IET MATERIALS, CIRCUITS AND DEVICES SERIES 65

MEMS Resonator Filters

Other volumes in this series

Volume 2	**Analogue IC Design: The current-mode approach** C. Toumazou, F.J. Lidgey and D.G. Haigh (Editors)
Volume 3	**Analogue–Digital ASICs: Circuit techniques, design tools and applications** R.S. Soin, F. Maloberti and J. France (Editors)
Volume 4	**Algorithmic and Knowledge-Based CAD for VLSI** G.E. Taylor and G. Russell (Editors)
Volume 5	**Switched Currents: An analogue technique for digital technology** C. Toumazou, J.B.C. Hughes and N.C. Battersby (Editors)
Volume 6	**High-Frequency Circuit Engineering** F. Nibler *et al.*
Volume 8	**Low-Power High-Frequency Microelectronics: A unified approach** G. Machado (Editor)
Volume 9	**VLSI Testing: Digital and mixed analogue/digital techniques** S.L. Hurst
Volume 10	**Distributed Feedback Semiconductor Lasers** J.E. Carroll, J.E.A. Whiteaway and R.G.S. Plumb
Volume 11	**Selected Topics in Advanced Solid State and Fibre Optic Sensors** S.M. Vaezi-Nejad (Editor)
Volume 12	**Strained Silicon Heterostructures: Materials and devices** C.K. Maiti, N.B. Chakrabarti and S.K. Ray
Volume 13	**RFIC and MMIC Design and Technology** I.D. Robertson and S. Lucyzyn (Editors)
Volume 14	**Design of High Frequency Integrated Analogue Filters** Y. Sun (Editor)
Volume 15	**Foundations of Digital Signal Processing: Theory, algorithms and hardware design** P. Gaydecki
Volume 16	**Wireless Communications Circuits and Systems** Y. Sun (Editor)
Volume 17	**The Switching Function: Analysis of power electronic circuits** C. Marouchos
Volume 18	**System on Chip: Next generation electronics** B. Al-Hashimi (Editor)
Volume 19	**Test and Diagnosis of Analogue, Mixed-Signal and RF Integrated Circuits: The system on chip approach** Y. Sun (Editor)
Volume 20	**Low Power and Low Voltage Circuit Design with the FGMOS Transistor** E. Rodriguez-Villegas
Volume 21	**Technology Computer Aided Design for Si, SiGe and GaAs Integrated Circuits** C.K. Maiti and G.A. Armstrong
Volume 22	**Nanotechnologies** M. Wautelet *et al.*
Volume 23	**Understandable Electric Circuits** M. Wang
Volume 24	**Fundamentals of Electromagnetic Levitation: Engineering sustainability through efficiency** A.J. Sangster
Volume 25	**Optical MEMS for Chemical Analysis and Biomedicine** H. Jiang (Editor)
Volume 26	**High Speed Data Converters** A.M.A. Ali
Volume 27	**Nano-Scaled Semiconductor Devices** E.A. Gutiérrez-D (Editor)
Volume 28	**Security and Privacy for Big Data, Cloud Computing and Applications** L. Wang, W. Ren, K.R. Choo and F. Xhafa (Editors)
Volume 29	**Nano-CMOS and Post-CMOS Electronics: Devices and modelling** Saraju P. Mohanty and Ashok Srivastava
Volume 30	**Nano-CMOS and Post-CMOS Electronics: Circuits and design** Saraju P. Mohanty and Ashok Srivastava
Volume 32	**Oscillator Circuits: Frontiers in design, analysis and applications** Y. Nishio (Editor)
Volume 33	**High Frequency MOSFET Gate Drivers** Z. Zhang and Y. Liu
Volume 34	**RF and Microwave Module Level Design and Integration** M. Almalkawi
Volume 35	**Design of Terahertz CMOS Integrated Circuits for High-Speed Wireless Communication** M. Fujishima and S. Amakawa
Volume 38	**System Design with Memristor Technologies** L. Guckert and E.E. Swartzlander Jr.
Volume 39	**Functionality-Enhanced Devices: An alternative to Moore's law** P.-E. Gaillardon (Editor)
Volume 40	**Digitally Enhanced Mixed Signal Systems** C. Jabbour, P. Desgreys and D. Dallett (Editors)

Volume 43 **Negative Group Delay Devices: From concepts to applications** B. Ravelo (Editor)

Volume 45 **Characterisation and Control of Defects in Semiconductors** F. Tuomisto (Editor)

Volume 47 **Understandable Electric Circuits: Key concepts, 2nd Edition** M. Wang

Volume 49 **Advanced Technologies for Next Generation integrated Circuits** A. Srivastava and S. Mohanty (Editors)

Volume 51 **Modelling Methodologies in Analogue Integrated Circuit Design** G. Dundar and M.B. Yelten (Editors)

Volume 53 **VLSI Architectures for Future Video Coding** M. Martina (Editor)

Volume 54 **Advances in High-Power Fiber and Diode Laser Engineering** Ivan Divliansky (Editor)

Volume 55 **Hardware Architectures for Deep Learning** M. Daneshtalab and M. Modarressi

Volume 58 **Magnetorheological Materials and Their Applications** S. Choi and W. Li (Editors)

Volume 60 **IP Core Protection and Hardware-Assisted Security for Consumer Electronics** A. Sengupta and S. Mohanty

Volume 64 **Phase-Locked Frequency Generation and Clocking: Architectures and circuits for modem wireless and wireline systems** W. Rhee (Editor)

Volume 67 **Frontiers in Securing IP Cores: Forensic detective control and obfuscation techniques** A Sengupta

Volume 68 **High Quality Liquid Crystal Displays and Smart Devices: Vol. 1 and Vol. 2** S. Ishihara, S. Kobayashi and Y. Ukai (Editors)

Volume 69 **Fibre Bragg Gratings in Harsh and Space Environments: Principles and applications** B. Aïssa, E.I. Haddad, R.V. Kruzelecky, W.R. Jamroz

Volume 70 **Self-Healing Materials: From fundamental concepts to advanced space and electronics applications, 2nd Edition** B. Aïssa, E.I. Haddad, R.V. Kruzelecky, W.R. Jamroz

Volume 71 **Radio Frequency and Microwave Power Amplifiers: Vol. 1 and Vol. 2** A. Grebennikov (Editor)

Volume 73 **VLSI and Post-CMOS Electronics Volume 1: VLSI and post-CMOS electronics and Volume 2: Materials, devices and interconnects** R. Dhiman and R. Chandel (Editors)

MEMS Resonator Filters

Edited by
Rajendra M. Patrikar

The Institution of Engineering and Technology

Published by The Institution of Engineering and Technology, London, United Kingdom

The Institution of Engineering and Technology is registered as a Charity in England & Wales (no. 211014) and Scotland (no. SC038698).

First published 2020

The Institution of Engineering and Technology
Michael Faraday House
Six Hills Way, Stevenage
Herts, SG1 2AY, United Kingdom

www.theiet.org

British Library Cataloguing in Publication Data
A catalogue record for this product is available from the British Library

ISBN 978-1-78561-896-3 (hardback)
ISBN 978-1-78561-897-0 (PDF)

Typeset in India by MPS Limited

Contents

Chapter 1

Introduction

Rajendra M. Patrikar[1]

Since the first wireless transmission in 1895 by Guglielmo Marconi, wireless communication has revolutionized the way how humans and society interact with one another. The radio frequency (RF) communication circuits technology is ubiquitous with its applications in mobile telecommunications technology, wireless local area network (WLAN), wireless sensor networks (WSN), wireless health monitoring, radio frequency identification (RFID), space communication, cordless computer interface devices, keyless entry, wireless personal area network (i.e., Bluetooth and ZigBee), broadcasting, a global positioning system (GPS), etc., to name a few [1].

Filters are an essential part of almost every electronic system and are inevitable in wireless communication systems. They can be broadly classified into two categories: the continuous-time filters or analog filters and digital filters. In this book, we focus is on continuous-time filters. Earlier, the design of filters typically involved passive components such as a resistor, capacitor, inductor, and active devices such as transistors. At very high frequencies, passive components can be realized by the metallic interconnects. Typically, the major part of the design process of filters is finding out the right parameters of these components and their right connections. However, this scenario has changed a lot with the entry of electromechanical components into this field and thus the addition of these components is playing an important role in today's electronics and in particular in wireless communications [2]. Since the size of the electromechanical components is in microns and decreasing further, these are called micro-electro-mechanical systems (MEMS). The filters designed using these components are called MEMS filters. Over the past few years, the MEMS-based on-chip resonators have shown significant potential for sensing, detecting, and high-frequency signal processing applications. MEMS resonators are being explored to fulfill the demands of the wireless communication circuits. In a radio transceiver, they are being utilized as oscillators and/or filters. Microresonators of the radiofrequency range are involved in a variety of commercial applications as a replacement of traditional bulky devices due to their small size, high frequency, high-quality factor, and most importantly complementary

[1]Centre for VLSI and Nanotechnology, Visvesvaraya National Institute of Technology, Nagpur, India

metal-oxide-semiconductor (CMOS) compatibility [2]. In this chapter, important issues that influence the development of MEMS filters are discussed.

Telecommunication associations all around the world are working on new cellular network technologies [2]. Most of the new solutions require advanced digital modulation schemes with higher spectral efficiency; multiple-input, multiple-output (MIMO)-based radio interfaces; and carrier aggregation (CA). A typical RF front-end of a smartphone comprises of power amplifiers, filters, duplexers, clocks, switches, and low noise amplifiers. The clock sets the frequency such that the window is opened at the right position in the RF spectrum, while the RF filter ensures the window has the right width to pass data at the needed bandwidth. Filters play a very important role in these instruments. For example, 3G networks used only about five bands, 4G LTE network is using 20 bands and with the advent of 5G, this number could rise to more than 40. A single smartphone, meant for international use, might need to filter transmit and receive paths for 2G, 3G, and 4G in up to 15 bands, apart from the filtering required for Wi-Fi, Bluetooth, and GPS. Such a phone might require as many as 30 to 40 filters. The situation is likely to become even more complex in the future and high-end smartphones could include 50 or more filters. The recently introduced 5G cellular technology has many improvements over 4G LTE which includes increased data capacity, lower latency, and low power consumption. 5G is projected to be 100 times faster than 4G and 10 times faster than wired fiber connections. The ability to achieve ubiquitous connectivity requires the RF front-end radio of a smartphone to handle these increased data rates and access the full bandwidth of both 4G and 5G wireless systems. Requirements of filters for 5G applications will include complex multiplexing, increasing integration, additional filters, and the capability to handle much higher frequencies. Overall, the system architecture for 5G RFFEs will be extremely complex and require a smaller footprint compared to the current technology since more than 100 RF filters will have to fit into a smartphone. All these enhancements depend on achieving efficiency in cost and power, reduction in the space needed for each filter, and the ability to manufacture filters in large quantities to meet fast-growing global demand [3].

The global demand for spectrum is leading governments around the world to reorganize the existing spectrum; this often results in the allocation of a new band that is adjacent to existing bands. The new bands usually have, small or none, guard bands. Not only that but the existing frequency bands for mobile communication are being extended or even newly defined by allocating additional frequency ranges. As a result, the spectrum suitable for mobile communication is becoming increasingly overcrowded. This requires RF front-end filters, which will separate these bands. Thus, the need for band-pass filters in the RF front end of mobile devices with steeper skirts and larger bandwidths (BWs) is increasing continuously. Usually, they are realized as single filters or multiplexers [2]. Besides steep skirts and large bandwidth, other requirements are low insertion loss (IL), high out-of-band suppression both close to the passband and also far away, power durability, electrostatic discharge stability, and good nonlinear behavior. All these parameters also come with power constraints and low power operation is essential.

Apart from mobile phones, most of the modern electronic systems are also increasingly becoming mobile and portable. These features of the modern systems require that they should have a small size and more important that they should consume low power. These features drive most of the specifications in today's scenario. Mobile communication is already an integral part of our lives, and its role is growing continuously. While early mobile phone functions were limited and concentrated around making landlines mobile, the capabilities of these phones initially improved to have the address book and text messaging of 80 characters. However, today's predominant smartphones and tablets support a large number of functions via third-party application software, bringing an insatiable hunger for data and higher download rates. In all these developments now, the telephone is just one of the applications. In the modern era not only human beings are using these communication aides but machines are also using the Internet and various communication methods, making the Internet of things (IoT) paradigm very popular where machines also need wireless interfaces. According to the market forecast, there will be several billion devices and equipment connecting through the low power wide area network (LPWAN), which depends on the standard of narrowband Internet of things (NBIoT) [4,5]. There is an increased demand for the interaction between physical objects and the real world. MEMS-based subsystems with versatile functionality are now widely adopted in most IoT and wireless sensor network (WSN) applications. IoT was coined in 1999 to describe the communication between versatile sensor nodes using modern wireless technology, later facilitating smart living for humans. Toward the goal of smart living, machine-to-machine communication (M2M) technology attracts significant attention for developing emerging IoT systems in different domains, including consumer electronics, automotive safety, transportation systems, energy grids, and healthcare facilities [4,5]. This machine-to-machine communication for IoT requires versatile filters with a varied range of center frequencies and bandwidths.

Until recently the stringent challenges of filter design were met with acoustic technologies. The kind of acoustic resonators we study herein are microelectromechanical devices. They experience acoustic wave propagation and eventually vibrate at a resonance frequency related to their dimensions and mechanical configuration when actuated. Resonance is a dynamic behavior that is observed when certain systems are excited properly. In general, these systems exhibit an amplified response to their input when the frequency of the excitation is equal to the resonant frequency of the system, with a more efficient transfer of the energy from the excitation source to the structure. The damping of the excitation energy is an important issue when considering the dynamic behavior of a system, whether the resonant response is desired or should be avoided. In the case of large scale mechanical structures, it is generally desired to avoid resonance as it often causes accelerated fatigue and eventually failure of the structure. However, they are useful when the amplitude of vibration is small and does not easily lead to a breakdown. Generally, any material is capable of supporting acoustic wave propagation including silicon. The piezoelectric properties of certain materials facilitate the wave propagation, thus improving the electromechanical energy conversion, so piezoelectrics are usually chosen as the acoustic layer of many acoustic-wave resonators [6]. They experience acoustic wave propagation and eventually vibrate at a resonance frequency related to their dimensions and mechanical configuration when driven with the appropriate conditions. Roughly, they are classified into two categories: surface acoustic wave (SAW) and bulk acoustic wave (BAW) resonators. Usually, a combination of both is normally found in the form of longitudinal, shear, mixed longitudinal-shear Rayleigh waves, Love waves, or Lamb waves [6]. For example, the waves we can see propagating on the surface of a lake after hitting the water mass with a stone illustrate the concept of a SAW. On the other hand, the sound waves traveling through the air until reaching our ears are of the BAW type. In these examples, the water and the air are the propagation media of the SAW and BAW, respectively; two types of the latter are found: the solidly mounted resonator (SMR) and the thin-film bulk acoustic wave resonator (FBAR). This is considered by the MEMS community to be a micromachined version of the conventional bulk acoustic wave resonators such as the quartz crystal. Undoubtedly, micromechanical resonators such as FBAR can perform as well as if not better than its bulky conventional counterparts and facilitate the miniaturization and power reduction of conventional RF systems.

Another technology that is promising and whose fabrication process is similar to FBAR technology is the contour-mode resonator (CMR) technology [7]. It has enabled the fabrication of multiple frequencies and high-performance resonators on the same silicon chip. The CMR technology can combine in a single device many important features that characterize existing resonators. The piezoelectric transduction enables simultaneous frequency scaling of the device and its direct interface to 50-Ω electronics.

In the literature, SAW BAW, FBAR, and CMR are treated as the same class of devices and often treated as micromachined electromechanical devices. BAW and FBAR are differentiated by its mounting technologies. In the literature, some other classes of devices, such as capacitive micromachined ultrasonic transducers (CMUTs), are also sometimes called BAW devices. In general, the devices are called MEMS if they are fabricated by using thin and thick film processes used in integrated circuits (IC) fabrication or popularly known as CMOS process. Today, SAW, SMR, FBAR, and CMR devices can be fabricated within standard IC technologies. Additionally, FBAR manufacturing entails micromachining steps, like MEMS resonator processes. On the other hand, FBARs resonate at far-from-fundamental acoustic modes, instead of purely mechanical modes. Both circumstances have thus created certain controversy regarding whether FBARs are considered as MEMS resonators. However, in this book, any microstructure used for filter applications are referred to as MEMS resonators. The distinct feature of all these components is high Q, indicating that less energy dissipated, helps in low power designs.

1.1 Low power issues

In all these systems, power consumption is an important issue since most of them would be working on battery. In general, not only battery-operated devices but the power consumption is very important in plugged equipment also due to environmental considerations. Thus, low-power design is of utmost importance. The power for signal processing and transmission has to be near-zero. In general [4], in conventional radio receivers, the power consumption can be reduced only at the expense of reduced sensitivity. The strategy of low-power design consists of decomposing the task to be solved in an intelligent fashion such that the rate of information processing is reduced as far as is possible without compromising the performance of the system. Intelligent decomposition of the task involves good architectural system decomposition, a good choice of topological circuits needed to implement various functions in the architecture, and a good choice of technological devices for implementing the circuits. MEMS resonators are good candidates for low-power designs because of their excellent quality factors. For example, wakeup receivers, which are often deployed to save power in the signal processing of unattended sensors can be built with MEMS resonators. Even if scheduling algorithms can be used to augment the efficiency of existing receivers, it can be shown that the lifetime of the sensor battery could be extended by several years by having an always-on RF receiver that practically consumes near-zero power (at levels comparable to the leakage rate of a battery). The main challenge associated with the use of ultra-low-power (in the order of 10 nW) in RF receivers is the limited sensitivity. It is also clear that radio architectures that rely on passive amplification and asynchronous demodulation of RF signals are the best candidates for achieving ultra-low power consumption [8,9].

In these systems front-end systems, "near-zero" power is achievable by performing signal processing functions in passive MEMS elements. MEMS devices that resonate at the frequencies of interest implement filtering, voltage step-up, and demodulation functions by leveraging their equation of motion and high-quality factor (Q). CMOS circuits then implement only the back-end receiver stages, resulting in much lower power consumption than in a system that realizes front-end and processing functions in CMOS. In addition to this advantage in power, MEMS devices can provide superior out-of-band rejection of signals far from the resonant frequency, a critical feature for receivers used in high-interference environments. Because low-power circuits tend to have high input impedance, minimizing parasitic capacitance at the circuit input node is essential to this design. CMOS-MEMS technology enables monolithic integration of various MEMS devices with CMOS circuits, by patterning the MEMS within the CMOS chip interconnects. The interconnect length and capacitance between MEMS and CMOS is therefore minimized, making this technology ideal for the MEMS demodulator [9].

Most of the above discussion is about long-distance communication where relatively high frequencies are used. However, interestingly such filter banks are also required for signal processing at low frequencies. An example of the ultra-low-power programmable analog bionic ear processor showed that such filter banks are necessary for acoustical signal processing. In this system, a microphone converts sound to electrical signals; and the signals are transmitted via a cable to a speech processor, the signals are processed by the speech processor to extract the logarithm of the spectral energy in 8–22 filter bands, and these energy coefficients are compressed to reduce the electrical dynamic range of stimulation. It is a necessity that filters would be extremely consuming low power and have a small size [10].

The feature "small size" is the major feature of all the forms of electronic systems today. The small size offers several advantages in these systems and efforts to make them as small as possible usually referred to as miniaturization.

1.2 Miniaturization

In general, all electrical/electronic systems have changed and are continuously changing because of the miniaturization of transistors. However, all other components are also miniaturized because the reduction of the size usually offers advantages. The major electronic system used worldwide is a cellular phone, which is using these technologies and becoming versatile with every generation. The trend is seen in almost every electronic component. Filter circuits, which are essential components in these systems also continuously becoming smaller and smaller and need to handle various frequency bands in modern technology. Thus, miniaturization had a tremendous impact on electronic system engineering. It is argued by Ray Kurzweil that imperative in any technology, which is as he put it as a natural process for any system development [11]. He has stated the Law of Accelerating Returns Evolution (sophistication) of life-forms or technology speeds up because they are built on their own recorded degree of order. Kurzweil calls this the Law of Accelerating Returns. This Law of Accelerating Returns gave us ever-greater order in technology, which led to computation—the essence of order. For life-forms, DNA provides the record. In the case of technology, it is the ever-improving methods to record information. Moore's law [12] (based on a temporary methodology, i.e., lithography) is only an example of the Law of Accelerating Returns. Beyond lithography, we may expect further progress in miniaturization based on DNA, quantum devices, AFM lithography, nanotubes, etc. All these areas are being explored and we surely will see some of these technologies will come in the market soon. One such area is quantum computing, which is expected to give acceleration to computation. In general, miniaturization offers the following advantages.

- Minimizing the energy and materials used in manufacturing:
  This is a very obvious advantage. In general, smaller devices will require less material and also the energy required for processing typically is proportional to the volume of the material that will also be reduced.
- Redundancy and arrays:
  It is easy to build an array with smaller components because it would require less area/volume. These units can be then connected in parallel configuration so that redundancy is built in the system. If anyone of these components fails, others can take over the operation and thus system continues to function.
- Integration with electronics, simplifying systems:
  The electronics that are becoming an integral part of the systems typically are called smart systems. Interfacing sensors and actuators with electronics usually used to increase the smartness of the systems. If these sensors and actuators are also small and can be integrated with transistors, then the systems do not face problems such as noise and mismatch.

- Reduction of the power budget:
  In general, smaller devices mean larger electric field; the supply voltage can be reduced and in turn, one can show that power requirement for the same functionality will be less for smaller devices.
- Faster devices:
  It can be shown that smaller devices have higher transconductance, which usually is an indication of how fast devices will respond to input signals. In general, faster operation can always be achieved by using higher power. That is the reason the figure of merit is usually a power delay product. Thus, faster devices and reduction in power budget have to be treated together to see their effectiveness.
- Increased selectivity and sensitivity:
  Because of higher transconductance, it is easy to design a system particularly amplifiers, which will have higher selectivity and sensitivity.
- Wider dynamic range
  Since smaller devices are faster and require less power budget, these properties can be translated to design the systems with a higher dynamic range. Thus, the response of the system will be linear for a larger range of input parameters.
- The exploitation of new effects through the breakdown of continuum theory in the microdomain:
  This phenomenon is seen in many areas of nanotechnology, for example, the gold nanoparticles exhibit paramagnetic behavior. In the case of electronics, different conduction mechanism is seen at nanolevel since the density of states of electron changes at the nanolevel.
- Improved reproducibility (batch concept):
  Most of the miniaturized components are fabricated in a batch and as a result cost of the fabrication per piece reduces with every miniaturization node. The batch fabrication also ensures the reproducibility since the process of fabrication usually can be optimized and affect all the components uniformly.
- Improved reliability:
  This is usually seen in the theory of reliability that the larger objects have more failure probability because there is a more probability that the defects may fall on this object or may get generated during the fabrication or operation. Thus smaller systems are inherently are more reliable.
- Cost/performance advantages:
  As discussed, fabrication cost of a smaller version of the component is less costly because of less material requirement as well as a reduction in the manufacturing cost. At the same time, the performance is higher for these devices. This is seen in the case of flash memories. The cost of the transistor is continuously going down; however, data transfer rates are continuously going up. This is true for all the miniaturized electronics. This is also an important part of this book.
- Minimal invasive:
  Miniaturized components take a small volume to be fit in any system and their presence is not noticed. The latest trend is to fit them into the human body.

Thus, it appears that low power and small size or (small area in case of planer technologies) are the major parameters with which specifications are buildup. In the case of filters as stated above, the requirements are becoming stringent. Using multiple filters for different bands will lead to a larger footprint. The digital design of the radio can be one of the solutions.

1.3 Tunable filters

As mentioned earlier, the spectrum of electromagnetic waves is strictly controlled by individual governments and accessible frequency bands and standards are determined by location, which means that multiband and multistandard operation is mandatory in current mobile phones and smartphones. This makes the design and fabrication of the RF front-end section of the transceiver very crucial and demanding. As a possible solution to ease this problem, the implementation of reconfigurable digital basebands called software-defined radio (SDR), combined with a flexible (tunable) RF front end, has received much attention. In an ideal SDR receiver, incoming RF signals are directly converted from analog-to-digital and all necessary signal processing functions, including band and channel selections, which are carried out in the digital domain. These efforts lead to moving antenna closure to the digital section [13]. Only a broadband low-pass filter (LPF) is necessary for antialiasing in the RF section. The operation mode can be changed and upgraded by uploading specific software. In an ideal SDR transmitter, outgoing RF signals are directly generated by a digital-to-analog converter (DAC) from the data stream. The major requirement of this system is the need for high speed, low-noise, high-resolution analog-to-digital converter (ADC) and a DAC. The implementation of the SDR is also being done on FPGA to maintain speed and configurability. However, it is clear that a tunable RF front end, especially tunable RF BPFs, is invaluable in realizing fully flexible SDRs [13].

The tunability of RF filters is a complex issue. For a filter to be truly tunable, the passband and stopband of the filter must slide along the frequency axis as a function of the control signal, which is usually a DC voltage. This DC voltage usually is set by the programmable digital system. The ideal situation would be a tunable filter in which a high-Q resonator scan be moved up and down continuously in frequency and processed further in a similar manner, which is usually digital signal processing. Circuits in which filters or parts of filters are reconfigurable with switches do not fall under the definition of tunable filters. The tuning range is the range between the lowest and highest center frequencies to which the filter can be set, and a truly tunable filter can be set to any number of frequency positions required to serve multiple bands within the tuning range.

Tunable RF filters that promise to simplify the RF front end dramatically by reducing filters and switch throw count have been a subject of research [14,15]. RF active filters have also been extensively investigated with various devices such as varactors (VC), fixed capacitor arrays. The cointegration of RF BAW filters and RF MEMS switches or VCs with CMOS circuitry seems to be a possible solution.

However, even though it is possible, it may not be practical. Achievable filter bandwidth, however, is narrow due to the small electromechanical coupling for the vibration mode. Recently, micromechanical RF channel-select filters capable of eliminating strong adjacent channel blockers directly after the antenna, which greatly reduces power consumption in RF front-ends have been designed and fabricated [16]. The combined 2.7-dB passband insertion loss and 50-dB stopband rejection of the demonstrated 206-element 0.1% bandwidth 223.4-MHz differential micromechanical disk filter represents a landmark for capacitive-gap transduced micromechanical resonator technology for tunable filters.

1.4 Integration with CMOS

CMOS technology is becoming pervasive and is being used in a large number of applications and MEMS resonators are becoming essential in many of them apart from the applications discussed above. Apart from the filter application of mechanical resonators, these devices also serve as timing (i.e., oscillator), and inertial detection (i.e., gyroscope), as well as mass sensing purposes. These other applications are also driving the research and development of resonators. Thus, integrating fabrication of resonators with CMOS processes has several advantages as discussed above. The major advantage here is the integration of digital and analog sections besides the packaging issues [4].

There are two major methods are being used for the integration of MEMS components with CMOS circuits known as (1) system in package (SiP) and (2) system-on-chip (SoC) methods. The SiP solution is available after fabricating these components their electrical connections are made through-wafer- or chip-level bonding techniques. Typical techniques used are through silicon via (TSV) and wire interconnection, thus a complete system with the necessary functionality is fabricated. From the requirement of system specification, this hybrid multichip approach allows the optimal design of the CMOS circuitry and MEMS devices using the appropriate and individual technology node. Even if the hybrid approach is a mainstream strategy from the commercial viewpoint owing to its high accessibility and flexibility for most MEMS designers and manufacturers, the nature of considerable bonding parasitics and cost for MEMS combo integration become the main challenges toward future wearable/IoT implementation. On the other hand, the SoC approach provides a generic solution to monolithically integrate the overall system by cofabricating MEMS and IC on the same substrate. As a consequence, this monolithic approach offers several merits, including fewer capacitance paths (i.e., undesired parasitics), smaller footprint, fast prototyping, and turnaround time only at the cost of restricted material properties and dedicated foundry sources. The technique with which the MEMS components are fabricated is also called surface micromachining. There are three possibilities in this approach, which can be explored. In the first approach, it is possible to fabricate the MEMS device first and then fabricate CMOS circuits along with appropriate interconnections. The second approach could be to use at least some process steps for both the MEMS and well CMOS circuits and complete integration can be done. Finally, MEMS device fabrication can be done after all the steps of CMOS circuit fabrication are over [4].

1.5 Inside the book

The organization of chapters in this book is as follows. In Chapter 2, we are first exploring the history of filters. The design of MEMS resonator-based filters needs more attention to resonator design. The circuit part becomes a bit trivial once the resonator is designed for intended functions. The design process of the resonator and filter is described in this chapter.

The operating frequencies of the resonators depend upon the physical parameters, the geometry of the resonators, and the transduction mechanism. In Chapter 3, the design of MEMS resonators that are operated at lower frequencies by using capacitive transduction is discussed.

The high-frequency counterparts of the MEMS resonators are very popular and are being explored for various applications including filters. With small size, high performance and complementary metal-oxide-semiconductor (CMOS) compatibility, RF MEMS resonators offer promising technology in contemporary RF front-end in wireless communication systems. Chapter 4 discusses the design issues of these resonators. The detailed methodology to design contour mode resonator (CMR) for GSM frequencies is discussed in this chapter.

The design of resonators and most of the MEMS devices use finite element modeling for analysis of the structures. This method is discussed in detail with examples in Chapter 5.

Fabrication of these components is typically done with microtechnology, which comprises of thin and thick film processing technologies and popularly used for CMOS circuit fabrication. It can be merged with on-chip signal processing systems and actuators, which can handle noise and many other issues effectively.

In Chapter 6, we are discussing fabrication issues of low-frequency MEMS resonators, which typically use capacitive transduction. In Chapter 7, the fabrication issues of high-frequency MEMS are discussed. One of the major requirements of these resonators is that it uses the piezoelectric layer and that usually is a different material than commonly used materials in CMOS fabrication processes. The deposition of thin piezoelectric material and top of that interdigitated electrode's deposition and patterning are important processes involved in the fabrication of CMR. Releasing of the device is also a challenge. These issues are elaborated in Chapter 7.

The fabricated components are to be used in the circuits, where design issues become very important particularly at higher frequencies. In Chapter 8, circuits used in wireless communication systems are discussed. In frequency control above the GHz band, piezoelectric materials are employed as a resonating body in SAW/BAW devices to obtain a sufficient electromechanical coupling coefficient since the amplitude of vibration decreases with increasing operating frequency.

The testing and verification of MEMS resonators are an important issue and the design optimizations can only be done after getting test results. The testing of these devices has to be done with special instruments and also with special layouts technique to take care of issues related to parasitic components associated with resonators and its connections. These details are discussed in Chapter 9.

The packaging of the MEMS devices, in general, a challenging issue since in many cases packaging materials cannot be used to seal it, unlike the integrated circuits. In the case of MEMS resonators for FBAR type, space should be available to vibrate (displacement) in vacuum preferably. There are several other packaging issues, which in general MEMS devices face are discussed in Chapter 10.

System reliability usually decides all commercialization issues. The system reliability is usually a product of reliabilities of all its components. For commercialization of MEMS resonators, reliability and sustainability are important issues for understanding the behavior of the device under various operating conditions and environment. Reliability of MEMS resonators is an important issue due to increasing complexity, integration level, and stability and to enhance desired performance in various applications. These issues are discussed in Chapter 11.

References

[1] Wu I.-T. "Integrated electrostatically- and piezoelectrically-transduced contour-mode MEMS resonator on silicon-on-insulator (SOI) wafer" (2014). Graduate Theses and Dissertations. https://scholarcommons.usf.edu/etd/5336

[2] Warder P., and Link A. "Golden age for filter design: Innovative and proven approaches for acoustic filter, duplexer, and multiplexer design." *IEEE Microwave Magazine*. 2015; **16**(7); 60–72.

[3] Patel M.S., Tweedie A., Harvey G., and Marra V. "OnScale simulation-driven optimization of 5G RF MEMS filters," OnScale White Paper series.

[4] Chen C.-Y., Li M.-H., and Li S.-S. "CMOS-MEMS resonators and oscillators: A review." *Sensors and Materials*. 2018; **30**(4): 733–56.

[5] Chen M., Miao Y., Hao Y., and Hwang K. "Narrow band Internet of things." *IEEE Access*. 2017; **5**(20): 557–77

[6] Campanella H. *Acoustic Wave and Electromechanical Resonators Concept to Key Applications*. Artech House; 2010.

[7] Piazza G., Stephanou P.J., and Pisano A.P. "Piezoelectric aluminum nitride vibrating contour-mode MEMS resonators." *Journal of Micro Electrochemical Systems*. 2006; **15**(6): 1406–18.

[8] Kochhar A., Galanko M.E., Soliman M., *et al.* "Resonant microelectromechanical receiver." *Journal of Microelectromechanical Systems*. 2019; **28**(3): 327–43.

[9] Colombo L., Galanko M.B., Abdelsalam H., *et al.* "Ultra-low-power and high sensitivity resonant micromechanical receiver." *IEEE Sensors: 2017 Conference Paper*. IEEE; 2017.

[10] Sarpeshkar R. *Ultra-Low Power Bioelectronics*. Cambridge University Press; 2010.

[11] Kurzweil R. "The law of accelerating returns." in Teuscher C. (ed), *Alan Turing: Life and Legacy of a Great Thinker*. Berlin, Heidelberg: Springer; 2004. Available at https://www.kurzweilai.net › the-law-of-accelerating-returns.
[12] DeBenedictis E.P. "It's time to redefine Moore's law again." *Computer*. 2017; **50**(2):72–75.
[13] Hashimoto K.-Y., Kimura T, Matsumura T., *et al.* "Moving tunable filters forward." *IEEE Microwave Magazine*. 2015;**16**(7).
[14] Gevorgian S.S., Tagantsev A.K., and Vorobiev A.K. *Tuneable Film Bulk Acoustic Wave Resonators*. London: Springer; 2013.
[15] Hashimoto K.-Y., Tanaka S., and Esashi M. Tunable RF SAW/BAW filters: Dream or reality? *Proceedings of the 2011 Joint Conference of the IEEE International Frequency Control and the European Frequency and Time Forum (FCS)*. IEEE.
[16] Akgul M., Ozgurluk A., and Nguyen C.T.-C. "RF channel-select micromechanical disk filters—Part II: Demonstration." *IEEE Transactions on Ultrasonics, Ferroelectrics, and Frequency Control*. 2019; **66**(1); 218–35.

Chapter 2
Filter design

Rajesh Junghare[1], *Raghvendra Deshmukh*[1] *and Rajendra Patrikar*[1]

Filters are essential parts of electrical/electronic systems. It is difficult to find any modern electronic system without filter/s. In some of the systems such as electronic communication systems, filter design becomes very crucial and many times decides the progress of these systems. Mobile communication systems are one of the examples. The filters by and large affecting the quality of systems and therefore it has been always a matter of research and development. As a result, a lot of work has been done in the last 100 years on the topic of filters. The electronic systems have grown in numbers almost exponentially in these years so is the scientific and engineering efforts on each component and subsystem and filters are no exceptions. Today, the number of mobile phones used is in the range of 5 billion and that means at least those many filters are being used. The effectiveness of filters can immediately be seen on the quality of the systems and thus industries also contributed a lot in these efforts. As seen during the evolution of the filters, there were a lot of scientific challenges that were also posed by the filters and thus filters remained an attractive area for researchers. There are so many names associated with the filers since the beginning and some of them will appear in this book also. The topic of filters, in general, is a vast one and has generated so many texts and handbooks apart from the research papers in journals and various articles in magazines. So even a review of all this work may turn out to be another book. This chapter is written to give a quick perspective of the complicity involved in designing and synthesizing the filters along with the brief history.

2.1 Brief history of filters

As happens in many electronic systems, the long-distance communication equipment was not so sophisticated, as M.I. Pupin commented, "One had to put his mouth very close to the transmitter, speak very slowly, very loud, and about very ordinary things, otherwise the meaning of the transmitted intelligence would not be

[1]Center for VLSI and Nanotechnology, Visvesvaraya National Institute of Technology, Nagpur, India

caught at all." Obviously, it was necessary to improve the properties of long-distance transmission lines. In 1886, Heaviside's brother Arthur, by then a leading post office engineer was experimenting with telephone lines in which the receivers were arranged in "bridge" or parallel circuits. To his surprise, he found that adding more telephones to a circuit actually improved the clarity of transmission. He turned for an explanation to Oliver, who soon showed that the leakage (filter?) of current through each telephone reduced the distortion, though it also weakened the signal.

Heaviside also found that adding more inductance to the circuit—for instance, by inserting coils at regular intervals along the transmission cable—would reduce the distortion even further. The extra inductance, he explained, would help carry the waves along in much the same way that loading a clothesline with birdshot makes it better able to convey transverse waves. He later joked that his name and inductive loading were "naturally and providentially connected. You heavify a line by the process of heavification" [1]. Whatever one called it, inductive loading offered a relatively cheap and easy way to improve telephone transmission, and AT&T and other companies later used it with great success. Heaviside never patented his idea, so he never made a penny from it.

Pupin found that a line, which contains coils at regular intervals can be represented by an equivalent uniform cable if the coils are spaced closely enough [2]. The equivalence decreases if the distance between two adjacent coils is increased, and disappears altogether if this distance is larger than half the wavelength of the signal that is propagated in the cable. Through his thorough mathematical and experimental research, Pupin found that the damping in cables for telegraphy and telephony can be substantially reduced by judiciously inserting these coils, which has resulted in the widespread use of these so-called 'Pupin lines' throughout the world.

The properties of these lines were further investigated by George A. Campbell. In 1903, he published some findings [3], among which a peculiar frequency-dependent effect of Pupin lines, namely that they have a well-defined critical frequency that marks a sudden change in the damping characteristics. Below this frequency, the damping is low and dependent only on the parasitic cable losses. If these losses are zero, the damping below the critical frequency is also zero. Above the critical frequency, the damping is high and almost independent of the cable losses. The transition at the critical frequency can be very sharp. The critical frequency itself is determined by the spacing of the coils and corresponds to a wavelength equal to twice the distance between them.

This effect was used to answer the question of how many coils are to be inserted in a given length of cable, but it was also immediately clear that this effect could be utilized, and Campbell pointed out that he used this effect to eliminate harmonics in signal generators. In fact, he used the cable as a low-pass filter, and he even mentioned the possibility of using the cable as a band-pass filter by replacing the coils by combinations of coils and capacitors.

A reel of cable is very large and therefore somewhat unwieldy as a filter, but the next step was so logical that it was undertaken independently in the same year

(1915) in Germany by Karl Willy Wagner [4], and in America by Campbell [5]. The line was simulated by a ladder construction of impedances, an instance of which is shown in Figure 2.1.

This step was also indirectly suggested by Pupin, and it gave rise to a filter that, due to its lineage, was referred to as an "electrical wave filter" in America and a "Wellensiebe" or, less frequently, a "Wellenfilter" in Germany. It was this invention that made the year 1915 to be usually regarded as the birth year of the electrical filter, although before that time simple frequency-selective circuits were also used.

In general, wave filters after Wagner and Campbell can consist of any number of sections, which consist of three impedances and can be of the T-type or of the Π-type. These types are shown in Figure 2.2.

The example of Figure 2.1 consists of two sections of the T-type. The sections in a wave filter were chosen identically so that the filter represented a homogeneous line.

Usually, for the impedances, reactance was chosen. This could result in a high pass, low pass, bandpass, or bandstop filters, or even filters with any number of disjoint passbands. Based on the frequencies of interest these were classified as lumped and distributed filters. Lumped filters or lumped networks are circuits whose elements (e.g., R, C, and L) are concentrated within their physical devices. The electrical and physical properties of the devices are defined at their terminals and the component connections are small compared to the wavelength of the highest signal frequencies, which are applied to the filter. Distributed filters are networks where the physical dimensions of the elements are in the same range as the signal wavelengths.

During this initial period of the circuit development, it was realized that getting near to ideal characteristics is difficult. For example, the low pass filter network required to pass all the frequencies below specified frequency with no attenuation

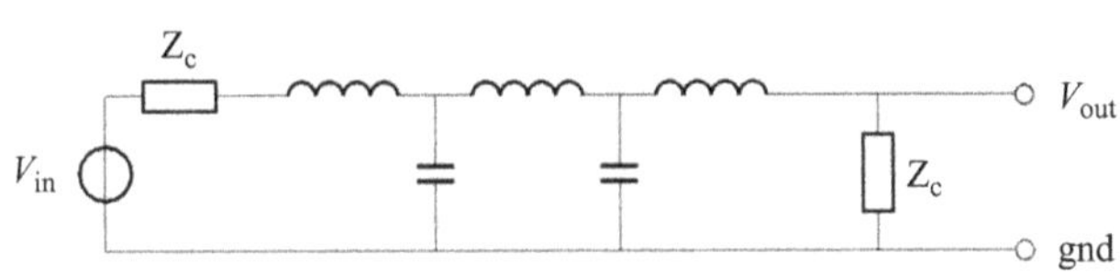

Figure 2.1 A lowpass electrical wave filter with a signal generator

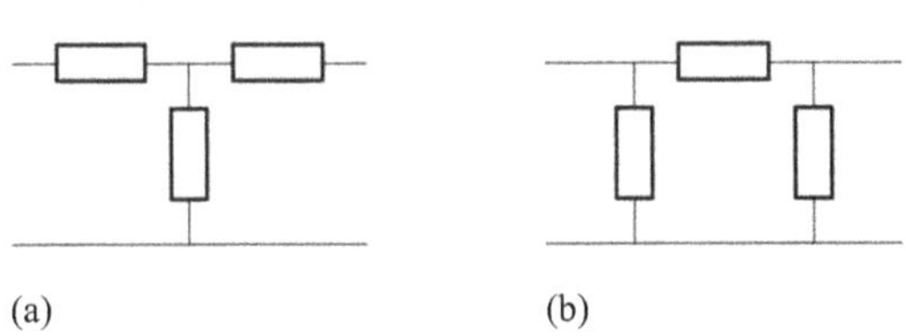

Figure 2.2 The sections of a wave filter can be of (a) the T-type or of (b) the Π-type

and frequencies above with infinite attention. The high pass network will have the complementary properties and that is no attenuation above the specified frequency and infinite attenuation for all the lower frequencies. However, these specifications are not achieved by a network based on physical components such as R, L, and C. The reason is that such ideal behavior demands the phase of the filter network to be linear function of frequencies. It can be shown that the impulse response of the network which gives such ideal response is noncausal time response and hence not realizable.

The solution to this problem was an important step in the development of the filter and modified the flow of the design. The after selection of filters specifications, the selection of realizable function was introduced as an important step. This step is also called an approximation problem. The filter specifications such as cut-off frequencies, passband gain, transition, bandwidth, and stopband attenuation are generally described in the frequency domain. The approximation method, a realizable rational function is selected so that it meets the specifications with prescribed tolerance. The procedure becomes highly algebraic and is used in the time domain or frequency domain which is more common.

A physical realization of an ideal filter transfer function is impossible. Hence, a frequency scheme is given, which is dependent on many system parameters and trade-offs. Within this frequency scheme, the real filter transfer function has to be located. Important selection criteria are for example a fast transmission from the pass-band to the stop-band or a minimum of filter distortion. The closer the ideal filter transfer function is approximated, the greater is the effort regarding a number of elements, power consumption, or costs. Depending on the latitudes various approximations for the ideal filter transfer functions are realizable. Important and popular approximation functions are Butterworth, Chebyshev, elliptic or Cauer, and Bessel approximations.

To design these filters, the filter impedances were to be constructed as combinations of inductances and capacitances. Design methods were developed by several people, under whom Otto J. Zobel has a prominent place because in 1923 he introduced a strategy that allowed the design of filters with an unlimited number of capacitances and inductances [6]. One of his inventions was the m-type derived filter, which simplified the design of complicated filters because these could be derived from relatively simple filters.

Darlington, in 1939, and Cauer, in 1940, both published the same theory to solve the set of problems just outlined and both displayed the same dazzling virtuosity in mastering a long sequence of thorny mathematical complications. The importance of the new method was not recognized immediately. It could be used to design better low-pass filters but it failed to provide such designs in practice because of the extremely heavy burden of computation required. It was not until the advent of cheap computation methods, in the 1950s, that Cauer–Darlington filters came into widespread use. So many computer-prepared designs have now been published that designing an elliptic-function filter involves little more work than copying numbers out of a book, and this technique is actually easier than the image-parameter method. Wilhelm Cauer designed passive filters with well-defined transfer functions [7], for which he used Chebyshev approximations. In a period that roughly extends from 1930 to 1940, Cauer published a great number of articles in which he developed a theory for designing filters with a defined attenuation behavior.

2.1.1 Active filters

Filters were realized as networks of inductors, capacitors, and resistors. It was recognized that the quality factor of the inductors was generally far less than of the capacitors, especially at low frequencies. In a number of applications, inductors with a sufficient quality factor were large and expensive, while high-quality capacitors were relatively small and cheap. The filter transfer functions that can be realized with capacitive and resistive elements only have their poles on the negative real axis of the complex Laplace plane. Complex poles are realizable if active circuits are added. This gave rise to the use of active RC filters [8].

A general design method for active RC filters has been proposed by R.P. Sallen and E.L. Key [9]. Although they indicated how sections of any order could be realized, in their synthesis procedure they proposed the realization of an active filter as a cascade of passive first-order sections and active second-order sections. It was possible to realize the second-order sections with just one active stage per section, which was important because active stages were expensive. The active elements used in filter design in various ways. For example, switch capacitor filters, tunable integrators became part of many systems. A breakthrough in high-frequency filtering is due to Bram Nauta, who proposed a transmittance-based integrator [10] with which 100 MHz filters can be designed.

With all these developments filters became third major development after radio and the electronic tubes in the electronics industry. Filters became essential parts of all electronic systems. In general, filters [11] direct, channel, integrate, separate, delay, differentiate, and transform all kinds of electric energy and information. The filter concept is now generalized and used in problems of mechanical and acoustical systems.

2.1.2 Electromechanical components

From the beginning, the principal objective of scientific filter design has been to find theoretical methods that would accommodate actual physical components and fit them exactly into optimized networks. The desire of the creative engineer has been to produce hardware with the performance that agrees exactly with the theoretical predictions. In L-C filters, for example, the inductors are often expensive toroids, whereas the capacitors are relatively inexpensive. The constant search for optima has produced an ingenious method for designing "minimum-inductance" filters that use fewer expensive components and more inexpensive ones. Since high-Q coils are bulkier and more expensive than those of moderate Q, various theories have enabled "predistorted" and "equalized" networks to maintain high performance with coils of lower Q. As described in Chapter 1, electronic systems began to use more and more of the available electromagnetic spectrum, the need for very narrowbands became urgent. It would have been difficult to find RLC components that would give the desired performance. At the same time, quartz crystals were explored for their possible applications in electronics. Filter design was designed developed to accommodate quartz resonators that offer practical Q's up to a quarter of a million and stabilities to match.

Quartz technology has been the focus of intense research and development since the first quartz crystal was demonstrated in 1922 [12]. The essential feature of frequency reference performance is that the frequency of the output signal should be constant over time; it should have good stability. Quartz crystal frequency reference products are sold in a dizzying array of sizes, shapes, and acronyms, but they can be classified into four broad categories based on the method of achieving stability [13]:

1. The basic crystal oscillator (XO). These are the cheapest and most common frequency references, with the lowest levels of performance.
2. Temperature-compensated crystal oscillator (TCXO). These are sold as a single component which includes the oscillator circuitry required to generate the frequency output with temperature compensation.
3. Microprocessor-controlled crystal oscillator (MCXO). These are oscillators whose output frequency is modified by a dedicated microprocessor, typically using a phase-locked-loop or digital frequency multiplier to adjust the output frequency as required to maintain output stability.
4. Oven-controlled crystal oscillator (OCXO). These devices are built inside a temperature-controlled enclosure for maximum temperature (and hence, frequency) stability.

The better performance of a quartz-based device requires increasing the size, power consumption, in turn, for the price of the device. Clever circuitry and manufacturing may modify the slope of this performance function somewhat, but it cannot be escaped entirely as long as the underlying technology is the same.

The quartz crystal, to a first approximation, is represented by a network of inductance and two capacitors that are inseparably given as a single unit. As can be seen, by discussions above, it was obvious the replacement of a few bulky and leaky components. However, a considerable amount of fresh theory was needed to treat quartz networks as components so as to reap the advantages of their enormous Q figures. But quartz crystals are expensive, so additional resourcefulness was needed to make use of the cheaper mechanical and ceramic resonators that have quality factors somewhere between those of L-C components and of quartz. This led to further developments in this area. In the beginning, researchers together with engineers had the dream to replace quartz by silicon and they started working on resonators designed for filter application in cell phones.

The answer to this quest is MEMS integration with the CMOS IC platform. The MEMS resonator compatible with CMOS integration can not only replace quartz crystals but also create new devices with superior frequency performance and order of magnitude improvement in size, cost, and power consumption. Such integration of MEMS devices with CMOS circuitry leads to develop new technologies. Now, quartz can completely replace by MEMS-based resonator as technology needs tiny devices compatible with silicon or CMOS platform. Hence quartz will be replaced mainly for two reasons: miniaturization and CMOS compatibility. The shrinking of technology requires ever smaller components and there are practical limits on the minimum size of quartz piece that can be used. New manufacturing and handling techniques for quartz have extended this limit for years, but it is impossible to keep up with the size reductions of digital technology, which uses batch fabrication techniques and follows Moore's Law [14]. The progress of modern technology is inextricably bound up with the progress of silicon and complementary-metal-oxide-semiconductor (CMOS) integrated circuit technology. Devices that can be manufactured with silicon technology are promoted because they can be manufactured cheaply using the existing silicon batch fabrication capacity, and digital circuitry can be integrated into them directly, enhancing their performance and function. For example, automobile dynamic stability systems were impossible to deploy on a commercial scale until silicon MEMS-based inertial sensing systems were developed. Optical projection systems are now made with an array of millions of separate mirrors on a silicon substrate, each smaller than the period at the end of this sentence. Such devices can only be manufactured using batch fabrication technology. Quartz crystals are not silicon-compatible, and so they become simply knocked out from the latest technology. It should be noted that there are some efforts being made to integrate quartz and CMOS-compatible silicon to create frequency references. However, the cost of combining the two separate technologies outweighs the benefits and it seems likely to remain so.

In 1855, surface acoustic waves (SAW) were mathematically founded by Lord Rayleigh [15]. In 1965, interdigital transducers (IDT) were invented by White and Voltmer, enabling the use of SAW in various filters [16]. At first, SAW filters were used in place of LC filters at the IF stage in TVs, and they were later also used for signal processing in radars. In 1977, just 12 years after the invention of IDT, Williamson released a list of 45 products that apply SAW devices. (Proceedings of 1977 IEEE Ultrasonic Symposium, pp. 460–468) In the list, 10 common uses at that time, such as IF filters for TVs, filters for CATV, and pulse compressors for radars, were introduced as well as distinctive characteristics. Researchers also investigated the use of bulk acoustic waves (BAW) at very high frequencies (more than 2 GHz) where the SAW is difficult to apply due to high absorption in the surface layer. The amplification of BAW at the frequency of 9.4 GHz by the supersonic drift of electrons was obtained by M Pomerantz in 1964 [17]. Free-standing membrane film bulk acoustic resonator (FBAR) was first demonstrated in 1980 by Gradkowski *et al.*, at United Technologies and independently that same year by Nakamura *et al.*, at Tohoku University Japan [18,19]. In 1982, a visionary paper was given by Lakin *et al.* of TFR [20], described the future potential of BAW (and FBAR) resonators. In particular, the paper emphasized the size. Size and performance (as we learned much later) are crucial for cell phone applications and are the enabling technologies that allowed both the shrinking of the early mobile handsets while allowing more components and functionality into the slimmer phones. The history of other MEMS resonators such as CMR for filter applications is not of more than 20 years.

2.2 MEMS resonator filter design

The design methodology has evolved over the years and the major impact also has come from digital computation technology. A lot of complexity of the design process is being handled by computers including documentation. But along with these digital filters also evolved and almost became a separate topic of filters. Again, there had been so much work done in digital signal processing and that is not being reviewed or discussed much in this chapter and in general in this book. In this book, MEMS-based RF filter design is illustrated which has a lot of challenges for its performance prediction. To help resolve these challenges, RF designers typically use several modeling and simulation tools to extract critical parameters prior to fabrication or testing. This simulation benchmark allows a designer to optimize filter performance based on resonator design, to select the appropriate number of resonators, and to reduce losses related to interconnect pads. Designers can also perform thermal analysis to account for environmental, packaging, and power handling issues. The parasitic losses, meal losses that account for the quality factor of resonators can also be optimized by electromagnetic modeling tools. Figure 2.3 shows a typical workflow that can be used to iterate on design concepts, using a combination of different modeling and simulation tools to converge to a design that meets the required filter performance. Typical tools used in this workflow include specialized computer-aided design (CAD) and finite element analysis (FEA) software.

In general, the design flow of MEMS resonator-based filter in CAD environment starts with attaining chosen resonance frequency. This resonance frequency is selected as per application; later is achieved by choosing proper geometry and dimension of geometry with suitable material. There is a fabrication limitation to selecting for geometry for higher frequencies. Also, the appropriate material is selected for the resonator which is compatible with the fabrication process and

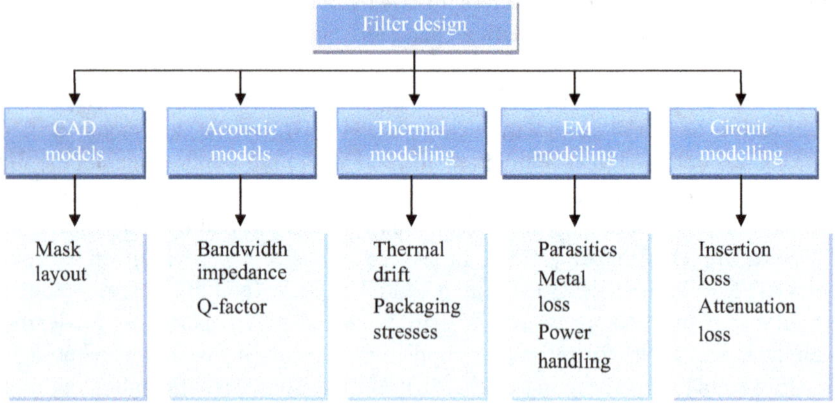

Figure 2.3 Typical modeling and simulation workflow used during RF filter design [21]

gives efficient performance. Once the structure is selected for desired resonance frequency, the suitable physics is used to model the operation of the resonator. There are various physics that can be incorporated in a model of resonator depending on the principle of resonator like electro-mechanics, thermal, and piezoelectric. In the model of resonator, one must consider the losses to get more accurate response of the resonating device. Figure 2.3 shows the various losses which can be encountered in the operation of resonator. In commercial tools like COMSOL, CoventorWare, or others, there is a feature to consider some of these, some can be taken care of by adding more parameters in the existing models. After considering all physics-based aspects of the resonator, various studies can be performed to understand the working of resonators. These studies are in time domain, frequency domain, or eigenvalue analysis. The eigenanalysis of structure gives the mode shape of structure which illustrates the deformation during resonance. Once the resonators are built with desired characteristics, they can be used as building blocks for designing the MEMS-based filter. In designing the filter, coupled resonators are used; hence the design methodology will moreover remain the same.

The other impact on filter technology definitely came from microelectromechanical components, which is the topic of this book. The implementation of this new technology not only requires usual network analysis but also requires careful design of the device which comes with so many options such as different materials and various physical geometry and size. The examples are given here to illustrate the methodology of these new types of filters. A similar methodology is used later in the next couple of chapters but targeted for different materials and applications and also fabrication.

2.2.1 RF MEMS filter

Over the past few years, extensive efforts have been devoted to replacing off-chip frequency-selective components (i.e., frequency references and filters) in telecommunication systems with on-chip silicon-micromachined MEMS resonators. Various fabrication platforms to realize low-power micromechanical components have been developed for future single-chip implementation of the wireless transceiver. Amongst these micromechanical components, one of the important devices, the modern MEMS-based filter, has much better performance compared to its SAW and BAW counterparts which have a bulky size that occupies more area on the board [22,62]. In RF receivers, the absence of filters that are capable of selecting single-channel bandwidths forces the front-ends of contemporary receivers to accept the unwanted signal. Thus, the available filers like FBAR or SAW are not efficiently operating which possess insufficient quality factor Q to manage practical RF channel selection without undue insertion loss (IL). Moreover, MEMS-based filter has a high-quality factor in comparison with SAW and acoustic filters. In order to achieve the desired selectivity, there have been attempts to design high order band-pass filters consisting of a number of coupled resonators [23–25]. The mechanical coupling technique, traditionally used for the implementation of high order filters from individual mechanical resonators, has been applied to micromechanical resonators for filter synthesis [26].

In this chapter, various design methods such as parallel resonators with electrical coupling, notching, and low-velocity coupling have also been described and demonstrated to realize a micromechanical disk-based array band-pass filter with smaller bandwidth while still retaining reasonable passband shapes [27–29]. With such a very small percent bandwidth filtering, it is possible to remove unwanted interference in the receive path of a communication handset. The percent of bandwidths achieved are small enough to make possible channel-selection much easier in a receive path chain, which could then greatly enhance the robustness and battery lifetime of future wireless transceivers [30,31]. Due to the very small size of the resonator element, to design a small bandwidth filter, the mechanical coupling will require submicron size of coupling elements (i.e., wires) that are critical to fabricate using optical lithography. In addition, filter characteristics are sharply dependent on the position and dimensions of the coupling elements [32].

High-order electromechanical filter design involves several elementary resonators interacting with each other in a way to generate a given passband function [33]. Customarily, the coupling of mechanical resonators is achieved with a soft mechanical beam connected between two resonators. However, the mechanical coupling approach presents several limitations concerning the geometry of the filter (in most cases only adjacent resonators can be coupled), the transfer function (only "all-pole" functions are achievable) and the tuning (the coupling strength can't be modified after fabrication). Thus, of resonators and coupling between them is important.

2.2.2 Advancement (evolution) of MEMS resonator

In literature, various shapes of capacitive micromechanical resonators have been reported to date, operating in different vibration modes and fabricated with different materials. The first of its kind of work is reported by Howe and Muller consisting of miniature cantilevers and doubly supported mechanical beams fabricated from polycrystalline silicon using the conventional MOS planar process [34]. This idea leads to the simplest MEMS resonator structure which is a clamped-clamped (CC) beam fixed to the substrate at both ends. Such thick clamped-clamped resonator which is made up of poly Silicon [32] is demonstrated by Bannon *et al.* and Lin *et al.* In Bannon's work, the resonators are demonstrated in vacuum and air ambiance where quality factor is degrading from 8,000 (vacuum) to 50 (air). Also, the single-crystal silicon (SCS) has also been used in making CC-beam resonators, as demonstrated by Pourkamali *et al.* [35]. They have used the *HARPSS* (high aspect ratio combined poly and single-crystal silicon) process, where the beam is of SCS, while the drive and sense electrodes are made of polysilicon. To enhance the resonant frequency, materials with higher Young's modulus are used in recent times. In 2011, Nabki *et al.* have reported 2-μm thick CC-beams fabricated using amorphous silicon carbide (SiC) with transducer gaps of 100 nm exhibiting resonant frequencies up to 26.2 MHz and Q in the order of 1,000 [36]. But continuous scaling of the dimensions of the beam to achieve higher frequencies degrades the performance of resonators. This is due to the higher losses, the high frequency via anchor, and substrate; eventually, it reduces quality factor to a few hundreds at higher frequencies. However, this issue can be moderately addressed in free-free beam resonators which are supported at flexural mode nodal locations using four quarter-wavelength support beams, with Q remaining high even with frequencies increasing past 100 MHz [37]. However, as the dimensions of beam resonators are further scaled down to achieve higher resonant frequencies, the attainable Q tends to decrease due to increased surface-loss and anchor-loss mechanisms.

Another widely used MEMS structure is comb-drive, first reported in 1989 by Tang *et al.*, which was fabricated with polysilicon with Q about 100 at a resonant frequency of 40 kHz [38]. The work in 2005 by Cioffi and Hsu reported a silicon comb-drive resonator using the SOI MEMS process with high Q around 50,000 [39]. In comb-drive resonators, pull-in effect between the electrodes is linear unlike the straight actuation, but this has limitation of scaling to achieve high frequency.

To achieve high frequencies and avoid disadvantages of smaller dimensions of beam geometry, bulk resonator is widely used. More commonly square or disk geometries are used to fabricate the bulk resonators. The work reported by Lee [40] has demonstrated a silicon extensional mode device vibrating at 2.18 MHz with Q of 1.16×10 6. These are excited through lateral capacitive gap drive electrodes on each side of the structure. The other more widely reported structure is a circular disk. The reported woks used circular disk as resonators with anchored at one or more nodal points on its periphery (for wine-glass mode), or with supported by a stem at its center (for radial-contour mode). The work reported by Lin *et al.* has achieved sub-100 nm capacitive gaps in their elliptic bulk-mode disk resonators of polysilicon, exhibiting a motional resistance of only 1.5 kΩ [37]. The work of Pourkamali [41] has reported HARPSS-based 18 μm thick SCS disk resonators with a Q of 46,000 at an elliptical mode resonance frequency of 150 MHz, exhibiting a resistance of 43.3 kΩ for 160 nm capacitive gaps. In 2009, a silicon resonator of similar type but with a larger radius by Lee and Seshia has shown an exceptional Q [42]. A radial contour mode disk of polysilicon has been reported by Clark *et al.* in 2005 with a Q of about 10,000 at 156 MHz [43]. However, there is limitation to continuous scaling of flexural mode resonators to achieve resonance frequency greater than 100 MHz. Also, the motional impedance grows rapidly. Due to this limitation, bulk acoustic mode resonators come to the rescue with the employment of resonant modes with much higher effective stiffness, increasing the operating frequency.

The other MEMS structure used for high-frequency resonators is Ring. The ring resonators consist of a circular ring supported by spokes emanating from a stem anchor at its center that has also been reported, with the ring expanding and contracting in width. In work CVD diamond ring resonators with balanced symmetrical quarter-wavelength support beams designed to suppress anchor loss, have achieved record-high Q's of 77,400 at 900 MHz and 42,900 at 3 GHz, the latter of which marks the highest f•Q of yet reported among room temperature acoustic devices [44]. The motional impedance is large as the size is small but can be reduced by reducing the transduction gap and using the array of resonator. One of the issues with this work is the fabrication process is not compatible with CMOS. The quality factor can also be increased with other geometries of resonator.

In reported work [45], an HF CVD diamond disk, the WGM has been shown to greatly enhance f•Q product by suppressing anchor losses, in particular posting a record-setting Q of 109,200 at 515 MHz, the highest Q reported to date in this frequency range for a room-temperature acoustic device. The higher modes improve the quality factor and increase the operating frequency at the same time it brings nonlinearity in sensing output

Recently, an elastic link between closely spaced resonators is introduced by an electric field, which is generated by the voltage applied between the resonators [31]. This approach offers several advantages, namely, voltage control of the elastic link (hence a good filter tunability) and the absence of a coupling spring. This technique has been used to realize a fourth-order filter composed of two clamped-clamped beam resonators coupled using this technique. However, the need to closely space the vibrating elements of the resonators is very difficult. From the geometrical shape of the resonators [31], it appears that the electrostatic coupling can seldom be used with three or more clamped-clamped beam resonators. Moreover, the technique can't be directly applied to resonators with nonrectilinear shapes (e.g., vibrating disk resonator). Structures in which an intermediate DC-isolated coupling electrode inserted in-between the vibrating elements have been presented [32]. The coupling electrode is placed close to the vibrating elements, in a way to yield two series-connected electrostatic transducers. It can have an arbitrary shape, thus liberating the designer from the geometrical restrictions mentioned above. The coupling strength is fixed at the design level by the value of the common node parallel-to-ground capacitance. The passband shape is tuned by adjusting the resonator bias voltage, which doesn't affect the coupling factor but only the natural frequency of resonators. This tuning technique increases the insertion loss and can only be used for small passband shape correction.

The two recent papers [46,47] introduces a design flow for micromechanical RF channel-select filters with tiny fractional bandwidths capable of eliminating strong adjacent channel blockers directly after the antenna, hence reducing the dynamic range requirement of subsequent stages in an RF front-end. In this work, mechanical system design flow described like VLSI transistor circuit design, hierarchical with a design stack built upon vibrating micromechanical disk building blocks capable of Q's exceeding 10,000. This paper had a detailed discussion about the different (half, quarter, and full) wavelength of coupling beam over operation of the filter.

In the design of the MEMS-based filter, the design of the resonator plays a vital role in achieving the desired characteristic of the filter. This section presents the detailed theoretical aspects of the resonator, disk resonator, coupled-resonator, and thermal stability of resonator for filter application. In this implementation of the filter, the disk resonator is chosen over the beam due to its high-frequency resonant mode. The design and operation of the disk resonator are discussed in Section 2.4 with the analytical electromechanical model. Also by considering the specification of filter to design, the disk resonator is designed in the FEM tool

COMSOL MultiPhysics®. The simulated results for the disk resonator have been shown in Section 2.4.

2.3 Theory of resonator

The concept of a micromechanical resonator can be easily understood from the simple macroscale example of a guitar string of a particular length and made of nickel and steel alloy which vibrates at a distinct resonance frequency when plucked. Hence, it mechanically selects a particular frequency with a Q of about a few hundred, which is far better than passive electronic resonators. Now if the dimensions of such a string are scaled down to the micron level, and fabricated with IC-compatible materials like silicon, polysilicon, etc., and excited electrostatically or piezoelectrically rather than mere plucking, a microscopic-guitar is created, named *clamped-clamped* (CC) beam resonator (as illustrated in Figure 2.4) in MEMS terminology, which can provide us with a resonance frequency of about 10 MHz at a Q of about 10,000 [13]. This is one of the simplest versions of MEMS resonators.

Such a MEMS resonator essentially consists of the following three components:

1. An input-transducer which converts the input electrical signal into a mechanical signal, that is, an electrostatic force (alternatively piezoelectric, magnetostatic, etc.).
2. A mechanical resonant structure which can vibrate in one or more modes due to the produced electrostatic force.
3. An output-transducer that senses the motion of the vibrating structure, hence converting the mechanical signal back to an output electrical signal.

Every mechanical structure, such as a disk or a beam, has several natural modes of resonance at which they will vibrate. This is analogous to wine glasses which vibrate at a distinct audible frequency when they are firmly hit. At such *macroscale*, these occur at very low frequencies—typically lower than a few kiloHertz. By shrinking the dimensions into the *microscale*, much higher resonant frequencies can be attained (e.g., several MHz or even up to GHz).

Mechanical structures are forced into resonance by applying forces onto them at specific frequencies, known as the resonant frequencies. Figure 2.4(a) shows a

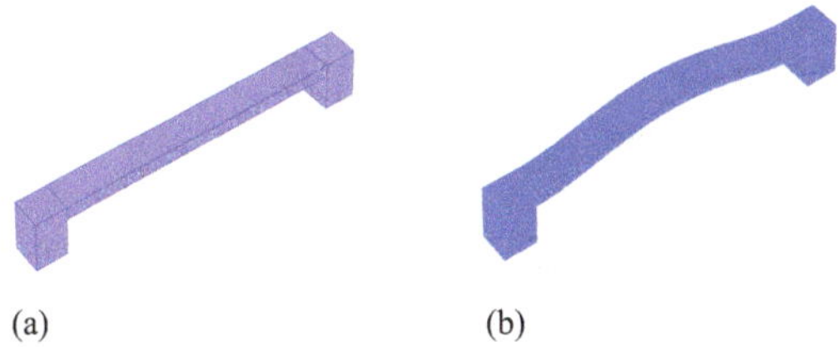

Figure 2.4 Clamped-clamped beam (b) 3D representation of the deformed beam at its natural eigenfrequency

beam having its movement constrained at both ends. This beam configuration is commonly known as "clamped-clamped" (c-c). Also Figure 2.4(b) shows are the deformed shapes for the "first" flexural-mode, which has the lowest resonant frequency and is labeled as "flexural" because the movement is in a direction parallel to the beam thickness. In Figure 2.4(b), the 3D representation of the deformed shape for an equivalent doubly anchored MEMS beam is shown, outlining the maximal displacement in the middle of the beam. The resonant frequencies of such beams are relatively high, because of their small size. As an example, a clamped-clamped beam that is 40 μm long would have a typical resonant frequency in excess of 8 MHz [48], depending on the structural material used. For smaller beams, the resonant frequency can reach values as high as ~100 MHz [24].

To provide the mechanical forces necessary to drive the resonator into resonance, transducers are needed. These transducers are principally operated by the thermal, magnetic, piezoelectric or electrostatic ways. The piezoelectric transducers are more efficient for the operation of resonance in comparison with other types of transducer. But the materials used in the piezoelectric transducer mostly limit their on-chip use in filters. However, due to the simplicity in fabrication and operation, electrostatic transducers are considered in this implementation of the filter. To understand the details of electrical and mechanical forces acting on the resonators requires a model. The design of the resonator can use this model for calculating physical parameters.

2.3.1 Mass damper model and electrical equivalent model

The behavior of a micromechanical resonator is modeled as a damped, driven harmonic oscillator. The micromechanical structure in Figure 2.4 (a doubly clamped oscillator) is identical to a simple mass-spring system (Figure 2.5) where the force pulling the suspended mass back toward the equilibrium position is provided by the stiffness of the material. In extensional mode resonators, the spring constant is determined by the elongation resistance (Young's modulus) in the resonator mass, while in flexural resonators, the spring is represented by the bending resistance of a membrane.

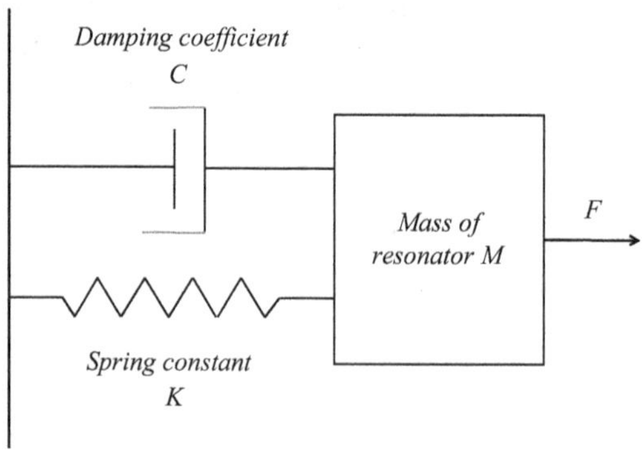

Figure 2.5 Mass damper model of vibrating structure

The mass-spring mechanical system is governed by a linear differential equation that balances forces according to Newton's Second Law. The forces acting on the mass are the external driving force, $F(t)$, the restoring force proportional to the displacement, and the damping force proportional to the speed:

$$m\left(\frac{d^2x}{dt^2}\right) + c\frac{dx}{dt} + kx = F(t) \tag{2.1}$$

An analogous system that obeys a similar differential equation is the inductor capacitor-resistor (LCR) tank circuit

$$L\left(\frac{d^2q}{dt^2}\right) + R\frac{dq}{dt} + \frac{q}{C} = V(t) \tag{2.2}$$

where charge (q) is the electrical equivalent of the mechanical displacement around the equilibrium position (x), the inductance (L) is related to the mass (m), the resistance (R) is the damping or loss coefficient (c), and the capacitance (C) is inversely related to the mass-spring system spring constant, (k).

Proceeding with the mechanical model, dividing out the mass and writing $c = \omega_0 m/Q$

And $k = \omega_0^2 m$, where Q is quality factor (2.1) becomes

$$\left(\frac{d^2x}{dt^2}\right) + \frac{\omega_0}{Q}\frac{dx}{dt} + \omega_0^2 x = \frac{F(t)}{m} \tag{2.3}$$

Solving for x, we get

$$x = \frac{F}{m(\omega_0^2 - \omega^2 + j\omega_0\omega/Q)} \tag{2.4}$$

From this, we can plot the magnitude and phase of displacement

$$|x| = |F| \times \frac{1}{\sqrt{m((\omega_0^2 - \omega^2)^2 - (\omega_0\omega/Q)^2)}} \tag{2.5}$$

And phase θ is given by

$$\theta = \tan^{-1}\left(-\frac{\omega_0\omega}{Q(\omega_0^2 - \omega^2)}\right) \tag{2.6}$$

Thus, we have relationships with electrical and mechanical domains; now we will calculate the actual effect of electrical forces on the mechanical moments.

2.3.2 Actuation

There are well-known methods for transforming electrical energy carried by a voltage signal to mechanical energy in an RF MEMS resonator like thermal, piezoelectric, magnetic, and electrostatic. In this implementation of the filter, electrostatic actuators, the most common energy conversion method, rely on the force between two charged plates

$$F = -\frac{1}{2}\frac{q^2(t)}{C \cdot d} \tag{2.7}$$

where q is the charge on each plate, C is the capacitance, and d is the gap distance between the two plates [49,50]. Most of the reported MEMS resonators, such as comb drives [40] doubly clamped beams [24], and BAW resonators [51], function with electrostatic actuation. Typically, one plate corresponds to the freely suspended MEMS resonator and the second plate is anchored to the substrate. When a voltage difference is applied across the two plates, the suspended plate will be drawn toward the fixed plate. The equation of motion follows the form of (2.1) replacing $F(t)$ with an electrostatic driving force,

$$m\left(\frac{d^2x}{dt^2}\right) + c\frac{dx}{dt} + kx = -\frac{1}{2}\frac{Q^2(t)}{\varepsilon \cdot A} \tag{2.8}$$

where A is the area of the plate and ε is the dielectric permittivity. The driving force can be rewritten as

$$\frac{1}{2}\frac{Q^2(t)}{\varepsilon \cdot A} = \frac{1}{2}\frac{\varepsilon A V_{\text{in}}^2}{(d-x)^2} = \frac{1}{2}\frac{\varepsilon A V_{\text{in}}^2}{d^2(1-x/d)^2} \tag{2.9}$$

where x/d is a very small quantity and can be neglected.

Additionally, V_{in} has both an AC and DC component and is written as

$$V_{in} = V_o + v_{ac}e^{j\omega t} \tag{2.10}$$

Substituting a linearized (2.9) and (2.10) into (2.8) and solving for the x, we find the displacement of the electrostatic resonator behaves as

$$x_{ac} = v_{ac}\frac{\varepsilon A V_0/d^2k'}{j\omega/Q\omega_0 + 1 - (\omega/\omega_0)^2} \tag{2.11}$$

where k' is the DC tunable spring constant

$$k' = k \cdot \left(1 - \frac{\varepsilon A V_0^2}{k d_0^3}\right) \tag{2.12}$$

Electrostatic actuation is advantageous since it is not dependent on material properties, unlike piezoelectric actuation. It is very effective at high frequencies, and the large impedance reduces power consumption. However, there are numerous disadvantages that electrostatic actuation must overcome like significant non-linearities are inherent to the driving force. Both nanometer-sized gaps (which arise critical fabrication challenges) as well as large DC voltages (often over 100 volts) are required to produce sufficient driving forces and detectable motion. Additionally, electrostatic fields generated in the transducer can be influenced by outside forces and can couple with the MEMS motion detection mechanism. Due to the capacitive nature of the actuation method, the input impedance is very high. Consequently, in order to interface a MEMS filter with a standard 50 Ω network, a lossy and bulky matching network is required if substantial signal reflection from the impedance mismatch is to be avoided, which is a part of the design process. The signal detection is again done by converting mechanical motion into electrical signals, which is discussed in Section 2.3.3.

2.3.3 Detection

One of the most common methods for detecting the mechanical motion of a MEMS device is by monitoring the capacitive characteristics of the electromechanical resonator. A micromechanical resonator can be configured to have a capacitive output port, that is, the resonator membrane forms one nonstationary side of the capacitor and the substrate or adjacent structure forms the static side of the capacitive output port. The current flowing through the MEMS capacitor is

$$i = \frac{dQ}{dt} \tag{2.13}$$

Since at resonance the gap is varying with time, (2.13) becomes

$$i = V_0 \frac{d(\mathrm{CV})}{dt} \tag{2.14}$$

which, following the notation of (2.9) and (2.10) and again rewriting in terms of $x/d0$, can be written as

$$i = \frac{d\left(\dfrac{\varepsilon A}{d-x}(V_0 - v_{\mathrm{ac}})\right)}{dt} = \frac{d\left(\dfrac{\varepsilon A}{d}\left(1+\dfrac{x}{d}\right)(V_0 - v_{\mathrm{ac}})\right)}{dt} \tag{2.15}$$

Thus, it has been seen from (2.15), the current depends on deformation occurred during resonance. Hence by measuring the displacement current produced by the vibrating resonator, the characteristics of the mechanical motion are to be determined. When monitoring the current, the MEMS resonator is viewed as a two-port system where vibrations are induced and measured on separate actuators. Like the electrostatic actuator, displacement current detection has disadvantages due to its capacitive (high impedance) nature, its susceptibility to electrostatic cross-talk, its nonlinear dependence on $d0$, and the necessity for small gaps. The electrostatic MEMS resonator is also be implemented as a one-port system, where motion is both induced and sensed with the same capacitive transducer.

Substituting x from (2.11) into (2.15) and solving for the impedance, $Z(\omega) = v/i$

$$Z(\omega) = \frac{\omega^2 - j\omega\dfrac{\omega_0}{Q} - \omega_0^2}{j\omega C\left(\omega^2 - j\omega\dfrac{\omega_0}{Q} + \omega_{AR}^2\right)} \tag{2.16}$$

where ω_0 is the resonant frequency and ω_{AR} is defined as the antiresonant frequency

$$\omega_{\mathrm{AR}}^2 = \omega_0^2\left(1 + \left(\frac{\varepsilon A V_0}{g_0^2}\right)\frac{1}{k'C}\right) \tag{2.17}$$

Depending on the frequency of stimulation, the impedance of the actuator will vary, enabling detection of the resonant motion.

The design process based on these models is illustrated with the case study for two geometries, disk, and ring resonators.

2.4 Case study: disk resonator

Micromechanical resonators can have different shapes like beams, square plates, circular disks, annular rings, comb, etc., and can again be classified according to their modes of operation, namely *flexural*, *torsional,* and *bulk* mode devices as shown in Figure 2.6 [37].

- Flexural mode of vibration is representative of transverse standing waves. In such devices, the displacement of the structures is orthogonal to the bending stress in the structure.
- In resonators vibrating in the torsional mode, the dominant stress is shear-stress and the displacement produced is rotational in nature.
- Bulk mode operation can be described in terms of standing longitudinal waves.

In general, bulk-mode vibration of microresonators is preferred for a high-frequency generation due to larger structural stiffness in comparison to other modes. Moreover, bulk-modes yield higher Q relative to flexural mode resonators of the same frequency. This is due to the fact that flexural modes have comparatively larger surface-to-volume ratios than bulk mode resonators, thus leading to increased losses from surface effects [37].

Hence, the bulk-mode operation is emphasized in this paper. A common example of a bulk-mode device is a circular-disk resonator which can vibrate in two distinct modes as illustrated in Figure 2.7(a) *radial-contour* (or, *extensional* or, *breathing*) mode where the shape of the disk expands and contracts equally in all

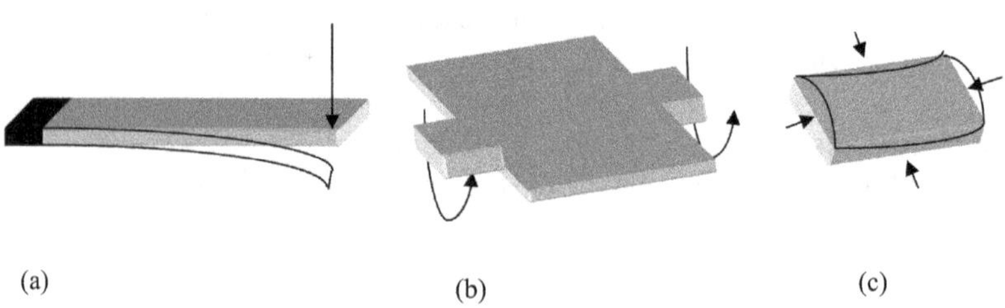

Figure 2.6 Modes of operation of resonator classified as (a) flexural mode (b) torsional mode, (c) and bulk mode (force is indicated by arrows)

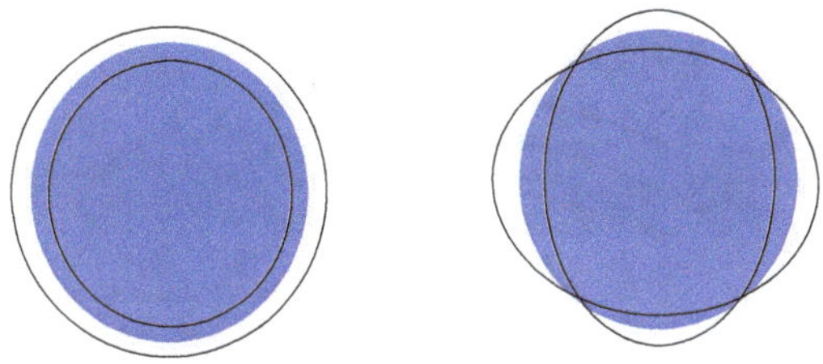

(a) (b)

Figure 2.7 Mode shape of circular disk resonator: (a) radial-contour mode or (b) wine-glass mode

the lateral surface, and Figure 2.7(b) *elliptical wine-glass* mode where disk expands along the circumference of the disk. For elliptical mode, the disk expands along one axis and contracts in the orthogonal axis forming two alternate and perpendicular ellipses per cycle of vibration with four nodal points at the perimeter. The wine-glass mode has expansion and contraction in-plane for higher modes. In this work, we consider disk in elliptical mode i.e. wine-glass mode of operation.

2.4.1 Design and operation

Figure 2.8 shows the perspective schematic view of disk resonator with radius R clamped at antinodal points. The electrodes with an arc length of θ_e are placed with a gap of d_d and d_s for driving and sensing electrodes, respectively. The principle operation of the resonator depends on DC-biased voltage (Vp) and excitation AC voltage (Vac). This device is operated with two driving and two sensing electrodes. Both driving electrodes are connected to ac drive voltage. With origin set at the center of the disk, the plane polar coordinates and is used in this work, as shown in Figure 2.8. The disk resonator vibrates in the in-plane elliptic bulk-mode illustrated by the dotted line, which involves both radial and circumferential displacements in the disk. This elliptic bulk-mode has four resonant nodes at the disk periphery, located 90° apart from one another, where the radial displacements diminish. In order to reduce the support loss in the disk, the support beam is located at one of these four resonant nodes, 45° away from the center of the drive electrode. In order to excite the elliptic bulk-mode with the maximum electromechanical coupling, the centers of the two electrodes are aligned with the line, where the radial displacements at the edge are maximum.

The signal is sensed by two different topologies, common-mode or differential mode configuration. The disk vibrates at resonance frequency when it is excited by driving electrodes. The disk is suspended by a support beam fixed at anchors. The placement of anchored points decides the mode of vibration. In this implementation, the disk is anchored at antinodal points of the elliptical mode of resonance. The support at antinodal points does not allow transferring large magnitude of acoustic vibration in substrate thereby yielding high-quality factors.

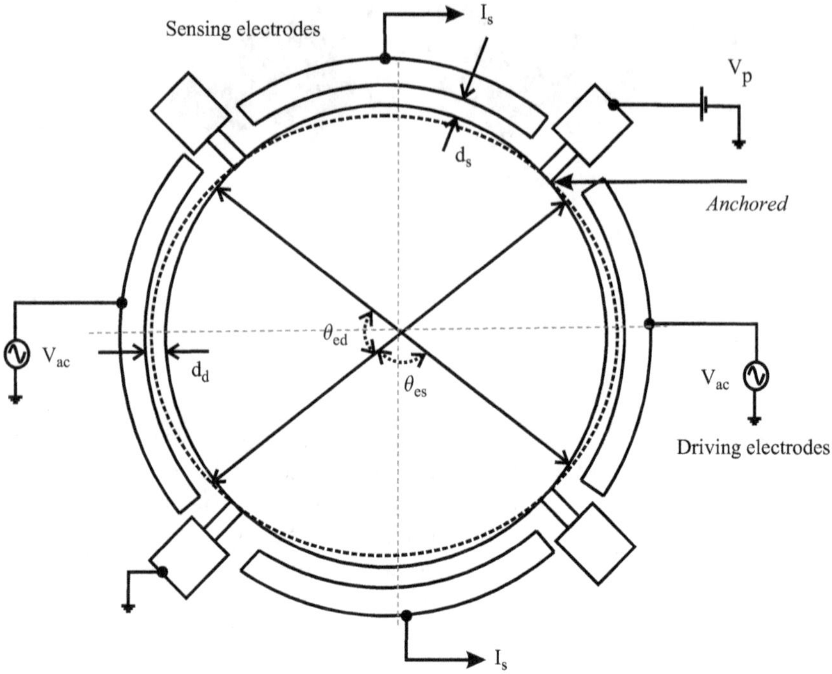

Figure 2.8 Schematic view of disk resonator

2.4.2 In-plane bulk mode resonance and modal shape estimation

In this section, a comprehensive derivation of the in-plane vibrations of a disk to obtain mathematical expressions for the mode shapes and resonant frequencies is discussed. This mathematical formulation is very useful in getting initial guess before staring FEM simulation of resonators. It is assumed that the vibration variables are independent of the thickness, and the support beam has a negligible effect on the in-plane vibrations of the disk. These assumptions are valid as long as the resonator thickness is much smaller than its diameter and the support beam size is much smaller than the disk size. Thus, the disk is modeled as a circular thin-plate with a free edge. For simplicity, it is assumed that the disk resonator is made of isotropic and homogeneous material. The 2D elastic theory governing the in-plane vibrations of a disk, in the absence of body forces, may be written in the following format [37]:

$$\frac{E}{1-v^2}\cdot\nabla(\nabla\cdot\vec{\mu})-\frac{E}{2\cdot(1+v)}\cdot\nabla\times\nabla\times\vec{\mu}=\rho\cdot\frac{\partial^2\vec{\mu}}{\partial t^2} \tag{2.18}$$

Where E, v, and ρ represents Young's modulus, Poisson's ratio, and mass density resonator material, respectively.

The displacement vector may be defined in terms of the pressure-wave (P-wave) scalar potential Φ and the shear-wave (S-wave) vector potential, $\vec{\Psi}$, via Helmholtz' theorem [52] as

$$\vec{\mu} = \nabla\Phi + \nabla \times \vec{\Psi} \tag{2.19}$$

The displacement variables are independent of the thickness of disk, that is, independent of z-coordinate.

By substituting (2.19) into (2.18) and taking the divergence and curl of (2.18), respectively, the elastic equations for P-wave and S-wave may be written as

$$\frac{\partial^2\Phi}{\partial t^2} = \alpha^2 \cdot \nabla^2\Phi \tag{2.20}$$

$$\frac{\partial^2\Psi}{\partial t^2} = \alpha^2 \cdot \nabla^2\Psi \tag{2.21}$$

where $\nabla^2 = (1/r) \cdot (\partial/\partial r) + (\partial^2/\partial r^2) + (1/r^2) \cdot (\partial/\partial\theta)$ in polar coordinates; $\alpha = \sqrt{E/\rho(1-\nu^2)}$ and $\beta = \sqrt{E/2\rho(1+\nu)}$ are the propagation velocities of the P-wave and S-wave, respectively.

The solutions to (2.20) and (2.21) expressed as

$$\Phi_m = A_m \cdot J_m(k_m r/R) \cdot \cos(m\theta).e^{j\omega_m t} \tag{2.22a}$$

$$\Psi_m = B_m \cdot J_m(h_m r/R) \cdot \sin(m\theta).e^{j\omega_m t} \tag{2.22b}$$

where J_m is Bessel functions of the first kind (J_m) and m is the mode order, it is equal to or greater than 2. The $m = 0$ corresponds to mode shapes that are independent of the circumferential direction, with displacements solely either in the radial direction (radial) or in the circumferential direction (torsional). And $m = 1$ involves a nonzero deformation at the center of disk (translational). For this disk resonator, we consider resonant modes at disk circumference.

In (2.22a) and (2.22b), ω_m is angular resonant frequency and A_m and B_m are the constants of the elastic waves, in the unit of square meter. And k_m and h_m are both dimensionless frequency parameters, respectively, expressed as

$$\begin{aligned} k_m &= \omega_m \cdot R/\alpha \\ h_m &= \omega_m \cdot R/\beta \end{aligned} \tag{2.23}$$

Substituting (2.21) into (2.19) leads to the radial (U) and circumferential (V) components of the displacement vector, respectively, expressed as

$$U_m = \left[A_m \cdot \frac{d}{dr} J_m(k_m r/R) + \frac{m}{r} B_m \cdot J_m(h_m r/R)\right] \cdot \cos(m\theta) \tag{2.24a}$$

$$V_m = \left[A_m \cdot \frac{-m}{r} J_m(k_m r/R) - B_m \cdot \frac{d}{dr} J_m(h_m r/R)\right] \cdot \sin(m\theta) \tag{2.24b}$$

For a disk with a free edge, the boundary conditions at $r = R$, radial normal stress (σ_r), and circumferential shear stress (τ_r) are 0. The expression for σ_r and τ_r both solves (2.24)

$$\sigma_r|_{r=R} = \frac{E}{(1-\nu^2)}\left[\frac{\partial U}{\partial r} + \frac{\nu}{r}\cdot\left(U + \frac{\partial V}{\partial \theta}\right)\right] = 0 \tag{2.25a}$$

$$\sigma_r|_{r=R} = \frac{E}{2.(1+\nu)}\left[\frac{\partial V}{\partial r} + \frac{1}{r}\cdot\left(\frac{\partial U}{\partial \theta} - V\right)\right] = 0 \tag{2.25b}$$

After solving this equation, we get A_{m} and B_m in matrix form, that is

$$\begin{bmatrix} a_{11} & a_{12} \\ a_{21} & a_{22} \end{bmatrix}\cdot\begin{bmatrix} A_m \\ B_m \end{bmatrix} = 0 \tag{2.26}$$

which is solely associated with k_m, h_m, and ν, while h_m can be expressed as $h_m = k_m \cdot \sqrt{2/(1-\nu)}$ from (2.23) and (2.24)

The solution of A_m and B_m obtained when the determinant of the matrix is zero. This determinant is zero only for eigenvalue which corresponds to the resonant frequency of in-plane vibrations of a disk. The eigenvalue equation for the resonant frequency can be expressed as follows:

$$\left[\frac{k_m \cdot J_{m-1}(k_m)}{J_m(k_m)} - m - \frac{h_m^2}{2(m^2-1)}\right]\cdot\left[\frac{h_m \cdot J_{m-1}(h_m)}{J_m(h_m)} - m - \frac{h_m^2}{2(m^2-1)}\right]$$
$$-m^2\cdot\left[\frac{h_m^2}{2(m^2-1)} - 1\right] = 0 \tag{2.27}$$

where J_m is Bessel function of the first kind, m is number of modes, k_m and h_m are nondimensional frequency parameters. The relation between h_m and k_m is $h_m = k_m \cdot \sqrt{2}/(1-\nu)$. This equation is used to find a frequency parameter. The resulting values of k_m and h_m could be verified with [52].

The resonance frequency is calculated as [37] (2.28)

$$f_m = \frac{k_m}{2\Pi \cdot R}\sqrt{\frac{E}{\rho(1-\nu^2)}} \tag{2.28}$$

where R is a radius of disk, E is Young's modulus, ρ is mass density per volume, and ν is Poisson's ratio.

From matrix:

$$\xi_m = \frac{B_m}{A_m} = \frac{J_m(k_m)}{J_m(h_m)}\cdot\frac{2\cdot\dfrac{k_m \cdot J_{m-1}(k_m)}{J_m(k_m)} + h_m^2 - 2m(m+1)}{\left[\dfrac{h_m \cdot J_{m-1}(h_m)}{J_m(h_m)} - (m+1)\right].2m} \tag{2.29}$$

The ratio gives the mode shape of resonating modes of a disk. Both the eigenvalue (2.27) and ratio (2.29) are solely functions of Poisson's ratio of the resonator structural material.

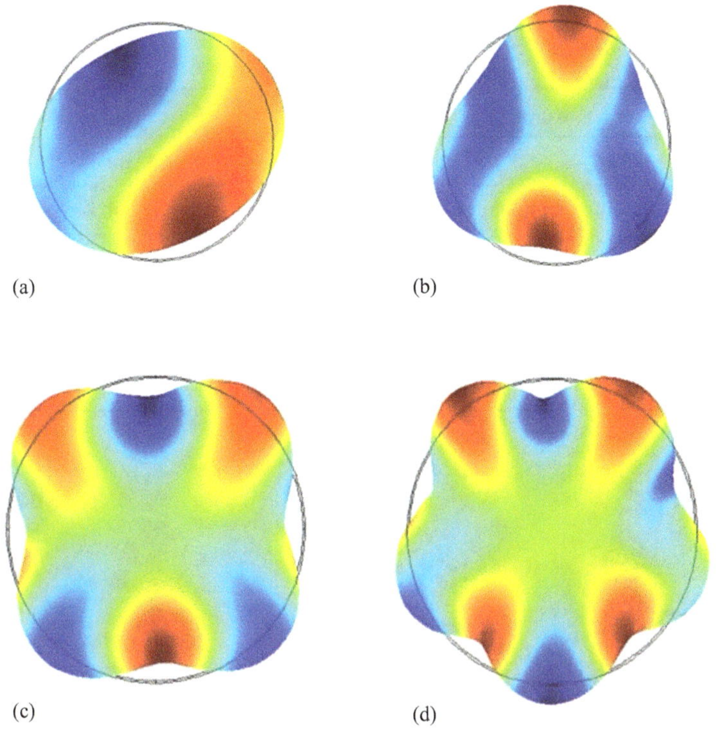

Figure 2.9 Bulk modes of resonance of disk in COMSOL FEM software: (a) elliptical bulk mode, that is, mode 2; (b) triangular bulk mode, that is, mode 3; (c) tetragonal bulk mode, that is, mode 4; and (d) pentagonal bulk mode, that is, mode 5

The modal shape for each mode of vibration of a disk is estimated using COMSOL MultiPhysics. Figure 2.9 shows the bulk modal response of disk simulated in COMSOL MultiPhysics. COMSOL FEM results are verified with numerical calculations performed with the above-mentioned formulation as shown in Table 2.1.

After studying the mechanical behavior of the disk we now will now look at the electromechanical aspect of the resonator.

2.4.3 Electromechanical model of disk resonator

A MEMS resonator consists of a resonant mechanical structure and transducers for energy conversion between electrical and mechanical domains, shown in Figure 2.10. The input transducer provides dc actuation and ac excitation to the resonator. The input voltage with dc with a small magnitude of ac is applied to the driving electrode. This electrostatic energy converts in equivalent mechanical force which transfers energy to the vibrating structure. The resonant mechanical structures, regardless of whether their shapes are that of beams, disks, rings, or plates, can all be modeled as spring-mass-damper systems. The equivalent mass at any

Table 2.1 Resonance frequencies for free, fixed disk, and theoretical calculation (frequencies are in GHz)*

	Dimensions		
Disk	**Radius = 1.75 μm, thickness = 0.4 μm**		
Mode	**Resonance frequency for free disk**	**Resonance frequency for the clamped disk at antinodal locations**	**Resonance frequency theoretically calculated**
Mode 2	2.4820	2.5052	2.4874
Mode 3	3.7672	4.0396	3.7711
Mode 4	4.8794	5.4796	4.8717
Mode 5	5.9271	6.3833	5.9011

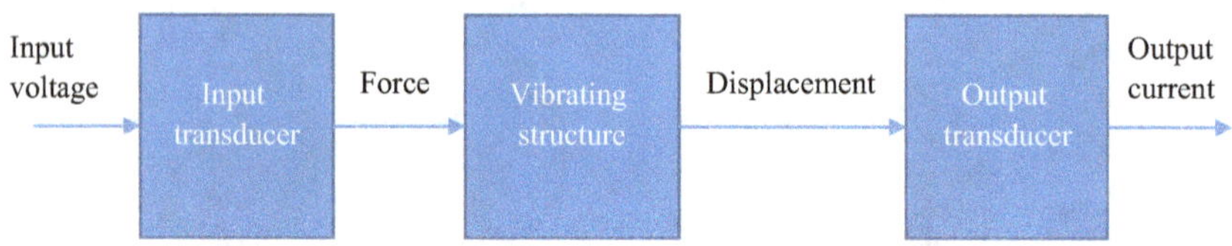

Figure 2.10 Block diagram of MEMS resonator

given point of the resonator is determined from the total kinetic energy of the resonator KE and the velocity at that point v(x,y). The output transducer senses the change in displacement at the sensing electrode and induces a current. This electrostatic energy comes from the mechanical vibration of the resonating structure. So, at output transducer kinetic energy of vibrating structure converts back into electrostatic energy.

Figure 2.11 shows the electro-mechanical model of resonator stating one to one equivalent for electrical and mechanical domains. The transduction coefficients, η_1 and η_2 model energy transduction efficiency between electrical and mechanical domains by considering the amount of mechanical force F generated from voltage input v_i and the output current i_o generated from mechanical displacement x, they are expressed as follows:

$$\eta_1 = \frac{F}{v_i} \eta_2 = \omega_0 \frac{i_0}{x} \tag{2.30}$$

The mass damper model for a disk is solved for further analysis of the disk resonator. The dynamic behavior of this infinitesimal element along the circumferential direction, θ, is described by the second-order equation of motion:

$$\left\{ m(\theta) \cdot \frac{d^2}{dt^2}[A_r \cdot U_R] + C_d \frac{d}{dt}[A_r \cdot U_R] + K(\theta) \cdot [A_r \cdot U_R] \right\} \cdot \cos(2\theta) = f_e(\theta) \tag{2.31}$$

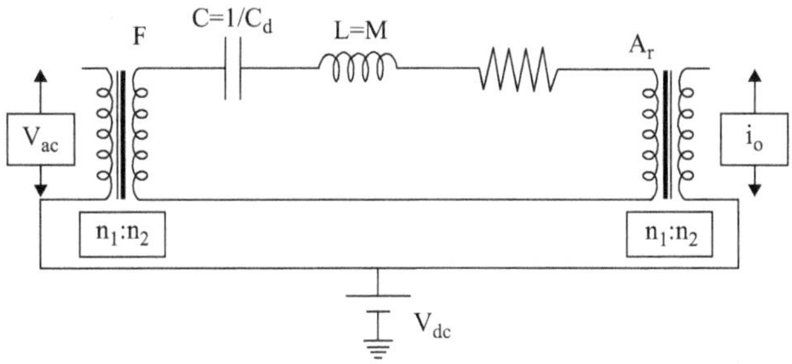

Figure 2.11 Equivalent electromechanical model of resonator

In this second-order difference equation, U_R represents the maximum radial displacement at the disk edge. Since the excitation and detection of this disk resonator are mainly through the gap variation along the radial direction, only the vibration variables along the radial direction are considered here. Through combining (2.24) and (2.29), the radial displacement at the location can be rewritten as

$$U = A_r \cdot \cos(2\theta) \cdot U_r \tag{2.32}$$

where $U_r = k \cdot J_1(kr') - \frac{2}{r'} \cdot J_2(kr') + \frac{2}{r'} \cdot \xi J_2(hr')$ and $r' = r/R$ denotes the dimensionless radial coordinate, normalized to disk radius R. The effective mass for an infinitesimal element, $d\theta$, can be expressed as

$$m(\theta) = \rho \cdot h \cdot R^2 \cdot \Sigma / U_R^2 \cdot d\theta \tag{2.33}$$

where $\Sigma = \int U_r^2 \cdot r' \cdot dr'$ is the integral for kinetic energy and $U_R = k \cdot J_1(k) - \frac{2}{r'} \cdot J_2(k) + 2 \cdot \xi J_2(h)$ is the dimensionless maximum radial displacement at the disk edge. As both Σ and U_R are functions of solely the Poisson's ratio of the material used. And C_d is the damping related coefficient for this element and $f_e(\theta)$ is the radial electrostatic force per unit radian from the drive and sense electrodes. Since the capacitive gap is extremely small compared with the disk radius, the capacitances for the drive and sense electrodes are to be calculated using a parallel-plate model. Thus, the electrostatic excitation force per unit radian from the drive and sense electrodes, respectively, are calculated as

$$f_d(\theta) = \frac{1}{2} \frac{\varepsilon h R}{d_d^2} \left[-2 \cdot V_{dc} \cdot v_{ac} + 2 \cdot A_r \cdot \frac{U_R \cdot \cos(2\theta)}{d_d} \cdot V_{dc}^2 \right] \tag{2.34}$$

$$f_s(\theta) = \frac{1}{2} \frac{\varepsilon h R}{d_s^2} \left[2 \cdot A_r \cdot \frac{U_R \cdot \cos(2\theta)}{d_s} \cdot V_{dc}^2 \right] \tag{2.35}$$

where d_d and d_s are the capacitive gaps for the drive and the sense electrodes, respectively, and ε denotes the permittivity of air.

Integrating this equation for θ from 0 to 2π gives a modified equation as follows:

By using this equation electrostatic stiffness and equivalent force for elliptic bulk-mode, respectively are expressed as

$$K_e = \varepsilon hR\left(\frac{\theta_e}{2} + \frac{\sin(2\theta_e)}{4}\right) \cdot V_{dc}^2\left(\frac{1}{d_d^3} + \frac{1}{d_s^3}\right) \tag{2.36}$$

$$F = \frac{\varepsilon hR}{d_d^2 \cdot U_R} \cdot (V_{dc} \cdot v_{ac}) \cdot \sin(\theta_e) \tag{2.37}$$

And the equivalent mass damper model is given by

$$M\left(\frac{d^2 A_r}{dt^2}\right) + C_d\frac{dA_r}{dt} + (K - K_e)A_r = F \tag{2.38}$$

where

$$M = \pi \cdot \rho \cdot h \cdot R^2 \cdot \Sigma / U_R^2 \tag{2.39}$$

$$K = M \cdot \omega^2 \tag{2.40}$$

2.4.4 Electrical model

The two-port electrical equivalent circuit model for the disk resonator can be developed by the derivation of its four-parameters (admittance parameters), which are defined as the ratio of the current measured at one port to the drive voltage at the same or the other port while the undriven port of the circuit is shorted to ground. For the micromechanical resonators, the admittance parameter can be further expressed in terms of the mechanical force-displacement transfer function and displacement-current transfer functions for the disk resonator $Z(j\omega)/F(j\omega)$, $Q(j\omega)/Z(j\omega)$, respectively. The electromechanical coupling at the input and output ports, $\eta_1, \eta_2, \eta_1', \eta_2'$. Here Z denotes the displacement amplitude of vibration A_r.

The input and output coupling terms are expressed as

$$\eta_1(j\omega) = \frac{F(j\omega)}{v_d(j\omega)} \tag{2.41a}$$

$$\eta_2(j\omega) = \frac{Q_s(j\omega)}{Z(j\omega)} = \frac{1}{j\omega} \cdot \frac{i_s(j\omega)}{Z(j\omega)} \tag{2.41b}$$

$$\eta_1'(j\omega) = \frac{Q_d(j\omega)}{Z(j\omega)} = \frac{1}{j\omega} \cdot \frac{i_d(j\omega)}{Z(j\omega)} \tag{2.41c}$$

$$\eta_2'(j\omega) = \frac{F(j\omega)}{v_s(j\omega)} \tag{2.41d}$$

where Q_s and Q_d are the charges going through the sense and drive electrode, respectively. While the electromechanical coupling from the drive electrode the sense electrode is denoted by η_1, η_2, the coupling from the sense electrode to the drive electrode is denoted by $\eta_1^{'}, \eta_2^{'}$.

Substituting (2.37) into (2.41a) and (2.41d) gives rise to the following expressions for the voltage-force transfer functions at the sense and drive electrodes:

$$\eta_1(j\omega) = \frac{\varepsilon hR}{d_d^2 \cdot U_R} \cdot (-V_{dc}) \cdot \sin(\theta_e) \tag{2.42}$$

$$\eta_2^{'}(j\omega) = \frac{\varepsilon hR}{d_s^2 \cdot U_R} \cdot (-V_{dc}) \cdot \sin(\theta_e) \tag{2.43}$$

The displacement-current transfer functions at the sense and drive electrodes can be written as

$$\eta_1^{'}(j\omega) = \frac{\varepsilon hR}{d_d^2} \cdot (-V_{dc}) \cdot \sin(\theta_e) \cdot U_R \tag{2.44}$$

$$\eta_2^{'}(j\omega) = \frac{\varepsilon hR}{d_s^2} \cdot (-V_{dc}) \cdot \sin(\theta_e) \cdot U_R \tag{2.45}$$

Substituting above equations in transfer function results in the transfer functions in the form of admittance of series *RLC* tanks with the equivalent inductance, capacitance, and resistance expressed, respectively, as

$$\begin{aligned} L_{21} &= \frac{K}{\eta_1 \cdot \eta_2 \cdot \omega^2} \quad & L_{11} &= \frac{K}{\eta_1 \cdot \eta_1^{'} \cdot \omega^2} \\ C_{21} &= \frac{\eta_1 \cdot \eta_2}{K(1 - K_e/K)} \quad & C_{11} &= \frac{\eta_1 \cdot \eta_1^{'}}{K(1 - K_e/K)} \\ R_{21} &= \frac{\sqrt{K \cdot M}}{Q \cdot \eta_1 \cdot \eta_2} \quad & R_{21} &= \frac{\sqrt{K \cdot M}}{Q \cdot \eta_1 \cdot \eta_1^{'}} \end{aligned} \tag{2.46}$$

The motional impedance is referred to as R_{21}.

It is worth mentioning that depending on the electrode configuration, the output current is in phase or 180° out of phase with respect to the input voltage. In case of in phase displacement of the resonator toward sense and drive electrodes, that is, confronting sense and drive electrodes, the current coming out of the device has 180° phase difference with the input voltage while for the four-electrode configuration.

2.4.5 FEM simulation of disk resonator

The structure for disk resonator with anchor and support beam is realized in COMSOL. Figure 2.12 shows the 3D view of the disk resonator. This structure is simulated for frequency domain analysis with solid mechanics as physics. The force of actuation is imparted on structure such that disk vibrates at only elliptical mode, that is, mode 2 for a disk. The force is imparted to disk periphery in differential configuration such that displacement and quality factor is higher for mode

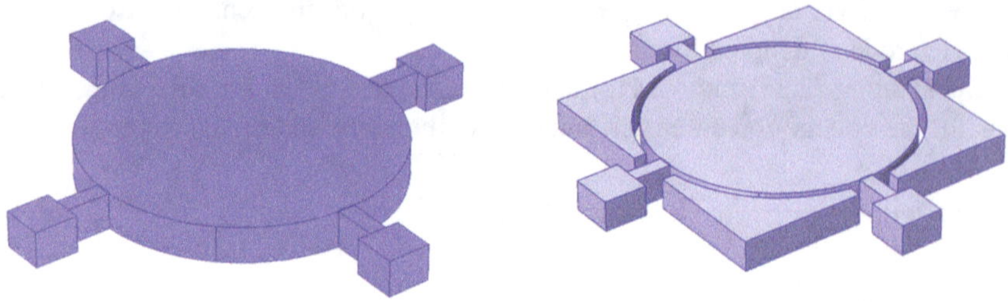

(a) (b)

Figure 2.12 The structure realized in COMSOL: (a) for solid mechanics physics and (b) for electro-mechanics physics

of operation, that is, elliptical mode. This type of actuation is known as differential actuation. So, force of actuation is provided by electrodes. Electrodes are connected to ac voltage. The dc actuation voltage is provided through anchor or electrodes.

2.4.5.1 Electrode shaping

The shapes of the drive and sense electrodes are critical to the operation of the disk resonator in the elliptic and tetragonal bulk-mode. First, the excitation of unwanted modes is avoided by shaping the driving electrode with respect to a particular vibration mode. As illustrated in Figure 2.8, the elliptical bulk-mode in operation is accompanied by its degenerate mode approximately 45° apart in the circumferential direction and other modes that are spaced in the same direction. Second, the shape of the drive electrode, symmetric to the line of resonance, contributes to proper capacitive transduction. Even though there are degenerate modes and higher modes are present but their amplitude of vibration and quality factor is low. Therefore, the span angle of the drive and sense electrodes should be maximized. Finally, both stronger electromechanical coupling and lower motional resistance are desirable from these resonators, requiring the span angle to be maximized.

2.4.5.2 Support beam

The dimensions and positions of support beam the resonance frequency and quality impact resonance frequency and quality factor of resonator. In this design of resonator, the radius of disk is chosen 1.75 μm which put fabrication limits on selecting dimension of support beam. Also dimensions are selected such that overall stiffness at anti nodal points is not lowered which might impact the quality factor. For calculated support beam, stiffness at anti nodal point is significantly lowered. Support beam are placed at anti nodal points so that acoustic waves are not transferred more to substrate which reduces quality factor. Hence, length of beam is selected such that even if beam carry waves their magnitude is less and it does not lowered stiffness at anti nodal points. For length of beam 0.12 μm at anti nodal points provides the stiffness equal to stiffness of disk. This dimension is not chosen for design because of the fabrication limitation, so length of beam is 0.5 μm is chosen.

The Figure 2.13 shows the displacement versus frequency plot for disk resonator which has excitation by semi elliptical-electrodes. The Figure 2.13 shows the spurious resonant modes which needs to avoid for an operation of resonator. The Figure 2.14 shows the displacement versus frequency plot for disk resonator which has excitation by quadratic elliptical-electrodes as shown in Figure 2.12(b). It is observed from the plot that the spurious resonant modes are eliminated and others high frequency resonant modes can be removed in differential mode of sensing.

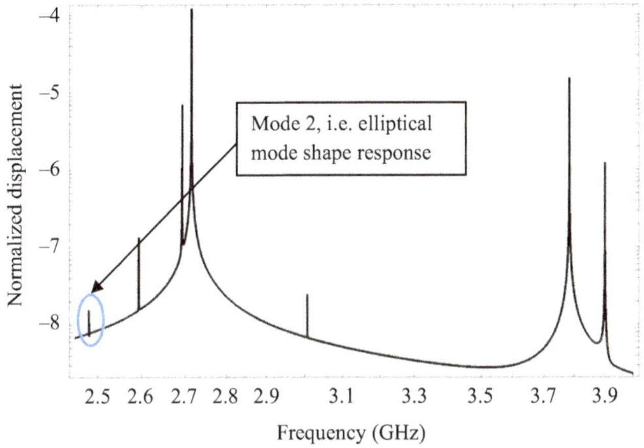

Figure 2.13 Displacement versus frequency plot for disk resonator with electrodes is not properly shaped. The circled response is the desired response for this work

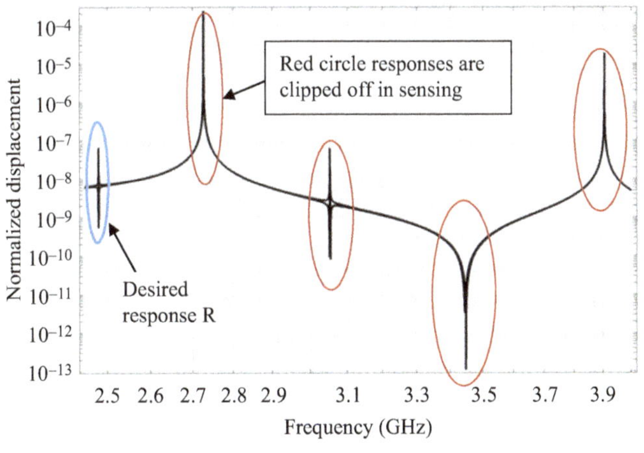

Figure 2.14 Displacement versus frequency plot for disk clamped by 4 support beam

2.4.6 FEM simulation of disk resonator with proposed fabrication flow

Silicon has been the dominant material in the microelectronics revolution of the twentieth century and has been the precursor to the microelectromechanical and nanoelectromechanical systems (MEMS/NEMS) revolution currently underway. It has been the material of choice for current MEMS devices, mainly because devices can be fashioned using standard microfabrication techniques. However, Si is not the best choice for devices where friction and wear are present since its poor mechanical and tribological properties limit its performance. Even for high-frequency applications, Si does not provide efficient results. Thus, new and robust materials with exceptional properties must be sought to meet the stringent demands that MEMS/NEMS devices require. Because of these limitations, other materials such as SiC, GaN, and diamond are now under investigation. Carbon in its various forms, specifically diamond, may become a key material for the manufacturing of MEMS/NEMS devices. The only ambiguity in using this material is its compatibility with the standard CMOS fabrication process.

The anchor losses occurred in resonators which can degrade the quality factor substantially, can be reduced by choosing different materials for fabricating the disk, stem, and I/O electrodes of disk resonators. The one such work has been reported by Wang *et al.* in 2004 in which disk is fabricated with poly-diamond and supporting central stem is made up of polysilicon [51]. This leads to impedance mismatch which consequently reduces the anchor losses.

The resonant frequency depends on the dimensions of a structure as well as on the properties of structural material. The resonant frequency of a MEMS device is proportional to the acoustic velocity, which is $\nu = \sqrt{E/\rho}$, where E and ρ are Young's modulus and density, respectively. It has been observed from the literature, silicon resonators (both poly and single-crystalline) have attracted the maximum attention due to well-established processing technology available in the IC industry and their compatibility with CMOS integration. Also, single-crystal silicon (SCS) has superior mechanical properties and hence, can offer resonators with lower internal losses as compared to those made from polysilicon. An SOI (silicon-on-insulator)-based process can also provide increased thickness of the structural layer, leading to increased transducer area; hence reduced motional resistance and higher power handling capability. However, as the conventional chemical vapor deposition (CVD) techniques cannot be used for depositing SCS, fabricating SCS resonators with ultrathin capacitive transduction gaps becomes difficult. The quest of achieving higher frequencies for resonance leads researcher to scale down dimension, and also to look for material with higher acoustic velocity and more superior than Silicon. The new materials must be compatible with existing CMOS process for batch fabrication. There are various reported work where alternate materials are used to fabricate MEMS resonator which is suited to CMOS process [53].

This section of chapter presents the diamond and derivatives material and its material properties. Diamond has the highest Young's modulus, hardness, and thermal conductivity and it is transparent from the UV to far IR region. Furthermore, its superior electronic properties make it suitable for use in heat sinks and radiation detectors [55]. Diamond is chemically inert, stable at high temperature (1,000 °C in a vacuum) and is suitable for operation in harsh environments [56,63]. Because of these characteristics, diamond is a very good candidate for realizing reliable, high power, and temperature-stable MEMS and microwave devices [56,57]. Thin-film diamond can be classified into a single crystal, microcrystalline (MCD), nanocrystalline (NCD), and ultra-nanocrystalline (UNCD) films. These films are grown on different substrates which are dependent on the respective application.

Table 2.2 compares the mechanical properties of NCD films with other materials used in microsystems technology. Along with positive mechanical attributes, NCD possesses low loss when used as a thin film at microwave frequencies.

Ultra-nanocrystalline diamond is among the best choice for bulk mode MEMS resonator. This material has the highest Young's modulus and low dissipation losses. Ultra-nanocrystalline grain structure provides several advantages over microcrystalline and nanocrystalline. UNCD can have Young's modulus and stiffness as high as a single-crystal diamond. Single crystal diamond has Young's modulus close to 1,200 GPa. The UNCD has Young's modulus close to 1,000 GPa. UNCD film's mechanical and electrical property depends on sp^3 bonding and losses are governed by sp^2 bonding which is mainly at grain boundaries. The nitrogen incorporation in plasma increases graphitization. This reduces resistivity which can minimize dissipation for UNCD film [55]. The new ultra-nanocrystalline diamond (UNCD) developed at Argonne National Laboratory [58] is emerging as one of the most promising forms of diamond with unique multifunctional properties.

The UNCD films are grown using a microwave plasma chemical vapor deposition (CVD) technique involving new CH4/Ar chemistry. The process yields films with extremely small grain size (3–5 nm), significantly smaller than nanocrystalline diamond films (30–100 nm grain size) produced by the conventional CH4/H chemistry [53,58]. The films possess many of the outstanding physical properties of the diamond, that is, they exhibit exceptional hardness, extremely low friction coefficient and wear, high thermal and electrical conductivity (the latter when deposited with nitrogen), high electrical resistance when grown with hydrogen addition to the CH4/Ar plasma, and high chemical inertness, optical transmittance, electrical carrier mobility, and dielectric breakdown strength. UNCD is characterized by a unique microstructure of sp^3-bonded grains and atomic grain boundaries (2–4 Å) with substantial sp^2 coordination. Preliminary results have shown that this unique microstructure results in outstanding mechanical properties (~97 GPa hardness and 967 GPa Young's modulus, which is similar to single-crystal diamond), unique

Table 2.2 Mechanical properties of diamonds

	Si	**SiC**	**Diamond**
Bandgap (eV)	1.12	2.2/2.9	5.5
Beak down field (10^6 V/cm)	0.5	3–6	10
Young's modulus (Gpa)	160	450	1,000–1,500
Fracture strength (Gpa)	1.37	—	10.3
Thermal conductivity (W/cmK)	1.47	4.9	22
Thermal stability (°C)	500		1,500

tribological properties (coefficient of friction of the order of ~0.02–0.0310), and field-induced electron emission (threshold voltage ~2–3 V μm−1 11).

The ultra-nanocrystalline diamond can be made highly conductive by the addition of nitrogen into the gas phase during deposition and this conductivity is n-type. However, this conductivity is not due to doping, but due to the manipulation of the nanostructure of the material, leading to enhanced sp^2 regions and mid-gap states. This leads to low mobility hopping-type conduction processes and impurity band conduction, with very high carrier concentrations and low mobility. Thus, this conductivity mechanism is of little use in active electronics, where the formation of depletion regions is crucial for modulation and rectification. However, UNCD can be highly useful where high carrier concentrations are required, such as in electrochemical electrodes, field emission, heterostructures, high-temperature stable ohmic contacts, etc.

Ultra-nanocrystalline diamond is among the best choice for bulk mode MEMS resonator. This material has the highest Young's modulus and low dissipation losses. Ultra-nanocrystalline grain structure provides several advantages over microcrystalline and nanocrystalline. UNCD can have Young's modulus and stiffness as high as the single-crystal diamond. Single crystal diamond has Young's modulus close to 1,200 GPa. The UNCD has Young's modulus close to 1,000 GPa. UNCD film's mechanical and electrical property depends on sp^3 bonding and losses are governed by sp^2 bonding which is mainly at grain boundaries. The nitrogen incorporation in plasma increases graphitization. This reduces resistivity which can minimize dissipation for UNCD film.

The chemical vapor deposition method can be used to form UNCD on Si substrate. This method uses CH_4 as a precursor in Ar plasma. In this new plasma chemistry, the diamond-growth species are gas-phase carbon dimers, derived either from collisional-induced fragmentation of gas-phase C in an Ar plasma or from methane via the reaction

$$2CH_4 \rightarrow C_2H_2 + 3H_2$$
$$C_2H_2 \rightarrow C_2 + H_2$$

in an atmosphere containing a small amount of hydrogen.

UNCD films can be grown using microwave plasma CVD (MPCVD) or hot filament CVD (HPCVD) process. HPCVD grown film has better quality than MPCVD [56]. Figure 2.15 shows the proposed fabrication flow to deposit UNCD films on Si substrate.

Prior to diamond growth, the wafer needs to go through a seeding step which aids in the growth of the thin film. Seeding is popularly done through three different techniques:

- Mechanical polishing of the wafer: In this technique, nanometer-sized diamond powder is sprinkled on the silicon wafer and the wafer is mechanically scratched. By this, the diamond powder is spread uniformly across the wafer and this acts as a seeding layer in the CVD system.
- Ultrasonication: Here a silicon wafer is suspended in a slurry of nanometer-sized powder with acetone or methanol for 20–30 min. Through this process,

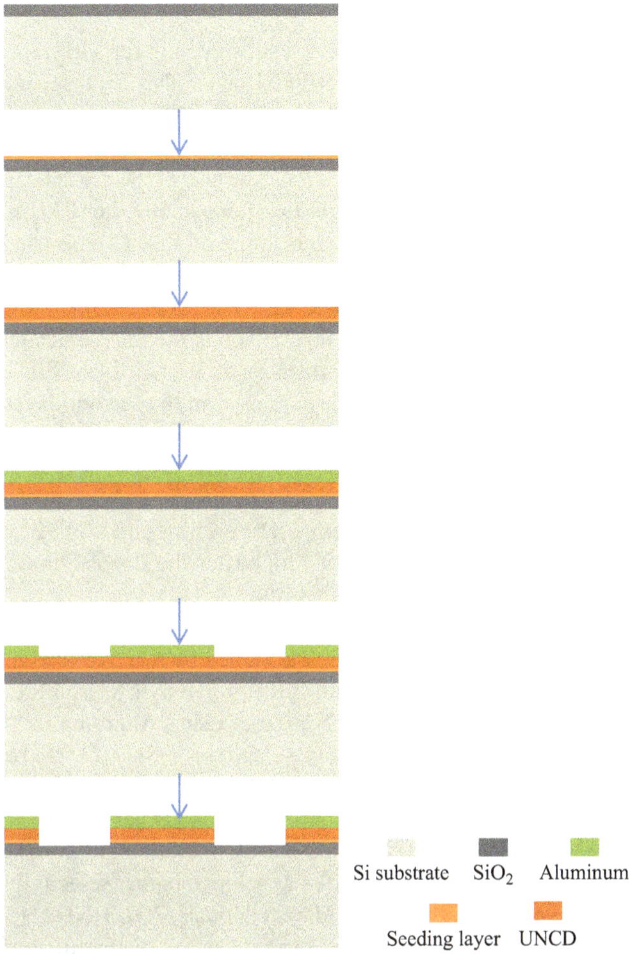

Figure 2.15 Proposed fabrication flow for the design of disk resonator

the surface of the wafer is damaged and seeded with the diamond powder for the subsequent growth process. Nucleation density of 106–1,010 cm^{-2} is achieved through this method [58].

- Bias enhanced nucleation (BEN): Although the first two processes are popular and result in good diamond films, nucleation density is best in the BEN process [56]. In the microwave plasma-enhanced CVD (MPECVD) process, prior to growth, in the BEN stage, the substrate is negatively biased at around 250 volts resulting in a starting current value of 10 mA.

The ultra-nanocrystalline diamond is among the best choice for this work of the high-frequency application. In the recent development of depositing UNCD at low temperatures, it is now possible that UNCD can be compatible with the CMOS process. The following steps deduced the proposed fabrication flow.

1. Thermal oxidation:
   UNCD films can be directly grown on Si substrate but for MEMS application it is advantageous to use SiO_2 as a sacrificial layer. This SiO_2 layer is grown on Si by thermal oxidation for the thickness of about 1 μm.
2. Cleaning and seeding of wafer:
   The substrate is then cleaned by methanol. The size of the crystal depends on the seeding layer and also affects the deposition temperature. For CMOS compatible processes, UNCD deposits at a low temperature of 400 °C. This can be achieved by depositing the thin (10 nm) Tungsten layer first. And then seeding layer of diamond nanoparticles (DNG) with dimethyl sulfoxide diluted with methanol solution is spin-coated on the SiO_2 layer. This is the ultrasonication of the Si-SiO_2 substrate. This decides the grain size of UNCD film. For UNCD film deposition, 4–12 nm of DNG particles are used in the seeding layer.
3. Deposition of UNCD:
   UNCD films can be deposited in the CVD process with CH_4(1%)/N_2(2–5%)/Ar Plasma, 1,200 W power at 2.45 GHz RF frequency, the pressure of 100 Torr. This deposition can be done at a temperature of around 400 °C to 800 °C to achieve 400 nm thickness of UNCD film. After the deposition of the UNCD film, the hard mask is deposited in this film.
4. Etching and patterning of UNCD:
   (a) UNCD film is hard to etch which is etched by reactive Ion etching (RIE). In this process, UNCD diamond films grown by CVD on SiO_2 sacrificial layers were etched in a CF_4/O_2 plasma using Aluminum (Al) as a hard mask. This etch was reasonably selective to SiO_2 (15:1), enabling the fabrication of diamond disk resonators. Al mask is deposited by RF sputtering on UNCD film at 350° C for UNCD patterning.
   (b) The Al film patterning is carried out by photolithography. Again, RIE is used with a mixture of Ar/HBr/Cl to achieve higher selectivity. Further to remove residual Al the wet acidic etchant (H_3PO_4, HNO_3, acetic acid) is used. These fabrication steps are to realize the disk structure.
5. Electrode deposition:
   Polysilicon for the electrode can be deposited by LPCVD. This layer is patterned and etched by lithography. And sacrificial layer SiO_2 is etched by HF. UNCD films are highly resistive to HF, this enables UNCD film and their structure to suspend easily.
6. Deposition of dielectric coating:
   The dielectric is deposited to improve the thermal stability of the device and also to improve motional impedance. For this work, silicon dioxide (SiO_2) and titanium oxide (TiO_2) are considered. The deposition for both layers is for a few nanometers. This deposition is possible to achieve by atomic layer deposition.

The commercial CAD tool CoventorWare FEM tool is used to simulate the disk resonator with the proposed fabrication flow in Figure 2.15. Figure 2.16 show the realized structure in CoventorWare and Figure 2.17 shows the plot of displacement versus frequency.

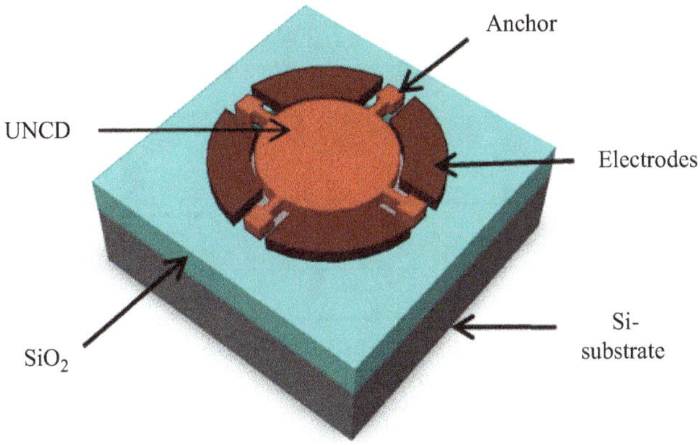

Figure 2.16 Disk resonator designed in CoventorWare FEM Software by basic steps of fabrication discussed in section 2.4.6

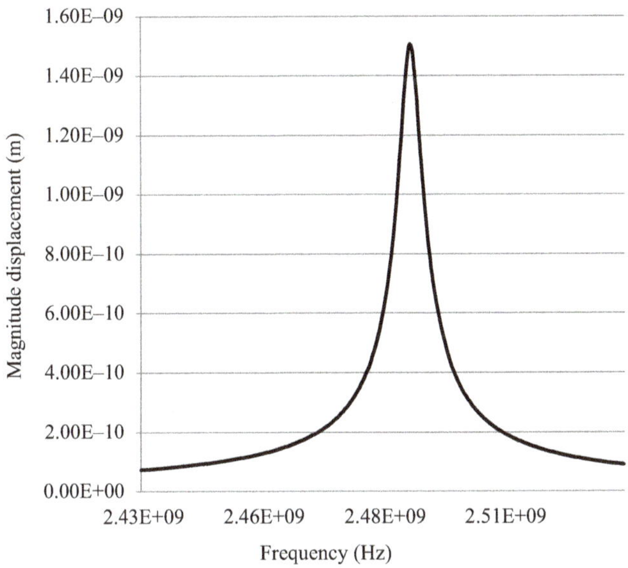

Figure 2.17 Displacement versus frequency graph plotted in CoventorWare FEM Software

2.4.6.1 Temperature compensation

The disk resonator with UNCD has temperature coefficient of frequency (TCF) −11 ppm/K calculated from data of Figure 2.18 simulated result in CoventorWare. The compensation of this shift in frequency is achieved by coating with dielectric. The dielectric coating provides two advantages for the performance of the disk

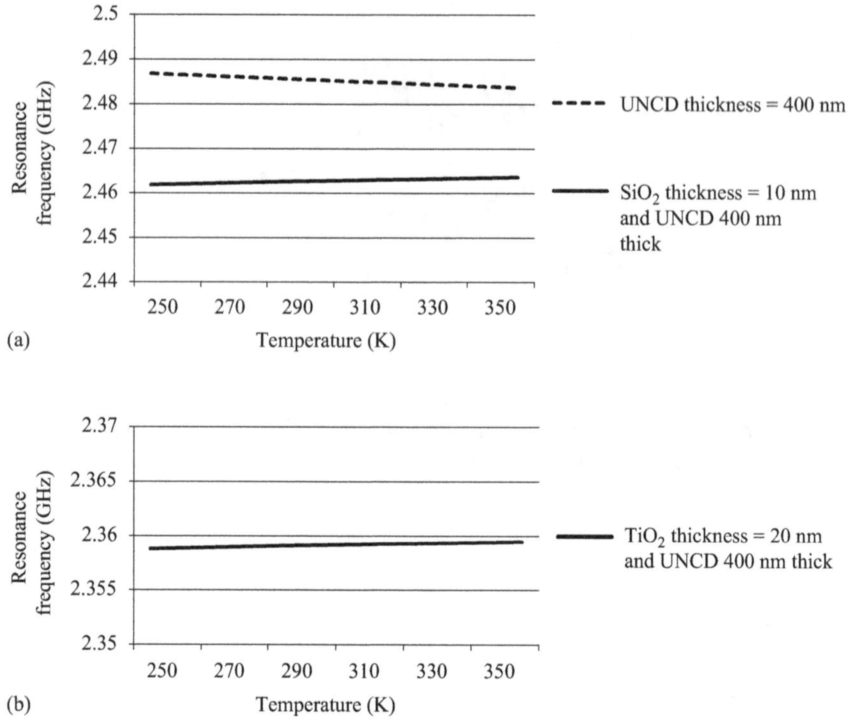

Figure 2.18 Plot of resonance frequency versus temperature for (a) UNCD without a coat and UNCD with SiO_2 10 nm coat and (b) UNCD with a TiO_2 coat

resonator. First, if this dielectric coating has the positive TCF of frequency, then shift in frequency can be compensated. Second, this dielectric coating also improves the motional impedance of resonator.

The oxides are the best choice as a dielectric because most of them have positive TCF. For a CMOS compatible process, SiO_2 is preferred due to ease of fabrication. The Young's modulus of SiO2 increases with increasing temperature. The temperature coefficient of expansion (α_L) is 0.5 ppm/K. The temperature of the coefficient of Young' modulus (TC of E) is around +185 ppm/K [54]. The dependence of shift in resonance frequency on α_L can be neglected. The 10 nm thick coating of SiO2 on UNCD provides the TCF of 8.5 ppm/K calculated for data of Figure 2.18(a). The other choice of metal oxide is tungsten oxide (TiO_2) which has a high dielectric constant. The influencing feature to select this material is because of adhesion with UNCD with high dielectric constant. The αL for TiO2 is in 8.4–11.8 ppm/K. And TiO_2 is the hardest metal oxide and it has excellent thermal-mechanical stability. For 20 nm thick TiO2 coating on UNCD, TCF of 2.5 ppm/K is achieved from Figure 2.18(b).

2.5 Coupled resonator

Till the last section, design of the resonator is explained in detail in keeping the aspect in mind of filter implementation. The targeted characteristics of filters are being achieved with coupling few resonators. The coupling of resonators is a crucial aspect in filter design which we will see in detail in this section.

Coupled-resonator filters are built from several identical resonators interacting through a reactive link [59,62]. Without coupling, such a system has two pairs of finite conjugated poles situated at the resonator natural resonance frequency. Nonzero coupling between the resonators results in poles splitting around the resonator natural frequency. For example, this happens when RLC resonators are magnetically coupled. In mechanical filters, the coupling link is elastic and generally realized with mechanical springs, as shown in Figure 2.19 for the case of a two resonator lumped-element system [60]. Input and output transducers needed to interface with electrical-domain signals are not shown. In the mechanical domain, the input value is the force acting on one of the resonators (or on a mass in the lumped parameter representation), the output value is the velocity or displacement of the other resonator (its lumped mass). Such a system has two conjugated pole pairs. If coupling strength increased, pole separation is observed. The frequency of one of the pole pairs increases whereas the pole pair frequency is constant and remains equal to the natural frequency of the resonators. Thus, the center frequency increases with the coupling. However, in the classical theory of coupled-resonator, the coupling link doesn't affect the center frequency, the latter remains equal to the resonance frequency of the isolated resonators. Nevertheless, a classical theory can be used here if we introduce the notion of the individual resonator in the context of filter which we will call filter individual resonators.

From Figure 2.19(b), the natural frequency of the filter individual resonators f_0 is different from the natural frequency f_R of the original resonators formed by m_0, $k_{0,}$ and μ_0:

$$f_0 = \frac{1}{2\pi}\sqrt{\frac{k_0 + k_c}{m_0}} = f_R\sqrt{1 + \frac{k_c}{k_R}} \tag{2.47}$$

In practical implementations, because planar IC processes typically exhibit substantially better *matching* tolerances then *absolute*, the constituent resonators in micromechanical filters are preferably designed to be identical, with identical spring

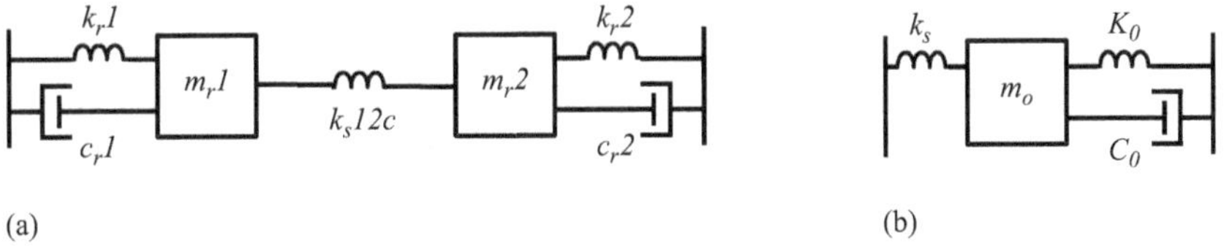

Figure 2.19 (a) Mass spring model for two coupled resonator and (b) equivalent filter individual resonator

dimensions and resonance frequencies. For such designs, the center frequency of the overall filter is equal to the resonance frequency of the resonators, while the filter passband (i.e., the bandwidth) is determined by the spacing between the mode peaks.

The relative placement of the vibration peaks in the frequency characteristic—and thus, the passband of the filter—is determined primarily by the stiffness of the coupling springs and of the constituent resonators at their coupling locations. In particular, for a filter with center frequency f_0 and bandwidth B, this stiffness satisfies the expression:

$$B = \left(\frac{f_0}{\alpha}\right)\left(\frac{k_s}{k_r}\right) \tag{2.48}$$

where α is the coupling coefficient and k_s is the spring constant of coupling spring, and k is the spring constant of the resonator at coupling location. One must take notice of (2.48) that filter bandwidth is not dependent on the absolute values of resonator and coupling beam stiffness, rather their ratio dictates bandwidth. Thus, the general procedure for designing a mechanical filter involves two main steps:

1. design of mechanical resonators with resonance frequencies at or near f_0 and with reasonable stiffness, and
2. design of coupling springs with appropriate values of stiffness to achieve the desired bandwidth.

2.5.1 Coupling beam design

As indicated in the ideal circuit of Figure 2.20(a), the function of the coupling beams is to implement appropriate values of stiffness to couple the resonators and achieve the needed bandwidth. The needed values of coupling spring constant are dictated by (2.48) and are obtained in large part by selecting appropriate coupling beam length and width dimensions assuming a given thickness defined by the process technology.

Designs using identical resonators are made possible by the fact that real coupling beams have finite mass, as well as stiffness, and thus actually function as acoustic transmission lines. As such, the reactance they present to adjoining resonators generally includes both mass and stiffness (i.e., inductive and capacitive) components, with values dependent upon both the dimensions of the couplers themselves and the frequency of operation. In particular, for frequencies within the filter passband, the lengths of the coupling beams correspond to specific wavelength-fractions that largely determine the impedances presented by couplers to their adjoining resonators.

For general wavelength fractions, the coupling beam can be modeled as shown in network of Figure 2.20(b) comprised of series mechanical impedances, and, and shunt impedance. The coupling beam dimensions yielding specific impedance values can be obtained through consideration of the impedance matrix for this flexural mode beam under fixed-fixed, sliding support boundary conditions.

This implementation presents the two different geometries for coupled-resonator, ring, and disk. In this work designing of coupling, the beam is being evaluated for two implementations of filters using ring and disk. In the case of filter design using a ring coupled resonator, the coupling beam dimensions play a different role than their role in filter design using in disk coupled resonator.

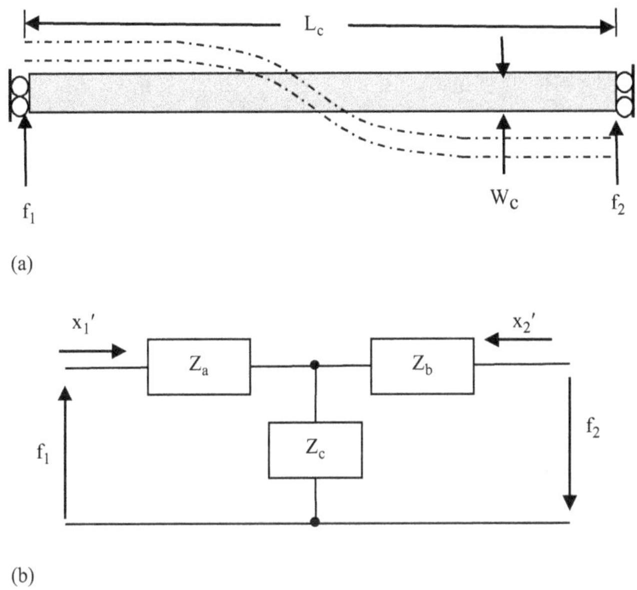

Figure 2.20 (a) Coupling beam under forces f_1 and f_2 with corresponding velocity responses and (b) general transmission line T—a model for the coupling beam

2.5.2 Case study: disk resonator-based filter

Figure 2.21 shows a mechanical circuit composed of two contour-mode disk resonators mechanically coupled to each other with resistors connected to both the input and the output ports. Figure 2.21(b) shows its equivalent mechanical circuits. The transformer turns ratios associated with the couplers (similar to Figure 2.11) $\eta c1$ and $\eta c2$, and model the mechanical impedance transformation realized by mechanically coupling one resonator to the other at a location different from the reference point based on which the electrical models of each resonator is calculated. Usually, the reference point is chosen as the location with the maximum mechanical displacement.

The coupling beam has two ways to couple disk resonators: one at low-velocity points and high-velocity points (low-velocity points and high-velocity points are similar to those shown in coupled ring resonators).

The coupling at low-velocity points gives the shift in center frequency. So, two couple disk resonator for filter application, it is advantageous to couple them at low-velocity points. The width of the coupling beam decides the shift in center frequency. For the narrow bandwidth filter, the coupling is at the low-velocity location. Figure 2.22 shows the modal shapes and displacement response for frequency. This coupling at nodal points gives the narrow bandwidth filters. The shift in frequencies decides mainly by the width of the coupling beam, as it affects the stiffness of the disk at the periphery.

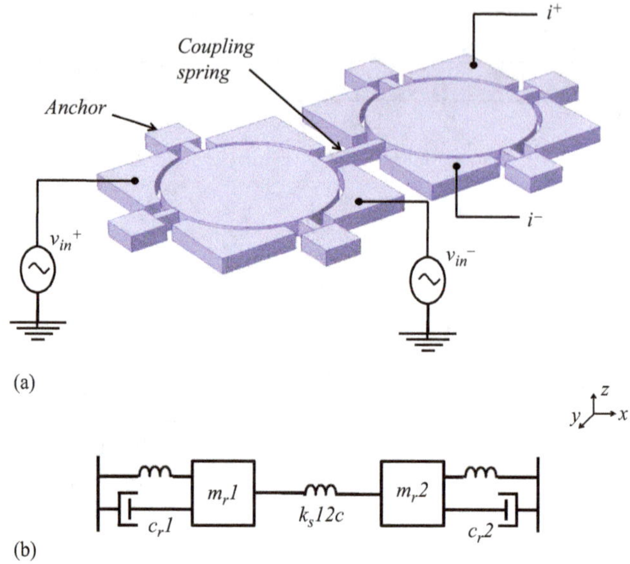

Figure 2.21 (a) Perspective view schematic of a two flexural-mode ring resonator micromechanical filter, along with the preferred bias, excitation, and sensing circuitry. (b) The equivalent mechanical circuit

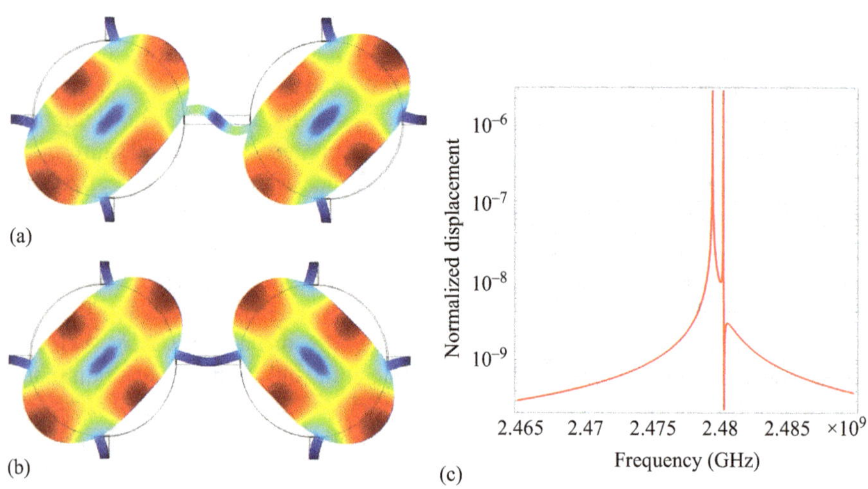

Figure 2.22 The two disk coupled at low velocity location with λ/4 (1.45 μm): (a) in phase frequency 2.479 GHz, (b) out of phase frequency 2.48 GHz, and (c) displacement versus frequency plot

The high-velocity coupling location is illustrated also for $\lambda/4$ and $\lambda/2$ coupling beam. For the $\lambda/4$ coupling beam, it acts as a transmission line that provides reactance. However, such a $\lambda/4$ coupling beam shifts the resonance frequency for larger bandwidth as compared to low-velocity coupling. This coupling is not useful for filter application as it degrades the performance of the desired mode and unwanted modes become more dominant. Figure 2.23 shows the disk coupled by the $\lambda/4$ (1.5 μm) beam. This $\lambda/4$ coupling beam contributes to reactance in the overall reactance of disk significantly which results in a large shift in resonance frequency.

The $\lambda/2$ coupling beam couples two resonators, which oscillate at the same frequency. By transmission line concepts, this beam does not provide in reactance which makes two resonators to vibrate at their natural frequency. Figure 2.24 shows two disks coupled by the $\lambda/2$ (3 μm) beam at the high-velocity location. It is observed that both resonators are vibrating at the same frequency.

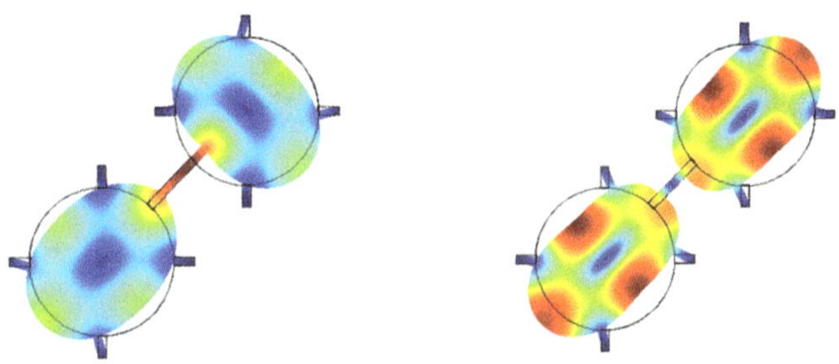

(a) (b)

Figure 2.23 The two disks coupled at high velocity location with $\lambda/4$ (1.5 μm): (a) out of phase frequency 2.41 GHz and (b) in phase frequency 2.52 GHz

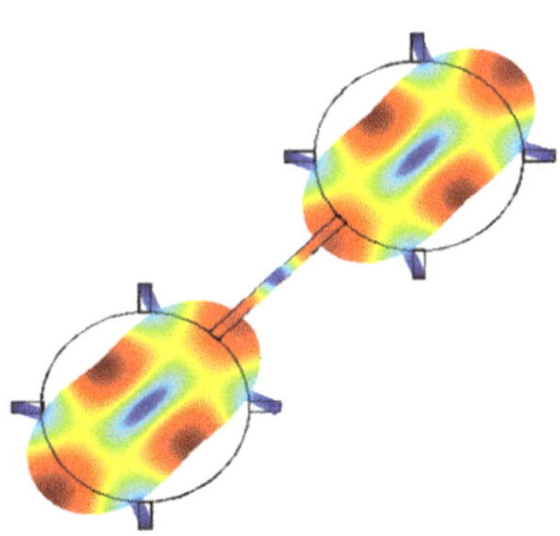

Figure 2.24 The two disk coupled at high velocity location with $\lambda/2$ (3¼m): (a) out of phase frequency 2.41 GHz and (b) in phase frequency 2.52 GHz

By using the above two couplings at a low-velocity location and at a high-velocity location, four disks coupled-resonator are realized. Figures 2.25–2.26 show mode shape and normalized displacement versus frequency, respectively. The coupling at high-velocity points by $\lambda/2$ coupling beam is making disks to vibrate at single frequency and coupling at low-velocity location shifting center frequency for desired bandwidth of the filter. The narrow bandwidth filtering is possible for low-velocity coupling as mentioned in Table 2.3.

The four disks are coupled together at high and low-velocity locations. This coupling is giving 1 MHz bandwidth with improved power handling, reducing motional impedance for RF application. The driving of resonator is carried out by

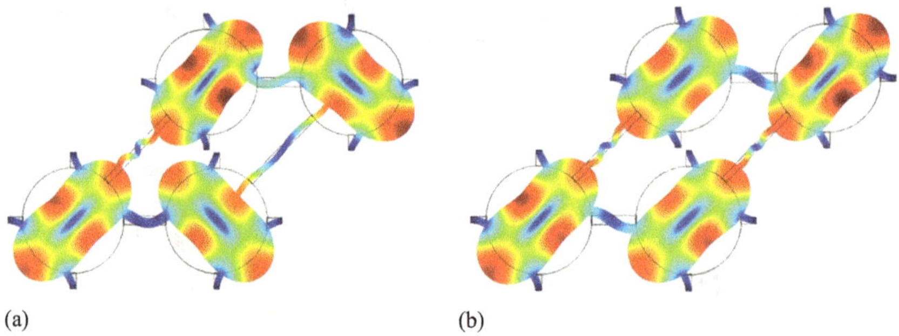

Figure 2.25 The four disk coupled at high velocity location with $\lambda/2$ (3 µm) and at low velocity location with $\lambda/4$ (1.5 µm and width = 0.3 µm): (a) out of phase frequency 2.488 GHz and (b) in phase frequency 2.489 GHz

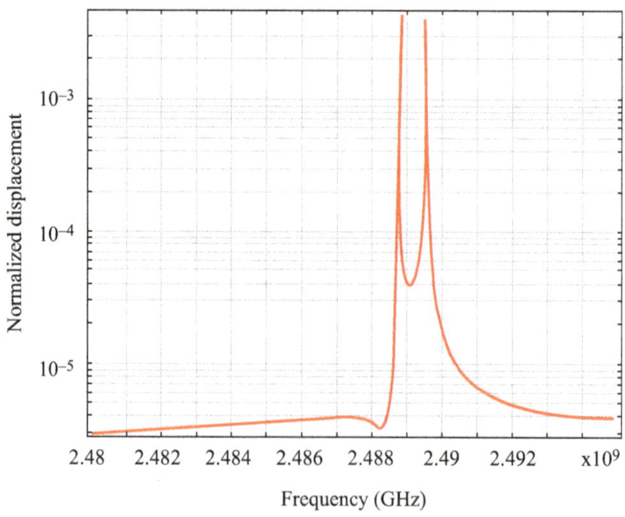

Figure 2.26 Normalized displacement versus frequency for four coupled disk resonator

Table 2.3 Summarized results of disk resonator-based filter

Resonance frequency	**Analytically calculated**		**Simulated in FEM COMSOL MultiPhysics**
	2.4871 GHz		**2.4820 GHz**
Temperature compensation of frequency (TCF)	UNCD	UNCD with SiO_2 coating	UNCD with TiO_2 coating
	11 ppm/K	8.5 ppm/K	2.5 ppm/K
Motional impedance for single resonator	58 MΩ	75 kΩ	24 kΩ
Band-pass filter with four disk resonator coupled	Bandwidth = 1 MHz and center frequency = 2.48 GHz		

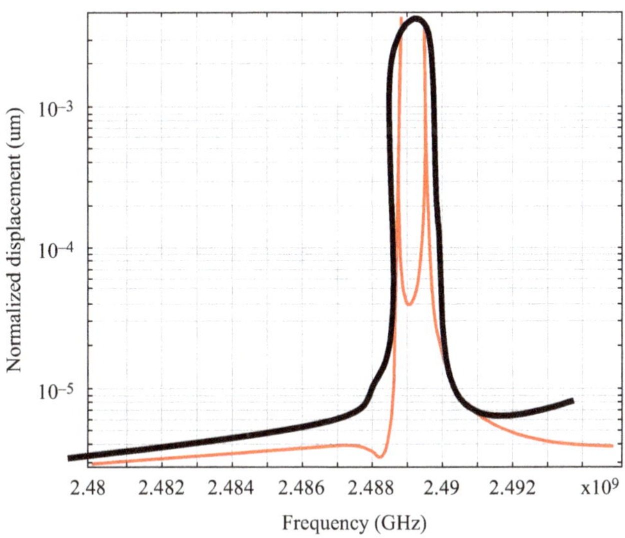

Figure 2.27 Four disks coupled-resonator filter response. Blackline shows the possible flat band response of filter

one pair of a resonator coupled in array mode and sensing of output is done by another pair of a resonator in array mode. Figure 2.27 shows filter response with 1 MHz bandwidth.

2.5.3 Case study: ring resonator-based filter

In this example of implementation of the filter, it is designed in a way that it is not required to use a coupling spring in a submicron dimensions, thereby having ease in fabrication and yet it realizes a small bandwidth by connecting a coupling spring to the location on the ring resonator where resonator has the highest stiffness (lowest velocity). It is relatively simple to access low-velocity location due to inherent symmetry in the ring resonator and therefore, it is not required to calculate a low-velocity (high stiffness) locations as compared to its clamped-clamped beam filter counterpart.

Figure 2.28 presents the perspective view schematic of a two-ring resonator filter, along with appropriate bias, driving, and sensing ports. To operate this filter, a DC-bias is applied to the suspended filter structure, and an AC input voltage is applied through a resistor to the input electrode (electrode 1), as shown in Figure 2.28. The application of this input creates a directed electrostatic force between electrode 1 and the conductive resonator that induces directed vibration of the input resonator when the frequency of the input voltage comes within the passband of the mechanical filter. This vibrational energy is imparted to the output resonator via the coupling spring, causing it to vibrate as well. The vibration of the output resonator creates a DC-biased, time-varying capacitor between the conductive resonator and output electrode, which then sources an output current.

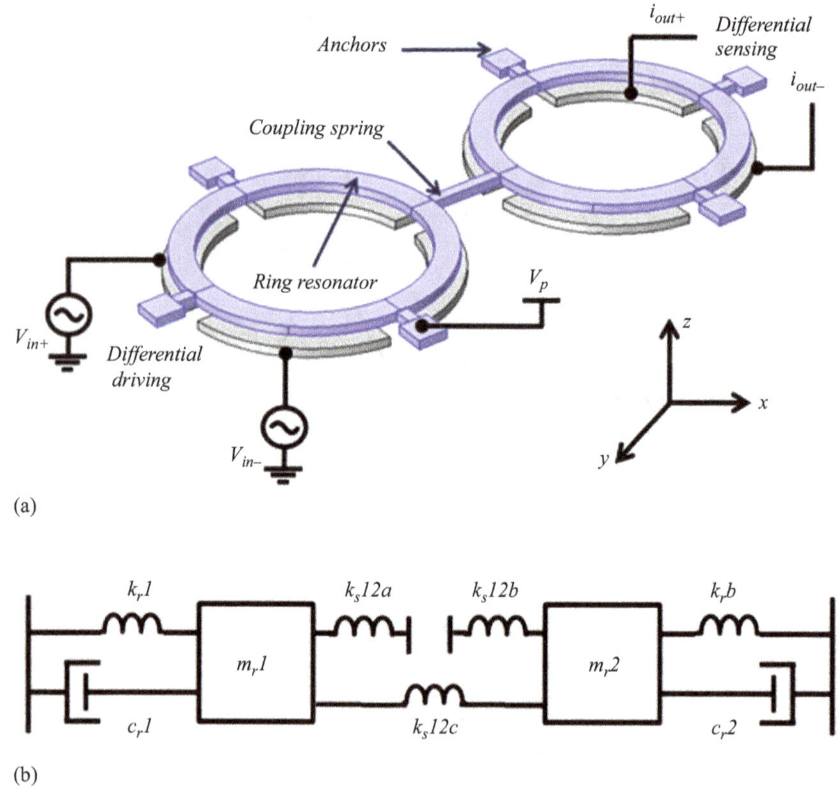

Figure 2.28 (a) Perspective view schematic of a two flexural-mode ring resonator micromechanical filter, along with the preferred bias, excitation, and sensing circuitry, and (b) the equivalent mechanical circuit

As shown, the filter consists of two identical flexural-mode ring resonators, coupled mechanically by a flexural-mode beam. The conducting electrodes for driving and sensing underlie the periphery regions of each resonator and serve as capacitive transducer electrodes positioned to induce out-of-plane resonator vibration. This coupled two-resonator system exhibited two mechanical resonance modes with closely spaced frequencies that define the filter passband. The center frequency of the filter is determined primarily by the frequencies of the constituent resonators, while the spacing between modes (i.e., the bandwidth) is determined largely by the stiffness of the coupling spring. Quarter-wavelength ($\lambda/4$) coupling is used on this microscale to alleviate mass loading effects caused by similar resonator and coupler dimensions [32]. In general, to operate this filter, differential AC input voltages v_i^+ and v_i^- are applied to the two input electrodes as shown in Figure 2.28(a). A DC bias V_p is then applied to the filter structure through anchors. The application of this input creates a z-directed electrostatic force between differential driving electrodes and the conductive resonator that induces z-directed vibration of the input resonator when the frequency of the input voltage comes within the passband of the mechanical filter. This vibrational energy is imparted to the output resonator via the coupling spring, causing it to vibrate as well. The vibration of the output resonator creates a DC-biased, time-varying capacitor between the conductive resonator and differential sensing output electrodes, which then sources output currents namely i_{out}^+ and i_{out}^- sensed in a differential way. Figure 2.28(b) depicts an equivalent mechanical circuit for filter structure. The following equations have been used to determine resonance frequency and mode shape of a homogeneous, uniform, and unsupported circular ring [30]:

$$f_0 = \frac{i(i^2-1)}{2\pi R^2}\left\{\frac{EI_x}{m\left(i^2+\frac{EI_x}{GC}\right)}\right\}^{\frac{1}{2}} \tag{2.49}$$

$$\begin{bmatrix} x \\ y \\ z \\ \theta \end{bmatrix} = \begin{bmatrix} 0 \\ \sin i\alpha \\ 0 \\ -\dfrac{i^2}{R}\left\{\dfrac{\dfrac{GC}{EI_x}}{1+i^2\dfrac{GC}{EI_x}}\right\}\sin i\alpha \end{bmatrix} \tag{2.50}$$

where i is the mode number; x,y,z, and θ are the mode shapes corresponding to deformations parallel to x,y,z axes, and rotation about z-axis, respectively; m is the mass per unit length of the ring; C is a torsion constant; E represents the modulus of elasticity; G is the shear modulus; I_x and I_y are the area moments of inertia about x and y axes, respectively; R represents the radius to mid-line of the ring; α is the angular position about the ring; μ is the mass density of ring material; and v is the Poisson's ratio. Equivalent mechanical elements, such as dynamic mass, spring, and damper, are also given below as equation (2.51), (2.52), and (2.53), respectively:

$$m_{re}(y) = \frac{\rho W_r h \int_0^{2\pi} \left[Z_{\text{mode}}(\theta')\right]^2 d\theta'}{\left[Z_{\text{mode}}(\theta)\right]^2} \tag{2.51}$$

$$k_{re}(y) = \omega_0{}^2 m_{re}(y) \tag{2.52}$$

$$c_{re}(y) = \frac{\sqrt{k_{re}(y) m_{re}(y)}}{Q} = \frac{\omega_0 m_{re}(y)}{Q} \tag{2.53}$$

Figure 2.29 shows a 3D simulated plot of vibrating mode shapes of a flexural-mode ring resonator mechanically coupled by $\lambda/4$ coupling spring. In the lower frequency mode, both resonators vibrate 180° out of phase at a frequency of 4.3961 MHz; and in the higher frequency mode, the resonators vibrate in phase at a frequency of 4.4324 MHz. Both high-velocity points (low-stiffness) and low-velocity points (high-stiffness) are indicated for the flexural-mode ring geometry in Figure 2.29. The support beams which play an important role are also highlighted.

The filter design is dominantly governed by the bandwidth equation [25,32]

$$BW = \frac{ks_{12}}{k_{12}} \frac{f_0}{k_{rc}} \tag{2.54}$$

where f_0 is the center frequency of the filter, it is also the resonant frequency of each of the resonators, k_{12} is the normalized coupling coefficient ($k_{ij} = 0.7225$ [32]), ks_{12} is

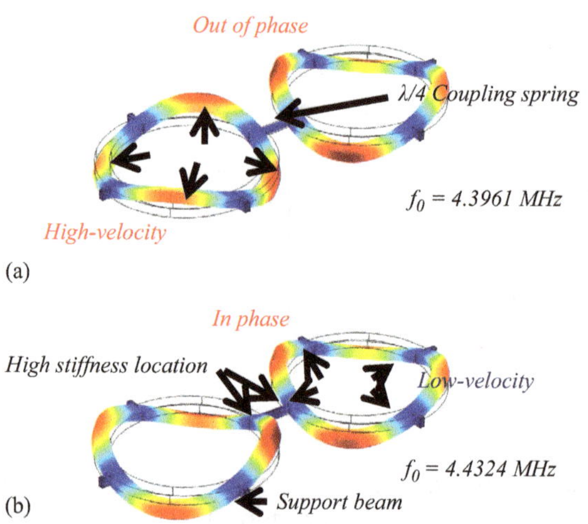

Figure 2.29 3D COMSOL simulated plot of vibrating mode shapes of a flexural-mode ring resonator mechanically coupled by λ/4 coupling spring: (a) out-of-plane mode shape and (b) in-plane mode shape

the coupling beam stiffness, and k_{rc} is the resonator effective stiffness at the coupling location. In our design, the effective stiffness is highest at the motionless nodal points of the resonator. From 2.54, it is obvious that the bandwidth depends only on the resonator stiffness, that is, the coupling location on the resonators and the coupling-beam stiffness. The other two parameters are constants. If the coupling is at a low-velocity location on the resonator, the effective mass increases and so is the effective stiffness; consequently, the filter bandwidth decreases [59]. In order to achieve a small bandwidth for the filter design, choosing the optimum coupling location plays an important role. It is relatively easy to determine the coupling location in this type of geometry due to the inherent symmetry of the ring structure where motionless nodal points (i.e., low-velocity points) are easy to locate. Moreover, the motionless nodal points do not change due to process variations as opposed to their clamped-clamped beam counterparts [32,61].

The interface between the mechanical and electrostatic building blocks was created in the electromechanics module of COMSOL. This model shows the response of a filter under an applied electrostatic load and computed the deformation of the ring due to the applied voltage. A bias voltage $Vp = 45$ V was applied to the ring structure via an anchor, as can be seen from the perspective view schematic of a two flexural-mode ring resonator mechanical filter, along with the bias, excitation, and sensing circuitry (Figure 2.28(a)). Figure 2.30 shows the simulated displacement versus frequency for two different configurations, common mode, and differential mode. The admittance plot was obtained as shown in Figure 2.31, which shows the resonance of two coupled ring resonators. To flatten

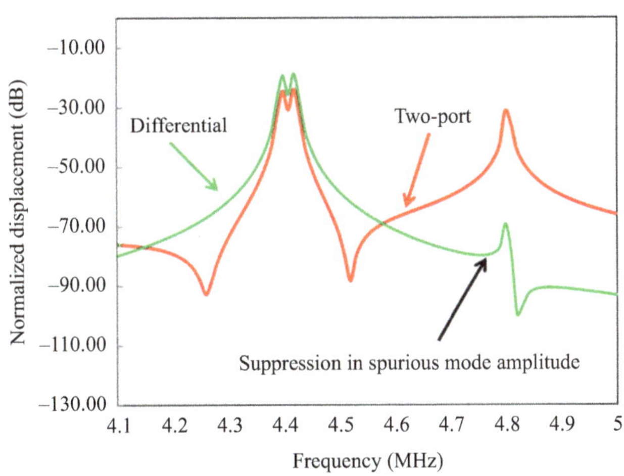

Figure 2.30 Simulated spectra of ring microresonator filter under two different driving and sensing schemes

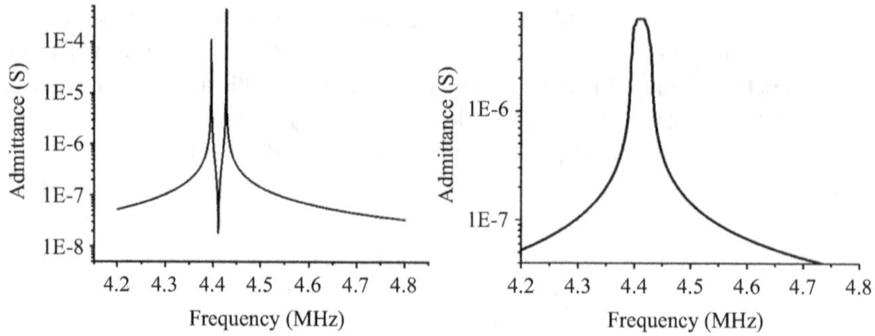

Figure 2.31 An admittance plot (COMSOL MultiPhysics) of ring microresonator filter under an applied voltage (a) without termination (b) with the termination of R_Q

the filter passband in Figure 2.31, the quality factors of the end resonators should be loaded via resistive termination with a value of R_Q given by [32]

$$R_Q = \left(\frac{Q_r}{qQ_f} - 1\right) \cdot R_x \approx \frac{Q_r}{qQ_f} \cdot \frac{m_r \omega_o}{Q\eta_e^2} \approx \frac{m_r BW}{q\eta_e^2} \tag{2.55}$$

where R_x is the motional resistance of a constituent end ring resonator; Q is the unloaded quality factor of the resonator; $Q_{\text{filter}} = f_o/BW$, BW the filter bandwidth; q the normalized parameter obtained from the reference [32]; m_r the dynamic mass of the ring resonator at its point of maximum displacement; and η_e the electromechanical coupling factor. For our design, the analytically determined value of the termination resistor is $R_Q \approx 240$ kΩ.

To attain low-loss capacitive resonators implemented as a filter, the considerable R_x for such a capacitive device can be significantly reduced by choosing appropriate values for the Q-factor, the electromechanical transduction factor, η_e, and the capacitive gap, thus relaxing the requirement of the termination resistance (R_Q) for the impedance matching of a given filter.

2.6 Summary

In this chapter, we have explored the history of filter development in the beginning. In general, classical filter, developments were along with the network theory. The filters based on the RLC components have limitations and achieving desirable characteristics is difficult largely because of its lower Q. The electromechanical components because of their high Q and small size quickly replaced these components to give required performance in the modern communication systems. The CMOS fabrication technology-based resonators, which are typically addressed as MEMS resonators, are showing promise for further development. The design of filters with MEMS requires material selection, process development, and the determination of resonator physical structure which is illustrated in this chapter with examples.

References

[1] http://quadrivium.nl/history/history.html by Gerrit Groenewold.
[2] Pupin M.I. "Wave transmission over non-uniform cables and long-distance air-lines." *Transactions of the American Institute of Electrical Engineers*. 1900; 17: 245–307.
[3] Campbell G.A. "On loaded lines in telephonic transmission." *The London, Edinburgh, and Dublin Philosophical Magazine and Journal of Science*, Series 6. 1903; 5(27): 313–30.
[4] Wagner K.W. Spulen- und Kondensatorleitungen. *Archiv für Elektrotechnik*. 1919; 8(2–3): 61–92.
[5] Campbell G.A. Electric wave-filter. U.S. Patent 1,227,113, July 15, 1915.
[6] Zobel O.J. "Theory and design of uniform and composite electric wave-filters." The Bell System Technical Journal. 1923; 2(1): 1–46.
[7] Cauer W. Siebschaltungen. Berlin: VDI; 1931
[8] Bowers J.L. R-C bandpass filter design. *Electronics*. 1947; 20(4):131–33,
[9] Sallen R.P., and Key E.L. "A practical method of designing RC active filters." *IRE Transactions-Circuit Theory*. 1955; 2(1): 74–85.
[10] Nauta B., and Seevinck E. "Linear CMOS transconductance element for VHF filters." *Electronics Letters*. 1989; 25(7): 448–49.
[11] Zverev A.I. "The golden anniversary of electric wave filters." *IEEE Spectrum*. 1966; 3(3): 129–131.
[12] Cady W.G. Piezo electric resonator. U.S. Patent Office, Patent No. 1,450,246, 1923; Method of Maintaining Electric Currents of Constant Frequency. U.S. Patent Office, Patent No. 1,472,583, 1923.
[13] Filler R.L., and Vig J.R. "Resonators for the microcomputer compensated crystal oscillator." in *Proceedings of the 43rd Annual Symposium on Frequency Control*, 1989, pp. 8–15.
[14] Moore G.E. "Cramming more components onto integrated circuits, Reprinted from Electronics, Vol. 38, no. 8, April 19, 1965, pp. 114 ff." in *IEEE Solid-State Circuits Society Newsletter*. 2006; 11(3): 33–35. DOI: 10.1109/N-SSC.2006.4785860
[15] Lord Rayleigh: "On waves propagating along the plane surface of an elastic solid." *Proceedings of London Mathematical Society*. 1885; 1–17(1): 4–11.
[16] White R.M., and Voltmer F. "Direct piezoelectric coupling to surface elastic waves." Applied Physics Letters. (1965); 7: 314. https://doi.org/10.1063/1.1754276
[17] Pomeranz M. "Amplification of microwave phonons in germanium." Physics Review Letters. 1964; 13: 208, 385.
[18] Grudkowski T.W., Black J.F., Reeder T.M., Cullen D.E., and Wagner R.A. "Fundamental-mode VHF/UHF miniature acoustic resonators and filters on silicon." Applied Physics Letters. 1980; 37(11): 993–95.
[19] Nakamura K., Sasaki H., and Shimizu H. "ZnO/SiO2-diaphragm composite resonator on a silicon wafer." Electronics Letters. 1981; 17(14); 507–09. DOI: 10.1049/el:19810355

[20] Ketcham R.S., Kline G.R., and Lakin K.M. "Performance of TFR filters under elevated power conditions." Proceedings of the 42nd Annual Frequency Control Symposium, Baltimore, MD, USA, 1988, pp. 106–11.

[21] "Simulation-Driven Optimization of 5G RF MEMS Filters." White Paper, July 2019. www.onscale.com

[22] Nguyen C.T.-C. "Vibrating RF MEMS overview: applications to wireless communications." Proceedings of Photonics West: MOEMS-MEMS 2005, San Jose, California, January 22–27, 2005, Paper No. 5715-201.

[23] Young M. *The Technical Writer's Handbook*. Mill Valley, CA: University Science; 1989.

[24] Wong A.-C., Clark J.R., and Nguyen C.T.-C. "Anneal-activated, tunable, 65MHz micromechanical filters." Digest of Technical Papers, 10th International Conference on Solid-State Sensors and Actuators, Sendai, Japan, June 7–10, 1999, pp. 1390–93.

[25] Wang K., and Nguyen C.T.-C. "High-order medium frequency micromechanical electronic filters." *Journal of Microelectromechanical Systems*. 1999; 8(4): 534–56. DOI: 10.1109/84.809070

[26] Greywall D.S., *et al.* "Coupled micromechanical drumhead resonators with practical applications as electromechanical bandpass filters." *Journal of Micromechanics and Microengineering*. 2002; 12(6): 925–38.

[27] Wang J., Ren Z., and Nguyen C.T.-C. "Self-aligned 1.14-GHz vibrating radial-mode disk resonators." *TRANSDUCERS'03 – 12th International Conference on Solid-State Sensors, Actuators and Microsystems. Digest of Technical Papers (Cat. No.03TH8664)*, vol.2. Boston, MA, USA, 2003, pp. 947–950.

[28] Clark J.R., Hsu W.-T., and Nguyen C.T.-C. "High-Q VHF micromechanical contour-mode disk resonators." *International Electron Devices Meeting 2000. Technical Digest. IEDM (Cat. No.00CH37138), San Francisco, CA*, USA, 2000, pp. 493–6. DOI: 10.1109/IEDM.2000.904363

[29] Abdelmoneum M.A., Demirci M.U., and Nguyen C.T.-C. "Stemless wine-glass-mode disk micromechanical resonators." The Sixteenth Annual International Conference on Micro Electro Mechanical Systems, 2003. MEMS-03 Kyoto. IEEE, Kyoto, Japan, 2003, pp. 698–701. DOI: 10.1109/ MEMSYS.2003.
1189845

[30] Li S., Lin Y., Ren Z., and Nguyen C.T.-C. "An MSI micromechanical differential disk-array filter." *TRANSDUCERS 2007 – 2007 International Solid-State Sensors, Actuators, and Microsystems Conference*, Lyon, 2007, pp. 307–11. DOI: 10.1109/SENSOR.2007.4300130

[31] Pachkawade V., Li C.-S., and Li S.-S. "A fully-differential CMOS-MEMS resonator integrated with an on-chip amplifier." *IEEE SENSORS*, Taipei, Taiwan, October 28–31, 2012, pp. 1–4.

[32] Bannon Ill F.D., Clark 1. R., and Nguyen C.T.-C. "High-Q HF microelecnomechanical Filters." IEEE Journal of Solid-Stare Circuits. 2000; 35(4): 512–26.

[33] Clark J.R., Wong A.-C., and Nguyen C.T.-C. "Parallel-resonator HF micromechanical bandpass filters." *Digest of Technical Papers, 1997 International Conference on Solid-State Sensors and Actuators*, Chicago, Illinois, June 16-19, 1997, pp. 1161–64.

[34] Howe R.T., and Muller R.S. "Polycrystalline silicon micromechanical beams." *Journal of The Electrochemical Society*. 1983;130(6): 1420–23.

[35] Pourkamali S., Hashimura A., Abdolvand R., Ho G.K., Erbil A., and Ayazi F. "High-Q single crystal silicon HARPSS capacitive beam resonators with self-aligned sub-100-nm transduction gaps." in *Journal of Microelectromechanical Systems*. 2003; 12(4): 487–96. DOI: 10.1109/ JMEMS.2003.811726

[36] Nabki F., Cicek P., Dusatko T.A., and El-Gamal M.N. "Low-stress CMOS-compatible silicon carbide surface-micromachining technology—Part II: Beam resonators for MEMS above IC." *Journal of Microelectromechanical Systems*. 2011; 20(3): 730–44. DOI: 10.1109/JMEMS.2011.2115130

[37] Hao Z., Pourkamali S., and Ayazi F. "VHF single-crystal silicon elliptic bulk-mode capacitive disk resonators – Part I: design and modeling." *Journal of Microelectromechanical Systems*. 2004; 13(6): 1043–53. DOI: 10.1109/ JMEMS.2004.838387

[38] Tang W.C., Nguyen T.-C.H., and Howe R.T. "Laterally driven polysilicon resonant microstructures." *IEEE Micro Electro Mechanical Systems, Proceedings, An Investigation of Micro Structures, Sensors, Actuators, Machines, and Robots*. Salt Lake City, UT, USA, 1989, pp. 53–59. DOI: 10.1109/MEMSYS.1989.77961

[39] Cioffi K.R., and Hsu W-T. "32KHz MEMS-based oscillator for low-power applications." *Proceedings of the 2005 IEEE International Frequency Control Symposium and Exposition*, 2005, Vancouver, BC, 2005, pp. 551–58. DOI: 10.1109/FREQ.2005.1573992

[40] Lee J.E.-Y., Bahreyni B., Zhu Y., and Seshia A.A. "A single-crystal-silicon bulk-acoustic-mode microresonator oscillator." *IEEE Electron Device Letters*. 2008; 29(7): 701–03. DOI: 10.1109/LED.2008.2000643

[41] Pourkamali S., Abdolvand R., and Ayazi F. "A 600 kHz electrically-coupled MEMS bandpass filter." *The Sixteenth Annual International Conference on Micro Electro Mechanical Systems, 2003. MEMS-03 Kyoto*. IEEE, Kyoto, Japan, 2003, pp. 702–05. DOI: 10.1109/MEMSYS.2003.1189846

[42] Lee J.E.Y., and Seshia A.A. "5.4-MHz single-crystal silicon wine glass mode disk resonator with a quality factor of 2 million." Sensors and Actuators A. 2009;156(1): 28–35.

[43] Clark J.R., Hsu W.-T., Abdelmoneum M.A., and Nguyen C.T.-C. "High-Q UHF micromechanical radial-contour mode disk resonators." in Journal of Microelectromechanical Systems. 2005; 14(6): 1298–1310. DOI: 10.1109/ JMEMS.2005.856675

[44] Naing T.L., Beyazoglu T., Lingqi W., Akgul M., Ren Z., Rocheleau T.O., and Nguyen C.T.-C. "2.97-GHz CVD diamond ring resonator with

Q >40,000." *2012 IEEE International Frequency Control Symposium (FCS) Proceedings*. Baltimore, MD, USA, May 21–24, 2012.

[45] Rocheleau T.O., Naing T.L., Ren Z., and Nguyen C.T.-C. "Acoustic whispering gallery mode resonator with Q > 109,000 at 515MHz." *2012 IEEE 25th International Conference on Micro Electro Mechanical Systems (MEMS)*, January 29–February 2, 2012, pp. 672–75.

[46] Ozgurluk A., Akgul M., and Nguyen C.T.-C. "RF channel-select micromechanical disk filters—Part I: Design." IEEE Transactions on Ultrasonics, Ferroelectrics, and Frequency Control (vol. 66, no. 1). IEEE, January 2019, pp. 192–217. DOI: 10.1109/TUFFC.2018.2881727

[47] Akgul M., Ozgurluk A., and Nguyen C.T.-C. "RF channel-select micromechanical disk filters—Part II: Demonstration." *IEEE Transactions on Ultrasonics, Ferroelectrics, and Frequency Control* (vol. 66, no. 1). *IEEE*, January 2019, pp. 218–35. DOI: 10.1109/TUFFC.2018.2883296

[48] Tilmans H.A.C., Elwenspoek M., and Fluitman J.H.J. "Micro resonant force gauges." Sensors and Actuators A. 1992; 30(1–2): 35–53.

[49] Legtenberg R., and Tilmans H.A.C. "Electrostatically driven vacuum-encapsulated polysilicon resonators. Part I: Design and fabrication." Sensors and Actuators A. 1994; 45: 57–66.

[50] Young M. The Technical Writer's Handbook. Mill Valley, CA: University Science, 1989.

[51] Jing W., Ren Z., and Nguyen C.T.-C. "1.156-GHz self-aligned vibrating micromechanical disk resonator." IEEE Transactions on Ultrasonics, Ferroelectrics, and Frequency Control (vol. 51, no. 12). IEEE, December 2004, pp. 1607–28. DOI: 10.1109/TUFFC.2004.1386679

[52] Bashmal S., Bhat R., and Rakheja S. "Frequency equations for the in-plane vibration of circular annular disks." Advances in Acoustics and Vibration. 2010, Article ID 501902, 8 pp.

[53] Ghodssi R., and Lin P. *MEMS Materials and Process Handbook*.

[54] Sandberg R., Svendsen W., Molhave K., and Boisen A. "Temperature and pressure dependence of resonance in multi-layer micro-cantilevers." Journal of Micromechanics and Microengineering. 2005; 15(8).

[55] Buja F., Kokorian J., Sumant A.V., and van Spengen W.M. "First adhesion measurements of conductive ultra-nanocrystalline diamond MEMS sidewalls." 2014 9th IEEE International Conference on Nano/Micro Engineered and Molecular Systems (NEMS), April 13–16, 2014, pp. 77–80. DOI: 10.1109/NEMS.2014.6908763

[56] Vlasov K I.I., Goovaerts E., Ralchenko V.G., Konov V.I., Khomich A.V., and Kanzyuba M.V. Vibrational properties of nitrogen-doped ultra-nanocrystalline diamond films grown by microwave plasma CVD. *Diamond and Related Materials*. 16(12): 2074–77.

[57] Lin C.-R., Liao W.-H., Wei D.-H., Tsai J.-S., Chang C.-K., and Fang W.-C. Formation of ultra-nanocrystalline diamond films with nitrogen addition. *Diamond and Related Materials*. 2011; 20(3): 380–84.

[58] Adiga V.P., Sumant A.V., Suresh S., *et al.* "Temperature dependence of mechanical stiffness and dissipation in ultra-nanocrystalline diamond." *Proceedings of SPIE*. 2009; 7318, 731–818.

[59] Shalaby M., Abdelmoneum M., and Saitou K. "Design of spring coupling for high Q, high frequency MEMS filters." *ASME 2006 International Mechanical Engineering Congress and Exposition Microelectromechanical Systems*. Chicago, Illinois, USA, November 5–10, 2006.

[60] Demirci M.U., and Nguyen C.T.-C. "A low impedance VHF micromechanical filter using coupled-array composite resonators." Digest of Technical Papers, the 13th International Conference on Solid-State Sensors & Actuators (Transducers'05). Seoul, Korea, June 5–9, 2005, pp. 2131–2134.

[61] Lopez J.L., Verd J., Uranga A., Giner J., Murillo G., and Torres F. "A CMOS-MEMS RF-tunable bandpass filter based on two high-Q 22-MHz polysilicon clamped-clamped beam resonators." *IEEE Electron Device Letters*. 2009; 30(7): 718–20.

[62] Li S.-S., Lin Y.-W., Ren Z., and Nguyen C.T.-C. "Disk-array design for suppression of unwanted modes in micromechanical composite-array filters." *Technical Digest, 19th IEEE International Conference on Micro Electro Mechanical Systems (MEMS'06)*. Istanbul, Turkey, January 22–26, 2006, pp. 866–69.

[63] Krauss A.R., Auciello O., Gruen, D.M., *et al.* "Ultra-nanocrystalline diamond thin films for MEMS and moving mechanical assembly devices." *Diamond and Related Materials*. 2001; 10(11): 1952–61.

Chapter 3

Microelectromechanical resonators design: low-frequency resonators

Amol Morankar[1]

3.1 Introduction

A majority of the current wireless communication systems used in radio frequency (RF) and intermediate frequency (IF) applications utilize several discrete off-chip resonant components, such as quartz-crystal metal packaged, quartz-crystal ceramic packaged, surface acoustic wave (SAW), and bulk acoustic wave (BAW) resonator filters. These components have demonstrated excellent performance in terms of quality factor (Q), dynamic range, thermal dependencies, larger tuning range, and aging stability but are bigger in size, fabricated using ultrafine micromachining technique and are not CMOS process compatible. A major disadvantage of these off-chip components is their high insertion loss that affects the performance and increases power dissipation, especially when additional amplification stages are required. Thus, it provides major interruption in the single-chip solution of a transceiver for wireless communication systems. Many functions in the RF frontends are passive and can be potentially replaced using RF MEMS devices [1–5]. The ability to reconfigure operating characteristics in real-time may result in a substantial reduction in the required number of discrete components for a given set of the wireless communication system. Thus, it may allow a single hardware component to emulate the behaviors of multiple discrete devices. Miniaturized multiband transceivers are essential in various wireless communication systems due to advancements in digital technology. Minimum insertion loss and power dissipation of passive elements such as capacitors, inductors, filters, resonators, and switches in RF transceiver are desirable. The ultimate power/bandwidth performance of the transceiver depends on the substrate parasitic [6]. RF MEMS technology with its versatility may provide two key resources on which wireless paradigm depends, that is, low power consumption and bandwidth. A large number of low/high-frequency MEMS resonator technologies are under investigation for efficient realization and replacement of quartz, SAW, and BAW.

[1]Riva Labs, CIVN, Visvesvaraya National Institute of Technology, Nagpur, India

MEMS resonators based on the capacitive transduction mechanism that can be made of integrated circuit compatible processes and materials are the forerunner. These capacitive resonators can be embedded in a variety of integrated circuit technologies that result in a higher level of integration and low-cost batch fabrication. The most important advantages of capacitive silicon resonators are the resonance frequency can be slightly tuned by changing the DC bias voltage that is required for their operation, postfabrication fine-tuning, and temperature compensation of the resonator. These resonators are smaller in size, consume practically zero DC power, and have the capability of frequency translation (mixing) and switching. Capacitive resonators with frequencies in the very high frequency (VHF) and ultra-high frequency (UHF) range have impedances on the order of tens or hundreds of kilo-ohms. To achieve low motional impedance of the capacitive resonators, the combination of large transduction area and the tiny electrode-to-resonator gap is preferred. Recently aluminum nitride contour mode resonator has demonstrated multifrequency operating characteristics on a single silicon chip. These resonators have demonstrated excellent filter characteristics but spurious modes are a major bottleneck obstructing the deployment of single-chip multifrequency resonators as a commercially viable solution for radiofrequency front-end filtering [6–8]. Also, MEMS resonators compatible with the CMOS process are mostly made of metal structures bearing low Q and poor temperature stability [6–8].

In this chapter, the design of MEMS resonator using Euler Bernoulli's beam theory for low-frequency applications is illustrated. Metal MEMS beam resonator exhibits various modes of vibration which depend on frequency and amplitude of input signal. These modes of vibration are exploited in the design of single and dual-frequency resonators using closed-form expressions. The mechanical and electrical coupling scheme to achieve the desired bandwidth and suppression of spurious/unwanted modes are illustrated respectively. Finally, temperature compensation using stress-induced frequency compensation through a mechanical coupling scheme is illustrated.

3.2 Low-frequency RF MEMS resonators

Electromechanical filters for signal processing applications are employed for several decades. Mechanical filters are explored where high Q, narrow bandwidth, low loss, and good aging stability are required. Also, mechanical filters were bigger in size, higher manufacturing cost and incompatible with the CMOS process. Due to advancement in the CMOS technology, mechanical filters were replaced with MOS capacitors and on-chip inductors that occupies a major portion of the integrated circuit. If these mechanical filters can be fabricated in smaller sizes using CMOS process compatible materials and methods, then the technological disadvantages of these filters would be eliminated. The advances in micromachining processes and microresonator materials have aided the feasibility of integrated micromechanical filters. The principles behind larger mechanical filters can be applied to micro-scale devices. These microdevices can be used to filter frequencies up to a few 100s of MHz and larger Q with proper packaging. Thus, a large number of high-frequency MEMS resonator technologies are under investigation for efficient realization and replacement of quartz, SAW, and BAW [9].

MEMS resonators are based on a capacitive transduction mechanism that can be made of integrated circuit compatible processes and materials. These capacitive resonators can be embedded in a variety of integrated circuit technologies that result in a higher level of integration and low-cost batch fabrication [10]. The most important advantages of capacitive silicon resonators are: the resonance frequency can be slightly tuned by changing the DC bias voltage that is required for their operation, post-fabrication fine-tuning, and temperature compensation of the resonator. These resonators are smaller in size, consume practically zero DC power, and have the capability of frequency translation (mixing) and switching [10]. Various MEMS resonators based on capacitive transduction mechanisms have been proposed in the literature bearing different shapes, vibration modes, structural material, and fabrication processes. Micromechanical resonators can have different shapes like beams, square plates, circular disks, annular rings, comb, etc., and can again be classified according to their modes of operation, namely flexural, torsional, and bulk mode devices [11].

- Flexural mode of vibration is representative of transverse standing waves. In such devices, the displacement of the structures is orthogonal to the bending stress in the structure.
- In resonators vibrating in the torsional mode, the dominant stress is shear-stress and the displacement produced is rotational.
- Bulk mode operation can be described in terms of standing longitudinal waves.

The majority of micromechanical circuits targeted for communication functions can be realized using micromechanical flexural-mode beam elements, as shown in Figure 3.1 with clamped-clamped (CC) boundary conditions [12]. This device consists of a beam anchored at both ends and an electrode underlying its central locations. Both the beam and electrode can be fabricated using materials, such as doped silicon, polysilicon, or metal. Surface micromachining can be the preferred method for the fabrication of micromechanical communication circuits since its ability to realize very complex geometries with multiple levels of suspension [12]. Experimental results for CC beam demonstrated high Q of 8,000 at frequency 8.5 MHz. Although high Q was achieved, anchor losses increase at a frequency above 30 MHz and therefore reduces Q.

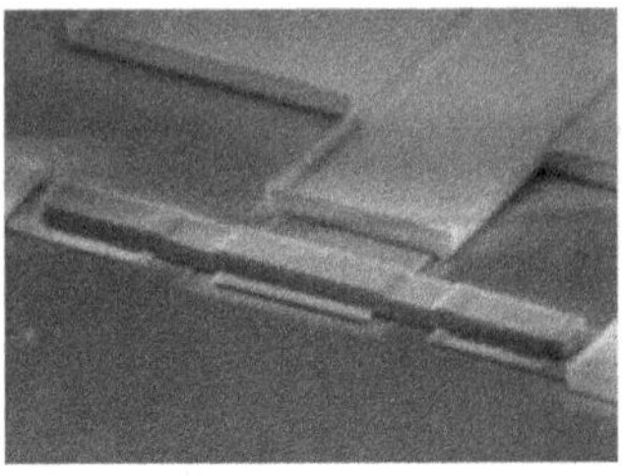

Figure 3.1 SEM image of flexural mode beam [12]

Two CC beam micromechanical resonators coupled by a soft flexural-mode mechanical spring was demonstrated [10]. The center frequency of a given resonator filter can be determined by the resonance frequency of the CC beam. The bandwidth can be determined by the coupling spring dimensions and its location between the resonators. These filters achieved a center frequency of around 8 MHz with Q's from 40 to 450 (percent bandwidths from 0.23 to 2.5%), insertion loss less than 2 dB and spurious-free dynamic ranges around 78 dB. Quarter-wavelength coupling was required on this micro-scale to avoid an increase in mass and reduction in resonant frequency caused by similar resonator and coupler dimensions. Figure 3.2 shows the SEM image of the resonator.

Polysilicon free-free beam micromechanical resonators based on MEMS technology operating in second and third-mode flexural vibrations have been demonstrated at frequencies in 102 MHz and high Q [13]. Higher resonant frequencies can be achieved by reducing the size of the resonator. Figures 3.3 and 3.4 show the SEM image of the second mode and third mode free-free beam resonator. These higher-mode free-free beam resonators provide several key advantages that include

- Lower series motional resistance Rx,
- Higher dynamic range,
- Higher power handling,
- Multiple ports that permit the 0° input-to-output phase-shift often preferred for high impedance micromechanical oscillators, and
- Invertible band-pass mixer-filters targeted for wireless applications.

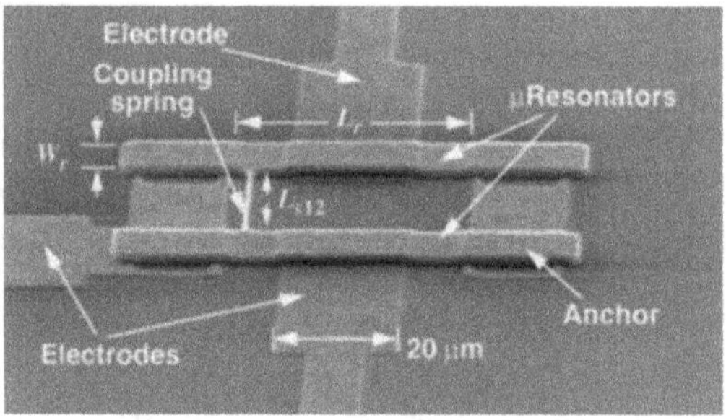

Figure 3.2 SEM image of the two CC beam resonator [10]

Figure 3.3 Second mode free-free beam [13]

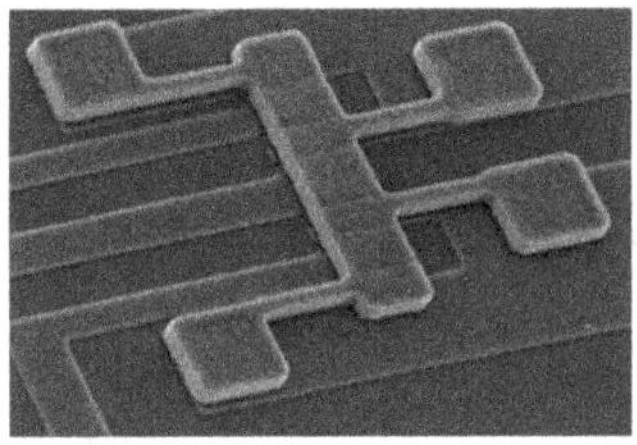

Figure 3.4 Third mode free-free beam [13]

It was observed that a free-free beam can be explored for future RF MEMS resonator due to its high Q and stiffness at higher frequencies. Also, it can provide better power handling capability with an increase in the number of beams [13].

Vibrating micromechanical system manufactured in silicon technologies usually employs a capacitive transduction mechanism that transforms the voltage to force and displacement to charge variation at the output electrode. More importantly, the efficiency of the transducer is decided by the gap between the transducer's electrodes. The transduction factor and the power efficiency of the transducer are inversely proportional to the second power of the gap width value [14]. Thus, for the case of the simple resonator (elementary second-order band-pass filter), the motional resistance is inversely proportional to the fourth power of this parameter [14]. Ideally, the value of the motional resistance should be low that determines the output signal level and insertion loss; therefore, it is necessary to have a transduction gap in the range of a few 100 nm. Practically, it is required to design a resonator with a lower motional resistance that is suitable for use as a filter in electronic signal processing circuits. The gap value should be less than 500 nm, which are often below the lithography resolution of conventional MEMS fabrication processes [14]. VHF 10 MHz micromechanical resonators fabricated in a thick-layer epipoly technology with an original postfabrication gap reducing method was demonstrated in [14]. Experimental results have proved the effectiveness of the method for gap adjustment. Thus, it demonstrates the possibility of designing micromechanical VHF filters in thick-layer silicon technology and the narrow gap that does not depend upon lithography resolution. Electrostatic motor, operating at only 30 V DC without any current consumption was utilized for gap reduction. The higher transmission was achieved with a lower transduction gap of 200 nm that experimentally validates the applicability of MEMS resonator for IF filter applications in wireless communication devices. Figure 3.5 shows the SEM image of the above-mentioned filter.

Laterally vibrating free-free beam micromechanical resonator was demonstrated in [15] that utilize the second mode flexural supports and optimal dc-bias application to suppress anchor dissipation and thereby attain Q's greater than 10,000 at 10.47 MHz. To suppress energy losses from the free-free beam to the anchors, the support beams were actually designed in such a way that two support beams were attached on opposite sides of the free-free beam as shown in Figure 3.6 [15]. The support beam from anchor to anchor was designed to vibrate in the second mode while a

Figure 3.5 SEM image of resonator with gap reduction electrostatic motor [14]

Figure 3.6 Laterally vibrating free-free beam resonator [15]

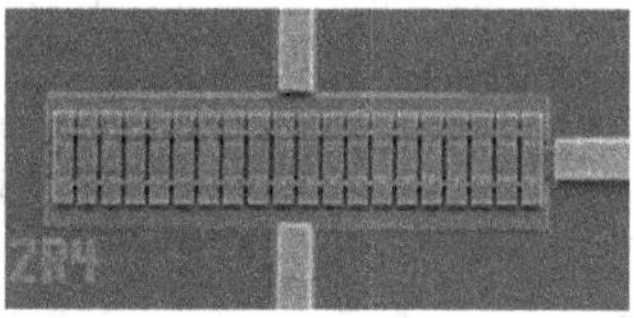

Figure 3.7 Free-free beam resonator array [16]

free-free beam to vibrate in fundamental mode frequency. It was observed that the resonance frequency and Q of this resonator were found to be sensitive to errors in support beam length. The use of metal electrodes in the capacitive transducers was found to be unfavorable due to the thermal expansion of the electrodes that made electrical stiffness a function of temperature. The thermal stability of –5 ppm/°C was measured [15].

Mechanically coupled array resonators were proposed in [16] for higher power handling capability and improved phase noise. A free-free beam resonator array was selected to validate the proposed concept. Figure 3.7 shows the SEM image of the fabricated devices. Although better power handling capability was demonstrated it did not produce the required phase noise that would be suitable for oscillators in GSM.

High-order micromechanical filters comprised of 3 to 4 mechanically coupled resonators, bridging between nonadjacent resonators was demonstrated using a polysilicon surface micromachining technology [17]. Sharper pass band to stop band roll-offs, larger stop band rejections, smaller filter shape factors, and very low insertion loss with more than 50 dB of stopband rejection was achieved. Bridging filter design techniques are expected to greatly benefit future communication systems. Figure 3.8 shows the FESEM image of the bridged micromechanical filter.

A square shape single crystal silicon micromechanical resonator with a quality factor of larger than one million and a resonant frequency of 2.18 MHz was demonstrated [18]. The resonator was excited in the square extensional bulk acoustic mode at 4 m, and it was fabricated in a foundry SOI MEMS process. Figure 3.9 shows the schematic of the square resonator.

A laterally vibrating radial contour mode micromechanical disk resonator was presented with a measured frequency and Q of 156 MHz and 9,400, respectively [19–21]. It combines polysilicon surface micromachining and metal electroplating to attain submicron lateral capacitive gaps. The geometric dimensions necessary to reach a given frequency are larger for this contour mode than for the flexural modes used by previous resonators. Thus, it makes this disk resonator a better candidate for IF and RF filter application stages in future miniaturized transceivers. Figure 3.10 shows the SEM image of the disk resonator.

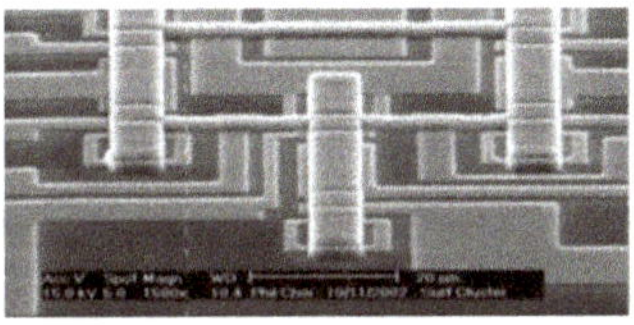

Figure 3.8 Bridged micromechanical filter [17]

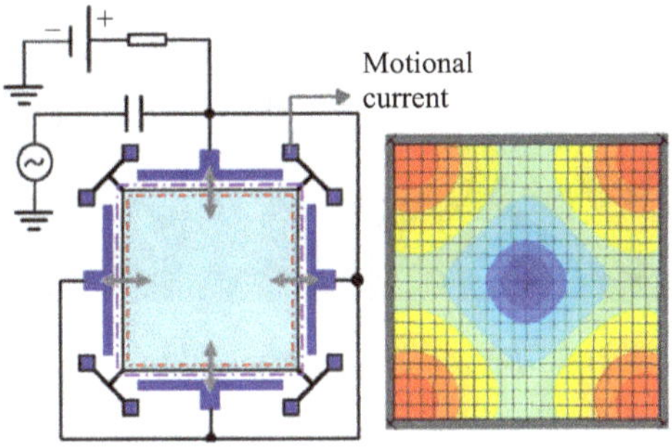

Figure 3.9 Schematic of square microresonator [18]

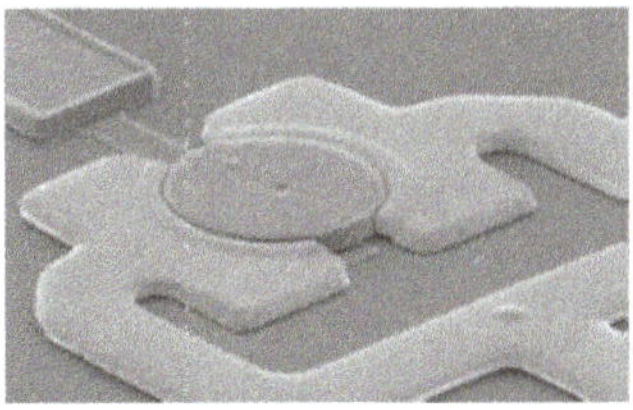

Figure 3.10 SEM image of the disk resonator [19]

A medium-scale integrated (MSI) vibrating micromechanical filter circuit that utilizes 128 radial-mode disk and mechanical link elements to achieve low motional resistance while suppressing unwanted modes and feed through signals was demonstrated in [22]. The ability to attain low insertion loss, small percent bandwidth on-chip and Qs >10,000 using capacitive transduced resonators were proved. The use of wavelength-optimized resonator coupling to affect a differential mode of operation that substantially improves the stopband rejection of the filter response and suppresses unwanted modes was validated [22,23]. Figure 3.11 shows the SEM image of the array disk resonator.

A new fabrication methodology that allows self-alignment of a micromechanical structure to its anchors was demonstrated in [24]. It was achieved using vibrating radial-contour mode polysilicon micromechanical disk resonators. Resonance frequencies up to 1.156 GHz and Qs at this frequency greater than 2,650 in both vacuum and air were measured. To satisfy power handling capability and phase noise for GSM standard wine glass disk resonator of thickness 3 μm supported with two beams was demonstrated in [25–27]. The 60-MHz wine glass disk oscillator exhibits an oscillator phase noise of 110 dBc/Hz at 1 kHz offset from the carrier and 132 dBc/Hz at far-from-carrier offsets [25]. Dividing down to 10 MHz for a fair comparison, these values equate to 125 dBc/Hz at 1 kHz offset and 147 dBc/Hz at far-from-carrier offsets which very nearly satisfy the GSM specification for wireless handsets [25]. Figure 3.12 shows the SEM image of the wine glass disk resonator.

Figure 3.11 SEM image of array disk resonator [22]

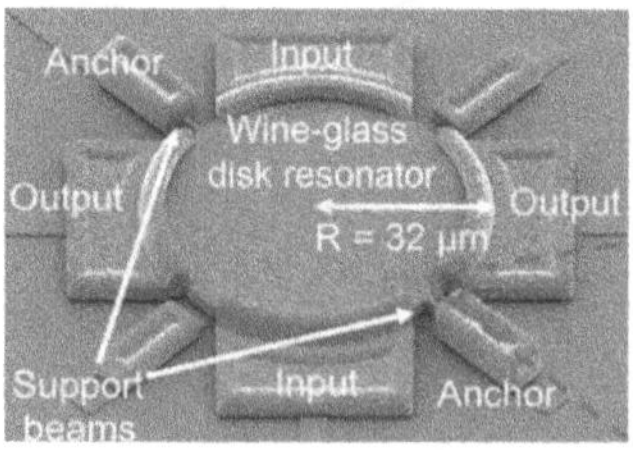

Figure 3.12 SEM image of wine glass disk resonator [25]

A reduction in phase noise by 13 dB was obtained over a 60 MHz surface micromachined micromechanical resonator oscillator by replacing the single resonator with a mechanically coupled array [28,29]. It also effectively raised the power handling ability of the frequency selective tank. Figure 3.13 shows the SEM image of the mechanically coupled wine glass disk resonator array.

Integrated CMOS-MEMS free-free beam resonator arrays operated in a standard two-port electrical configuration with low motional impedance and high power handling capability was demonstrated in [30]. Frequency centered at 10.5 MHz was demonstrated using the combination of pull-in gap reduction mechanism and mechanically coupled array design. The mechanical coupling elements using short stubs that connect each constituent resonator of an array to its adjacent ones at the high-velocity vibrating locations were demonstrated to enhance the desired mode and reject all other spurious modes. A single second-mode free-free beam resonator with quality factor $Q > 2{,}200$ and motional impedance Rm < 150 kΩ has been used to achieve mechanically coupled resonator arrays. Performance improvement of approximately 10x on motional impedance and power handling as compared with that of a single resonator was achieved [30]. Figure 3.14 shows the SEM image of a free-free beam resonator array with meander spring for gap reduction. Three different kinds of two-port flexural resonators, with both clamped and free ends, and with nominal resonance frequencies between 5 MHz and 50 MHz, were demonstrated in [31]. Novel free-free third mode resonator, as well as a tunable free-free resonator, was designed to maintain a high-quality factor despite its tunability. Simulations showed that third mode free-free resonators have the best frequency stability behavior [31].

The structure proposed in [32] was based on two CC beam resonators, designed with an identical resonant frequency of 26 MHz and mechanically coupled at a low-velocity point. A novel U-shape spring was chosen as the coupling

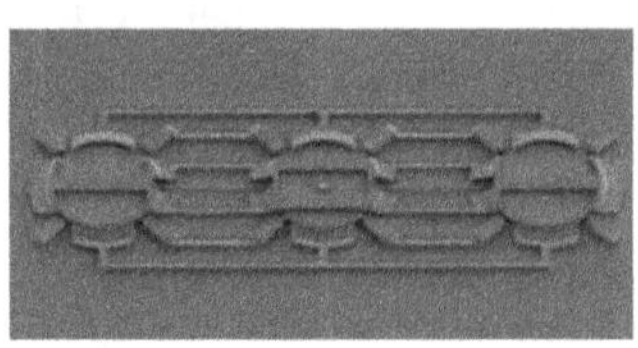

Figure 3.13 Mechanically coupled wine glass disk resonator [29]

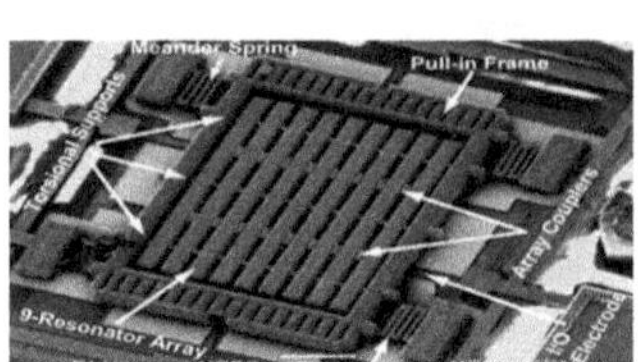

Figure 3.14 SEM image of the free-free beam resonator array [30]

element in order to set the frequency separation of both mechanical modes and therefore the bandwidth of the filter. The fabricated band-pass filter has shown very low passband distortion (less than 0.1 dB) without using any specific resistive termination, a central frequency of 26.7 MHz, and a bandwidth of 120 kHz (with a bias voltage of 18 V_{dc}) [32]. Novel filter coupling scheme that combines the merits of mechanically and electrically coupled methods to enable a well-defined narrow bandwidth and a decent stopband rejection in a CMOS-MEMS band-pass filter was proposed in [33]. Design strategies, such as oxide-rich structure, free-free beam arrayed design, and deep submicrometer transducer's gap result in featuring small motional impedance, low insertion loss, bandwidth control, and greater stopband rejection. Figure 3.15 shows the SEM image of the proposed free-free beam array resonator. Narrowband filter centered at 8.58 MHz with the bandwidth of 35.1 kHz and insertion loss of 17.4 dB was successfully demonstrated.

Dynamic behavior of an electrically actuated CC shallow arch to realize band-pass filters were investigated in [34]. Simulation results were shown demonstrating various scenarios of dynamic snap-through motion near the first and the third natural frequencies. Results could be also promising to build a band-pass filter with a sharp roll-off from the pass-band to stop-band, a flat bandwidth, and a high center frequency. This kind of filter is very simple, easy to fabricate, and small in size compared to other MEMS filters. Resonators based on the mechanical coupling of two or more vibrating microstructures usually suffer from mistuning problems and challenges in fabrication. The damping ratio can be used to control the bandwidth and the center frequency of such a filter [34]. Linear, undamped, and unforced vibration problem of micromechanical filters and closed-form expressions for their natural frequencies and mode shapes were obtained in [35]. The model described in this work treats the filter as a distributed-parameter system. For a micromechanical filter made of two CC beam resonators connected via a coupling beam, it was solved using boundary-value problem (BVP) composed of five equations and 20 boundary conditions for its natural frequencies and mode shapes. Due to the fact that the coupling beam is weak, the natural frequencies of the single resonator are split into two close frequencies for the filter: one frequency corresponds to an in-phase mode and the other corresponds to an out-of-phase mode. The bandwidth is very sensitive to the coupling location, whereas the center frequency is insensitive [35]. Finally, the design and characterization of a dual-frequency oscillator that consists of a reliable seesaw-shaped tungsten resonator

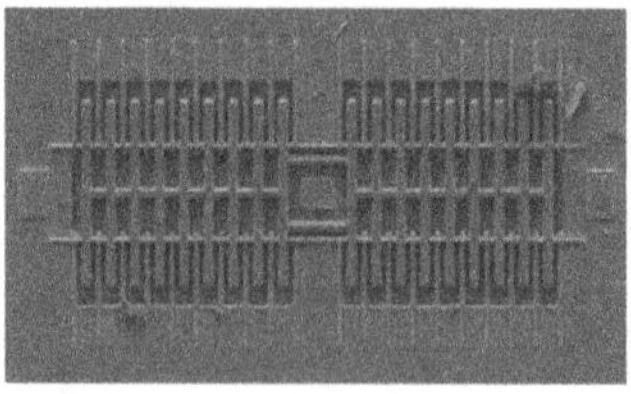

Figure 3.15 Free-free beam array structure [33]

integrated with the back end of a line of a standard 0.35-μm complementary metal-oxide-semiconductor (CMOS) technology [36,37] was demonstrated. The seesaw-shaped tungsten resonator built in a reliable and robust VIA3 platform with negligible curling or residual stress is electrostatically actuated applying an AC voltage to the driving electrode plus a DC bias voltage to the resonator. These voltages combine to produce a force that can excite two mechanical modes, the torsional (TM) and the vertical mode (VM) with natural frequencies of 553 and 906 kHz, respectively. Figure 3.16 shows the FESEM image of the seesaw resonator.

A large number of MEMS resonators are required if a wide range of operating frequencies need to be incorporated into the system. Quartz crystal and SAW resonators are unable to provide a feasible solution owing to limited maximum operating frequencies and large size. Recently aluminum nitride contour mode resonator has demonstrated multifrequency operating characteristics on a single silicon chip. These resonators have demonstrated excellent filter characteristics, but spurious modes are a major bottleneck obstructing the deployment of single-chip multifrequency resonators as a commercially viable solution for radiofrequency front-end filtering. MEMS resonators based on the capacitive transduction mechanism that can be fabricated using CMOS compatible process and materials result in a higher level of integration and low-cost batch fabrication. The most important advantages of capacitive silicon resonators are the resonance frequency can be slightly tuned by changing the DC bias voltage that is required for their operation, postfabrication fine-tuning and temperature compensation of the resonator. These resonators are smaller in size, consume practically zero DC power and have the capability of frequency translation (mixing) and switching. Capacitive resonators with frequencies in the very high frequency (VHF) and ultra-high frequency (UHF) range have impedances on the order of tens or hundreds of kilo-ohms. To achieve low motional impedance of the capacitive resonators, the combination of large transduction area and the tiny electrode-to-resonator gap is preferred.

MEMS resonators have also recently emerged as a compelling solution for the implementation of multifrequency resonators on the same substrate, owing to their high $f \cdot Q$ product and the fact that their center frequencies can be lithographically defined. These MEMS vibrating structures can be used according to their operational resonance modes which depend on the frequency range of applications. The power handling capability is limited by the small size single beam resonators,

Figure 3.16 Seesaw resonator [36]

therefore, resonator array is preferred. The advantages of the resonator array design are large transduction area and higher power handling capability. A large transduction area effectively reduces the required DC-bias voltage and motional impedance. The band-pass filters are often realized through several coupling approaches such as electric and mechanical coupling. The mechanical coupling enables the identification of the desired bandwidth through the physical dimension and position of the coupling beam. Electrical coupling lessens the parasitic capacitance that appears across the input and output ports that dominate the electrical measurements in capacitive transduced resonators.

MEMS resonators compatible with the CMOS process are mostly made of metal structures bearing low Q and poor temperature stability. To address thermal stability of MEMS resonators based on capacitive transduction mechanism passive temperature compensation techniques, such as the stress-induced frequency compensation [38], composite structures consisting of silicon dioxide [39], and degenerate doping [40], without consuming excessive power [41] or requiring complicated active compensation parabolic-type control circuits [42,43] are demonstrated. Composite structures consisting of silicon dioxide have been recently used for temperature compensation since it is easy to process and has a temperature coefficient of expansion (TCE) of opposite sign as compared to most other materials commonly used in MEMS. However, the design and simulation of complicated composite structures to achieve a low-temperature coefficient of frequency (TCF) may be computationally expensive and time-consuming. Therefore, indeed it is necessary to address the issues of suppression of spurious modes, multifrequency operation, narrow bandpass, low insertion loss, better stopband rejection, thermal stability, and CMOS process compatibility. Two-port electrical configurations, coupling schemes, and array structure are the keys to address low insertion loss, better stopband rejection, and narrow bandpass. MEMS resonators can be enabled for filter characteristics through a combination of larger transduction area, small transduction gap, electrostatic actuation, and low loss material. RF MEMS resonator finds applications in the implementations of the following circuits:

- Timing reference in oscillators,
- trans-impedance amplifier,
- Pierce oscillator, and
- Band-pass filters using either mechanical or electrical or both coupling schemes.

3.3 Actuation mechanism

MEMS often involves movable mechanical parts that require microactuators. This micro actuation requires mechanical energy so as to obtain a vibrating or translating or rotating motion based on the requirement for the MEMS device. The various types of mostly used actuation techniques are electrostatic, magnetic, thermal, and piezoelectric. Thermal actuation is rarely applied due to high power dissipation that makes it unsuitable for many electronic communication applications.

3.3.1 Electrostatic actuation

An electric charge is created around the electric field due to the potential difference between two conductors. This electric field generates a force between these two conductors. This widely-known principle has been actively used in electrostatic actuators for MEMS. Electrostatic actuation is mostly applied in a dielectric medium such as air or vacuum that can sustain electric fields across the electrodes. On a microscopic scale, this is a huge advantage, because most of the structures have a very low aspect ratio (i.e., width and length are large as compared to thickness and gap in the z-direction), so the distance between bodies is very small. In spite of all the available actuation techniques, the electrostatic actuation technique is preferred because of its ease in control, large force, and low power operation. The advantages of electrostatic actuation include relatively large displacement, ease of fabrication, controllable linearity of actuation, and most importantly convenience of integration with the integrated circuit processes to form a single-chip system. The applied actuation voltage is sufficient to control mechanically developed forces for the actuation of the moving parts. Since the system is nonlinear, instability of the pull-in often results that leads to failures including stick, wear, dielectric changing, and breakdowns. Application of electrostatic actuation mechanism needs careful studies in effect of voltage and temperature on spring softening, material selection, and reasonable design. In industry, it is used in microresonators, switches, micromirrors, accelerometers, etc. Almost every kind of microactuator has one or more electrostatic actuation-based version. Electrostatic actuation introduces nonlinear behavior such as hysteresis, jump, and dynamic instabilities. This nonlinear behavior is employed in many devices such as ultrasensitive mass-sensors, switches with low actuation voltage, etc.

3.3.2 Piezoelectric actuation

Piezoelectric materials show a small strain when they're placed under an electric field. This small strain is not so useful at the macro scale, but at a microscale, it can be a very effective source. The piezoelectric effect is realized as the linear electromechanical interaction between the mechanical and the electrical state in crystalline materials with no inversion symmetry. The piezoelectric effect is a reversible process in those materials exhibiting the direct piezoelectric effect. The internal generation of electrical charge resulting from an applied mechanical force exhibits the reverse piezoelectric effect and mechanical force resulting from an applied electrical field. Actuators based on piezoelectric ceramic material prime movers (or piezo-actuators) are finding broad acceptance in applications where precision motion and/or high-frequency operation is required. It can produce smooth continuous motion with resolution levels at the subnanometer level. Piezoelectricity is found in useful applications such as the production and detection of sound, generation of high voltages, electronic frequency generation, microbalances, and ultrafine focusing of optical assemblies.

3.3.3 Magnetic actuation

A plate is supported by a torsional hinge structure of embedded conducting wires, constituting multiwindings positioned at different locations. The conducting wires of these windings are therefore of different lengths. Two permanent magnets are placed on the side of the plate, such that the magnetic field lines are parallel to the plane and orthogonal to the torsional hinges. When current passes through the coils, Lorentz forces will develop and cause rotational torque on the plate. The direction of the torque depends on the direction of input currents. A MEMS magnetic Actuator is a device that uses the MEMS process technology to convert an electrical signal (current) into a mechanical output (displacement) by employing the well-known Lorentz force equation or the theory of magnetism. When a current-carrying conductor is placed in a static magnetic field, the field produced around the conductor interacts with the static field to produce a force. This Force can be used to cause the displacement of a mechanical structure.

3.4 Design of low-frequency MEMS resonator

Euler Bernoulli's beam theory exactly predicts the frequency of the resonant modes that completely depends on the material properties and beam dimensions. The angular frequency of the *n*th resonant mode of a CC beam [44] is given as

$$\omega_n = \frac{\pi^2 \alpha_n^2}{L^2} \sqrt{\frac{EI}{\rho A}} \tag{3.1}$$

where A is area, I is the moment of inertia, α_n is mode coefficient, L is the length of the beam, E is Young's modulus, and ρ is the density of the structural material, respectively. The dimensions of the anchor structure are assumed to be much larger than beam resonator and both are made of the same material. Beam theory can be used to determine the vibration amplitude and frequency of a beam resonator since zero displacements of the beam resonator at the clamped end can be assumed.

3.4.1 Clamped-clamped beam design

For the most practical design, the length (L) of the CC beam depends on the desired resonance frequency. Width (w) of the CC beam depends upon the ratio of length (L) to width (w) that satisfy Euler Bernoulli's beam theory. The thickness (t) of the beam completely depends upon the process constraints. For a CC beam uniform cross-section in the absence of an electromechanical coupling, the Euler Bernoulli equation for the resonance frequency f is given using (3.2) [32]

$$f = \frac{1}{2\pi\sqrt{12}} \beta_n^2 \sqrt{\frac{E}{\rho} \frac{t}{L^2}} \tag{3.2}$$

where E is Young's modulus and ρ is density of the structural material, respectively. β_n is mode coefficient whose values for the first four modes are given as 4.73, 7.853, 10.996 and 14.137 [32]. Euler Bernoulli's equation (3.2) is applicable only if the length to width and length to thickness ratios are large (greater than 10). Length of the CC beam can be determined by the desired resonance frequency when operated in mode 1, 2, 3, etc. and width of the CC beam can be chosen to

Table 3.1 Characteristics of CC beam resonator

Mode	Mode number (N)	Mode coefficients (β)	f_n/f_1
Fundamental	1	4.73	1.0
First harmonic	2	7.853	2.57
Second harmonic	3	10.996	5.404
Third harmonic	4	14.137	8.932
Fourth harmonic	5	17.279	13.34

satisfy $L/w > 10$ and minimum manufacturable value. Characteristics of the flexural mode CC beam resonator and ratios of the higher modes to the fundamental mode f_n/f_1 are shown in Table 3.1.

Careful design of the micromechanical resonator device includes a selection of geometrical parameters of CC beam and slim mechanical couplers. Desired frequency characteristics and bandwidth of the resonator device can be addressed using electrical coupling through precise placement of input-output electrodes and precise placement of mechanical couplers at node points, respectively. A large transduction area can be obtained through a mechanically coupled array structure consisting of more than two CC beams and output electrodes. Thus, design ensures narrow passband, better stopband rejection, desired bandwidth and suppressed spurious mode associated with the complex design. Low insertion loss or motional resistance can be addressed through a large transduction area and submicrometer transduction gap.

3.4.2 Mechanical coupler design

The bandwidth of the resulting filter is given using (3.3) [10]

$$B = \frac{f}{k_r(x)} \frac{k_c}{k_{12}} \tag{3.3}$$

where B is the required bandwidth, k_c is the stiffness of the coupler, $k_r(x)$ is the stiffness of the resonator at point x from the anchor, and k_{12} is the normalized coupling coefficient for the given filter type. Once the dimensions of the CC beam have been chosen for desired frequency, stiffness of the coupling beam can be obtained using (3.4) [45] and stiffness of CC beam at point x from the anchor can be obtained using simulation.

$$k_c = \omega w_c t \sqrt{E\rho} \tag{3.4}$$

where width (w_c) of the couplers that can be chosen to satisfy desired bandwidth (B) for the filter. The length of the mechanical coupler (L_c) can be chosen that satisfies the expression

$$\cos(\alpha)\sinh(\alpha) + \sin(\alpha)\cosh(\alpha) = 0 \tag{3.5}$$

where $\alpha = \left(\frac{L_c \rho w_c t \omega^2}{EI}\right)^{0.25}$ and $I = \frac{w_c t^3}{12}$ [46]

Alternatively, coupling beam length corresponding to a quarter wavelength at the frequency of operation can be chosen that satisfies (3.5) and it is given by (3.6)

$$L_c = \frac{1}{4f}\sqrt{\frac{E}{\rho}} \tag{3.6}$$

Desired bandwidth and Q depends upon the location of mechanical coupling between CC beams and couplers, that is, point x from the anchor in (3.3). The proposed mathematical model in [35] exactly predicts Q depending upon the location of coupling called as node points. Node points can be separately identified based on desired bandwidth when operated in mode 1 and mode 3. Figure 3.17 shows the node points obtained through (3.3) when operated in mode 1 and mode 3. When the CC beam is operated in mode 3, the beam observes zero displacements at points (one-third of the beam length) as shown in Figure 3.18, thus the coupling at those points would result in maximum Q. Exactly these node points were identified

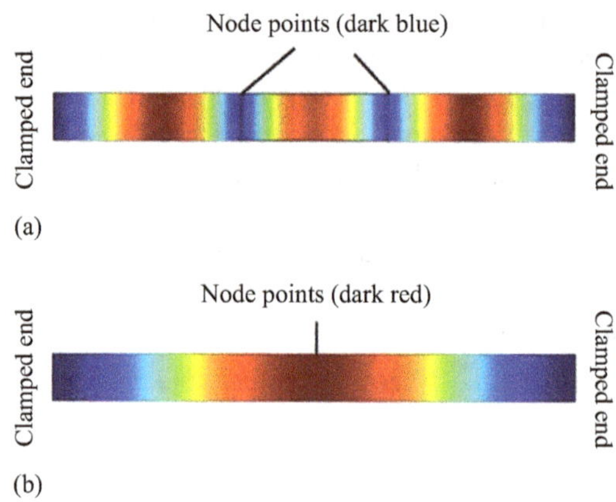

Figure 3.17 CC beam with node points [47]: (a) node points when operated in mode 1 resonating frequency and (b) node points when operated in mode 3 resonating frequency

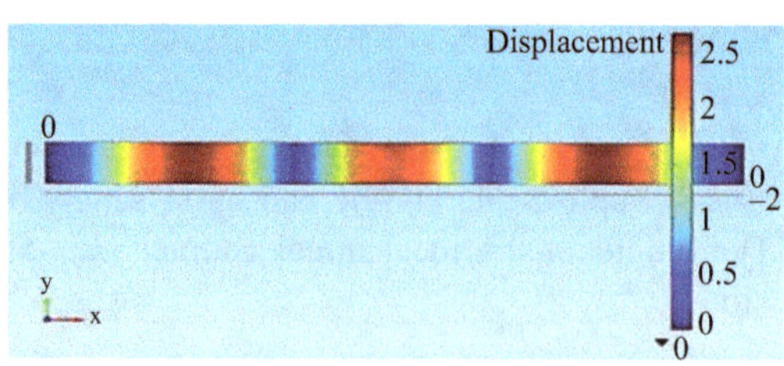

Figure 3.18 CC beam with zero displacement (dark blue) when operated in mode 3 [47]

to obtain desired bandwidth in mode 1. Similarly, to obtain desired bandwidth in mode 3, the node point was identified at the 1/3rd of the X. It was assumed that X is equal to 1/3rd length of the beam when operated in mode 3. Thus, equal displacement under electrostatic actuation can be obtained by identifying node points for operation in respective modes as above [47]. Node points were identified at 1/3rd of the beam length to achieve equal Q (around 100) in both the modes of operation.

3.4.3 Electrical coupling scheme

The response using an electrical coupling scheme can be obtained through a combination of opposite motional signals from two separate systems using a readout amplifier [48] or capacitive coupling. Desired passband and better stop-band rejection can be obtained by operating in-phase around the desired center frequency and out-of-phase for outside the passband, respectively. Thus, spurious responses which are out-of-phase signals get subtracted and cancellation of feed-through current results in better stopband rejection. Mixed electrical and mechanical coupling scheme results in desired bandwidth, better stopband rejection, low insertion loss, and sharp roll-off. Feedthrough cancellation can be obtained using two output electrodes that are precisely placed under the CC beam [47]. Figure 3.19 shows the placement of input-output electrodes in the MEMS resonator. When the resonator is excited with mode 1 and mode 3 resonating frequency signals, all the constituent resonators will move away from the output electrodes. The input current I_i moves in the opposite direction of the output currents I_{o1} and I_{o2} illustrated in Figure 3.20(a), 3.20(c), and 3.20(e).

When the resonator is excited with mode 2 and mode 4 resonating frequency signals, the constituent resonators move toward output electrode 1 and away from output electrode 2. The output current I_{o1} moves in the opposite direction of the output current I_{o2} as illustrated in Figure 3.20(b), 3.20(d), and 3.20(f). Further I_{o1} and I_{o2} are out-of-phase with 180° phase difference that would result in cancellation at mode 2 and 4. Finally, the total output current I_o can be obtained by summing the I_{o1} and I_{o2} together representing the entire filter spectrum.

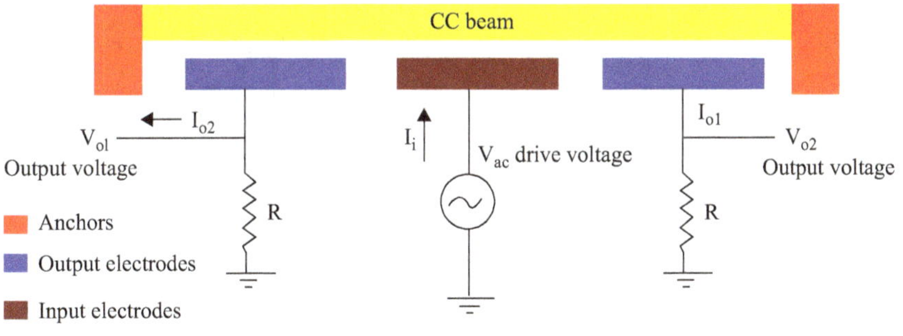

Figure 3.19 Electrical coupling scheme using two output electrodes and single input electrode (not to scale) [47]

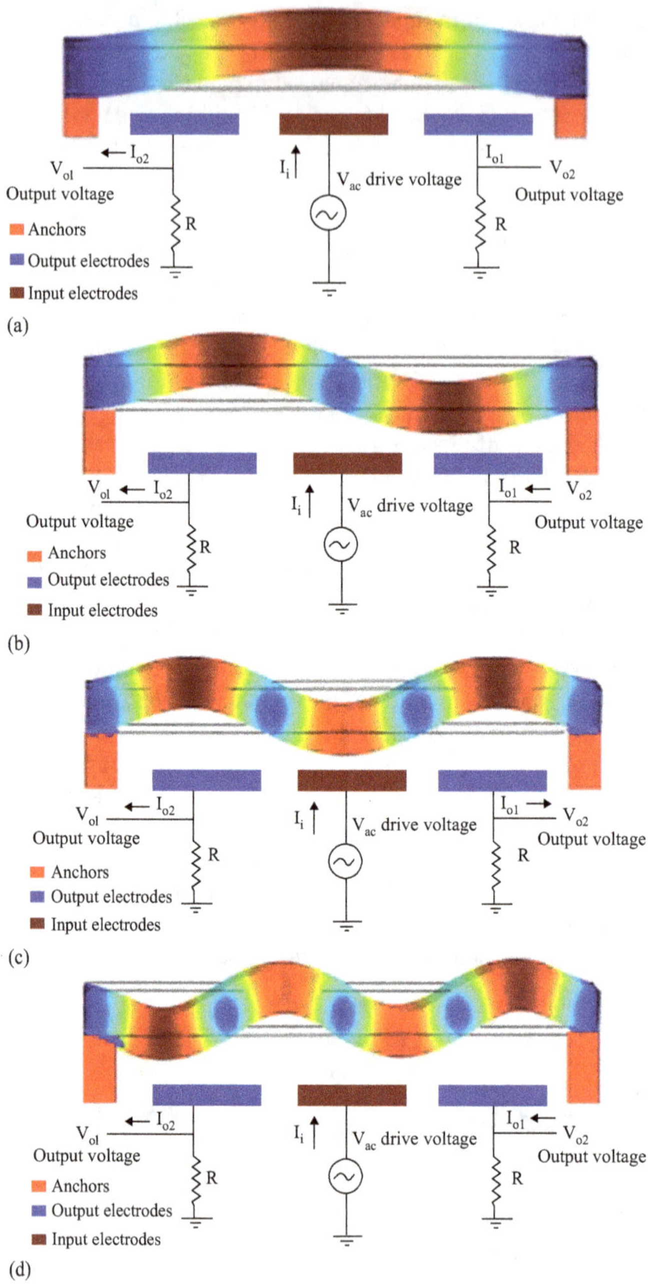

Figure 3.20 Description of the in-phase and out-of-phase mode (not to scale) [47]: (a) mode 1, (b) mode 2, (c) mode 3, (d) mode 4, (e) currents at mode 1 and 3, and (f) currents at mode 2 and 4

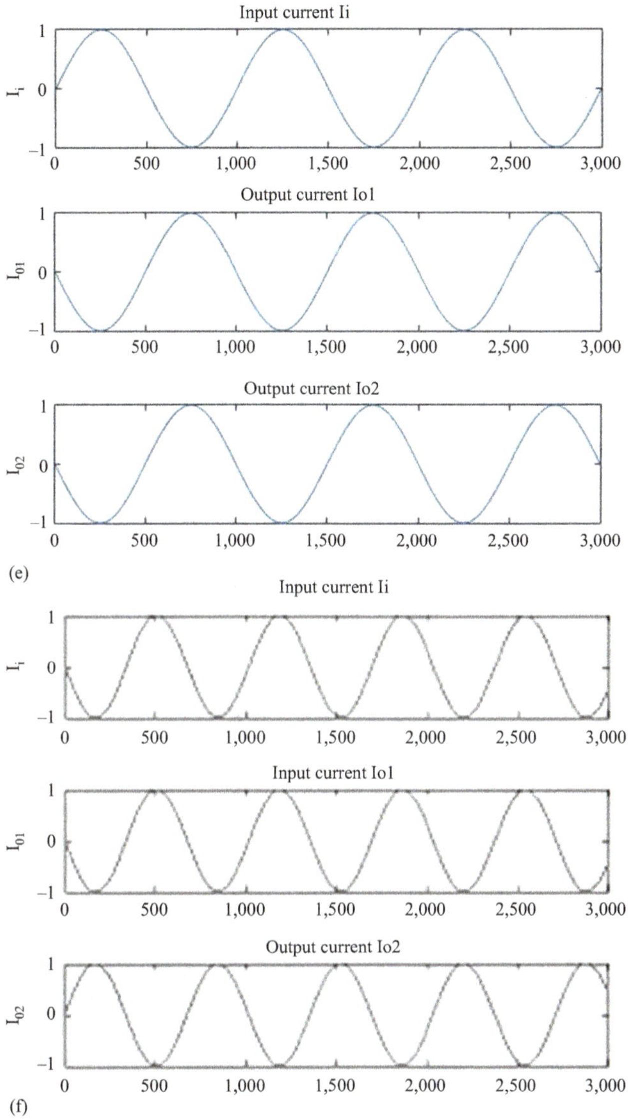

Figure 3.20 (Continued)

3.4.4 Suppression of spurious responses

The most fundamental equation that governs the behavior of beam structures when subjected to change in temperature is given by (3.7)

$$\varepsilon_{\text{total}} = \varepsilon_{\text{mechanical}} + \varepsilon_{\text{thermal}} \tag{3.7}$$

Unavoidable thermal strains (ε) are induced in beams through heating. These strains take the form of thermal expansion to an increased length and curvature.

CC beams cannot expand in length. Let us consider when the CC beam is subjected to a uniform rise in temperature (ΔT). This temperature change generates internal load and if temperature attains a value of first critical temperature and the first critical buckling load is achieved the structure suffers a transversal displacement. Thus, the frequency values of a given vibration mode decrease until the buckling in that mode is achieved. Practically, CC beams in an array experience temperature variation due to heating. That results in changes in vibration shapes and natural frequencies due to compressive thermal loads and expansion for each CC beam.

To address the thermal stability of MEMS resonators based on capacitive transduction, passive temperature compensation techniques, such as the stress-induced frequency compensation, composite structures consisting of silicon dioxide, and degenerate doping, without consuming excessive power or requiring complicated active compensation, parabolic-type control circuits are demonstrated. Suppression of spurious responses due to change in temperature of the beam is desirable to achieve single-chip and single/dual-band frequency response solutions. Therefore, it is necessary to provide temperature compensation with the help of stiff mechanical couplers. Temperature compensation in metal MEMS resonator can be explored using a mixed mechanical and electrical coupling scheme. Mechanical coupler forces all the beams to vibrate at uniform resonant frequencies. Whereas placement of mechanical couplers at low and high-velocity points ensures minimum variations in mode shapes irrespective of the variations in temperature of CC beams. These coupler forces all the beams to vibrate at uniform resonant frequencies irrespective of the variations in temperature of CC beams. It facilitates to suppress generation of spurious modes due to heating. Thus, the placement of mechanical couplers at low and high-velocity points ensures minimum variations in mode shapes and frequencies. Better stop-band rejection can be achieved through electrical coupling. It achieves nonaddition of unwanted and spurious responses and also reduces parasitic capacitance across the input-output ports [12,49]. Figure 3.21 and 3.22 show the COMSOL simulated

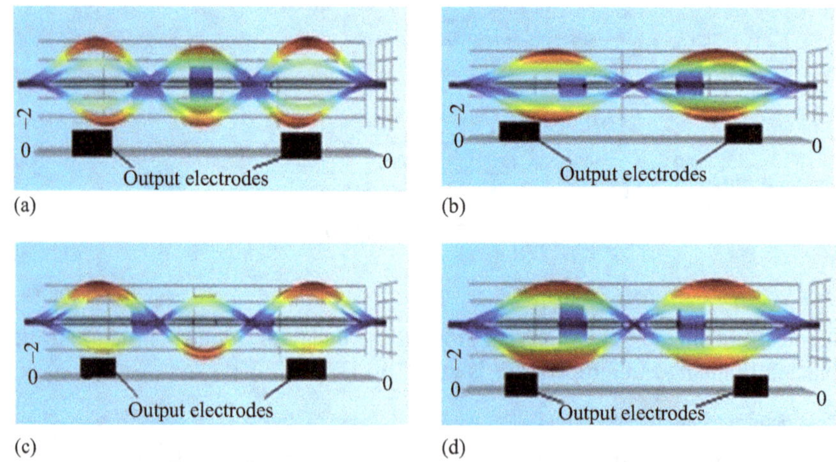

Figure 3.21 Spurious modes at different frequencies clearly indicating nonaddition of responses at output electrodes [47]

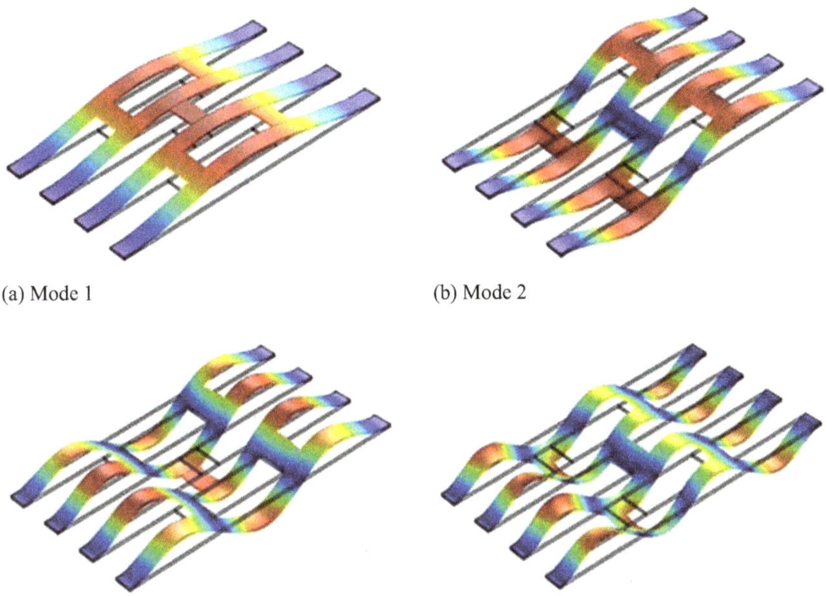

(a) Mode 1 (b) Mode 2

(c) Mode 3 (d) Mode 4

Figure 3.22 COMSOL simulated mode shapes of device

mode shapes of spurious modes, it also clearly indicates the nonaddition of responses at the output electrodes.

3.5 Summary

Low-frequency MEMS resonator device design using a mixed mechanical and electrical coupling scheme is illustrated in this chapter. The closed-form expressions are easier to handle and provide automated analysis before fabrication. The electrical coupling scheme is also useful in suppressing spurious modes due to an increase in the operating temperature of the device by no addition to the responses at the output electrode. Mechanical couplers at equal velocity point aids in minimum variations in mode shapes and resonant frequencies. Thus, mixed electrical and mechanical coupling scheme enhances operation in desired modes and suppresses undesired and spurious modes. The device can be designed using Euler Bernoulli's beam theory. Desirable center frequency and bandwidth can be obtained using well-defined equations and physical parameters. Although resultant frequency and bandwidth can be achieved, other filter parameters such as insertion loss, output power, and stopband rejection which is the function of the transduction gap, the width of the CC beam, and the number of CC beams need careful modeling of the device. A comprehensive model is required that would give complete design automation before fabrication of the device under electrostatic actuation.

Filter characteristics, suppression of spurious modes, and thermal stability can be obtained using CC beam array that is anchored to the substrate using rigid contact anchors. These anchors are responsible to dissipate large amounts of energy and reduce Q. Although it is known that a free-free beam achieves high Q. But dual-frequency characteristics, desired bandwidth, and suppression of unwanted responses may be difficult to achieve using a free-free beam. Hence it is required to adopt techniques that would reduce anchor losses and enhance Q. RF MEMS resonator filters to be deployed in a transceiver for wireless communication systems needs to satisfy all the stringent specifications given to the filter designer. One of the most important among them is thermal stability. Metals MEMS resonator filter has poor thermal stability hence composite structure using silicon dioxide. To improve further the thermal stability of the device, we suggest fabricating the composite structure of the device using metals and silicon dioxide. Fabrication of the low-frequency MEMS resonator devices is illustrated in Chapter 6.

References

[1] Basu J., and Bhattacharyya T.K. "Microelectromechanical resonators for radio frequency communication applications." *Microsystem Technologies.* 2011;**17**(10–11):1557–80.

[2] Héctor J., and Santos D.L. *RF MEMS Circuit Design for Wireless Communications*. Artech House; 2002.

[3] Hilbert J.L. "RF-MEMS for wireless communications." *IEEE Communications Magazine*. 2008; **46**(8): 68–74.

[4] Randy J.R., Héctor J., and Santos D.L. "MEMS for RF/microwave wireless applications: The next wave." *Microwave Journal.* 2001.

[5] Helfenstein M., and Moschytz G.S. *Circuits and Systems for Wireless Communications*. Kluwer Academic, 2002.

[6] Nguyen C.T.C., Katehi L.P.B., and Rebeiz G.M. "Micromachined devices for wireless communications." *Proceedings of IEEE*. 1998;**86**(8): 1756–68.

[7] Nguyen C.T.-C. "RF MEMS for wireless applications." *Conference Digest, Device Research Conference*. Santa Barbara, California, June 24–26, 2002, 9–12.

[8] Zou Q., Bi F., Tsuzuki G., Bradley P., and Ruby R. "Temperature-compensated FBAR duplexer for band 13." In: *Proceeding of IEEE International Ultrasonics Symposium (IUS),* Prague, Czech Republic, 2013, 236–38.

[9] Yen T.T., Pisano A.P., and Nguyen C.T.C. "High-Q capacitive-piezoelectric ALN lamb wave resonators." In: *IEEE International Conference on MEMS*, 2013.

[10] Bannon F.D., Clark J.R., and Nguyen C.T.C. "High-Q HF microelectromechanical filters." *IEEE Journal of Solid-State Circuits*. 2000;**35**(4):512–26.

[11] Chandorkar S.A., Agarwal M., Melamud R., Candler R.N., Goodson K.E., and Kenny T.W. "Limits of quality factor in bulk-mode micromechanical resonators." In *Proceedings of the 21st IEEE International Conference on MicroElectroMechanical Systems*, Tucson, Arizona, 2008, 74–77.

[12] Nguyen C.T.C. "Vibrating RF MEMS for low power communications (invited)." *Proceedings of MRS Fall Meeting*, Boston, Massachusetts, December 2–6, 2002, pp. J12.1.1–J2.1.12.
[13] Demirci M.U., and Nguyen C.T.C. "Higher-mode free-free beam micromechanical resonators." *Proceedings of IEEE International Frequency Control Symposium*, Tampa, Florida, May 5-8, 2003, 810–18.
[14] Galayko D., Kaiser A., Buchaillot L., Legrand B., Collard D., and Combi C. "Design, realization, and test of micromechanical resonators in thick-film silicon technology with post-process electrode-to-resonator gap reducing." *Journal of Micromechanics and Microengineering*. 2003;**13**(1–7).
[15] Hsu W.-T., Clark J.R., and Nguyen C.T.-C. "Q-optimized lateral free-free beam micromechanical resonators." *Digest of Technical Papers, the 11th International Conference on Solid-State Sensors & Actuators (Transducers'01)*, Munich, Germany, June 10–14, 2001, 1110–13.
[16] Lee S., and Nguyen C.T.C. "Mechanically-coupled micromechanical arrays for improved phase noise." *Proceedings of IEEE International Ultrasonics, Ferroelectrics, and Frequency Control*, Montreal, Canada, August 24–27, 2004, 280–28.
[17] Li S.-S., Demirci M.U., Lin Y.-W., Ren Z., and Nguyen C.T.C. "Bridged micromechanical filters." *Proceedings IEEE International Frequency Control Symposium and Exposition*, 2004, 280–86.
[18] Lee J.E.Y., Bahreyni B., Zhu Y., and Seshia A.A. "A single-crystal-silicon bulk acoustic mode microresonator oscillator." *IEEE Electron Devices Letter*. 2008; 29(7): 701–03.
[19] Clark J. R., Hsu W.T., and Nguyen C.T.C. "High-Q VHF micromechanical contour-mode disk resonator." *Technical Digest, IEEE International Electron Devices Meeting*, San Francisco, California, December 11–13, 2000, 493–96.
[20] Basu J., Chakraborty S., and Bhattacharyya T.K. "Micromechanical radial-contour mode disk resonator for a CMOS-MEMS oscillator." *Annual IEEE India Conference (INDICON)*, 2010, 1–4.
[21] Demirci M.U., and Nguyen C.T.C. "Single-resonator fourth-order micromechanical disk filters." *Proceedings of 18th International IEEE Micro Electro Mechanical Systems Conference*, Miami, Florida, January 3–February 3, 2005, 207–10.
[22] Li S.S., Lin Y.W., Ren Z., and Nguyen C.T.C. "An MSI micromechanical differential disk-array filter." *Digest of Technical Papers, in 14th IEEE International Conference on Solid-State Sensors & Actuators (Transducers'07)*, Lyon, France, June 11-14, 2007, 307–11.
[23] Li S.S., Lin Y.W., Ren Z., and Nguyen C.T.C. "Disk-array design for suppression of unwanted modes in micromechanical composite-array filters." *Technical Digest, 19th IEEE International Conference on Micro Electro Mechanical Systems (MEMS'06)*, Istanbul, Turkey, January 22–26, 2006, 866–69.
[24] Wang J., Ren Z., and Nguyen C.T.C. "1.156-GHz self-aligned vibrating micromechanical disk resonator." *IEEE Transactions On Ultrasonics, Ferroelectrics, And Frequency Control*. 2004; **51**(12): 1607–28.

[25] Lin Y.W., Lee S., Li S.S., Xie Y., Ren Z., and Nguyen C.T.C. "Series resonant VHF micromechanical resonator reference oscillators." *IEEE Journal of Solid-State Circuits*. 2004;**39**(12); 247– 91.

[26] Lin Y.W., Lee S., Li S.S., Xie Y., Ren Z., and Nguyen C.T.C. "60-MHz wine glass micromechanical disk reference oscillator." In: *IEEE International Solid-State Circuits Conference*, San Francisco, California, February 2004, 322–23.

[27] Lee S., and Nguyen C.T.C. "Phase noise amplitude dependence in self-limiting wine-glass disk oscillators." *Technical Digest, 2002 Solid-State Sensor, Actuator, and Microsystems Workshop*, Hilton Head Island, South Carolina, June 6–10, 2004, 33–36.

[28] Lin Y.W., Li S.S., Ren Z., and Nguyen C.T.C. "Low phase noise array-composite micromechanical wine-glass disk oscillator." *Technical Digest, IEEE International Electron Devices Meeting*, Washington, DC, December 5–7, 2005, 287–90.

[29] Lin Y.W., Hung L.W., Li S.S., Ren Z., and Nguyen C.T.C. "Quality factor boosting via mechanically-coupled arraying." *Digest of Technical Papers, the 14th International Conference on Solid-State Sensors & Actuators (Transducers'07)*, Lyon, France June 11–14, 2007, 2453–56.

[30] Li M.-H., Chen W.-C., and Li S.-S. "Mechanically coupled CMOS-MEMS free-free beam resonator arrays with enhanced power handling capability." *IEEE Transactions on Ultrasonics, Ferroelectrics, and Frequency Control*. 2012; **59**(3); 346–57.

[31] Paci D., Mastrangeli M., Nannini A., and Pieri F. "Modeling and characterization of three kinds of MEMS resonators fabricated with a thick polysilicon technology." *Analog Integrated Circuits Signal Processing*. 2006; **48**: 41–47.

[32] Giner J., Uranga A., Torres F., Marigo E., and Barniol N. "Fully CMOS integrated bandpass filter based on the mechanical coupling of two RF MEMS resonators." *IEEE Electronics Letters*. 2010; **46**(9).

[33] Chen C.-Y., Li M.-H., Chin C.-H., and Li S.-S. "Implementation of a CMOS-MEMS filter through a mixed electrical and mechanical coupling scheme." *Journal of Microelectromechanical Systems*. 2016; **25**(2): 262–74.

[34] Ouakad H.M. "An electrostatically actuated MEMS arch band-pass filter." *Hindawi Journal of Shock & Vibration*. 2013; 809–819.

[35] Hammad B.K. "Natural frequencies and mode shapes of mechanically coupled microbeam resonators with an application to micromechanical filters." *Hindawi Journal of Shock & Vibration*. 2014.

[36] Riverola M., Sobreviela G., Torres F., Uranga A., and Barniol N. "Single-resonator dual-frequency BEOL-embedded CMOS-MEMS oscillator with low-power and ultra-compact TIA core." *IEEE Electron Device Letters*. 2017;**38**(2): 273–76.

[37] Riverola M., Sobreviela G., Uranga A., and Barniol N. "Intrinsic feed through current cancellation in a seesaw CMOS-MEMS resonator for integrated oscillators." In: *IEEE International Frequency Control Symposium*, 2016, 1–4.

[38] Hsu W.-T., and Nguyen C.-C. "Stiffness-compensated temperature-insensitive micromechanical resonators." In: *15th IEEE International Conference on Micro Electro Mechanical Systems*, January 2002, 731–34.

[39] Tabrizian R., Casinovi G., and Ayazi F. "Temperature-stable silicon oxide (SilOx) micromechanical resonators." *IEEE Transaction on Electron Devices*. 2013;**60**(8):2656–63.

[40] Ng E., Ahn C., Yang Y., *et al.* "Localized, degenerately doped epitaxial silicon for temperature compensation of resonant mems systems." *Transducers Eurosensors XXVII: The 17th International Conference on Solid-State Sensors, Actuators, and Microsystems*, 2013, 2419–22.

[41] Hopcroft M., Melamud R., Candler R.N., *et al.* "Active temperature compensation for micromachined resonators." In: *Technical Digest Solid-State Sensor, Actuator and Microsystems Workshop*, 2004, 364–67.

[42] Sundaresan K., Ho G., Pourkamali S., and Ayazi F. "Electronically temperature compensated silicon bulk acoustic resonator reference oscillators." *IEEE Journal of Solid-State Circuits*. 2007;**42**(6):1425–34.

[43] Ho G., Sundaresan K., Pourkamali S., and Ayazi F. "Micromechanical IBARs: Tunable high-q resonators for temperature-compensated reference oscillators." *Journal of Microelectromechanical Systems*. 2010;**19**(3): 503–15.

[44] Wang K., Yu Y., Wong A.C., and Nguyen C.T.C. "VHF free-free beam high Q micromechanical resonator." *12th IEEE International Micro Electro Mechanical Conference*, Florida, 1999, 453–58.

[45] Amar A.B., Bahloul D., Gagnon F., and Kouki A.B. "MEMS filter's design and modeling based on width-extensional mode plate resonator for wireless applications." *Microsystem Technology*. 2015; **21**:1567–76.

[46] Bannon F.D., Clark J.R., and Nguyen C.T.-C. "High frequency microelectromechanical IF filters." *Technical Digest IEEE International Electron Devices Meeting*, 1996, 773–76.

[47] Morankar A., and Patrikar R.M. "Dual frequency MEMS resonator through mixed electrical and mechanical coupling scheme." *IET Journal of Circuits, Devices & Systems*. 2018;**12**(1); 88–93.

[48] Chen C.-Y., Li M.-H., Chin C.-H., Li C.-S., and Li S.-S. "Combined electrical and mechanical coupling for mode-reconfigurable CMOS MEMS filters." In: *Proceedings of 27th IEEE International Conference on MEMS*, January 2014, 1249–52.

[49] Morankar A., and Patrikar R.M. "Effective method for temperature compensation in dual band metal MEMS resonator." *22nd International Symposium on VLSI Design and Test (VDAT – 2018)*. Thiagarajar College of Engineering, Madurai, India, June 28–30, 2018. *Springer Communication in Computer & Information Science*. 2019; **892**:233–41.

Chapter 4

Microelectromechanical resonator design for high frequency

Prasanna Deshpande[1] *and Rajesh Pande*[2]

4.1 Introduction

Electronics have become an essential part of human life. Sir Nikola Tesla said it, science proved it, it is a known fact that everything including our own bodies is made up of energy vibrating at different frequencies. The conventional microelectromechanical systems (MEMS) technology converts energy from mechanical to the electrical domain or vice versa—sensors and actuators play an irreplaceable role in our modern life and are offered by many suppliers. In contrast to their unique function, radio frequency microelectromechanical systems (RF MEMS) process electrical signals using mechanically vibrating structure and have replaced on-chip electrical RF devices to provide frequency control functions due to their extraordinary performance compared to on-chip electrical counterparts. Frequency selective elements such as resonators are being increasingly employed in applications related to timing and frequency control, and as building blocks in micro/nanofabricated oscillators and/or filters. With small size, high performance, and complementary metal-oxide-semiconductor (CMOS) compatibility, RF MEMS resonator offer promising technology in contemporary RF front-end in wireless communication systems.

Many wireless communication systems develop a certain degree of miniaturization. On one side, miniaturization is making the system lighter and more efficient, while the growing IC manufacturing technology can produce large quantities of small product at a lower cost. The term RF MEMS refers to electronic devices with a moving submillimeter-sized part which provides RF functionality [1]. RF MEMS provides a class of new devices and components that display superior high-frequency performance relative to conventional semiconductor devices. A key advantage of the RF MEMS devices compared to semiconductor counterparts is the electromechanical isolation, means the RF circuit does not couple significantly to the actuation circuit. Lower power consumption is another advantage. However, the low switching speed is a serious drawback of RF MEMS devices [2]. Wireless communication technologies such as CDMA (code division multiple access), GSM (global system for mobile

[1]Department of Electronics and Communication Engineering, Shri Ramdeobaba College of Engineering and Management, Nagpur, India
[2]Department of Electronics Engineering, Shri Ramdeobaba College of Engineering and Management, Nagpur, India

communication), 3G, and 4G provide us a voice, data, and broadband communication. In order to maintain the quality and reliability of these technologies, the specifications given to a design engineer are more rigorous. RF MEMS switches, varactors, inductors, and resonators are ideal for reconfigurable systems possess low insertion loss and very high-quality factor even up to tens of GHz range of frequencies. [3]. With quality factor more than 100 times higher than their on-chip electrical counterparts, RF MEMS resonators offer promising building blocks to synthesize low-loss band-pass filters/oscillators used in contemporary wireless communication systems.

4.2 Motivation and challenges

The emerging demands of a highly integrated wireless communication system have raised the needs for the fine frequency selection purpose on a single substrate. Most of the microwave hardware runs at 50–75 Ω characteristics impedance system. At ac signals, the maximum power transfers from the source to load occurs when load impedance Z_L equals to source impedance Z_S. The RF transceivers serve as a medium to match characteristics impedance between the source and the load. Improvements in the mismatching of these characteristics impedance bring low reflections between the interconnected circuit elements.

Similarly, the system performance and higher order integration could also be benefited from the replacement of discrete crystal/filter components with thin-film MEMS. Higher frequencies for the carrier waves permit the higher rate of data transfer. Therefore, a resonator with high-quality factor (Q) operating at the high resonance frequency (fr) and low motional resistance (Rm) is a challenge [1,2].

Among all the transduction mechanisms, researchers have focused on piezoelectric transduction for its better frequency scaling and its capability to interface directly with 50–75 Ω electronics. This is an extremely important advantage for RF applications, which is not easily achievable with electrostatically transduced resonators. Similarly, the actuation voltage requirement in the piezoelectric resonator is less than the electrostatic resonator [4]. Another advantage of the piezoelectrically transduced resonator is its low motional resistance (Rm) in the range of several ohms at high frequency which is unlike in electrostatically transduced resonators. Therefore, for high frequency application with high Q and low Rm, we have selected piezoelectric resonators for our design and fabrication.

The purpose of this chapter is to explain the transduction mechanisms used in resonators. The literature overviews on the different kinds of resonators; their characteristics are explained. The details of high frequency piezoelectrically actuated resonator and its design and development with the use of piezoelectric material are also addressed along with the optimization technique.

4.3 High frequency resonators

The piezoelectric resonators utilize electromechanical coupling capabilities to transform electrical signals into mechanical vibration at the frequency of interest.

The piezoelectric resonators have lower motional resistance and higher power handling capabilities and are capable of generating acoustic waves compared to capacitive actuation [5,6]. The three primary piezoelectric materials used in piezoelectric MEMS devices for various applications are zinc oxide (ZnO), aluminum nitride (AlN), and lead zirconium titanate (PZT). The highest piezoelectric constant and electromechanical coupling coefficient, as well as higher acoustic wave attenuation and lower sound velocities, are reported in PZT. However, poor stability and piezoelectric properties affected by operating temperature and less mechanical deformation for a long duration due to brittleness are the major issues of PZT. Due to the presence of lead, PZT is restricted in medical electronics and other applications. AlN films have higher phase velocity and chemical stability and have lower piezoelectric coupling. It is more difficult to deposit also [7,8].

This chapter focuses on the details of laterally vibrating one port contour mode ZnO MEMS resonator. The design of resonator using piezoelectric ZnO thin film has been employed due to its better coupling coefficient, low deposition temperature, excellent bonding, unique semiconducting, and optical properties [9,10]. Among all piezoelectric materials, zinc oxide is considered as a pollution-free green piezoelectric material. The ZnO is highly tensile and may undergo huge mechanical deformation for a long duration unaffected by temperature variation. The ZnO thin film can be deposited at room temperature and a variety of acidic etchants are also available [11,12]. A low toxic, biocompatible/biodegradable, and the stable material property of ZnO may deploy the resonator for medical application. The device design and analysis are presented with FEM-based software COMSOL^{TM} Multiphysics.

4.4 Literature survey

4.4.1 Introduction

Sir Guglielmo Marconi has demonstrated the first wireless transmission in 1895. It was a great deal to interact with one and another. The wireless networks are used in the verity of applications like in satellite transmission, radio and television broadcasting, sensor networks, a global positioning system (GPS), global system for mobile communications, and, most importantly, in today's smartphones. Day to day growth in wireless communication devices is looking for new ways to further miniaturize the RF front-ends with reducing cost and power consumption. The best quality wireless transceivers and subsystems have been relying on high-quality factor (Q) and low motional resistance (Rm) to implement resonators, hence oscillators/filters. A great deal of research effort has been devoted by various researchers to the development of miniaturized resonators that are open to direct integration with an integrated circuit (IC).

The integration of mechanical elements, sensors, actuators, and electronics on a single silicon substrate through microfabrication technology is referred to as MEMS. While the electronics are fabricated using IC-process sequences, the micromechanical components are fabricated using "micromachining" processes, for example, deposition, lithography, and etching. Silicon parts are selectively etched away or new structural layers are added to form the mechanical and electromechanical devices.

The MEMS show extraordinary advantages which include the on-chip integration of electromechanical systems and the circuitry used to control them, allowing further miniaturization. Its fabrication technologies allow parallel fabrication of thousands of systems by leveraging the parallel fabrication techniques of the integrated circuit industry. This may lead to a reduction in the manufacturing cost and improvement in reliability [13]. The fabrication methods for MEMS devices are bulk micromachining, surface micromachining, and molding.

The bulk micromachining is accomplished by removing material from a substrate to create holes, cavities, channels, or other desired shapes. Early bulk micromachining was accomplished using isotropic or anisotropic wet etching of silicon or glass substrates. In particular, several chemicals such as KOH (potassium hydroxide) or TMAH (tetramethylammonium hydroxide) etch a silicon substrate preferentially depending on the crystalline planes in the direction of etching. The etch rate for these chemicals is tens to hundreds of times faster in the <100> crystalline plane compared to the <111> plane. This effect has been used to create a wide variety of features using simple wet etching. Another common technique of bulk micromachining uses a deep reactive ion etch (DRIE) plasma etcher. Using this technique, silicon, as well as some other materials, can be etched very quickly and anisotropically, making possible very thick structures with small widths [14].

In surface micromachining, films are deposited on the substrate and patterned using photolithography to create micromechanical devices. The films normally alternate between structural and sacrificial layers with the MEMS parts being made from the structural layers. The sacrificial layers serve to support the structural components during fabrication. After the structural layers are patterned, the sacrificial material is removed, probably using wet chemical etching [14].

Finally, MEMS parts are often made by creating a mold, which may then be filled to create the desired part. Molds have been made from a variety of polymers, including some types of photoresist, metal and deep-etched silicon wafers. Photolithography is normally used to define the mold pattern. The mold may be filled by electroplating, for the metal parts. Polymer parts may be created by pouring or pressing the precursor into the mold. After the part has been molded, it may be removed from the mold by either etching the mold away, or, if the mold is to be used again, by peeling away the mold. Micromolding, also referred to as LIGA, was first performed in Germany. The LIGA process required an X-ray source to fully expose thick layers of photosensitive material, but many molding techniques have since been developed that use visible or ultraviolet light sources [14].

The MEMS are classified into RF (radio frequency) MEMS, Bio-MEMS, optical MEMS, and energy and power MEMS. The MEMS has been commercialized in various areas like automotive, defense, biomedical, communication, and aerospace industries. Today, MEMS applications cover a wider range. Some examples listed in Table 4.1.

Table 4.1 Examples of applications in the wide field of MEMS [15]

Category	Application
Inertia sensors	Accelerometer (airbag) and gyroscopes
Pressure sensors	Blood pressure, car tire pressure, microphone, and IOP
BioMEMS/microfluidics	Micro-bio-analysis, DNA chips, and inkjet printer head
Optical MEMS	Optical fiber switch and adaptive optics
RF MEMS	Switches, antenna, filter/oscillator, and high Q inductor
Energy and Power MEMS	Energy harvester and high power switches

4.5 Fundamentals of MEMS resonator

Resonance is a phenomenon where a system shows a selective response at a specific frequency. The resonance may occur in different domains like mechanical, electrical, electromagnetic, optical, acoustic, etc., when the system is capable of storing energy and transferring energy from one mode to another. In a mechanical domain, the system may vibrate with greater amplitude with some frequencies than the other. In the electrical domain, the series/parallel R-L-C circuit acts as a resonating part. The two modes of storing energy are the electrical field when the capacitor is charged and the magnetic field when the current flows through the inductor. The system oscillates when the energy is transferred continuously between the two. When the resonance occurs, the impedance of capacitor and inductor are equal but opposite hence they cancel each other. Thus, at the resonance, the circuit will have minimum impedance and the frequency at which this phenomenon occurs is known as resonant frequency [3,16,17].

4.5.1 MEMS resonator

A MEMS resonator consists of a resonant mechanical structure and transducers for energy conversion between electrical and mechanical domains, shown in Figure 4.1. The resonant mechanical structures are reported so far as beams, disks, rings, or plates can all be modeled as spring-mass-damper systems [17,18].

Figure 4.1 (a) shows the block schematics of MEMS resonator and 4.1(b) is its generic electrical model. The electromechanical transducer in the input port converts input electrical energy (i.e., actuation voltage) to mechanical energy (i.e., strain or stress). On the output port, the transducer converts the mechanical energy (i.e., displacement/deformation) back to electric energy (i.e., output currents). The efficiency of such energy transduction is quantified by the electromechanical coupling coefficient [18].

The quartz crystal resonators as a reference and clock circuit offering large quality factor (Q) $> 100{,}000$, high frequency stability, and reliability and hence are in high demand, but having some limitations, that is, costly manufacturing process, severe levels of shock and vibration, poor compatibility with CMOS technology, etc. The on-chip tank circuits with monolithic inductors and capacitors suffer from very poor Q-values even < 10 [3,19]. The MEMS resonator can offer very high Q

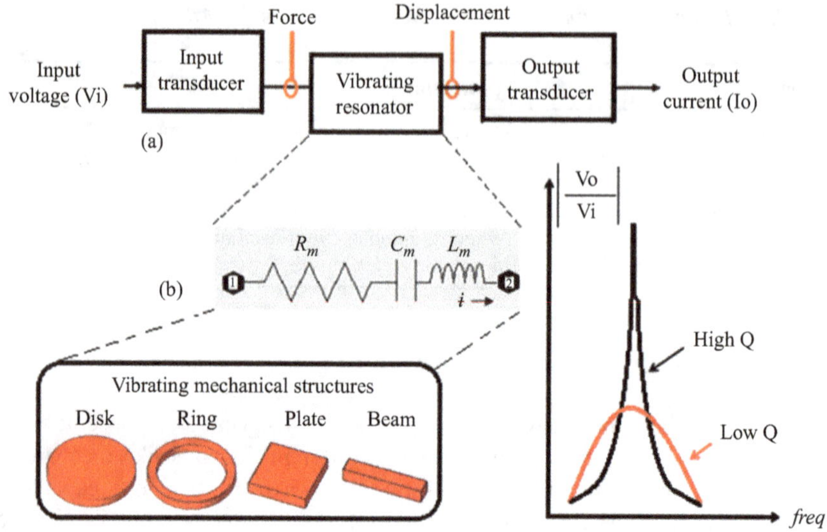

Figure 4.1 (a) Block diagram of a MEMS resonator and (b) a general electric model for a MEMS resonator [18]

Table 4.2 The basic comparison of quartz resonator and SiTime silicon MEMS resonator [3,23]

Features	Quartz resonator	MEMS resonator
Size	2–5 mm	400 µm
Frequency	1–80 MHz	1–50 MHz
Quality factor ($\times 10^3$)	100–200	75–150
CMOS compatibility	No	Yes
Shock/vibration immunity	Poor	Good
Cost	High	Low

value closer to quartz in vacuum as well as in air operating at VHF, UHF, and RF frequency range. MEMS resonators consume less power and having excellent CMOS compatibility with better temperature stability. In addition, these resonators are robust to shock and vibration. Due to the fact that mechanical resonators have a much high Q than their electrical counterparts, they have become alternative components to quartz crystals for modern communication applications [3,16,19–22]. The comparison of a SiTime MEMS resonator with respect to the quartz resonator is given in Table 4.2.

4.5.2 History of MEMS resonator

The year-wise development in MEMS resonator in diverse areas of engineering and science is listed in Table 4.3.

Table 4.3 Years and MEMS resonator developments

Year	Development
1965	Westinghouse research labs developed resonant gate transistor
1960s	Development of piezoelectric-on-silicon resonators
1970s	Continued development of silicon micromachining
1982	Kurt Petersen's paper in proceedings of the IEEE
1984	Polysilicon surface micromachining
1985	Review literature on resonant silicon sensors by Langdon (GEC-Marconi)
1988	First resonant gyroscopes were discussed
1989	Comb drive resonator
1990s	Efforts on vacuum encapsulation and temperature compensation initiated
1994	First fully integrated CMOS-MEMS oscillator
1994	Agilent develops film bulk acoustic resonators (FBAR)
1998	Review paper on surface micromachining in Proceedings of the IEEE
1998	Bosch developed a micromachined gyroscope
2000	Polysilicon bulk mode resonators
2001	Formation of first start-up commercializing silicon MEMS resonator technology
2002	Analog devices introduced first commercial integrated MEMS gyroscope
2000s	Development of epi-poly vacuum encapsulation technology by Bosch/Stanford
2004	First GSM specification MEMS oscillator
2004	VTT paper on nonlinear effects of silicon MEMS resonators
2004	Formation of SiTime with a license from Bosch on vacuum encapsulation technology
2005	Development of AlN resonator technology
2000s	Efforts on passive and active temperature compensation of MEMS resonators
2007	Review paper on MEMS for timing and frequency control
2010	Acquisition of Silicon Clocks by Silicon Labs
2012	Timing products were demonstrated by NXP Semiconductors
2012	Acquisition of VTI by Murata
2013	Acquisition of Discera by Micrel
2014	Acquisition of SiTime by MegaChips
2015	Acquisition of Sand9 by Analog Devices and SiT5022 by SiTime
2016	SiT8021μPower oscillator
2017	SiT1252 embedded MEMS resonator
2018	SiT5357 Elite Platform™ Precision Super-TCXOs

Source: Seshia Ashwin A, Cambridge University (UK) Nano-Science Centre, Department of Engineering, Power point presentation at IITB, India during GIAN workshop in 2016, www.sitime.com, www.analog.com

4.5.3 MEMS resonators—modes of vibration

A simple example of a micromachined resonator is a cantilever beam fixed at one end, called resonant gate transistor (RGT) as shown in Figure 4.2 with input transducer (driving), mechanical resonant structure (which can vibrate in one or more modes), and an output transducer (sensing) [24].

Micromechanical resonators can have different shapes like beams, rings, disks, square plates, comb, etc., and again be classified according to their modes of operation, namely flexural mode, torsional mode, and bulk mode devices [3,25,26], shown in Figure 4.3.

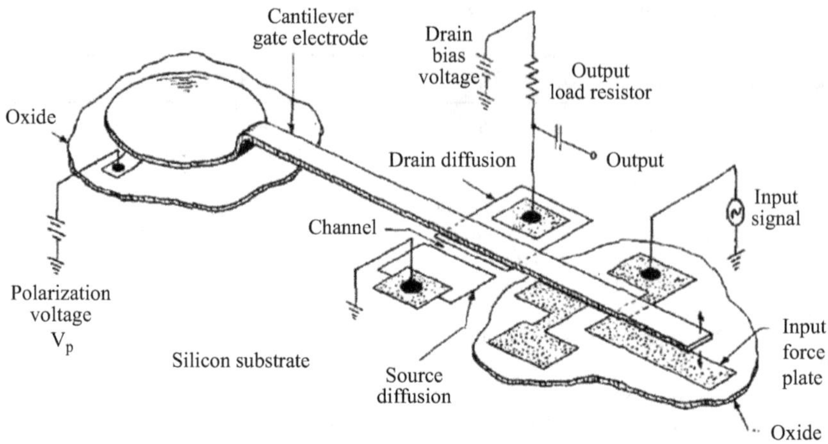

Figure 4.2 Resonant gate transistor [24]. Reprinted with permission. © 1967 IEEE

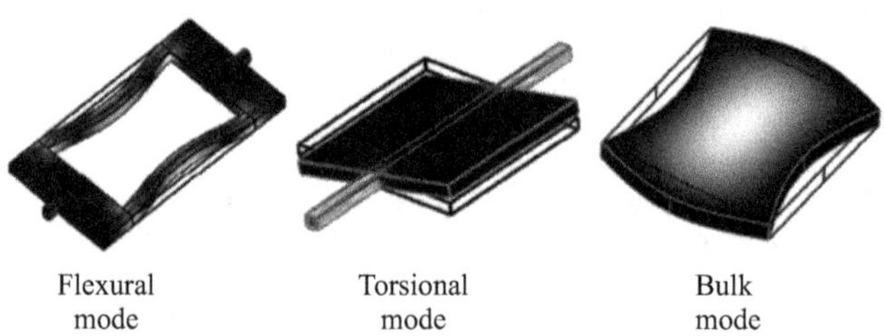

Figure 4.3 Vibration modes of mechanical resonators [3,26]. Reprinted with permission. © 2008 IEEE

Flexural mode represents the formation of transverse standing waves. The dominant stress is bending stress. Flexural mode resonators are suitable for low frequency operations, hence are more prone to losses from surface effects [3,16,25,26].

Torsional mode is the representation of sheer stress. The displacement produced is rotational in nature. These resonators exhibit lower anchor losses and lower squeeze film damping and hence having very high Q [3,16,25,26].

Bulk mode represents the formation of longitudinal standing waves. Because of their larger structural stiffness, the bulk mode is preferred for high frequency and high Q operations [3,16,25,26].

Figure 4.4 shows the commonly used vibrational modal shapes for square plate bulk mode devices: extensional (contour), wine glass, and lame [3,26].

4.5.4 Analogy between mechanics and electronics

Irrespective of the structure, the resonator can be mechanically represented by using mass, spring, and damper arrangement, shown in Figure 4.5.

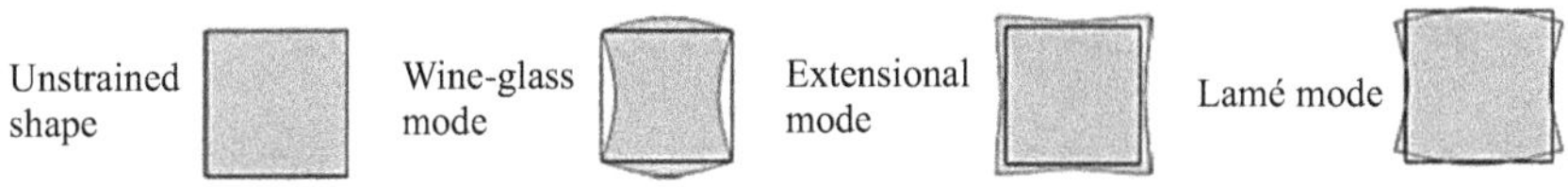

Figure 4.4 Mode shapes for bulk mode devices [3,26]. Reprinted with permission. © 2008 IEEE

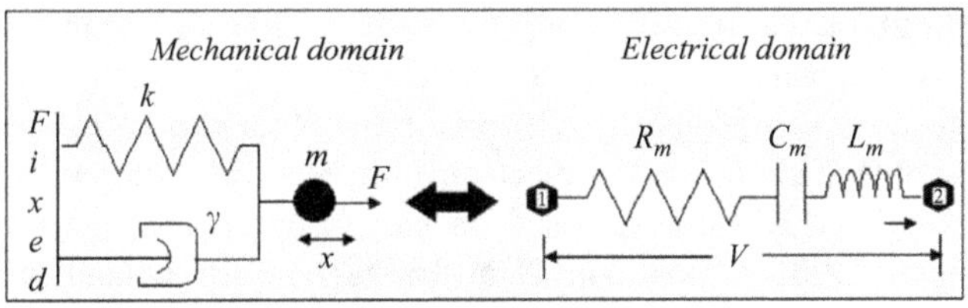

Figure 4.5 The mechanical and electrical analogy [27]

Table 4.4 Direct electromechanical analogy [16]

Mechanical parameter	Electrical parameter
Force (F)	Voltage (V)
Velocity (v)	Current (I)
Compliance ($1/K$)	Capacitance (Cm)
Damping (γ)	Resistance (Rm)
Mass (m)	Inductance (Lm)
Displacement (x)	Charge (q)

Mechanical variables can be made correspond to equivalent electrical variables in order to model the behavior of a mechanical resonator with standard RLC parameters, shown in Table 4.4.

The analogies are the result of the similarities of the equations governing the behavior of electrical and mechanical systems. For example, Newton's second law of motion relating the force, F and velocity, v for a rigid mass, m is

$$F = m\frac{dv}{dt} = m\frac{d^2x}{dt^2} \tag{4.1}$$

Equation (4.1) is mathematically analogous to the constitutive equation of an electric inductor:

$$V = L\frac{di}{dt} = L\frac{d^2q}{dt^2} \tag{4.2}$$

In this analogy, the force, F plays the same role as the voltage V, the velocity v, as the current i, and the displacement x as the charge q. The mass m in the mechanical system corresponds to the inductance L in electrical circuits [16,27].

4.6 Transduction mechanism of MEMS resonators

There have been several transduction mechanisms explored at the MEMS level. The dominant ones are electrostatic (validated in the production of low frequency oscillator), electrostrictive (dielectric or internal transduction, research only), piezoelectric (validated in the production of high frequency filters and high Q MEMS oscillators, requires less actuation signal) [4], thermal or thermo-elastic (piezoresistive sensing), optical transduction, and magnetic (the magnetic field is generated around the mechanical structure and ac current is made to flow) (Table 4.5).

MEMS resonators are typically the time base generators or references whose operating principle is similar to the mechanical tuning fork. Many micromachined resonators have been researched based on electrostatic transduction. The comb drive resonator in 1986 by William Tang has become the fundamental platform now of many microfabricated devices. For the first time, large lateral displacement was demonstrated by him. This comb structure performs well in the kHz range of frequencies. It was difficult to operate these devices in the UHF range since the comb fingers increase the mass of resonator [29,30]. Cioffi and Hsu in 2005 have reported a silicon comb drive resonator with Q~50,000 [31]. Another type of MEMS resonator studied is beam-type devices. Different types of beams such as clamped-free, clamped-clamped, free-free etc. [32–34] with the different material, including single-crystal silicon [35], poly-silicon [32], poly-diamond [36], poly-silicon carbide [37], poly-silicon germanium [38], etc. of the resonating structure have also been investigated. By shrinking their sizes, the beam resonators have been reported the resonance frequency in RF range with considerable large quality factor about 10,000. The biggest issue in these beam resonators is their high motional resistance in the range of 100 KΩ [3]. The drawback of shrinking in size

Table 4.5 Transduction mechanism in MEMS resonators [28]

Excitation	Detection	Advantages	Drawback
Piezoelectric	Piezoelectric	Large dynamic range, acoustical isolation technologies available	Needs IC compatibility development
Electrostatic	Capacitive	Surface micromachining	Complex process if only bulk micromachining is available
Electrothermal	Piezoresistive	Effective and simple, bulk micromachining is available	Heat generation power, bandwidth
Optothermal	Optical	Electrical isolation, immunity to most electromagnetic noise	Hard to integrate
Magnetic	Magnetic	Effective and simple, variety of excitation mechanism	Magnetic materials or magnets are required, magnetic fields in the system

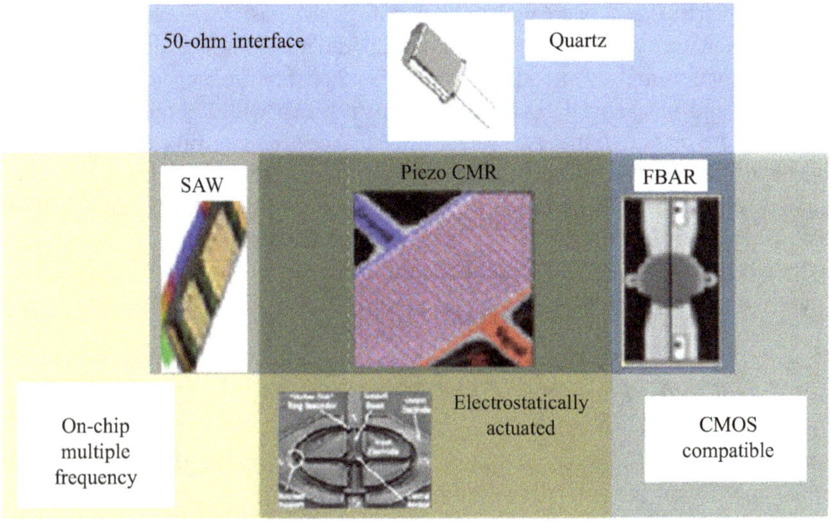

Figure 4.6 Comparison of piezo CMR with other resonator types [4]

required in beam resonators to be operated in the UHF range has been overcome by disk resonators. The disk resonators could reach the frequency in the UHF band without shrinking in size due to its high spring constant. Since disk resonators are much larger than beam type, they are much more robust [3].

The performance of MEMS resonators has been continuously improved. The *f-Q* products demonstrated from in-plane mode electrostatically transduced resonators have exceeded that of some best quartz crystal resonators. However, the motional resistance of these devices is still in the order of 100 KΩ, complicating the interfacing of the resonators with RF front-end. The objective of this research is to develop MEMS resonator featuring low motional resistance (*Rm*) with the high *f-Q* product.

The piezoelectric transduction with contour mode resonator (CMR) technology has proved as one of the most promising techniques for multiple frequency devices fabricated on the single chip with high *Q* and low *Rm* [4,30,39–41]. With the piezoelectric material, the fundamental frequency is defined by the in-plane dimensions, then the resonator is said to be contour mode. CMR is the combination of many important features which existing resonators have. In contour mode, the vibrations of the device are parallel to the major surfaces. If the structure vibrates along its length, called length-extensional mode (LEM) and if the structure vibrates along with its thickness, called thickness extensional mode (TEM) [42]. Figure 4.6 shows the comparison of piezoelectric contour mode resonator with other resonators.

4.7 Acoustic microresonator technologies

Acoustic microresonators are meant for the reduction in size and power consumption of mobile radio equipment and sensing systems that the telecommunication and sensors industries have been undertaking during the past few years. The kind of acoustic resonators is microelectromechanical devices. They experience acoustic wave propagation and eventually vibrate at a resonance frequency related to their dimensions and mechanical configuration when driven with the appropriate conditions. They are classified into two categories: surface acoustic wave (SAW) and bulk acoustic wave (BAW) resonators. Two types of the latter are found: the solidly mounted resonator (SMR) and the thin-film bulk acoustic wave resonator (FBAR). SAW, SMR, and FBAR devices can be fabricated within standard IC technologies. On the other hand, FBARs resonate at far-from-fundamental acoustic modes, instead of purely mechanical modes [28].

4.7.1 The concepts and the working principle of acoustic wave propagation

Many kinds of acoustic resonators exist. Musical instruments like drums, guitars, pianos, and organs amplify the vibration of a string or a shock by using resonant cavities or pipes. The instrument produces sound waves of specific tones regarding the size of the acoustic cavity. In this chapter, the focus is given on microelectromechanical acoustic resonators and, more specifically, on piezoelectric-based acoustic resonators [28].

An acoustic wave is a disturbance in an elastic medium that propagates in space and time, thus transferring the energy supplied by an excitation source along with the medium in the form of oscillation or vibration. Acoustic wave propagation entails elastic deformation of the medium along the propagation axis or in other axes. In contrast to electromagnetic waves, acoustic waves do require a medium to propagate, and their propagation speeds depend on the mechanical properties of the wave-supporting material. Virtually any material is capable of supporting acoustic wave propagation, including silicon. Nevertheless, the piezoelectric properties of certain materials facilitate the wave propagation, thus improving the electromechanical energy conversion, so piezoelectrics are usually chosen as the acoustic layer of many acoustic-wave resonators. Also, the acoustic-wave phase velocities are slower than those of the electromagnetic wave traveling in the same medium. In a first approach, there exist two types of acoustic waves: surface acoustic waves (SAW) and bulk acoustic waves (BAW). A combination of both is normally found in the form of longitudinal, shear, mixed longitudinal-shear Rayleigh waves, Love waves, or Lamb waves. The waves we can see propagating on the surface of a lake after hitting the water mass with a stone illustrate the concept of a SAW, meaning the propagating media here is water. On the other hand, the sound waves traveling through the air until reaching our ears are of the BAW type, meaning the propagating media is air [28].

As illustrated in Figure 4.7(a), the surface particles move in ellipses in planes normal to the surface and parallel to the direction of wave propagation. The particle displacement is significant at a depth. This movement is retrograde at the surface and thin depths and becomes prograde at greater depths. This phenomenon was defined by Lord Rayleigh in 1885.

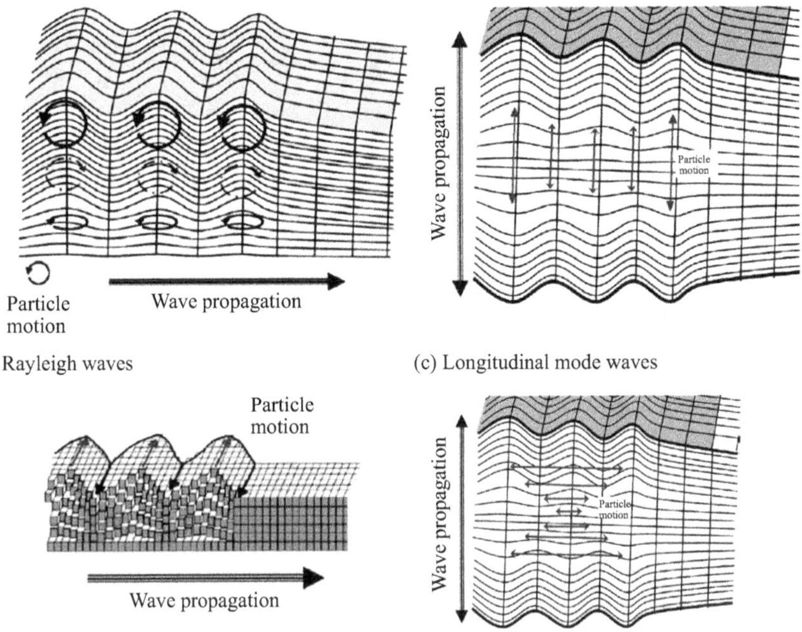

(a) Rayleigh waves

(b) Love waves

(c) Longitudinal mode waves

(d) Shear or transverse mode waves

Figure 4.7 The differences between Rayleigh waves, Love waves, longitudinal, and shear acoustic waves: (a) Rayleigh waves, (b) Love waves, (c) longitudinal mode waves, and (d) shear or transverse mode waves [28]

Love waves travel faster than Rayleigh waves. The particle motion of a Love wave, depicted in Figure 4.7(b), forms a horizontal line perpendicular to the direction of propagation, creating horizontally polarized shear waves (SH waves). Moving deeper into the material, motion alternately increases and decreases as one examines deeper layers of particles. The amplitude, or maximum particle motion, decreases rapidly with depth, and it decays with the square root of the distance traveled by the wave [28].

Bulk acoustic waves are longitudinal, shear-mode, or a combination of both. Longitudinal waves travel through the medium parallel to the same axis of the oscillations or vibrations of the particles in the medium, that is, in the same or opposite direction as the motion of the wave as shown in Figure 4.7(c).

A transverse or shear-mode wave propagates and transfers its energy in the direction perpendicular to the oscillations occurring in the medium. If the shear wave moves in the positive x-direction, for example, particles in the medium oscillate in the y-z plane, as represented in Figure 4.7(d). Shear-mode resonance occurs at longer wavelengths than longitudinal-mode vibrations [28].

Acoustic resonators are microelectromechanical devices exerting acoustic wave propagation and vibrating at a resonance frequency of interest. In some sense, the resonator behaves like an acoustic cavity trapping the wave in the medium. To do that, transmission and reflection of the wave are promoted by the appropriate

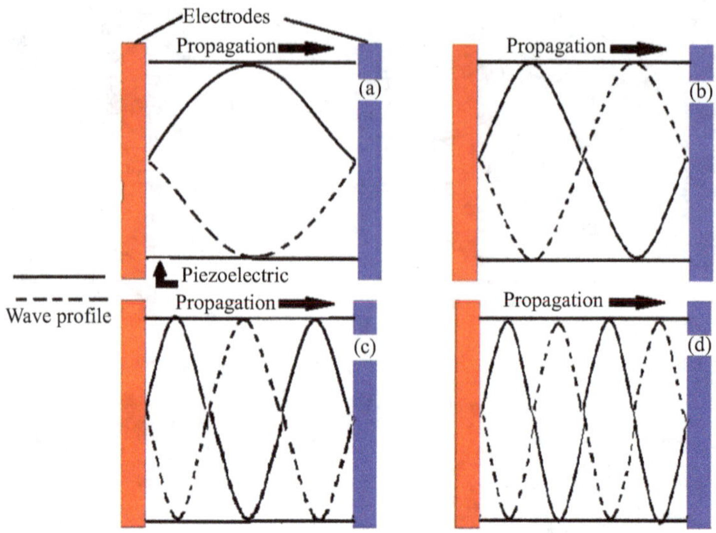

Figure 4.8 Propagation of longitudinal-mode waves inside λ/2 resonators [28]

means, such as electrodes and acoustic layers functionally designed of the type and frequency of the acoustic wave. The amplitude of the wave is maximum when the transmitted and reflected waves have λ, $\lambda/2$, or $\lambda/4$ phase shifting, according to the separation of the electrodes, dimensions of the acoustic layer, and acoustic mode. In longitudinal waves, wavelength λ is propagating along with the bulk of a $\lambda/2$-long resonant cavity. The transmitted and reflected waves are in-phase, henceforth, the interference between them reinforces the energy inside the cavity. If the incident and reflected waves are out of phase, they are suppressed after destructive interference. This happens not only for the fundamental wavelength λ but also for the shorter waves of wavelength λ/n equal to an entire fraction n of the fundamental wavelength λ [28].

The sequence of Figure 4.8(a)–(d) depicts the propagation of longitudinal modes through the $\lambda/2$-long resonator. Silicon and other materials have been used to manufacture acoustic resonators. But, for the high-frequency requirements, and the need for miniaturization, the thin-film piezoelectric technologies are suitable for new generation thin-film acoustic wave resonators. For these requirements, new fabrication techniques and materials were introduced. Thin-film aluminum nitride (AlN) and zinc oxide (ZnO) became the standard for the new kind of SAW and BAW resonators [28].

4.8 The piezoelectric theory

A crystal acquires the charge when being compressed, twisted, or distorted and is said to be piezoelectric. Piezoelectricity—a find from the Greek word "piezein," meaning "to press." Piezoelectricity is the property of a material to deform after electric field excitation or, alternatively, to undergo electrical displacement when an external strain is applied to the crystallographic structure (Figure 4.9).

The variables are in the mechanical (stress or strain) and electrical (electric displacement or electric field) domains. In the case of piezoelectric materials, the acoustic wave is about five times lower than electromagnetic waves. The magnitude of generated electric energy is directly proportional to the magnitude of aggregate stress. If the applied force is large enough, the transduction of mechanical energy to electrical energy will be taking place and this phenomenon is referred to as piezoelectric effect constitutive by piezoelectric equations:

$$\begin{aligned} S_i &= S_{ij}^E T_j + d_{li} E_l \\ D_m &= \varepsilon_{mn}^T E_n + d_{mk} T_k \end{aligned} \tag{4.3}$$

For $i, j, k = 1,2\ldots6$ and $l, m, n = 1,2,3$, where T is applied mechanical stress, E is the applied electric field, d is a stain, ε_{mn}^T is the permittivity under constant stress, D is electric displacement, S_i is the mechanical strain, d is piezoelectric coefficient, and S_{ij}^E is the compliance tensor under constant E [9,43].

Piezoelectric transduction requires metal electrodes directly on the thin film layer to apply an electric field. Several piezo coefficients available ($d_{31}(e_{31})$, d_{33} (e_{33}), d_{15} (e_{15})) can be exploited to excite the desired mode of vibration (Figure 4.10). Flexural, contour-mode, thickness-extensional and shear resonators have been reported. Piezo generates a body force. Piezoelectric coupling is strong and provides for effectively large kt^2 (electromechanical coupling coefficient) in general at least one order of magnitude greater than other transduction mechanisms. Quartz has low coupling but characterized by high Q. Shear MEMS resonators approaching GHz have shown the f-Q product of 1×10^{13}. Lateral, in-plane vibrations can be excited in piezoelectric films at the high frequency. The high-quality factor over the entire frequency range has also been demonstrated with low impedance value [43].

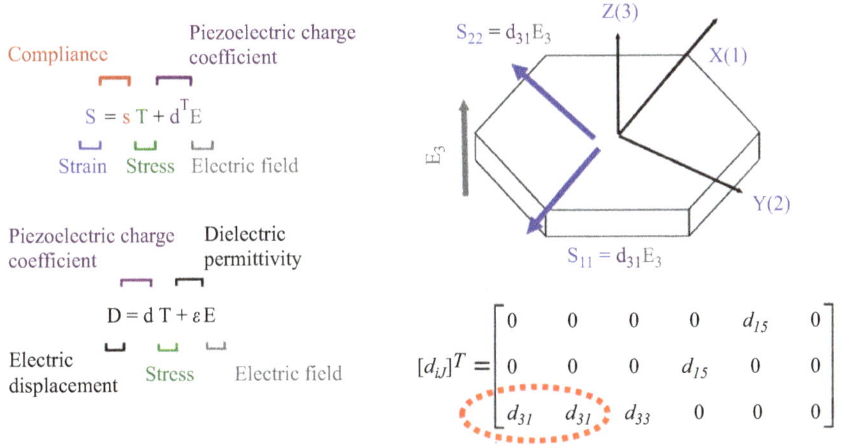

Figure 4.9 Relation between stress-strain and electric displacement [43]

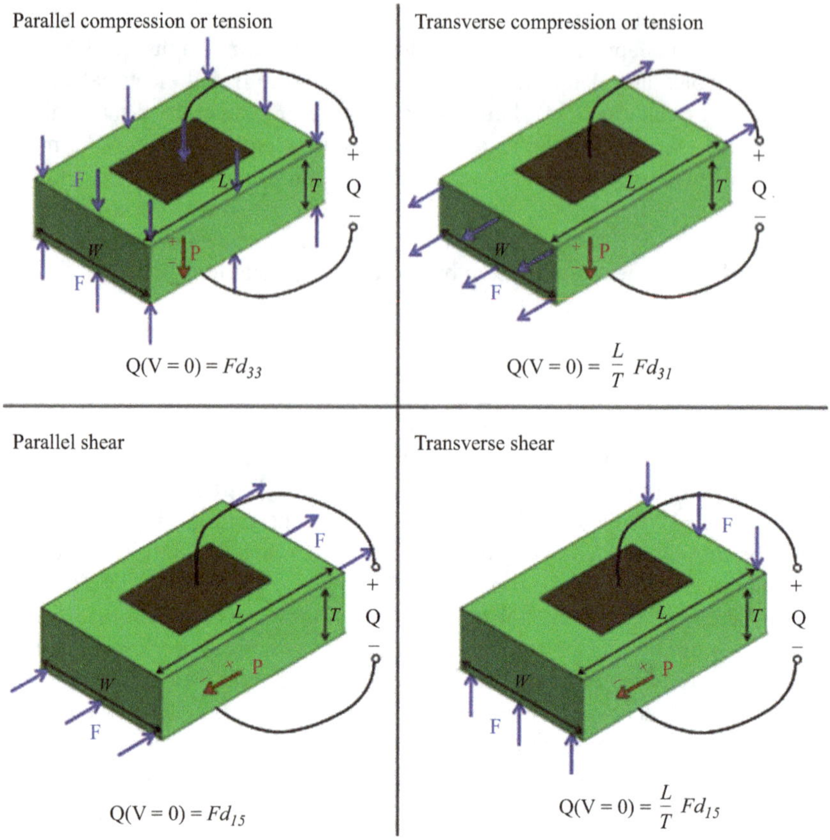

Figure 4.10 A piezoelectric slab with different piezoelectric actuation modes d_{33}, d_{31}, and d_{15} actuation with different pole and force direction [44]

The piezoelectric resonators utilize electromechanical coupling capabilities to transform electrical signals into mechanical vibration at the frequency of interest. The piezoelectric resonators have lower motional resistance, higher power handling capabilities, and are capable of generating acoustic waves compared to capacitive actuation [5,6]. The three primary piezoelectric materials used in piezoelectric MEMS devices for various applications are zinc oxide (ZnO), aluminum nitride (AlN), and lead zirconium titanate. Table 4.6 shows a comparison of basic piezoelectric materials AlN, ZnO, and PZT.

4.8.1 Piezoelectric resonator modes and associated frequency

The classification of the piezoelectric resonators is usually according to their mode of vibration. They offer a wide range of modes that cover a vast frequency range from several kHz to tens of GHz. Table 4.7 describing [45] the various frequency ranges

Table 4.6 Comparison of the basic piezoelectric materials [9,46]

Property	AlN	ZnO	PZT
Longitudinal acoustic velocity (m/s)	10,400	4,655	3,300
Young's modulus (GPa)	330	112	49–65
Density (kg/m^3)	3,300	5,680	7,600
Intrinsic loss	Very low	Low	High
Poisson's ratio (σ)	0.24	0.36	0.31
Dielectric constant (F/m)	8.2–8.9	9.2	80–400
d_{31} piezoelectric strain coefficient (pC/N)	–1.8	– 4.7	–130
d_{33} piezoelectric coefficient (pC/N)	5.6	12.4	200
CMOS compatibility	Yes	Yes	No

Table 4.7 Piezoelectric resonator modes, associated frequency, and applications [45]

Modes	Associated frequency	Application area
Flexural	1 kHz	Piezoelectric buzzer and ISM equipment
Length	10 kHz	Filter/resonator/inductive application
Area expansion	100 kHz	Filter/resonator
Thickness shear	1 MHz	Filter/resonator and AM radio
Thickness expansion	10 MHz	Filter/resonator, FM radio, cordless phone, and radio-controlled cars
Surface acoustic wave	100 MHz	Wildlife tracking collar and cell phone
BGS/SH wave	10 MHz	Filter/resonator, FM radio, cordless phone, and radio-controlled cars
Bulk acoustic wave	1 GHz	Air traffic control, GPS, GSM, Bluetooth, etc.

associated with different piezoelectric vibration modes. Researchers have been fabricated the devices based on these modes of oscillation and their required application.

4.9 Piezoelectric MEMS resonator

4.9.1 SAW resonator

In 1885, Lord Rayleigh first described the acoustic waves with longitudinal and vertical shear components and named them surface acoustic waves (SAW). These discoveries remained scientific research topics until the breakthrough invention of using interdigital transducer (IDT) on piezoelectric substrates by R. White and F. Voltmeter in 1965. Since then, this technology has dominated the market for decades. Figure 4.11 shows commonly used SAW resonator topologies and the images of the propagating surface acoustic waves. This type of acoustic wave only propagates near the surface of a semi-infinite solid and the amplitude decays

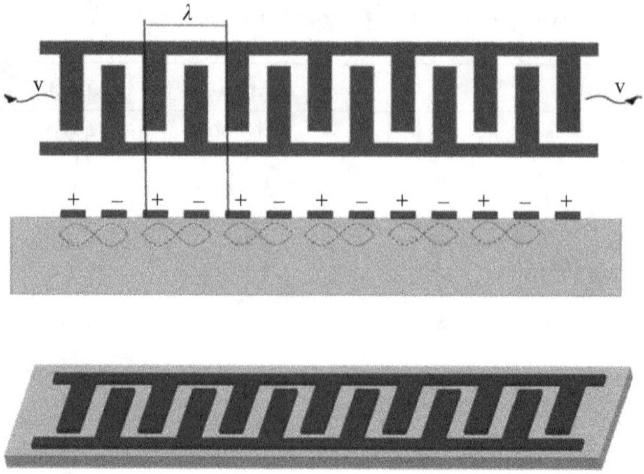

Figure 4.11 SAW resonator [28]

exponentially with depth into the substrate. They can be confined by reflector structures or gratings at either end of the IDT array and form standing waves. Besides, SAW can couple with any media in contact with the surface and is therefore widely used in nondestructive defect testing. One of the earliest successful applications was the intermediate frequency (IF) band-pass filter for television receivers. Single crystalline quartz, lithium niobate (LiNbO3), and lithium tantalite (LiTaO3) are commonly used substrate materials for SAW devices [46].

The frequency of SAW devices is determined by the IDT electrode pitch which in turn is limited by the resolution of the lithography tool. For higher frequency applications, substrates with higher acoustic velocity are necessary. SAW resonators have been a commercial used for radiofrequency applications, especially for filter and oscillator implementations. Their impact has made possible considerable reductions in the size and power of the chipsets of mobile devices. More modest, but also important, has been the impact of SAW resonators in the mass detector and pressure sensor devices, with application in bioparticle detection [28].

4.9.2 BAW resonator

BAW devices experience acoustic wave propagation through the bulk of its active layer structure, hence their name. This feature differentiates BAW and surface acoustic wave (SAW) devices, the acoustic waves propagating in longitudinal or shear-transversal modes. In both cases, the acoustic wave causes deformation of the active layer, which is typically a piezoelectric material made with thin-film technology. As in SAW resonators, piezoelectric and the inverse piezoelectric effects are the actuation and detection mechanisms involved in BAW device operation [28].

According to these principles, a voltage applied to the resonator's electrodes induces strain of the acoustic layer and vice versa; after a mechanical strain of the acoustic layer, a voltage can be read out the electrodes. Another difference between BAW and

SAW devices is their physical layout. In BAW, the acoustic layer is a component element of a stacked structure in which the acoustic wave is confined [28] (Figure 4.12).

A couple of metal layers acting as electrodes complete the structure of BAW resonators. Therefore, the BAW device is fabricated on top of a carrying substrate, typically silicon, and the acoustic layer and electrodes are located on top of said substrate. In SAW, the acoustic layer may be the carrying substrate by itself, and the IDT electrodes are located on top of it, both in the same plane. Typical operation frequencies for SAW range from 30 MHz to 1 GHz, while the center frequency of BAW can be found in the 1 to 10 GHz band. The resonance frequency of a BAW resonator operating in fundamental, longitudinal mode is mainly determined by the thickness *t* of the acoustic layer [28].

BAW resonators are classified as film bulk acoustic resonator (FBAR) and solidly mounted resonator (SMR).

4.9.2.1 FBAR

FBAR is a device consisting of a piezoelectric material sandwiched between two electrodes and acoustically isolated from the surrounding medium. AlN and ZnO are two common piezoelectric materials used in FBARs. Micromachining is needed to release the FBAR, thus providing acoustical isolation between the device and the substrate in order to achieve high-quality factors. The result of the process is a structure with an air gap, a cavity, or a membrane underneath the resonator's structure. Front-side or back-side surface and bulk micromachining are the technological options for device releasing [28] (Figure 4.13).

4.9.2.2 SMR

The device structure and working principle of SMRs are basically the same as those of FBARs. Instead of the air gap in FBARs, SMR devices implement a stack of $\lambda/4$-thick acoustically mismatched layers, which are placed underneath the resonator to provide acoustic isolation from the substrate. The large acoustic-impedance

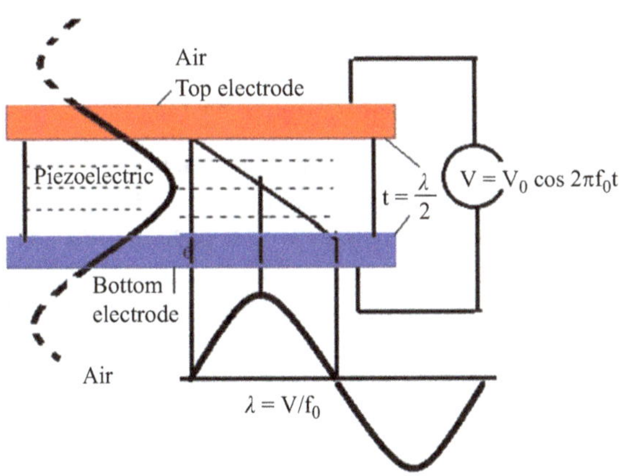

Figure 4.12 Electric charge displacement and poling in a BAW resonator due to an electric potential applied to its electrodes [28]

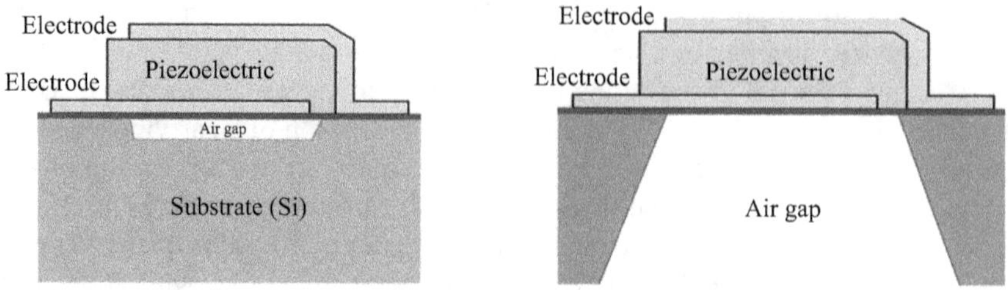

Figure 4.13 (a) Surface micromachined and (b) bulk micromachined [28]

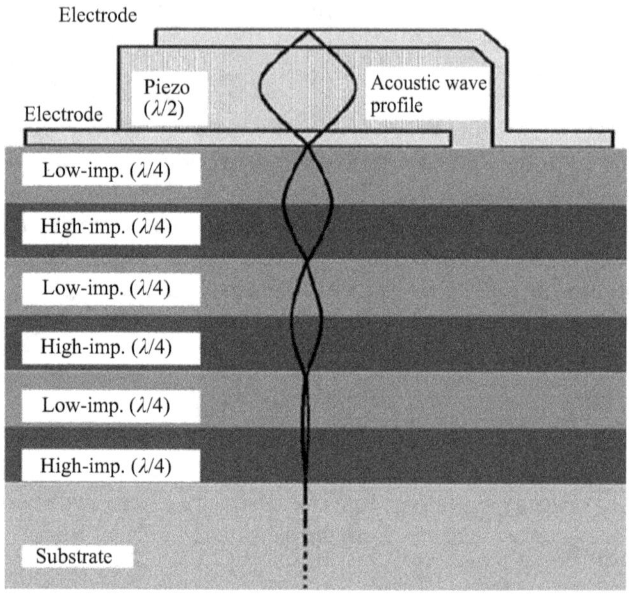

Figure 4.14 Wave propagation through SMRs and the Bragg reflector [28]

mismatching between the alternating λ/4-thick materials causes the BAW to be systematically reflected to the resonator, thus confining the energy in the acoustic layer of the device. For this reason, the stack is called a reflecting mirror, or Bragg reflector. SMRs exhibit good robustness, low stresses, and practically no risk of mechanical damage in dicing and assembly. On the other hand, FBARs exhibit higher Q factors and thermal isolation due to the air gap. However, the Q factor of SMRs can be tailored to achieve design specifications as a function of the number of reflector periods N (Figure 4.14).

4.9.2.3 FBAR and SMR applications

FBAR finds application in a variety of systems ranging from radio frequency (RF) to sensing components. Although the telecommunication industry has been the very first engine stimulating the development of FBAR applications, new sensing applications have been demonstrated in the past few years. The first RF applications of FBARs were thus devoted to supply fully passive components, able to compete with SAW and ceramic technologies, such as filters and duplexers. In FBAR-based sensor applications, one or more FBAR devices are the constituent elements of a system operating under piezoelectric actuation or detection mechanisms. Mass sensors and biochemical, liquid, or gas detectors are some examples [28].

Filter realization using BAW (FBAR or SMR)

BAW resonators (FBAR or SMR) are arranged side by side to realize a filter. This category of BAW filters does not allow mode conversion, from symmetric mode to differential mode, or impedance transformation. Two main architectures of filters are ladder filters "Π" or "T" as shown in Figure 4.15(a) and lattice as shown in Figure 4.15(b). It is necessary to have two types of resonators having different resonant frequencies to build these filters (Figure 4.16). The frequency shift can be obtained by adding an additional layer on top of the standard resonator. In Figure 4.16, the transmission coefficient of a ladder structure is presented with the

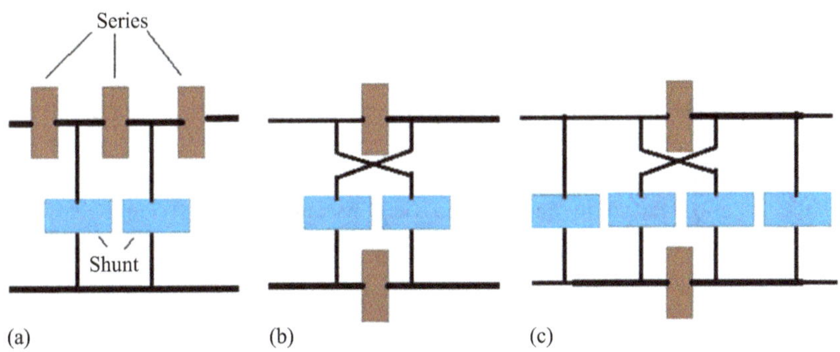

Figure 4.15 Topologies: (a) ladder, (b) lattice, and (c) mixed [47]

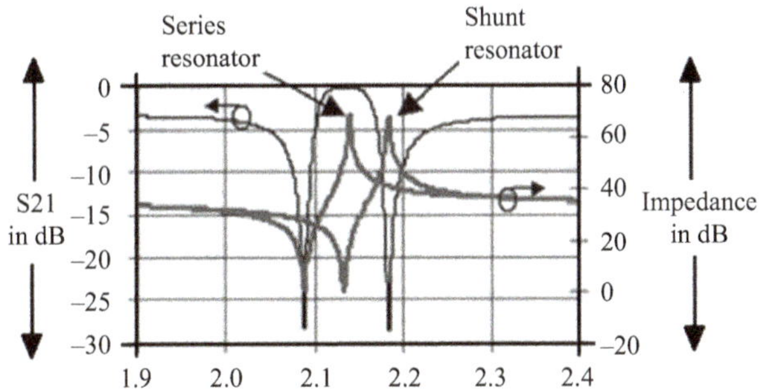

Figure 4.16 Transmission coefficients of a ladder structure with two resonators [47]

impedances of both resonators. The resonance of the series resonator gives the lower limit of the passband, while the resonance of the shunt resonator gives the upper limit of the passband [47].

4.10 Some more piezoelectric MEMS resonators by different researchers

- The first known demonstration of a piezoelectric resonator was made by R. Besson in 1977 in Besancon, France. He demonstrated 5 MHz electrodeless hemispherical quartz resonators, called "B.V.A." resonators, having Qs of 3 million and aging stabilities of $\sim$ 8 ppb per day [48].
- G. Piazza and his team (2006) were demonstrated plate and ring-shaped contour-mode AlN piezoelectric resonators for low motional resistance ranging from 50 to 700 Ω. The reported Q in this research was 4,300 at 230 MHz in air. The center frequency of these piezoelectric resonators was set by lithographic techniques. They have suggested the parallel arrays arrangement of small resonators to maintain overall structural stiffness and low motional resistance at high frequency [40] (Figure 4.17).
- In 2010, at UC Berkeley, Hung and Nguyen fabricated electrodeless piezoelectric resonators, called "capacitive-piezoelectric" resonators due to the introduced capacitive gaps, with much higher frequencies on-chip using AlN as the piezoelectric material for 1.2 GHz ring resonator with a Q of 3,100 and a motional resistance of 889 Ω with effective coupling coefficient of 0.3% [49,50] (Figure 4.18).
- Yen, Pisano, and Nguyen in 2013 demonstrated the first capacitive-piezo Lamb Wave Resonator (LWR) at 1 GHz with an improved quality factor of 5,000 [51] (Figure 4.19).

  The figure presents a perspective view illustration and cross-section of the capacitive-piezoelectric AlN Lamb wave resonator. To minimize damping and

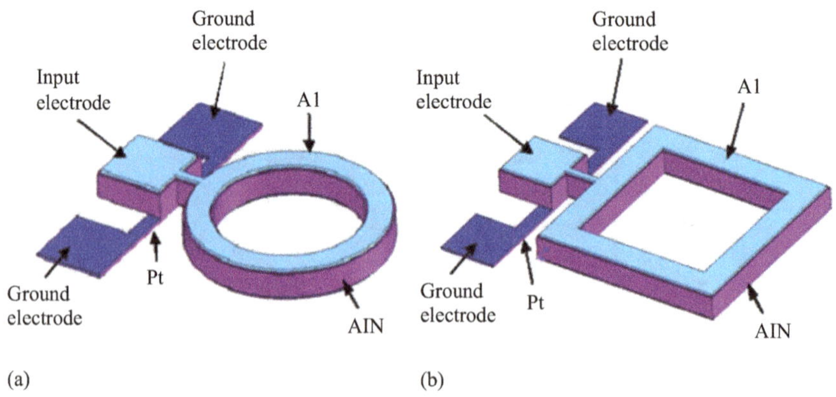

Figure 4.17 AlN ring-shaped contour-mode micromechanical resonators: (a) one-port circular ring and (b) one-port square-shape ring [40]. Reprinted with permission. © 2006 IEEE

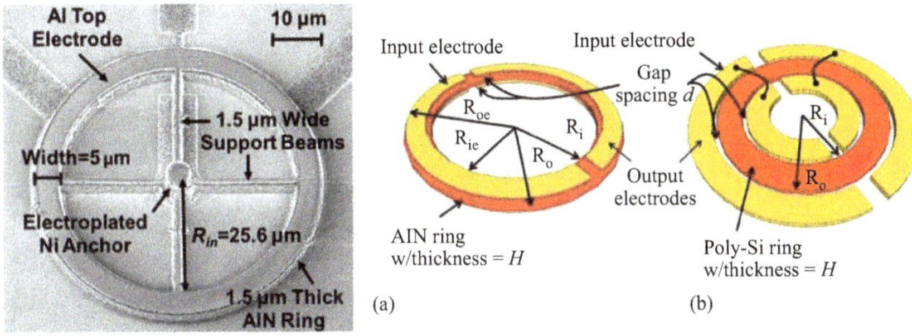

Figure 4.18 *Fabricated 1.2-GHz capacitive-piezo AlN ring resonator. Electrode arrangements for (a) the capacitive-piezo AlN ring resonator and (b) the capacitive-gap poly-Si ring resonator discussed in the text and to be compared in both devices operated in the same radial mode shape at 1.2 GHz [49,50]. Reprinted with permission from author*

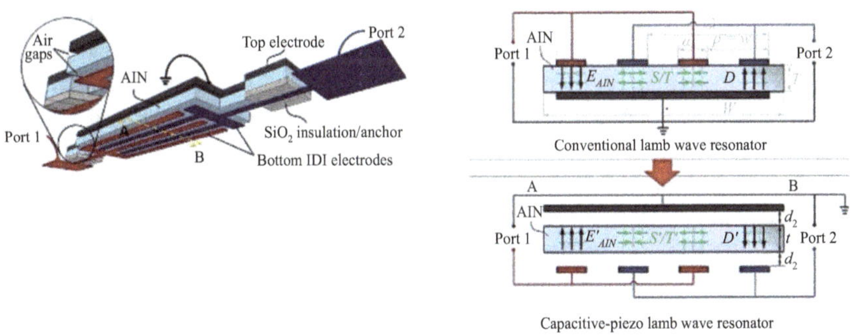

Figure 4.19 *(a) Perspective-view illustration of a capacitive-piezo Lamb wave resonator and (b) its cross-section alongside that of a conventional device with contacting electrodes. The capacitive-piezo AlN resonator structure is separated from its top and bottom electrodes by distance d_1 and d_2, respectively [51]. Reprinted with permission. © 2013 IEEE*

interface strain caused by metal electrodes, the Lamb wave resonators demonstrated use piezoelectric AlN as the resonator structure, but incorporate capacitive electrode to resonator gaps to a couple in electric fields, thereby effectively realizing a combined "capacitive piezo" transducer. The introduction of air gaps between the piezoelectric resonator structure and its metal electrodes eliminates metal to AlN interface losses and metal damping to enable high *Q*, while the submicron spacings of the gaps still allow strong electric fields across the piezoelectric resonator, preserving high electromechanical coupling and adequate motional impedance. To operate this device, an AC voltage applied across the input electrodes generates an electric field across the piezoelectric layer that induces mechanical stress and strain via the reverse piezoelectric effect. When the input AC signal matches the resonance frequency of the device, an acoustic vibration with half-wavelength equaling the IDT electrode pitch is excited, generating a Lamb wave propagating through the AlN plate. At the output electrode, by means of the direct piezoelectric effect, the mechanical stress and strain caused by Lamb wave mode vibration induce an electrical displacement current that generates opposite AC charges on opposing AlN surfaces, which are then collected by the output electrodes to deliver an output signal proportional to the displacement [51].

- The design and experimental verification of thin-film super high frequency laterally vibrating piezoelectric aluminum nitride (AlN) MEMS resonator suitable for the narrowband filter at a frequency above 3 GHz with two-port arrangements have been demonstrated by Matteo Rinaldi and his team with the structure Si/Pt/AlN/Au. These devices were employed to synthesize the highest frequency reported MEMS filter of 3.7 GHz (till the year 2009) based on AlN contour-mode resonator technology which attained electromechanical coupling in excess of 1.5% with Q of 500 in average [4,52].
- Gryba *et al.* presented high frequency ZnO/Si one port resonator in 2010. They compared the ZnO structure with the AlN structure using the same geometrical dimensions. And concluded more specifically, the effective coupling coefficient of ZnO was found higher than AlN structure, which is necessary to build an RF ladder filter. Very less variation in motional resistance and quality factor with these two-different piezo material structures have been noted by them [53] (Figure 4.20).
- Cremonesi *et al.* (2014) have provided an experimental and numerical relation of the quality factor of AlN CMRs with respect to the size of the resonator. They reported the quality factor and resonance frequency of 2,700 and 270 MHz, respectively, for the active area of the device of 2,440 square micrometer. They reported that the proper optimization of this area results in less leakage of acoustic energy that leaks into the substrate through the device anchors and, consequently, hence maximizes Q [54].

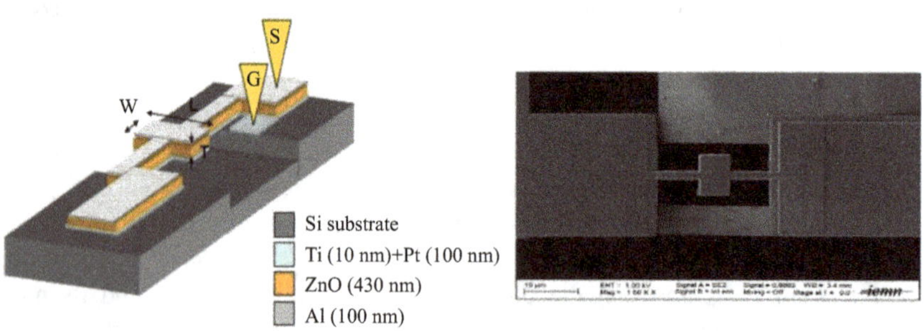

Schematic of one port ZnO piezoelectric contour-mode resonator.

SEM of one – port 10 x 5.6 μm ZnO rectangular plate.

Figure 4.20 Schematic of one port ZnO piezoelectric contour-mode resonator. SEM of one port 10 × 5.6 μm ZnO rectangular plate [53]

- Atsushi Isobe *et al.* in 2009 has been working on a tuning fork type resonator with the two radial extensional resonators. The figure of merits was 71 at the resonant frequency and 94 at the anti-resonant frequency, which was the highest values of the AlN-film contour-mode resonators [55].
- A new design approach for improving the anchor loss of the Lamb wave resonators has been presented by J. Zou in 2014 at the University of California, Berkeley. A conventional square plate and the butterfly-shaped AlN plate he has designed. With the butterfly-shaped device, the energy loss via tethers was reduced, hence high *Q* of 2,433 at 863.6 MHz center frequency compared to 1,916 of a quality factor at 862.0 MHz center frequency of conventional plate resonator. But, the electromechanical coupling coefficient and the motional resistance were degraded using a butterfly arrangement [44].
- A rigorous overview of the MEMS CMR technology using aluminum nitride as basic piezoelectric material, its main structure, working principle, etching of piezo film, and methods of performance optimization was presented by H. Yunhong and his team in 2016. The presented resonator has the ability to integrate multifrequency devices on a single chip for applications in the next generation of wireless communications [56].
- A thin film zinc oxide piezoelectric on substrate (TPoS) microresonator has been successfully demonstrated for the narrowband filtering application by M. Ossama in 2015. Highest reported measured frequency for these devices was 767 MHZ for minimum feature size (finger pitch) of 4 μm. The measured unloaded quality factor of this first prototype was 1460 and motional resistance of 21 Ω. The reported electromechanical coupling coefficient was 2.2% [57] (Figure 4.21).
- Gavin K. Ho *et al.* in 2008 reported the design, fabrication, and characterization of piezoelectrically-transduced micromechanical single-crystal-silicon resonators operating in their lateral bulk acoustic modes to address the need for high-*Q* frequency-selective components. Resonators were fabricated on

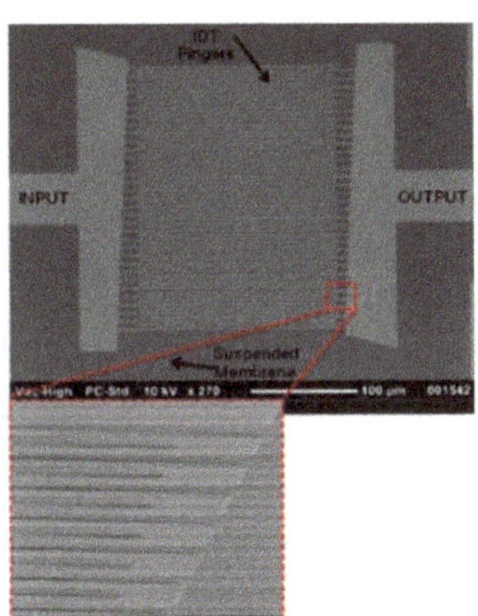

Figure 4.21 Thin film ZnO piezoelectric micro-resonator [57]. Reprinted with permission. © 2014 IEEE

SOI substrates with a 300 nm zinc oxide film. A bulk acoustic mode was observed with 600 Ω impedance ($Q = 3{,}400$) at 90 MHz [58] (Figure 4.22).

- Author of [59] demonstrated the piezoelectric MEMS resonators with high-quality factors in liquids can be used to monitor the change in grape must during wine fermentation, which is a direct quality indicator of the fermentation process. Yu *et al.* [60] presented MEMS gyroscope through the use of the acoustic Sagnac effect, which measured the phase difference between two sound waves traveling in opposite directions in a circular MEMS structure actuated by PMUTs [61].
- O. Mortada *et al.* in 2017 reported the design, simulation, fabrication, and test results of ZnO-based contour-mode microresonators integrating piezoelectric zinc oxide (ZnO) layers. The inter-digitated (IDT) type microresonators were fabricated on ZnO films and suspended top of 2 μm thick silicon membranes. They analyzed several possibilities of increasing the quality factor (Q) and the electromechanical coupling coefficient (kt^2) of the devices by varying the numbers and lengths of the IDT electrodes and using different thicknesses of the ZnO layer. They fabricated IDTs of different finger numbers with different lengths for three different thicknesses of ZnO films. The measured Q factor confirms that reducing the length and the number of IDTs fingers enables to reach better electrical performances at resonant frequencies around 700 MHz. The extracted results for an optimized microresonator device showed a Q of 1,180 and a kt^2 of 7.4% with a resonant frequency of around 700 MHz [62] (Figure 4.23).
- S. Ghosh and G. Piazza (2016) demonstrated the use of elasto-optic modulation in an integrated MEMS Lamb wave structure. The device consists of a photonic resonator embedded into a laterally vibrating piezoelectric resonator for the strain to be transferred. They observed optical modulation at the device resonance frequency of 843 MHz. They predicted the device to be used for signal modulation in RF-photonic systems [63] (Figure 4.24).

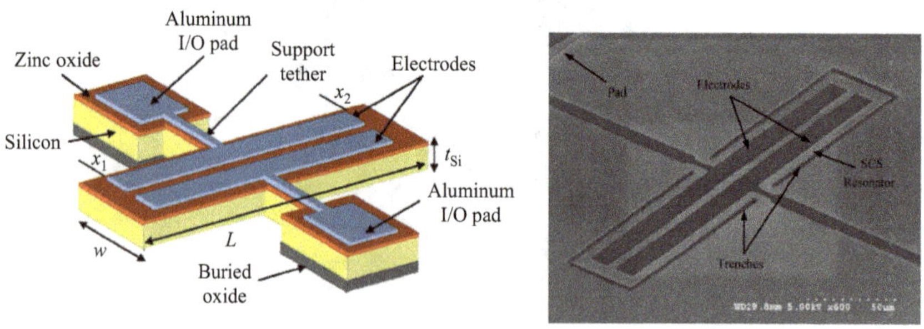

Figure 4.22 Piezoelectric-on-silicon lateral bulk acoustic micromechanical resonator (left). SEM of a 240 μm × 40 μm × 5 μm piezoelectric-on-silicon lateral bulk acoustic resonator (right) [58]. Reprinted with permission. © 2008 IEEE

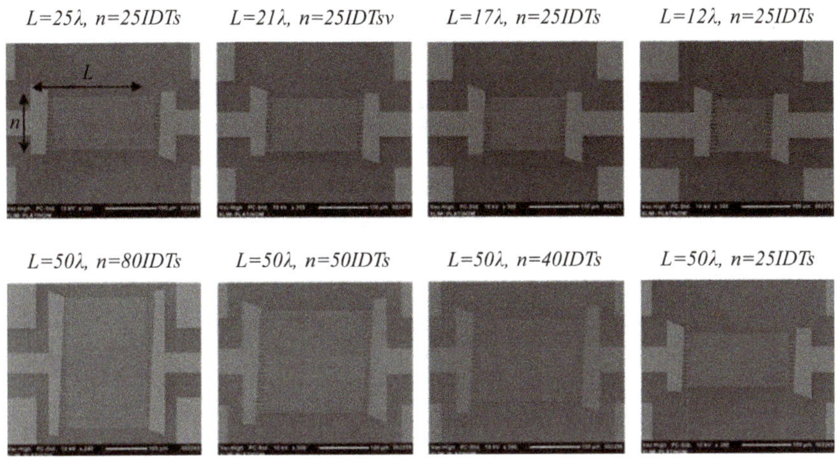

Figure 4.23 SEM picture of the fabricated devices: eight microresonators having different lengths L of IDT electrodes and different numbers n of IDT electrodes [62]

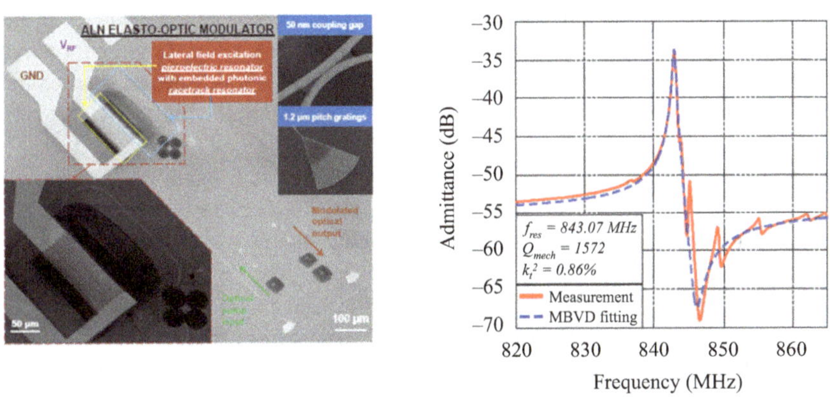

Figure 4.24 SEM images highlighted device core consisting of racetrack resonator arm embedded in MEMS resonator (bottom left) and integrated photonic elements including coupling gap and input/ output gratings (top right inset). Electromechanical device response for the piezoelectric resonator fit to the MBVD model [63]. Reprinted with permission. © 2016 IEEE

- M. Zadehsafari *et al.* (2017) reported a new structure for AlN Lamb wave disk-shaped resonator with several electrodes configurations. Electrodes are structured into different forms to study wave propagation with different electrodes configuration. Simulation results are presented in the paper and also discussed wave propagation [64] (Figure 4.25).

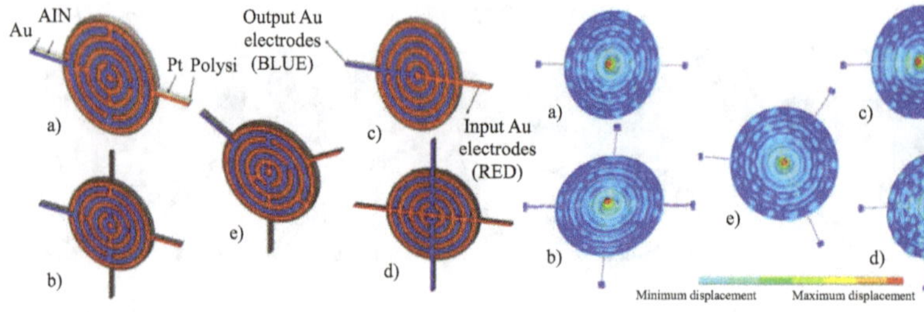

Figure 4.25 Several configurations for MEMS resonator. Simulation results for different structures [64]. Reprinted with permission. © 2017 IEEE

4.11 Subject of investigation

It is impractical to think of human life without energy, frequency, and vibrations. We can connect with each other any time through a radio frequency communication network even though we are not at the same place. RF communication is working efficiently in various fields like defense, biomedicine, optical, etc. It is observed that MEMS resonators of RF range have been designed, fabricated, and demonstrated by many of the researchers with great success. The target parameters, the structure geometry, input/output components, feasible design parameters, materials to be used, fabrication process, associated applications, etc. are very crucial to select and to be optimized properly to have a good quality product. The use of MEMS resonators in RF applications exists but its limitations concern with respect to its slow improving quality factor and motional resistances to transfer maximum power between transceivers compared to the quartz resonators. The high *f-Q* product, good bonding of electromechanical coupling, and low motional resistance (*Rm*) are the necessary means to design and fabricate realistic resonator. Experimentation at the fabrication level is a very costly affair, hence the device design and its rigorous analysis need to be performed using finite element machine (FEM) tool before fabrication.

4.12 Design and modeling of MEMS resonator

Modeling is a fundamental task in the production cycle of MEMS resonators. Modeling allows the prediction of the resonator's static and dynamic responses. Modeling is performed for the design process, when the decisions of choosing materials, geometries, and dimensions are to be made. Besides aiding the designer in defining the fabrication technology, modeling allows the analysis of the performance of a working device, so we can fit the actual frequency response to given models in order to extract their parameters. The design of modern system-on-a-chip blocks composed of resonators and integrated circuits is largely based on modeling activity [28].

The design of resonators is addressed by different modeling approaches, each having its own context and usefulness. Analytical models are based on mechanical,

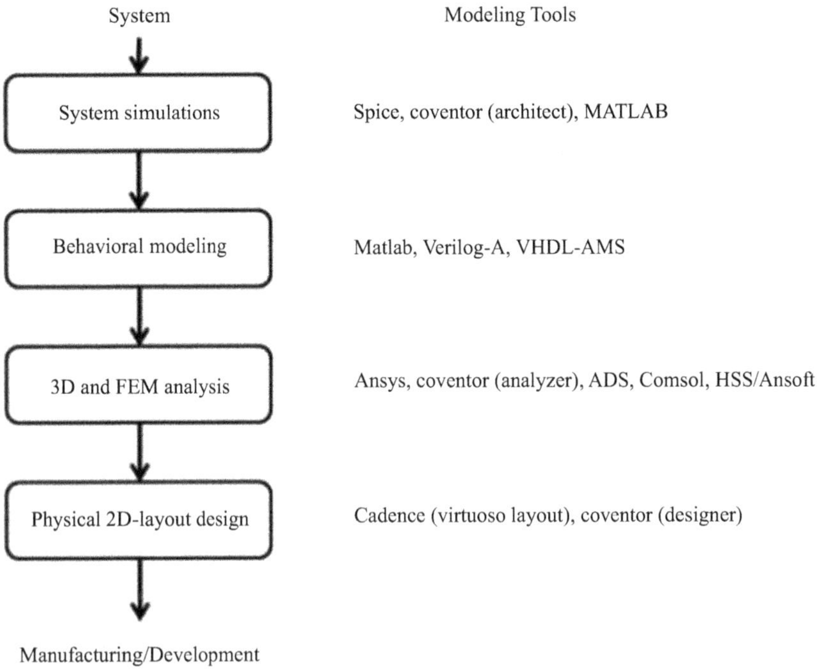

Figure 4.26 Modeling hierarchy of MEMS resonator [28]

electromechanical, and purely electrical equivalent-circuit representations of the resonator physics [28] (Figure 4.26).

4.12.1 Finite element modeling

The finite element modeling (FEM) analysis field spans from the formulation of numerical analysis and computational efficiency to the operational aspects of FEM commercial tools. Computer-aided FEM is a powerful numerical-analysis tool that allows accurate prediction of the static and dynamic responses of a multiple-domain physical system. FEM analysis reproduces the geometry and forces interaction of complex systems whose analytical formulation is unfeasible. Starting from the structural model of the system, FEM analysis couples the structural physics with electrostatic, magnetostatic, piezoelectric, thermal, optic, fluidic, and electromagnetic domains, among others. Some commercial software tools like ANSYS (Ansys Inc., Canonsburg, Pennsylvania), Coventor, IDEAS, and COMSOL Multiphysics are available in the market for FEM analysis [28].

The FEM of FBAR, MEMS, or NEMS resonators is a system constituted of the following components:

1. The resonator itself;
2. The environment—air and vacuum;
3. The degrees of freedom or boundary conditions; and

4. The excitation sources—atmospheric or relative pressure, gravity or another inertial force, mechanical acceleration, fluidic pressure, magnetic field, electric potential, and so on (Figure 4.27).

The fundamental component of a FEM is the *element*: the model geometry is structured by a group of elements, typically thousands of them. When modeling the complete system, the FEM analysis software implements a set of equations pertinent to the physical domains involved in the interaction. These equations are applied and solved at the element level, and the results are stored in the system memory and used as an input parameter for the next element to be analyzed. In the end, all these results are scaled up and superposed to evaluate the global system response, whose quality will depend on the accuracy, correction, and complexity of the built model [28].

In FEM analysis, first, the system characteristics and physical interactions between the system components are defined. The model of the system is built, by using a compiler or preprocessor. First, we define the materials and element types; then we generate the geometry, and next, we mesh this model with a number of elements that can be specified by the designer. Once the model is made and meshed, the degrees of freedom (DOF) or boundary conditions of the system are applied to it. This establishes the physical reference frame, and it involves initial force definition, initial charging, and clamping of the structure, among others. Once this task is completed, we proceed to solve the model according to the simulation settings defined by the user. Structural, modal, and harmonic analyses can be carried out, with or without initial stresses or loads included in the analysis. Linear and nonlinear analysis options are also available in commercial tools. When the analysis finishes, the postprocessing system of the FEM tool allows the evaluation of the simulation results, as they are provided in the form of graphs, charts, tables, or 3D plots [28].

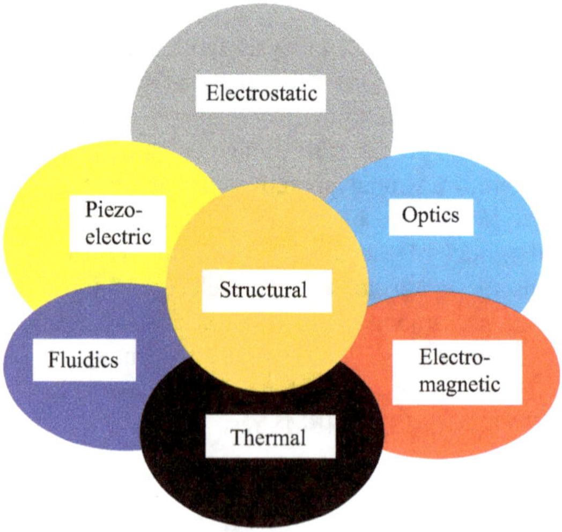

Figure 4.27 Physical domain in MEMS modeling [28]

4.13 One port lateral field excited contour mode piezoelectric MEMS resonator

4.13.1 Introduction

A high-quality resonating device depends upon its physical and mechanical parameters and their properties. The dimensions of the design directly affect its performance parameters like quality factor (*Q*), resonance frequency (*fr*) and motional resistance (*Rm*), inductance (*Lm*), capacitance (*Cm*, *C0*), etc. [4]. To obtain better performance, the resonator must have high resonance frequency and high-quality factor and low motional resistance. At present, the methods of improving these performance parameters mainly focus on shrinking the geometric size of the resonator, such as narrowing the width and space of the comb-shaped electrodes (interdigitated). It may be noted that the size scaling is greatly restricted by the lithography technology [56]. Therefore, the research of contour-mode resonator (CMR) focuses attention on its design parameters to be optimized properly. Thickness field excitation (TFE)-based resonators are two-port resonators consisting of piezo material film sandwiched between the top and bottom electrodes, hence increasing fabrication step and in lateral field excitation (LFE) which is a one-port device and there is no bottom electrode, shown in Figure 4.28. The top layer in both types is in an interdigitated structure form. Our work presents the design of one port MEMS ZnO contour mode resonator based on lateral field excitation (LFE).

We have used COMSOL™ Multiphysics FEM tool to analyze the resonance frequency, mode shapes, deflection, von mises stress, electric potential, admittance, quality factor, etc. of one port LFE contour mode MEMS resonator before proceeding to costly fabrication. Multiple trials of fabrication are not practically acceptable because of the involvement of time and cost, hence validation through numerical analysis and simulation through the efficient tool is very essential. The geometrical design for the targeted parameters is of prime importance to achieve good performance and also to understand reliability issues to build models. The parameter optimizations are done through the Taguchi method.

When an AC signal is applied to LFE/TFE-based resonator, the electric field induces corresponding mechanical strain (also stress) which causes the dilation or contraction (vibration) of the resonator in contour mode direction as shown in Figure 4.29 [65]. If the signal frequency coincides with the natural frequencies of

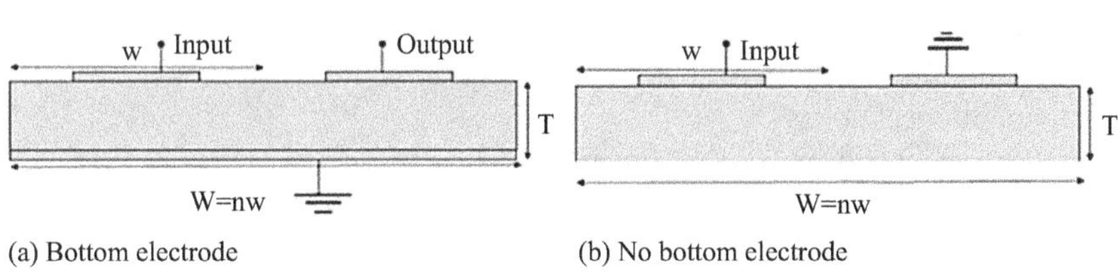

Figure 4.28 (a) Thickness field excitation and (b) lateral field excitation

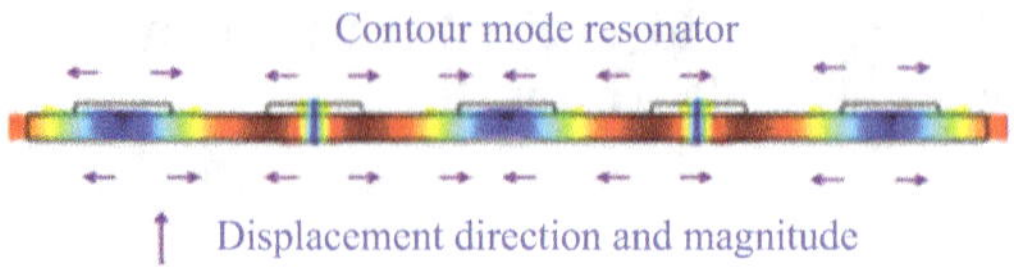

Figure 4.29 Displacement of LFE-based CM resonator

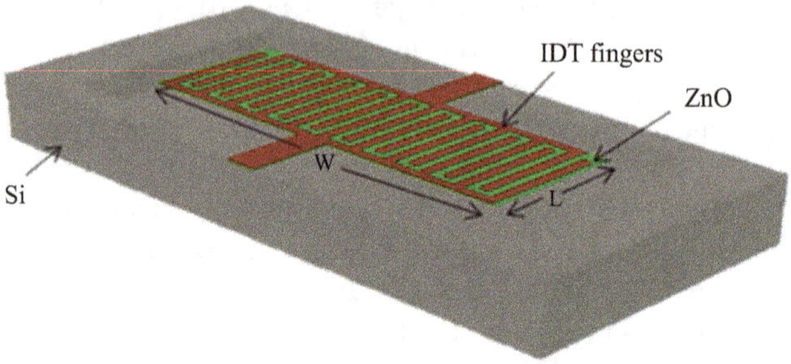

Figure 4.30 Schematic representation of designed one-port contour mode resonator [65,66]

the structure, the specific vibration mode will be excited. The static capacitance *C0* is coming from the dielectric property of the material contained between electrodes and is in parallel to the electromechanical resonator.

4.13.2 Design and analysis of contour mode resonator

The quality of a resonating device depends upon its physical and mechanical parameters. For optimizing these parameters in the resonator design, we have to try a number of combinations of physical dimensions and thus a large number of combinations are to be tried. This is a factorial design of an experiment. To solve this problem, the Taguchi method of design of experiments was used, which is explained in Section 4.16.1. The dimensions of the design directly affect its performance parameters like quality factor(*Q*), resonance frequency (*fr*) and motional resistance (*Rm*), inductance (*Lm*), capacitance (*Cm, Co*), etc. [51]. To obtain better performance, the resonator must have desired resonance frequency and high-quality factor as well as low motional resistance. At present, the methods of improving these performance parameters mainly focus on shrinking the geometric size of the resonator, such as narrowing the width and space of the comb-shaped electrodes (interdigitated fingers). It may be noted that the size scaling is greatly restricted by the lithography technology.

Design and analysis of the structure were carried out using COMSOL^{TM} simulations. This software uses FEM to solve for eigenfrequencies, admittances, displacement, etc. at each node created for the model. Interdigitated electrode (IDE or IDT) structure has been designed and analyzed in this work as shown in Figures 4.30 and 4.31.

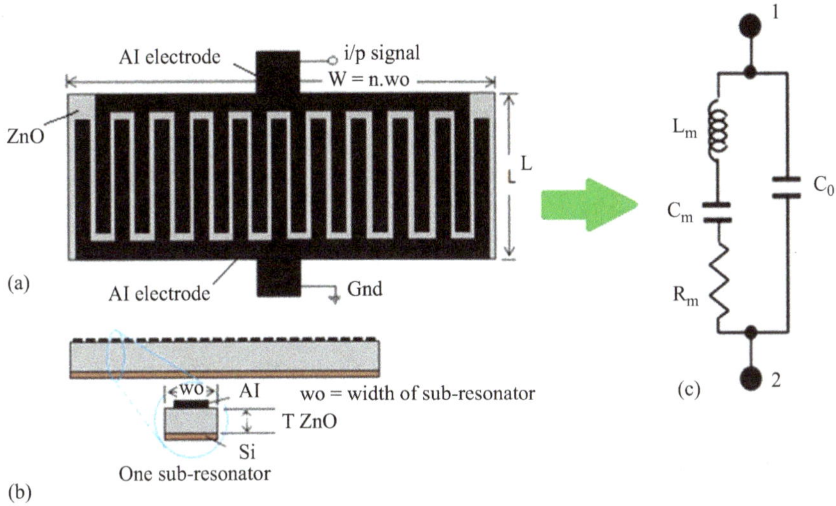

Figure 4.31 (a) One-port 21 interdigitated electrode CMR [65], (b) one subresonator in magnifying form [65], and (c) equivalent electrical Butterworth–Van Dyke model [66]

Table 4.8 Parameters used in design of resonator [66]

Parameters used in design
No. of IDT fingers (*N*)
Length of the resonator (*L*) in μm
Width of top Al electrode (*we*) in μm
Spacing/gap between two IDT fingers (*s*) in μm
Total width of the resonator ($N \times wo$) in μm
Thickness of piezoelectric material (*TznO*) in nm
Thickness of top Al metal (*Tal*) in nm
Width of subresonator (*wo or wr*) in μm
Young's modulus of the piezoelectric material *I* in GPa
The density of the piezoelectric material (ρ) in kg/m^3

A one-port ZnO contour mode resonator (CMR) and its equivalent Butterworth-Van Dyke (BVD) model are presented in Figure 4.31. The resonator consists of ZnO thin film placed between the silicon substrate and top aluminum (Al) interdigitated electrodes. The top electrode is IDT fingers to excite lamb waves propagating in the ZnO film. In our case total, 21 subresonators (IDT fingers) are connected in parallel. The resonance frequency is primarily defined by the IDT width, *wo* and is effectively decoupled from the ZnO film dimension. The parameters used to design the resonator are as per the Table 4.8.

The equivalent electrical circuit of the resonator is represented by its modified Butterworth-Van Dyke (BVD) model, shown in Figure 4.31(c). The motional resistance, *Rm* comes from the mechanical damping transformed into the electrical domain. *Cm* (motional capacitance) and *Lm* (motional impedance) are used to represent, with *Rm*, the induced components by the mechanical motion of the device. The capacitor C_0 is added to account for the feedthrough between two electrodes; it is a function of subresonator width (*wo*), length (*L*) of resonator, and thickness (*T*) of piezo material. All equivalent electrical parameters can be expressed as a function of the resonator capacitance, resistance, and inductance, is represented in (4.4) [53,66,67].

$$
\begin{aligned}
Rm &= \frac{1}{N}\frac{\pi}{8}\frac{T}{L}\frac{peq^{\frac{1}{2}}}{Eeq^{\frac{3}{2}} d_{31}^{2} Q} \\
Lm &= \frac{1}{N}\frac{\pi}{8}\frac{woT}{L}\frac{peq}{Eeq^{2} d_{31}^{2}} \\
Cm &= N\frac{8}{\pi^{2}}\frac{woT}{L} Eeq d_{31}^{2} \\
C0 &= N\,\varepsilon 0\,\varepsilon 33\frac{woL}{T}
\end{aligned}
\tag{4.4}
$$

where *N* is the number of subresonators (IDT fingers), $\varepsilon 0$ is the free space permittivity, $\varepsilon 33$ is the dielectric constant of ZnO along the c-axis, *Eeq* is the equivalent Young's modulus and ρeq is the mass density, and d_{31} is the piezoelectric constant [53,66,67].

The design of a contour mode resonator starts from the selection of the targeted resonant/center frequency (*fr*). The width of the subresonator plays an important role to achieve *fr*. In-plane geometry is selected with piezoelectric material ZnO placed between the silicon wafer and aluminum interdigitated (IDT) fingers. Following are the design parameters of a typical contour mode resonator.

(a) Resonance frequency (*fr*)

The frequency at which the resonator oscillates with maximum amplitude is known as the resonance frequency. The resonance frequency, ωr in terms of the mass of the resonating structure (*m*) and the spring arrangement of the resonator (*k*) is given by (4.5) [4].

$$
\omega r = \sqrt{\frac{k}{m}} \tag{4.5}
$$

If the mass is subjected to an external force, it oscillates. If the mass is driven at its natural frequency, it vibrates with a large amplitude. An electromechanical device in which electrical and mechanical energies are reciprocally converted to each other at a certain frequency called the resonance frequency is the basic principle of the piezoelectric resonator. In order to finalize the application and dimensions of the resonator, the desired resonance frequency was computed using (4.6) [4].

$$
fr = \frac{1}{2wo}\sqrt{\frac{Eeq}{\rho eq}} \tag{4.6}
$$

where *wo* is the width of subresonator, *Eo* (in GPa) is the equivalent elasticity modulus of the ZnO/Al layer. Equations (4.7) and (4.8) were used for the calculation of *Eo and ρo* respectively, described in [3,4,66]:

$$Eo = \frac{(EZnO)(TZnO) + (EAl)(TAl)}{TZnO + TAl} \tag{4.7}$$

where, *EZnO* and *EAl* are elasticity moduli of zinc oxide and aluminum, respectively, and *TznO* and *Tal* are the thicknesses of zinc oxide and aluminum, respectively. Similarly, *ρo* (in kg/m^3) is the equivalent mass density of the ZnO/Al layer:

$$\rho o = \frac{(\rho ZnO)(TZnO) + (\rho Al)(TAl)}{TZnO + TAl} \tag{4.8}$$

where *ρ ZnO* and *ρ Al* are mass densities of zinc oxide and aluminum, respectively.

(b) Quality factor (*Q*)

The quality factor *Q* can be approximated by the ratio of the center frequency and the 3dB bandwidth of the resonator [68] (Figure 4.32):

$$Q = \frac{fr}{3dB \text{ bandwidth}} \tag{4.9}$$

The series resonance frequency (*fs*) is obtained at maximum conductance and the parallel resonance frequency (*fp*) is obtained at the maximum resistance [68]. The mechanical quality factor of a resonator describes the ratio of energy stored (vibration stored in the resonator) to the energy dissipated per cycle of vibration is a well-known fact:

$$Q = 2\pi \frac{\text{Energy stored}}{\text{Energy dissipated per cycle of oscillation}} \tag{4.10}$$

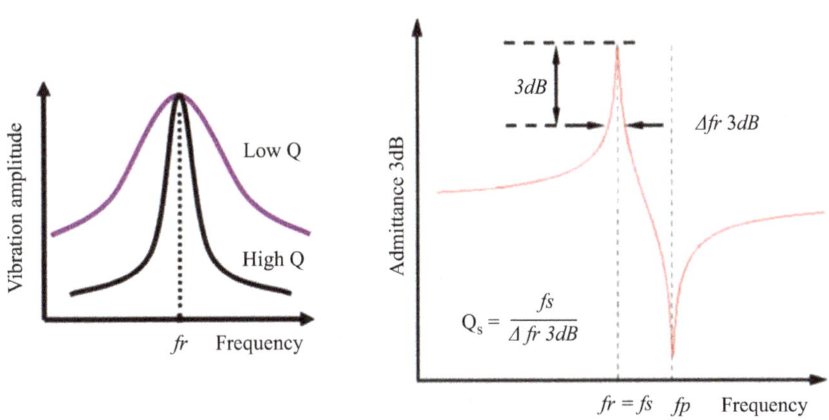

Figure 4.32 A typical graph of the quality factor (Q) [68]

Q is a dimensionless parameter suffers a large number of losses like a material loss, anchor loss, loss due to air damping, etc. So, higher is the quality factor of a resonator means lesser losses are there in the structure:

$$Q = \left(\frac{1}{Q_{\text{material}}} + \frac{1}{Q_{\text{anchor}}} + \frac{1}{Q_{\text{air}}}\right)^{-1} \tag{4.11}$$

Air damping losses can be prevented by operating the resonator in a vacuum, Anchor losses are because of the radiations of acoustic energy from the resonating structure through anchors and the material losses relates to the irreversible transformation of acoustic energy to thermal energy [68].

(c) Motional resistance (*Rm*)

Most of the RF systems use a 50 Ω transmission line to transfer maximum power from the source to load, and the serial motional resistance of the resonator must be designed at a reasonable level for it to be coupled with a practical system. Piezoelectrically actuated resonator is having less motional resistance compared with the electrostatically actuated resonator. *Rm* is a critical parameter because it directly relates to the insertion loss of the filter and the white noise of the resonator [69].

(d) Electromechanical coupling coefficient (kt^2)

The ability of piezoelectric material to transform electrical energy into mechanical energy and vice versa represents the electromechanical coupling coefficient (kt^2). Thus, it is defined as the ratio of electrical energy to mechanical energy. The stored electrical and mechanical energies are calculated by

$$kt^2 = \frac{\pi^2}{4}\frac{f_p - f_s}{f_p} \approx \frac{\pi^2}{8}\frac{Cm}{Co} \tag{4.12}$$

In order to improve the effective coupling of a resonator, the piezoelectric materials with a higher piezoelectric coupling coefficient are desired. Even though, its high electrical resistance, high breakdown voltage, and small dielectric constant, AlN has a slightly smaller coupling coefficient compared to ZnO [56].

4.14 Finite element simulations using COMSOL™ Multiphysics

The structure shown in Figures 4.30 and 4.31(a) was designed and simulated using COMSOL™ Multiphysics. The ZnO-based IDT structure was validated for functional verification, support numerical calculations, etc. The dimensions of the resonator affect resonance frequency, quality factor, motional resistance, etc. Change in the geometrical parameters will help to decrease motional resistance and an increase in resonance frequency and quality factor. The Al and ZnO material properties are tabulated in Table 4.9. The device design parameters with dimensions are tabulated in Table 4.10.

Table 4.9 Material properties [4,66]

Material	Density (kg/m^3)	Young's modulus (Pa)	Poisson's ratio
Al	2,730	73E + 09	0.334
ZnO	5,660	112E + 09	0.32

Table 4.10 Device design dimensions [66]

Parameters	Value	Parameters	Value
Number of IDT fingers (N)	21	Width of subresonator (wo or wr)	2.72 μm
Width of top electrode (we)	1.38 μm	Total width of resonator ($W = N \times wo$)	57.12 μm
Length of the resonator (L)	18.52 μm	Thickness of ZnO ($TZnO$)	250 nm
Spacing between fingers (S)	1.34 μm	Thickness of Al (TAl)	100 nm

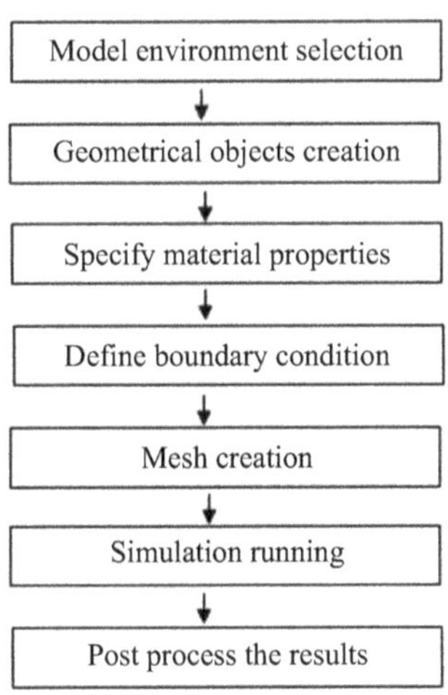

Figure 4.33 Stepwise structure design methodology in COMSOL™ [66]

The effective model implementation enables the design and device optimization efficiently. Several simulations for various inputs and device dimensions were addressed and prominent simulation results were considered. The following stepwise methodology has been adopted to design a CMR structure in COMSOL™ Multiphysics (Figure 4.33).

2D and 3D model was used for simulation. Physical boundary conditions like the fixed end and the free end are defined. Then meshing is done into meshable and nonmeshable parts. The active part of the model is the ZnO layer. Base structure and anchors are skipped in order to reduce the computational load. The resonator structure design and sample mesh model are shown in Figures 4.34 and 4.35, respectively.

The setting of the mesh before computations determine the resolution of the finite element mesh used to discretize the model. Tetrahedron meshing is the

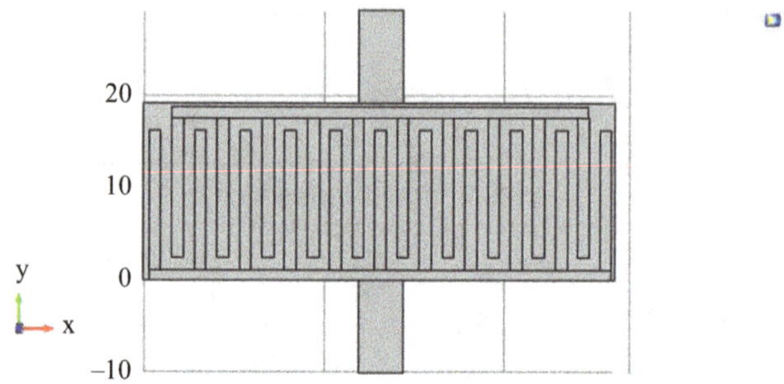

(a)

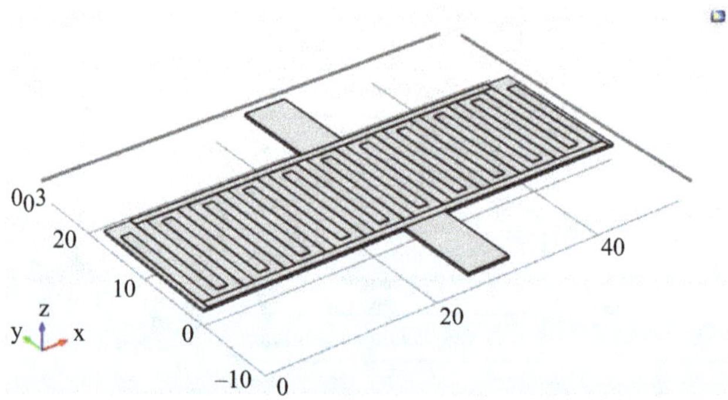

(b)

Figure 4.34 (a) 2D model and (b) 3D model of LFE contour mode resonator

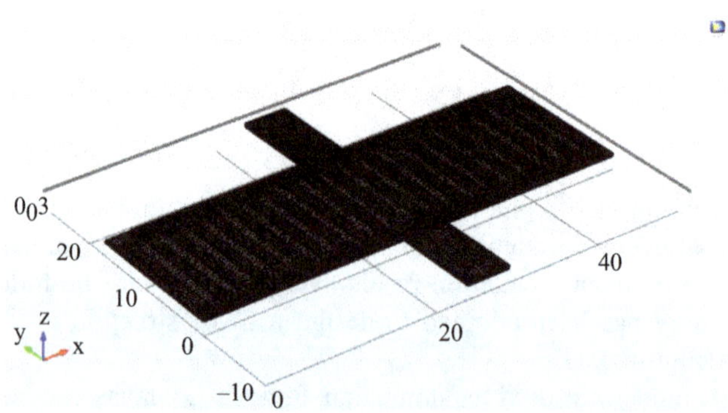

Figure 4.35 Meshed structure of LFE contour mode resonator

default element for most physics within the software. If the geometry contains small thin edges and faces, then slightly finer mesh than the default settings may be incorporated to get a more accurate result. This refinement in the mesh size definitely improves the computational accuracy but contains some sacrifice in speed and requires memory usage [66].

4.15 Mode shapes for lateral vibrating contour mode one-port resonator

For the selected dimension of width (Table 4.10), the mode shape analysis is done for different numbers of IDTs in the resonator as shown in Figure 4.36. Different numbers of IDT fingers (N = 3,5,9,11,15,17,19,21) were chosen. The simulated resonance frequency for eight designs with varied IDTs was found to be ~900 MHz in every case. But, it was observed that the increase in numbers of IDT fingers increases in the admittance (S) means reducing the motional impedance, hence reducing losses and improvement in Q, which are the key performance parameters of the resonator. It is clear that the admittance is increasing when the number of IDT fingers is increased; hence by keeping the active area of the resonator in mind, to avoid more complexity during fabrication, and for the targeted performance parameters like resonance frequency, quality factor, and motional resistance, the design has been optimized for 21 IDT fingers [66].

Figures 4.36(a) shows the resonant modes for the determination of resonance frequency. The first three modes are identified for our resonator from its frequency response with frequency range spanning from 750 MHz to 1 GHz. The resonant characteristics of these modes were explored by COMSOL™. The simulation result shows that the admittance of 0.0102 (S) at 901 MHz is obtained in the first mode as depicted in Figure 4.36(b). A sharp peak in figure is due to the lower resistance at resonance. It can be seen that the simulation result of the structure provides a resonance frequency in the GSM range [66].

The eigenvalue solver of COMSOL™ evaluates the eigenfrequency analysis where the quality factor of ~340 is attained at eigenfrequency of 9.0082E8 + 1.1678E5i (Figure 4.37). This study helps to validate the design with the numerical analysis of the target resonance frequency of GSM range. To meet required functional parameters of the device, contour mode-based lateral field excited ZnO piezoelectric resonator has been designed, optimized, and analyzed using the FEM tool. The design parameters have been optimized by the Taguchi method of design of experimentations. The FEM tool is useful to predict the device performance and adjust the parameters accordingly for better results before going for fabrication [66].

4.16 Parameter optimization of one port contour mode MEMS resonator

The reliability of the fabricated device depends upon various input geometrical parameters and the targeted results which we expect after execution. Therefore,

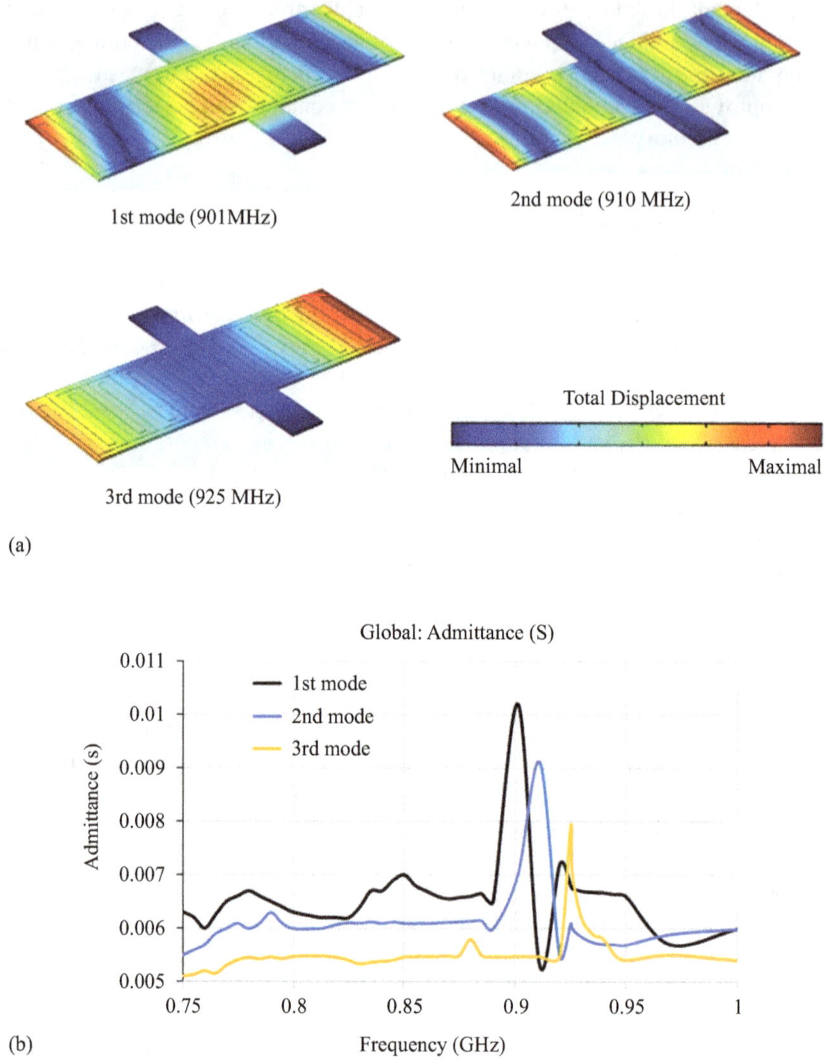

Figure 4.36 *(a) COMSOL^{TM} simulation shows a 3D example of the resonant modes of contour mode resonator for which 21 parallel IDT fingers were considered. Shading indicating the relative magnitude of displacement. (b) Equivalent admittance curve and comparison between frequency characteristics for 1st, 2nd, and 3rd contour resonant modes [66]*

these parameters must be carefully studied. For optimizing the number of parameters in our resonator design, the numbers of combinations of control factors (input factors) have been tried. Every combination with respect to each and every control parameter is difficult to interlink and consume a big time with the waste of

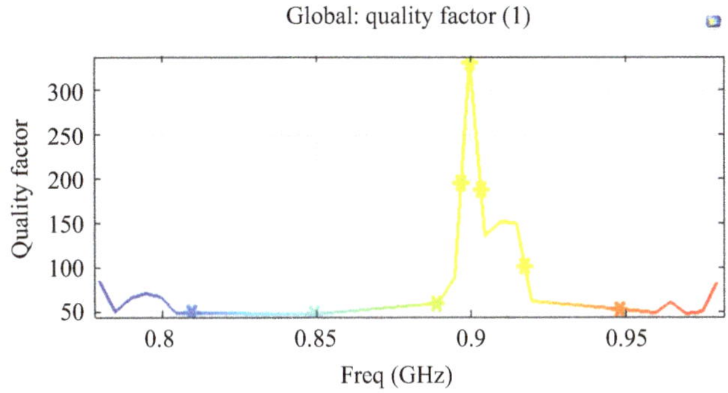

Figure 4.37 COMSOL™ simulation result—quality factor versus resonance frequency graph

resources. So, for enhanced optimization with lesser efforts, Taguchi method of design of experiments (DOE) has been implemented followed by an analysis of variance (ANOVA) to get our vital parameters [70–75].

4.16.1 Taguchi method

To produce a high-quality product at a low cost to the manufacturer is the basic objective of Taguchi method. It is the best suitable method where there are intermediate numbers of variables and their interactions between other variables with only a few variables are contributing significantly. The Taguchi method involves two powerful elements. First, it is a disciplined way of developing a product or investigating complex problems. Second, it provides a means to cost-effectively investigate the available alternatives. Although Taguchi's method was built upon well-developed concepts of optimization through the design of experiments [76], the experimental design proposed by Dr Genichi Taguchi involves using orthogonal arrays to organize the parameters affecting the process and the levels at which they should be varied. Instead of having to test all possible combinations, Taguchi method tests a pair of combinations. This allows for the collection of the necessary data to determine which factor affects the product quality most with a minimum amount of experimentations, thus saving time and resources with a high degree of certainty [70]. Figure 4.38 shows a comparison of the general practice and the Taguchi approach.

The general steps while performing Taguchi analysis are

- Identification of process objectives/target parameters. In our case, these may be resonant frequency (*fr*), motional resistance (*Rm*), and quality factor (*Q*).
- Finding design parameters affecting the process. In our case, these may be the length, width, thickness, gap, number of IDT fingers, the material used (E,ρ), etc.
- Creation of orthogonal array based on a number of parameters and levels of variation of each parameter.

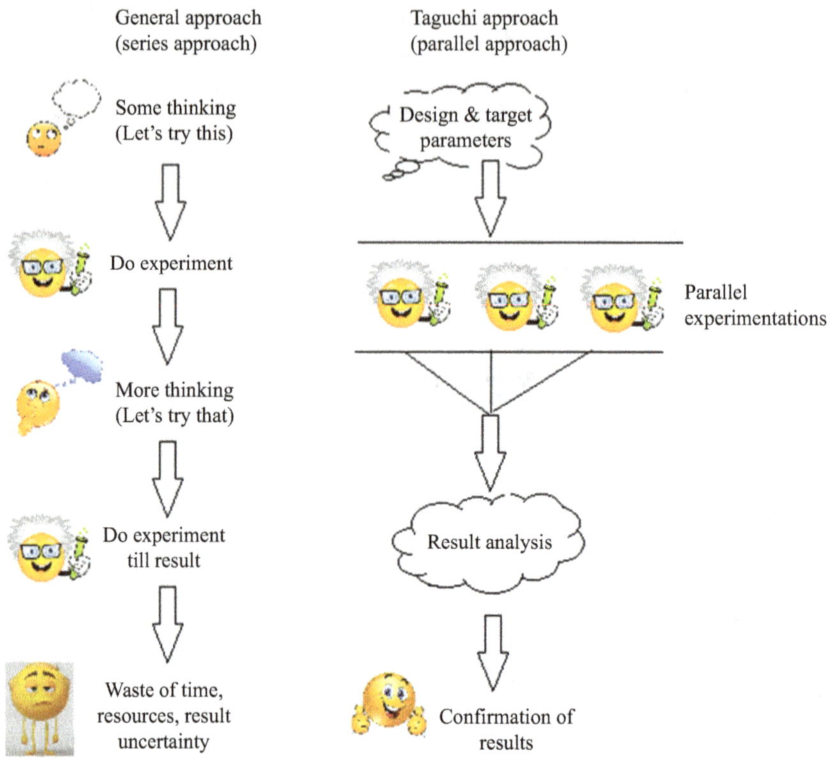

Figure 4.38 A comparison of the general approach and the Taguchi approach

- Experimental analysis.
- ANOVA performs complete data analysis to determine the effect of the different parameters on the performance.

4.16.2 ANOVA statistics

Analysis of variance or ANOVA is a collection of statistical models and its associated procedure. ANOVA performs a statistical test and complete data analysis to determine the different parameters on the performance measures [70,72,74].

To optimize the design of our microresonator, we have selected ten control factors that decide the important design parameters as shown in Table 4.11 and three levels are identified for each variable control factor as shown in Table 4.12.

The trial runs are calculated according to the L27 orthogonal array [70,73,76] for the ten control factors with the three-factor levels to determine the motional resistance (*Rm*), the resonant frequency (*fr*), and the quality factor (*Q*) as target parameters shown in Table 4.13.

Now, it is time to apply ANOVA analysis to the trial run result of Table 4.13. In the Taguchi method, the results are statistically analyzed using ANOVA to determine the percentage contribution of individual parameters to the response [70].

Table 4.11 Control factors [75]

Variables	Control parameters
p1	No. of IDT fingers (*N*)
p2	Length of the resonator (*L*) in μm
p3	Width of the top electrode (*we*) in μm
p4	Spacing/gap between two subresonators (*s*) in μm
p5	Total width of the resonator ($N \times wr$) in μm
p6	Thickness of piezoelectric material (*T*) in nm
p7	Thickness of top Al metal (*T top*) in nm
p8	Width of subresonator (*wo or wr*) in μm
p9	Young's modulus of the piezoelectric material (*E*) in GPa
p10	The density of the piezoelectric material (ρ) in kg/m^3

Table 4.12 Factor levels [75]

Factor levels	p1	p2	p3	p4	p5	p6	p7	p8	p9	p10
1	15	15	0.8	0.8	40	0.25	0.05	2	63	7,660
2	21	17	1.2	1.2	50	0.2	0.1	2.5	112	5,606
3	27	20	1	1	60	0.3	0.15	3	310	3,260

Through ANOVA analysis, we were able to find out the percentage contribution of each factor as shown in Figures 4.39–4.41.

So, we can see from the above charts that the material properties (E, ρ) play the crucial role for *Rm, fr, and Q*. It is clear that the quality factor and resonance frequency are less affected by the number of IDT fingers (subresonators), motional resistance is dependent on *N*. The width of subresonator is an important factor in deciding the performance of all the three target parameters. We cannot ignore the thickness *T*, which plays an important role in the motional resistance and the resonance frequency. Similarly, quality factor and resonance frequency can be adjusted by changing the value of the width of the IDT finger. The thickness of the top aluminum electrode only affects the resonance frequency as well as the length of the resonator is also contributing less to change the value of motional resistance, hence quality factor.

It is concluded from all the above three graphs that Young's modulus and the density of the material are the major percentage contributor for the three target parameters. The response of other percentage contributors, *viz.* length of the resonator, the thickness of the piezoelectric slab, and width of the sub-resonator for respective calculations can be seen from the graphs shown in Figure 4.42.

From the responses of Figure 4.42, we state that, as the length of the resonator increases, motional resistance decreases whereas resonance frequency and quality factor are degrading. So, the proper balance has to be maintained in deciding the length of the resonator for our design. As the piezoelectric thickness is increased,

Table 4.13 Experimental trial runs [75]

Expt.	p1	p2	p3	p4	p5	p6	p7	p8	p9	p10	Rm (Ω)	fr (GHz)	Q
1	15	15	0.8	0.8	40	0.2	0.05	2	63	7,660	106.89	0.72	719.79
2	15	15	0.8	0.8	50	0.25	0.1	2.5	112	5,606	38.980	0.89	893.95
3	15	15	0.8	0.8	60	0.3	0.15	3	310	3,260	4.261	1.63	1,625.25
4	15	17	1	1	40	0.2	0.05	2.5	112	5,606	27.515	0.89	893.95
5	15	17	1	1	50	0.25	0.1	3	310	3,260	3.133	1.63	1,625.25
6	15	17	1	1	60	0.3	0.15	2	63	7,660	142.59	0.72	716.96
7	15	20	1.2	1.2	40	0.2	0.05	3	310	3,260	2.130	1.63	1,625.25
8	15	20	1.2	1.2	50	0.25	0.1	2	63	7,660	101.00	0.72	716.96
9	15	20	1.2	1.2	60	0.3	0.15	2.5	112	5,606	35.082	0.89	893.95
10	21	15	1	1.2	40	0.25	0.15	2	112	3,260	12.953	1.47	1,465.35
11	21	15	1	1.2	50	0.3	0.05	2.5	310	7,660	5.959	1.27	1,272.32
12	21	15	1	1.2	60	0.2	0.1	3	63	5,606	84.478	0.56	558.72
13	21	17	1.2	0.8	40	0.25	0.15	2.5	310	7,660	4.382	1.27	1,272.32
14	21	17	1.2	0.8	50	0.3	0.05	3	63	3,260	65.019	0.73	732.67
15	21	17	1.2	0.8	60	0.2	0.1	2	112	5,606	15.723	1.12	1,117.44
16	21	20	0.8	1	40	0.25	0.15	3	63	3,260	46.055	0.73	732.67
17	21	20	0.8	1	50	0.3	0.05	2	112	5,606	20.047	1.12	1,117.44
18	21	20	0.8	1	60	0.2	0.1	2.5	310	7,660	2.980	1.27	1,272.32
19	27	15	1.2	1	40	0.3	0.1	2	310	5,606	2.714	1.86	1,859.06
20	27	15	1.2	1	50	0.2	0.15	2.5	63	3,260	31.841	0.88	879.21
21	27	15	1.2	1	60	0.25	0.05	3	112	7,660	35.508	0.64	637.30
22	27	17	0.8	1.2	40	0.3	0.1	2.5	63	3,260	42.142	0.88	879.21
23	27	17	0.8	1.2	50	0.2	0.15	3	112	7,660	25.065	0.64	637.30
24	27	17	0.8	1.2	60	0.25	0.05	2	310	5,606	1.995	1.86	1,859.06
25	27	20	1	0.8	40	0.3	0.1	3	112	7,660	31.957	0.64	637.30
26	27	20	1	0.8	50	0.2	0.15	2	310	5,606	1.357	1.86	1,859.06
27	27	20	1	0.8	60	0.25	0.05	2.5	63	3,260	29.851	0.88	879.21

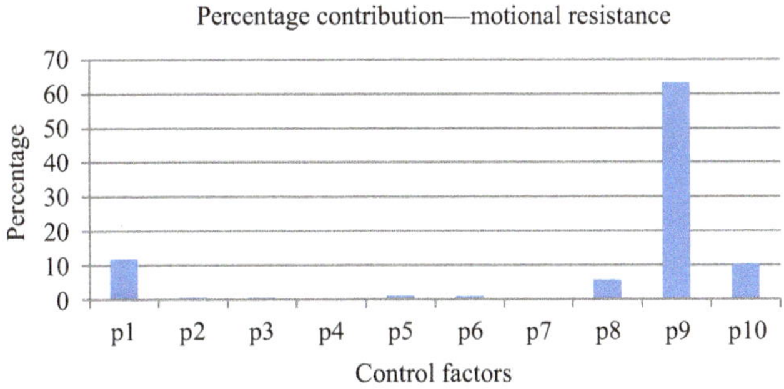

Figure 4.39 Percentage contribution Rm [75]

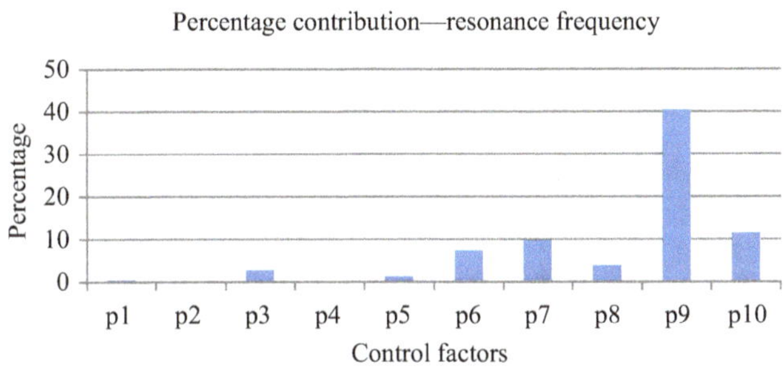

Figure 4.40 Percentage contribution fr [75]

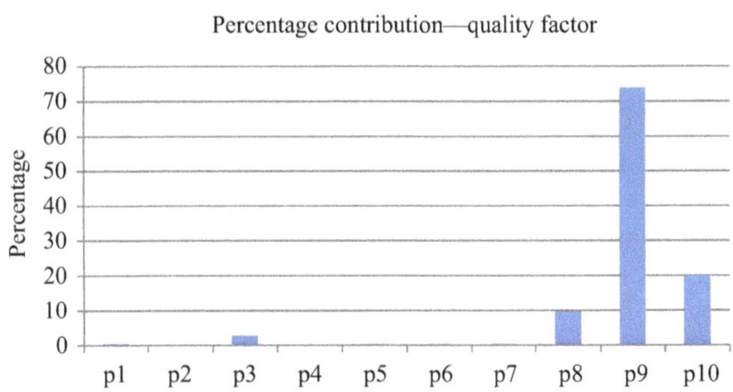

Figure 4.41 Percentage contribution Q [75]

then *Rm* increases whereas *fr* and *Q* are decreased and if the width of the sub-resonator is increased, *Rm* and *fr* are decreased and negligible change in *Q* occurs. As a number of IDT fingers are more, then the value of the motional resistance can be obtained less and we can achieve more value of the resonance frequency with high *Q*.

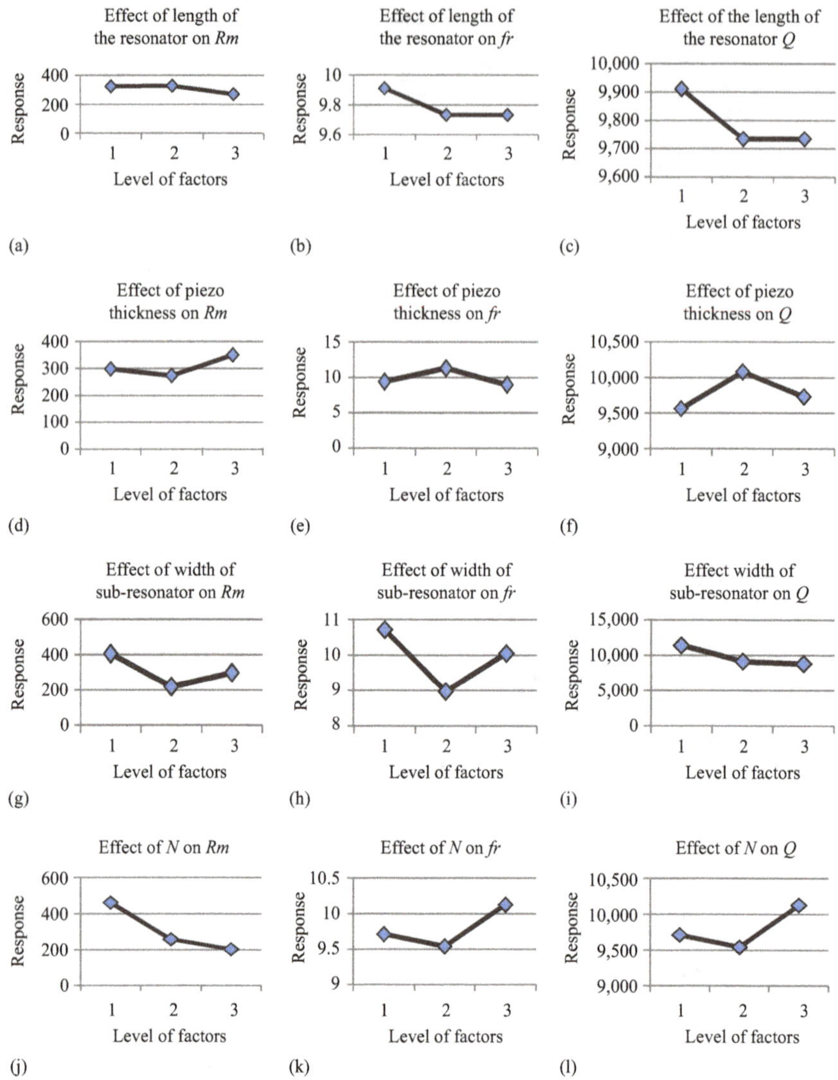

Figure 4.42 Effect of (a) length of resonator on Rm, (b) length of resonator on fr, (c) length of resonator on Q, (d) piezo thickness on Rm, (e) piezo thickness on fr, (f) piezo thickness on Q, (g) width of subresonator on Rm, (h) width of subresonator on fr, (i) width of subresonator on Q, (j) N on Rm, (k) N on fr, and (l) N on Q [75]

By considering all the responses and the effects of each and every control parameter, the final dimensions of the design were adjusted and finalized. With these optimized values of the parameters (Table 4.9), the simulations have been performed using COMSOL[TM] Multiphysics before proceedings toward fabrication.

4.17 Summary

This chapter introduced the fundamentals of MEMS resonators and their transduction mechanisms. We discussed surface and bulk acoustic wave propagations and examined some of the types of acoustic waves, including Rayleigh, longitudinal, and shear-mode waves. We paid attention to SAW and BAW devices as well. The device design and modeling have also been addressed along with the details of one port contour mode MEMS resonator. The parameter optimization through Taguchi analysis has also mentioned.

Acknowledgements

We acknowledge Indian Nanoelectronics User Program (INUP), IITB, Mumbai, India and Center for VLSI & Nanotechnology, VNIT, Nagpur, India. We are thankful to Dr. S. Balpande, Dr. D. Khushalani, and Ms. P. Talekar of RCOEM, Nagpur, India.

References

[1] Van Caekenberghe K. "Modeling RF MEMS devices." *IEEE Microwave Magazine*. 2012. DOI: 10.1109/MMM.2011.2173984.

[2] Azarnaminy S.F., "Modeling, design and fabrication of MEMS filters for RF applications." A thesis for the degree of Master of Applied Science at Concordia University, Montreal, Quebec, Canada, 2005.

[3] Basu J., and Bhattacharya T.K. "Microelectromechanical resonators for radio frequency communication applications." *Microsystem Technologies*. 2011; 17(10–11):1557–1580. DOI: 10.1007/s00542-011-1332-9

[4] Rinaldi M., Zuniga C., Zuo C., and Piazza G. "Super-high-frequency two-port AlN contour-mode resonators for RF applications." *IEEE Transactions on Ultrasonics, ferroelectrics, and Frequency Control*. 2010;57(1).

[5] Hamzah M.H., Karim J., Ralib A.A.M., and Nordin A.N. "Design and analysis of a boosted pierce oscillator using MEMS SAW resonators." *Microsystem Technologies*. 2017. https://doi.org/10.1007/s00542-017-3627-y

[6] Karim J., and Nordin A.N. "Implementation of CMOS oscillator for CMOS SAW resonator." *Symposium on Design, Test, Integration, and Packaging of MEMS and MOEMS*. 2016;7–11.

[7] Mengwei L., Junhong L., Jun M., and Chenghao W. "Design and fabrication of a MEMS Lamb wave device based on ZnO thin film." *Journal of Semiconductors*. 2011;32(4). DOI: 10.1088/1674-4926/32/4/044006

[8] Bhatia D., Sharma H., Meena R.S., and Palkar V.R. "A novel ZnO piezoelectric microcantilever energy scavenger: Fabrication and characterization." *Sensing and Bio-Sensing Research*. 2016. http://dx.doi.org/10.1016/j.sbsr.2016.05.008 2214-1804/© 2016

[9] Balpande S.S., Pande R.S., and Patrikar R.M. "Design and low cost fabrication of green vibration energy harvester." *Sensors and Actuators A: Physical*. 2016; 251(1): 134–41.

[10] Singh S, and Chakrabarti P. "Simulation fabrication and characterization of ZnO based thin film transistors grown by radio frequency magnetron sputtering." *Journal of Nanoscience and Nanotechnology*. 2012;12: 1880–85.

[11] Zhang Y., Nayak T.R., Hong H., and Cai W. "Biomedical applications of zinc oxide nanomaterials." *Current Molecular Medicine*. 2013; 13(10): 1633–45.

[12] Kołodziejczak-Radzimska A, and Jesionowski T. "Zinc oxide-from synthesis to application: A review." *Materials*. 2014;7: 2833–2881; DOI: 10.3390/ ma7042833

[13] Yao J.J. "RF-MEMS from a device perspective," *Journal of Micromechanics and Microengineering*. 2000; R9–R38.

[14] https://compliantmechanisms.byu.edu/content/introduction-microelectro-mechanical-systems-mems

[15] MEMS and Nanotechnology Exchange. http://www.memsnet.org

[16] Abdolvand R., Behreyni B., Lee J.E.-Y., and Nobki F. "Micromachined resonators: A review." *MDPI Micromachines*. 2016;7: 160. DOI: 10.3390/ mi7090160

[17] Sutagundar M, Sheeparamatti B.G., and Jangamshetti D.S. "Research issues in MEMS resonators." *International Journal of Engineering and Science*. 2014; 4(8): 29–39.

[18] Hung L.-W., and Nguyen C. "High-Q low-impedance MEMS resonators." Electrical Engineering, and Computer Sciences. University of California at Berkeley Technical Report No. UCB/EECS-2012-218. 2012. http://www.eecs.berkeley.edu/Pubs/TechRpts/2012/EECS-2012-218.html.

[19] Lin Y.W., Lee S., Li S.S., Xie Y., Ren Z., and Nguyen C.T.C. "Series-resonant VHF micromechanical resonator reference oscillators." *IEEE Journal of Solid-State Circuits*. 2004;39(12): 2477–91.

[20] Nguyen C.T.C. "MEMS technology for timing and frequency control." *IEEE Transaction on Ultrasonics Ferroelectrics Frequency Control*. 2007; 54: 251–70.

[21] Yole emerging MEMS: Technologies and markets. Report. Yole Development. France, 2010.

[22] Najla K. "Modeling and characterization of RF MEMS resonators." Thesis. The University of Texas at Arlington; 2009.

[23] Kim H.C., and Chun K. "RF MEMS technology." *IEEJ Transactions*. 2007; 2: 249–61.

[24] Nathanson H.C., Newell W.E., Wickstrom R.A., and Davis J.R. "The resonant gate transistor." *IEEE Transactions on Electron Devices*. 1967; 14(3): 117–33.

[25] Taylor J.T., and Huang Q. *CRC Handbook of Electrical Filters*. Boca Raton, FL: CRC Press; 1997.

[26] Chandorkar S.A., Agrawal M., Melamud R., Candler R.N., Goodson K.E., and Kenny T.W. "Limits of quality factor in bulk mode micromechanical resonators." *Proceedings of the 21st IEEE International Conference on Micro Electro Mechanical Systems*. Tucson, Arizona. 2008; 74–77.

[27] Tilmans A.C.H. "Equivalent circuit representation of electromechanical transducers: I. Lamped-parameter systems." *Journal of Micromechanics and Microengineering*. 1996;6: 157–76.

[28] Campanella H. "Acoustic Wave and Electromechanical Resonators: Concepts to Key Applications". *Artech House*. Norwood, MA 02062, ISBN-13: 978-1-60783-977-4

[29] Tang W.C., Nguyen T.-C., and Howe R.T. "Laterally driven polysilicon resonant microstructures." In *Technical Digest, IEEE Micro Electro Mechanical Systems Workshop*, February 1989, 53–59.

[30] Tang W.C., Nguyen C.T.-C., Judy M.W., and Howe R.T. "Electrostatic-comb drive of lateral polysilicon resonators." *Sensors and Actuators*. 1990; A21–23:328–31.

[31] Hsu W.T., and Nguyen C.T.C. "Stiffness-compensated temperature-insensitive micromechanical resonators." In *Proceedings of the 15th IEEE International Conference on Micro Electro Mechanical Systems*, Las Vegas, Nevada, 2002, 731–34.

[32] Yang Y.T., Ekinci K.L., and Huang X.M.H. "Monocrystalline silicon carbide nano electromechanical systems." *Applied Physics Letters*. 2001;78: 162–64.

[33] Quevy E.P., Bhave S.A., Takeuchi H., King T.J., and How R.T. "Poly-Si high- frequency resonator based on lithographic definition of nano-gap lateral transducers." In: *Proceeding of Solid-State Sensor, Actuator, and Microsystems Workshop*, Hilton Head Island, SC, 2004.

[34] Roukes M.L. "Nanoelectromechanical systems." In: *Proceeding of Transducer*, 2001, 658–61.

[35] Wang K., Yu Y., Wong A.-C., and Nguyen C.T.-C. "VHF free-free beam high Q micromechanical resonators." In: *Proceedings 12th International IEEE Micro Electro-Mechanical Systems Conference*, 1999, 453–58.

[36] Xie Y., Li S.S., Lin Y.W., Ren Z., and Nguyen C.T.-C. "UHF micromechanical extensional wine-glass mode ring resonators." In: *Proceedings of IEEE the 16th Annual International Conference on Micro Electro Mechanical Systems*, MEMS2003, 2003, 698–701.

[37] No S.Y., and Ayazi F. "The HARPSS process for fabrication of nano-precision silicon electromechanical resonators." In: *Proceedings of IEEE 2001 Conference on Nanotechnology* (IEEE NANO01), 2001, 489–94.

[38] Kruse P.W. *Elements of Infrared Technology Generation, Transmission, and Detection.* New York: Willy; 1963.

[39] Weinstein D., and Bhave S.A. "Internal dielectric transduction of a 4.5 GHz silicon bar resonator." *IEEE International Electron Devices Meeting*. 2007, 415–18.

[40] Piazza G., Stephanou P.J., and Pisano A.P. "Piezoelectric aluminum nitride vibrating contour-mode MEMS resonator." *Journal of Microelectromechanical Systems*. 2006; 15(6): 1406–18.

[41] Abdolvand R., Ho G.K., Butler J., and Ayazi F. "ZnO-on-nano-crystalline-diamond lateral bulk acoustic resonators." In: *Proceedings of the 20th IEEE International Conference on Micro Electro Mechanical Systems*, Kobe, Japan, 2007, 795–98.

[42] Patni R., Joshi M., Mehta S., and Mohan A. "Design of piezoelectric aluminum nitride MEMS resonator." *Proceedings of the World Congress on Engineering and Computer Science*, 2011, Vol. 1, WCECS 2011, San Francisco, USA.

[43] Piazza G. "MEMS resonators for frequency control and sensing applications." Penn Micro and Nano Systems Laboratory (PmaNS Lab), Department of Electrical and Systems Engineering, University of Pennsylvania Philadelphia, PA, USA, http://pmans.ese.upenn.edu/

[44] Zou J. "High-quality factor lamb wave resonators." M.S. Degree Thesis in Engineering – Electrical Engineering and Computer Sciences, University of California, Berkeley, 2014.

[45] Frederick A.A. "Analysis and fabrication of MEMS tunable piezoelectric resonators." MS Thesis, University of Pittsburgh, 2006.

[46] Yen T.-T. Experimental study of fine frequency selection techniques for piezoelectric aluminium nitride lamb wave resonators. MS Thesis. Department of Electrical Engg. & Computer sciences, University of California at Burkeley, 2003.

[47] Benech P., and Duchamp J.-M. "Piezoelectric materials in RF applications." http://dx.doi.org/10.5772/63125. Book chapter 9-201-227.

[48] Besson R.J. "A new 'electrodeless' resonator design." In: *31st Annual Symposium on Frequency Control*, 1977, 147–52.

[49] Hung L.-W., and Nguyen C.T.-C. "Capacitive-piezo transducers for higher Q contour mode AlN resonators at 1.2 GHz." In: *Proceedings of the 2010 Solid-State Sensor, Actuator, and Microsystems Workshop*, Hilton Head Island, SC, 2010.

[50] Hung L.-W., and Nguyen C.T.-C. "Capacitive-piezoelectric transducers for high- Q micromechanical AlN resonators." *Journal of Microelectromechanical Systems*. 2015; 24(2): 458–73.

[51] Yen T.-T., Pisano A.P., and Nguyen C.T.-C. "High-Q capacitive-piezoelectric AlN Lamb wave resonators." In: *2013 IEEE 26th International Conference on Micro Electro Mechanical Systems (MEMS)*. IEEE, 2013; 114–17. http://ieeexplore.ieee.org/xpls/abs_all.jsp?arnumber =6474190

[52] Rinaldi M., Zuniga C., Zuo C., and Piazza G. "AlN contour-mode resonators for narrow-band filters above 3 GHz." Department of Electrical and Systems Engineering, University of Pennsylvania, Philadelphia, 2009

[53] Gryba T., Carlier J., Wang S., Zhao X.Z., Guo S., and Lefebvre J.-E. "One port contour-mode ZnO piezoelectric MEMS resonator." *Microelectronic Engineering*. 2011; 88: 3003–10.

[54] Cremonesi M., Frangi A., Cassella C., and Piazza G. "Enhancement of the quality factor of AlN contour mode resonators by acoustic reflection: Numerical design and experimental investigation." Eurosensors 2014, the XXVIII Edition of the Conference Series. *Procedia Engineering*. 2014; 87: 468–71.

[55] Isobe A., Kengo Asai (Hitachi Ltd., Central Research Laboratory). "Contour-mode AlN resonator with high coupling factor." *Proceedings of Symposium on Ultrasonic Electronics*. 2009;30: 271–72.

[56] Yunhong H., Meng Z., Guowei H., Chaowei S., Yongmei Z., and Jin N. "A review: Aluminum nitride MEMS contour-mode resonator." *Journal of Semiconductors*. 2016; 37(10).

[57] Ossama M., Pierre B., Aurelian C., Matthieu C., and Orlianges J.-C. "A zinc dioxide-on-silicon MEMS resonator for narrowband filtering." *21st IEEE International Conference on Electronics, Circuits, and Systems (ICECS)*, 2014.

[58] Ho G.K., Abdolvand R., Sivapurapu A., Humad S., and Ayazi F. "Piezoelectric- on-silicon lateral bulk acoustic wave micromechanical resonators." *Journal of Microelectromechanical Systems*. 2008;17(2): 512–20.

[59] Pfusterschmied G., Toledo J., Kucera M., *et al*. "Potential of piezoelectric MEMS resonators for grape must fermentation monitoring." *Micromachines*. 2017; 8.

[60] Yu Y., Luo H., Chen B., Tao J., Feng Z., Zhang H., Guo W., and Zhang D. "MEMS gyroscopes based on acoustic Sagnac effect." *Micromachines*. 2017; 8.

[61] Schmid U., and Schncider M. "Editorial for the special issue on piezoelectric MEMS." *Micromachines*. 2018; 9:237. DOI: 10.3398/mi9050237.

[62] Mortada O., Zahr A., Orlianges J.-C., Crunteanu A., and Chatras M. "Analysis and optimization of acoustic wave micro-resonators integrating piezoelectric zinc oxide layers." *Journal of Applied Physics, American Institute of Physics*. 2017;121(7).

[63] Ghosh S., and Piazza G. "Elasto-optic modulator integrated in high-frequency piezoelectric MEMS resonator." *IEEE 29th International Conference on Micro Electro Mechanical Systems (MEMS)*. 2016. DOI: 10.1109/MEMSYS.2016.7421544

[64] Zadehsafarii M., Zolfagharii P., Akbari A., and Ghavifekri H.B. "A Piezoelectric lamb-wave disk-shape MEMS resonator for RF applications." *4th International Conference on Electrical and Electronic Engineering (ICEEE)*. 2017. DOI: 10.1109/ICEEE2.2017.7935787

[65] Deshpande P.P., Talekar P.M., Khushalani D.G., and Pande R.S. "Piezoelectric aluminum nitride micro electromechanical system resonator for RF application." *International Journal of Applied Engineering Research*. 2018;13(6): 4263–67.

[66] Deshpande P.P., Pande R.S., and Patrikar R.M. "Fabrication and characterization of zinc oxide piezoelectric MEMS resonator." *Microsystem Technologies*. 2019. https://doi.org/10.1007/s00542-019-04509-w

[67] Yen T.-T. "High-Q aluminum nitride RF MEMS lamb wave resonators and narrowband filters." Ph.D. Thesis, University of California, Berkeley, 2012.

[68] Collaboration. "IEEE standard on piezoelectricity." Institute of Electrical and Electronics Engineers, New York, NY, 1988. IEE ANSI/IEEE Std 176-1987.

[69] Yan L. "Piezoelectrically transduced low impedance MEMS resonators." Ph.D. Thesis, University of California, Irvine, 2005.

[70] Khushalani D.G., Dubey V.R., Bheley P.P., Kalambe J.P., Pande R.S., and Patrikar R.M. "Design optimization and fabrication of microcantilever for switching application." *Sensors and Actuators A: Physics. 2015;*225(2015); 1–7.

[71] Mahamuni N.N., and Adewuyi Y.G. "Application of Taguchi method to investigate the effects of process parameters on the transesterification of soybean oil using high-frequency ultrasound." Chemical Engineering Department, North Carolina Agricultural and Technical State University, Greensboro, North Carolina, Energy Fuels, 2010.

[72] Hwang I.-H., and Lee J.-H. "Self-actuating biosensor using a piezoelectric cantilever and its optimization." *Journal of Physics*, International MEMS Conference 2006.

[73] Ross, P. *Taguchi Techniques for Quality Engineering*. New York: McGraw-Hill; 1988.

[74] Kshirsagar A., Apte P., and Dattagupta S.P., "Optimization of pull-in voltage and contact force for MEMS series switch using Taguchi method." *ICSE Proceedings*, Melaka, Malaysia, 2010.

[75] Deshpande P.P., Talekar P.M., Khushalani D.G., and Pande R.S. "Design optimization and fabrication issues of MEMS resonator." *International Journal of Technical Innovation in Modern Engineering & Science*. 2018; 4(5): 1372–79.

[76] Roy R.K. *A Primer on the Taguchi Method*. Society of manufacturing Engineers; 2010.

Chapter 5

Finite-element modeling of RF MEMS resonators

Ravi Solanki[1], Sakthi Swarrup J[2] and Ashutosh Mahajan[2]

Radio frequency microelectromechanical system (RF MEMS) devices are potential candidates for replacing the off-chip complementary metal-oxide-semiconductor (CMOS)-based reference oscillators, filters, and switches working at high frequency in advanced communication systems [1,2]. The piezoelectric MEMS resonators has been introduced for Ka band (26.5–40 GHz) that is now opened for 5G communication systems [3].

The present resonator technologies broadly classified into electrostatically actuated resonators and piezoelectric material-based resonators. Under electrostatically actuated resonators, different structures such as beam, ring, and disk are made to vibrate using electric field generated between the electrode present on the resonating structure and the fixed electrode. While, piezoelectric material-based resonators have a piezoelectric material layer sandwiched between two electrodes. This category includes surface acoustic wave (SAW) devices which are based on traveling acoustic waves along the resonator surface, thin film bulk acoustic resonators (FBAR) in which vibration along thickness is utilized, and contour-mode resonators in which length or width extension mode is excited for resonator operation. Additionally, the contour mode resonators can be designed for either lateral excitation or transverse excitation of the modes.

The key requirements for the ideal resonator is high Q-factor, desired resonance frequency, and low motional resistance, so that interfacing with the 50 ohm electronics can be achieved. Based on the application and key requirements, a suitable configuration (electrostatic or piezo-based), resonator geometry (disk, beam, plate, etc.) and resonator materials, needs to be correctly chosen. This process of designing and optimizing every single unit of resonator demands accurate physical modeling and simulation strategies.

Modeling of the resonator is the first step where the behavior of the RF MEMS resonator can be known before the fabrication and make the optimum design possible. Mainly, there are two types of modeling approaches used for RF MEMS resonators: mechanical model of resonator structure along with its electrical

[1]Centre for VLSI and Nanotechnology, Visvesvaraya National Institute of Technology, Nagpur, India
[2]Centre for Nanotechnology Research, Vellore Institute of Technology, Vellore, India

equivalent model and the computational approach in which the governing physical equations are solved using the suitable numerical technique. Since RF MEMS resonators can have arbitrary shapes, finite element method (FEM) is best suited for determining the static and dynamic response of the resonating structure.

In this chapter, the quick review of different structures, modes, and excitation mechanisms is given in the beginning. The mechanical model for the RF MEMS resonator is discussed and ways to extract the important device parameters are given along with the electrical equivalent model. The importance of the physical modeling of a MEMS device is established and various governing partial differential equations (PDEs) are discussed. A brief introduction to the FEM for solving PDEs is given to give the reader a glimpse of the FEM machinery running behind the screens of commercial simulation tools. We discuss a finite element assembly for the Poisson's equation and its solution for demonstration purpose followed by the details of a few commercial tools such as CoventorWare, Intellisuite, and COMSOL Multiphysics with specific examples.

5.1 Classification of RF MEMS resonators

The RF MEMS resonators can be classified based on various parameters such as device structure, the shape of the resonating structure, modes of vibration, actuation and coupling mechanisms.

5.1.1 Structure

RF resonator structure mainly consists of a substrate which gives mechanical strength to the structure, resonating structure i.e. beam, and the support structure for the resonating beam i.e. anchor, as seen in Figure 5.1. Classification based on the boundary condition at the ends of the resonating structures are as follows:

1. Clamped-clamped beam: In this configuration both the ends of the beam are clamped and results in zero displacement at the end. The example of this configuration is a guitar string.
2. Clamped-free beam: The beam in which one end is free to move while the other is clamped can be referred as clamped-free beam. This configuration is also known as cantilever.

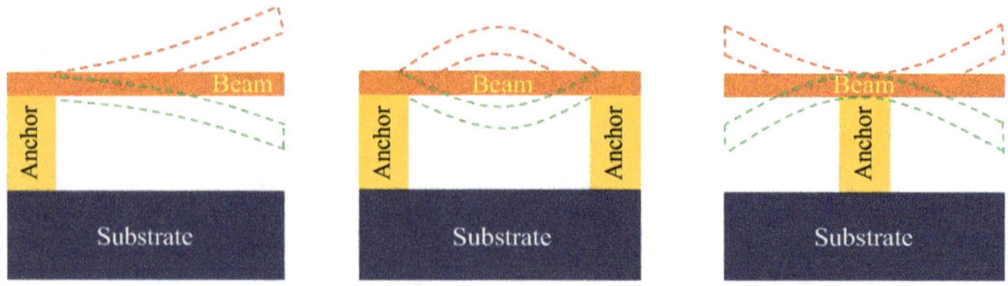

Figure 5.1 (a) Clamped-free, (b) clamped-clamped, and (c) free-free structure

3. Free-free beam: In this configuration, the resonating beam rests on the support beam such that both its ends are free to move, and the zero displacement comes at the point of support.

5.1.2 Shapes

The shape of the resonating structure is not limited to the beam but other structures such as the square plate, circular disk, ring shape, lateral comb, and membrane also acts as a resonating structure. The boundary conditions mentioned above can also be seen for these structures as well.

5.1.3 Vibration modes

The resonating structure can vibrate in different modes given as flexural, torsional, and bulk. The modes which will be excited depends on boundary conditions, anchor locations, and resonators shape and size. Each mode will vibrate with specific resonance frequency.

5.1.4 Actuation mechanisms

The RF MEMS resonator consists of three components: input transducer which converts input signals such as electrostatic, piezoelectric, or magnetostatic into mechanical signals; mechanical resonant structure which can vibrate in one or more modes; and the output transducer which converts the mechanical signal back to electrical [4]. For RF operation, piezoelectrically actuated MEMS resonators are more suitable compared to electrostatically actuated structures because of their low motional resistance.

5.1.5 Coupling mechanisms

Coupling between individual mechanical resonant structures can also be done with mechanical or electrostatic coupling or the combination of both schemes.

The classification establishes the scope of modeling required for RF MEMS resonators. The modeling and simulation approach should have the capability to address any possible configuration and structure mentioned above.

5.2 Modeling of RF MEMS resonators

Modeling involves the process of obtaining the mathematical equations whose solutions accurately represents the device working. Such a model can be used to predict the response of the device under any situation without actually performing the experimental measurements.

It is evident that modeling can significantly reduce device development cost and time since there is no need to perform a real experiment during device design and optimization. With the increasing computational power available in the present computers, more accurate models can be used and more refined simulation studies can be performed using advanced TCAD tools.

The challenge in RF MEMS modeling is the involvement of the multiple physical domains during its working. Also, as mentioned in the classification, the resonant frequency is highly shape- and size-dependent.

Various studies that can be performed using the modeling are:

1. Modal analysis: Modal analysis is carried out to identify the natural or eigen-frequency of the resonator under no excitation or actuation.
2. Static analysis: This analysis is performed by applying static actuating condition, that is, applying DC voltage on the RF MEMS resonator structure.
3. Dynamic analysis: The change of resonance frequency with the change of excitation that is, by applying time-varying voltages is studied under dynamic analysis. This analysis is also termed as harmonic analysis or frequency response analysis of the resonant structure in which the response of the resonator is analyzed for a wide frequency band.

Modeling can be done mainly in two ways: mechanical along with its electrical equivalent modeling and physical modeling using finite element analysis.

5.2.1 Mechanical model

The RF MEMS resonator is a mechanical oscillator that can store energy, which continuously exchanges from potential to kinetic form. However, this conversion is not perfect and there is some loss or energy dissipation per conversion cycle. The simplest model for such a device can be a system of spring-mass-damper as seen in Figure 5.2 in which spring-mass component stores energy through continuous conversion from potential to kinetic energy, and the damper component represents all the loss mechanisms.

The equation of motion for the spring-mass-damper system can be written using Newton's law of motion as

$$M_{\mathrm{eff}}\frac{\partial^2 x}{\partial t^2}+\zeta_{\mathrm{eff}}\frac{\partial x}{\partial t}+K_{\mathrm{eff}}x=F_{\mathrm{in}} \tag{5.1}$$

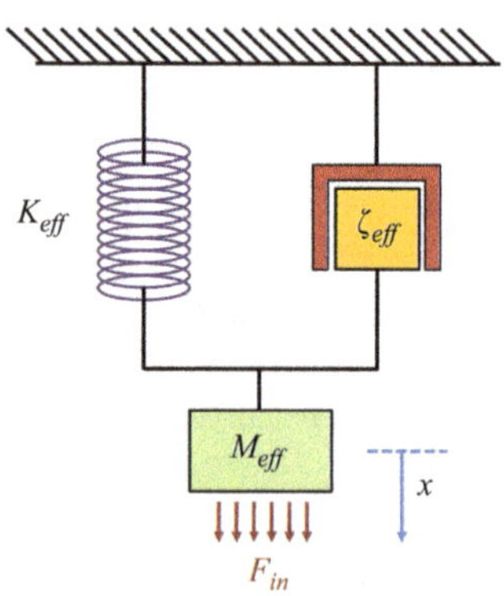

Figure 5.2 Schematic of spring-mass-damper system as a model for RF MEMS resonator

where M_{eff} is the effective mass for fundamental frequency f_0 which differs from the physical mass of the structure, ζ_{eff} is the effective damping factor which represents all the energy loss mechanisms, K_{eff} is the effective spring constant which signifies the restoration force, and F_{in} is an applied force [5]. All these effective parameters are frequency- and time-independent. The two main parameters required for RF MEMS resonator analysis are the natural frequency and quality factor or Q-factor.

The calculation of undamped natural frequency which states as modal analysis of the structure is given as

$$f_0 = \frac{1}{2\pi}\sqrt{\frac{K_{\text{eff}}}{M_{\text{eff}}}} \tag{5.2}$$

by setting input excitation F_{in} and ζ_{eff} in equation (5.1) to be zero.

For the static analysis, constant force $F_{\text{in}} = \hat{F}$ is applied and dynamic analysis is performed by applying time-varying force $F_{\text{in}} = \hat{F}\cos\omega t$.

The Q-factor by definition is the ratio of maximum energy stored to the energy dissipated in one cycle. The relation of damping factor ζ_{eff} and Q-factor is given as

$$\zeta_{\text{eff}} = \frac{2\pi f_0 M_{\text{eff}}}{Q} \tag{5.3}$$

The resonator frequency in the presence of damping is

$$f_r = f_0\sqrt{1 - \frac{1}{2Q^2}} \tag{5.4}$$

The spring-mass-damper model being simple, however, can only be applied for analyzing the system with single resonating frequency, or multiple frequencies without any coupling between the two frequencies. Moreover, the determination of the effective mass, spring constant, and damping coefficient for complex geometry is difficult to calculate in this model.

5.2.2 Electrical equivalent model

The RF MEMS resonator device consists of the mechanical resonating structure, excitation circuit for actuating it, and sensing circuit for taking output. The performance analysis of the final device requires modeling of the resonating structure along with its electrical circuits. As we have already seen, the spring-mass-damper system models the mechanical resonating structure; however, since the actuating and sensing circuits are in the electrical domain, it would be advantageous to use the electrical equivalent model for the resonating structure as well. The electrical equivalent for the resonating structure is given as series *R-L-C* circuit. The strength of the coupling between resonator structure with the input and output transduction mechanisms is given by η that will scale the R, L, and C components of the equivalent electrical circuit. The relation between mechanical domain quantities and the electrical domain considering transduction efficiency is given in Table 5.1 [6] and the equivalent electrical model for the complete device is given in Figure 5.3.

Table 5.1 The relation between mechanical and electrical domain parameters [6]

Mechanical domain	Electrical domain
Force, F_{in}/η	Voltage, V_{in}
Velocity, $\eta\dot{x}$	Current, I
Displacement, ηx	Charge, q
Compliance, η^2/K_{eff}	Capacitance, C
Mass, M_{eff}/η^2	Inductance, L
Damping, $\zeta_{\text{eff}}/\eta^2$	Resistance, R

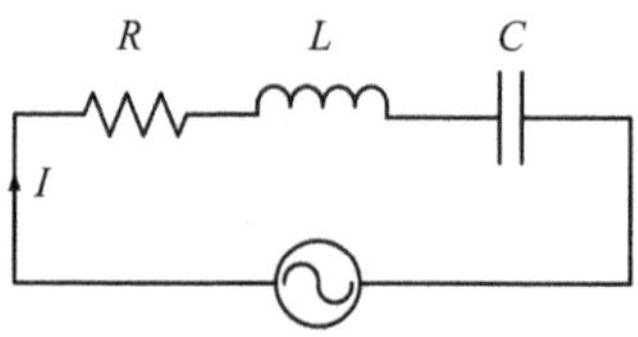

Figure 5.3 The electrical equivalent model of the RF MEMS resonator

A series electrical resonance circuit equation is obtained after putting the electrical parameters in equation of motion for spring-mass-damper system as

$$L\frac{d^2q}{dt^2} + R\frac{dq}{dt} + \frac{q}{C} = u \cdot \cos\,\omega t \tag{5.5}$$

The static, modal, and dynamic analysis of the complete RF MEMS resonator device along with transduction circuit can be performed by solving the equation 5.5.

5.2.3 Numerical simulation

The application of mechanical modeling and its electrical equivalent is limited due to the following reasons: the analysis of the single resonant mode of resonating structure, the difficulty in the calculation of effective parameters such as mass, damping coefficient, and spring constant for any arbitrary geometry, and missing the physical modeling of the transduction mechanism.

Analytical modeling of the complex micromachined system can be carried out by simplifying assumptions and is not fully accurate. Moreover, analytical modeling is not feasible for the MEMS resonators due to the multiphysics nature of the system of equations describing its behavior. Under this situation, a full numerical scheme is needed to check the validity of the analytical model and obtain the system performance accurately. A detailed computation simulation is essential when the complexities of the micromachined system cannot be analytically simplified.

The popular technique for simulation of microsystems of any arbitrary geometrical shape is FEM. The FEM is a numerical technique to solve a partial differential equation (PDE) with known boundary conditions. It allows solving PDE which governs the physical mechanism of the RF MEMS resonators for any geometry and material. This gives scope to make a generic software tool for simulating and optimizing RF MEMS devices. The details of the governing equations related to the important physical domains concern with RF MEMS resonators are discussed in the section 5.3 followed by the details of FEM implementation for solving PDEs.

5.3 Governing PDEs

The physical laws of any domain like mechanical, electrical, thermal, fluid, etc., and various conservation laws like mass, energy, momentum, etc. are governed by a PDE along with its boundary conditions. For any dynamical system for example MEMS, different physical domains coexist or coupled during its working. The dynamics of that complete system can then be modeled by the set of PDEs related to the involved physical domains and is referred to as the system's governing equations. In that case, the PDEs of the governing equations need to be solved simultaneously to understand the behavior of the dynamical system. Such type of analysis is called a coupled-domain analysis. Various physical domains involved for RF MEMS resonators are shown in Figure 5.4 with the corresponding PDEs which are discussed below.

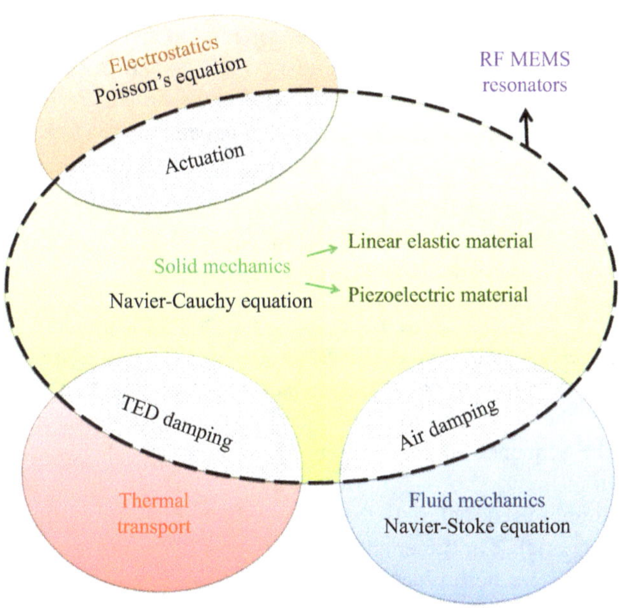

Figure 5.4 Multi physical-domains for RF MEMS resonator with corresponding equations

5.3.1 Beam mechanics

The most common structure for RF MEMS resonators is a slender beam. The slender beam is a structure for which beam length (L) is much larger than the cross-section dimensions. The slenderness (s) of the beam is defined as $s = L\sqrt{(A/I)}$, where (A) and (I) is the constant cross-section area along beam axis and its second moment, respectively. Bending of the slender beam can be calculated by solving the one-dimensional (1D) beam mechanics equation. For $s > 100$, Euler–Bernoulli beam theory can be applied [7].

The equation of transverse displacement (u) of the beam under Euler–Bernoulli equation is given as [8]

$$\frac{\partial^2}{\partial x^2}\left(EI\frac{\partial^2 u}{\partial x^2}\right) + \rho A\frac{\partial^2 u}{\partial t^2} = f(x,t) \tag{5.6}$$

where E is the modulus of elasticity, I is the moment of inertia, ρ is the mass density, and f is the transverse loading force, which can be obtained from electrostatic force or stationary mass on the tip.

The underlining assumption with the theory is that during deflection the cross-section should always be normal to the deflection axis. For the thick beams as well as a slender beam where the transverse shear deformation dominates, Timoshenko beam theory can be applied.

5.3.2 Solid mechanics

The branch of physics that deals with the deformation of any three-dimensional (3D) structure under the action of any force is solid mechanics or structural mechanics. These forces can be applied intentionally for the actuation of the structure, which is commonly electrostatic. For structural analysis, three types of equations need to be solved:

1. Equilibrium equations relating stress σ to the applied force,
2. Kinematic equations relating strains ε to displacement gradients, and
3. Constitutive relationship or equation of state relating stress to strain. This relationship describes the behavior of the material.

In the above equations, the stress and strain are denoted in the form of tensors and can be defined as matrix σ_{ij} and ε_{ij}. The indices j denote direction of applied force and i denote the direction of normal to the surface. The depiction of various stress components on the application of body force in x and z directions is shown in Figure 5.5. The sequence of these three equations to get displacement from the force is shown in Figure 5.6.

The commonly used materials for RF MEMS resonators are linear elastic and piezoelectric material. The kinematic equation, which gives strain–displacement vector relationship, for both type of material is

$$\varepsilon_{ij} = \frac{1}{2}\left(\frac{\partial u_i}{\partial x_j} + \frac{\partial u_j}{\partial x_i}\right) \tag{5.7}$$

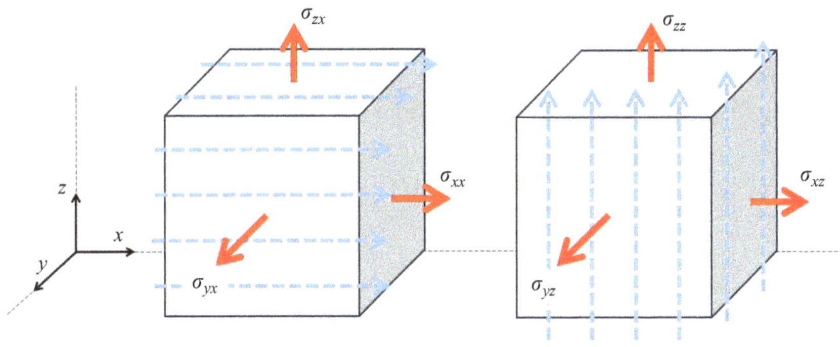

Figure 5.5 Representation of various stress components on the application of body force represented by blue arrows in x and z directions

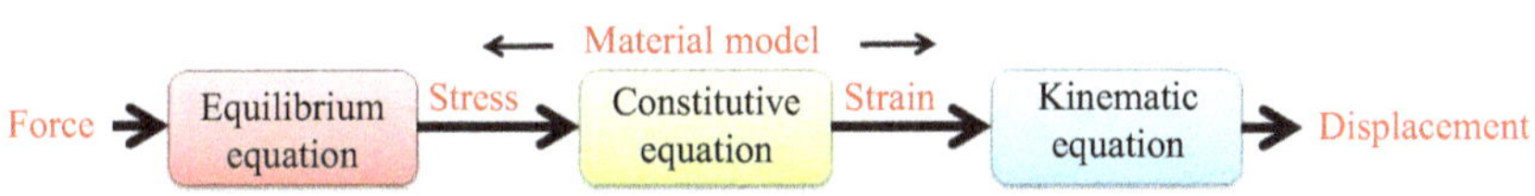

Figure 5.6 Relation between force and displacement through the three equations

The equation of equilibrium for the materials derived from Newton's law is given as

$$\nabla \cdot \sigma + f = \rho \frac{\partial^2 u}{\partial t^2}, \tag{5.8}$$

where f denotes body force per unit volume and ρ as mass density of the solid. The body force f, for example can comes from electrostatic force during electrostatic actuation or temperature-induced force during thermoelastic damping analysis.

The constitutional relation which is material dependent is defined individually for linear elastic and piezoelectric material.

Linear elastic material

The constitutive relationship for linear elastic materials, for example, metals (isotropic) or semiconductors (anisotropic), is generalized Hooke's law. For isotropic material, stress–strain relationship is given as

$$\sigma_{xx} = \frac{E}{(1+\nu)(1-2\nu)}[(1-\nu)\varepsilon_{xx} + \nu(\varepsilon_{yy} + \varepsilon_{zz})] \quad \sigma_{xy} = \frac{E}{(1+\nu)}\varepsilon_{xy} \tag{5.9}$$

$$\sigma_{yy} = \frac{E}{(1+\nu)(1-2\nu)}[(1-\nu)\varepsilon_{yy} + \nu(\varepsilon_{xx} + \varepsilon_{zz})] \quad \sigma_{yz} = \frac{E}{(1+\nu)}\varepsilon_{yz} \tag{5.10}$$

$$\sigma_{zz} = \frac{E}{(1+\nu)(1-2\nu)}[(1-\nu)\varepsilon_{zz} + \nu(\varepsilon_{xx} + \varepsilon_{yy})] \quad \sigma_{zx} = \frac{E}{(1+\nu)}\varepsilon_{zx} \tag{5.11}$$

where E and ν are the material constant termed as Young's modulus and Poisson ratio. Stress and strain are second-order symmetric tensors. The relationship in the matrix form can be written as $\sigma = D\varepsilon$, where D is the 6×6 elasticity matrix.

Using kinematic and constitutive relations, the equation of equilibrium can be written as

$$\frac{E}{2(1+\nu)}\left(\frac{1}{(1-2\nu)}\nabla(\nabla \cdot u) + \nabla^2 u\right) + f = \rho\frac{\partial^2 u}{\partial t^2} \tag{5.12}$$

The above equation is called Navier–Cauchy equation for solid mechanics which can be solved for any 3D structure to calculate the deformation caused by any force.

Piezoelectric material

Piezoelectric material is a class of dielectric in which applied electric field develops mechanical strain inside. The external field changes the alignment of the polarization domain inside the dielectric material.

The piezoelectric constitutive equations that couple the mechanical quantities such as stress and strain with the electrical quantities in the piezoelectric material are given as

$$\begin{aligned} \{\sigma\} &= [C^{\mathrm{E}}]\{\varepsilon\} + [\mathbf{e}]^{\mathrm{T}}\{\nabla\phi\} \\ \{D\} &= [\mathbf{e}]\{\varepsilon\} - [\varepsilon^{\mathrm{S}}]\{\nabla\phi\} \end{aligned} \tag{5.13}$$

where C^E is the elastic stiffness matrix (evaluated at the constant electric field), $\mathbf{e}$ is the piezoelectric matrix and the superscript T denotes the matrix transpose, and ε^S is the dielectric permittivity matrix (evaluated at constant strain). The ϕ is electric potential which can be written in terms of electric field $\vec{\xi} = -\nabla\phi$ in the constitutive relation.

The constitutive relation for piezoelectric material, along with kinematic and equilibrium equation, when solved with the mechanical and electrical boundary conditions imposed on the entire boundary of the structure completely determine the motion of the piezoelectric material under applied field. This constitutes the basic simulation engine for the piezoelectric MEMS resonators.

5.3.3 Electrostatics

The physics that deals with the stationary charges and its field is termed as electrostatics. The PDE for electrostatics is obtained using Gauss law, which is $\nabla \cdot \mathbf{D} = \rho_q$, and the relation between electric flux density and electric field, which is $\mathbf{D} = \varepsilon\xi$. The resultant equation is termed as Poisson's equation

$$\nabla \cdot \varepsilon\xi = \rho_q \tag{5.14}$$

where ρ_q is the volume charge density inside the enclosed region. Poisson's equation is a generalized form of Laplace equation, which has $\rho_q = 0$, and is used to calculate the potential distribution in material with known charge density. The voltage applied on the electrodes for actuation acts as a boundary condition for the electrostatic equation.

5.3.4 Thermal domain

The energy stored in the resonators is in the form of elastic phonons. When the elastic vibration period is comparable to the thermal transport time constant, those elastic phonons interact with thermal phonons, which move randomly and cause phonon–phonon scattering, through diffusion [9]. This results in conversion of vibration energy into thermal energy, and this irreversible loss mechanism is termed as thermoelastic damping. In a material with positive thermal coefficient, temperature increases at the point of compression and decreases at the point of tension. Due to this temperature gradient created in resonator structure during vibration, the diffusive thermal transport occurs from hot to cold region and causes energy loss [10].

The PDE that governs this transport is termed as heat diffusion equation, which gives the temporal distribution of heat in the structure, and is given as

$$\rho c_p \frac{\partial T}{\partial t} - \nabla \cdot (k \nabla T) = \dot{q} \tag{5.15}$$

where k is thermal conductivity, c_p is the specific heat capacity, and $\dot{q}$ is the heat generation rate which can come from strain-induced heat flow.

5.3.5 Fluid domain

During the movement of RF MEMS resonator structure, the surrounding air moves in and out of the structure. Due to the large surface to volume ratio at the micro-scale regime, the airflow exerts force on the resonator structure. This results in energy loss and termed as air damping. The presence of air modifies the spring constant of the system and thus affects its frequency response. Thus, the modeling of air motion becomes important for suspended structures. The motion of air is a 3D phenomenon and requires 3D flow simulation. The PDE which describes the flow of any fluid is Navier–Stokes equation

$$\rho\left(\frac{\partial \vec{U}}{\partial t} + (\vec{U} \cdot \nabla)\vec{U}\right) = -\nabla \vec{P} + \rho \vec{g} + \mu \nabla^2 \vec{U} + \frac{\mu}{3}\nabla(\nabla \cdot \vec{U}) \tag{5.16}$$

where $\vec{U}$, $\vec{P}$, and ρ are the fluid velocity, pressure, and density. μ is the viscosity coefficient and $\vec{g}$ is the acceleration due to gravity. The Navier–Stoke equation should be solved along with continuity equation that guarantees the conservation of flow

$$\frac{\partial \rho}{\partial t} + \nabla \cdot (\rho \vec{U}) = 0 \tag{5.17}$$

For large perforations in the MEMS resonator, solving Navier–Stoke equation for the 3D flow is computationally expensive and the problem in that case can be reduced into 2D and solved using perforation profile Reynolds equation [11]

$$\nabla \cdot \left(D_{\mathrm{h}} \frac{h^3 Q_{\mathrm{ch}}}{12\eta} \nabla p\right) - C_{\mathrm{h}} \frac{j\omega h p}{P_{\mathrm{A}}} - Y_{\mathrm{h}} p = v_{\mathrm{z}} \tag{5.18}$$

where h is the static air gap height, η is the viscosity coefficient, Q_{ch} is the relative flow rate, P_A is the ambient pressure, $v_z(x,y)e^{j\omega t}$ is the surface velocity in the z-direction, $p(x,y)e^{j\omega t}$ is the pressure variation to be solved from the equation, $D_h(x,y)$ is the relative diffusivity, $C_h(x,y)$ is the relative compressibility, and $Y_h(x,y)$ is the perforation admittance profiles.

5.3.6 Coupled-domain analysis

The physical domains are said to be coupled when force term defined in one domain results in system response in another domain. The analysis of RF MEMS resonators requires to deal with multiple physical phenomena concerned with different domains. For example, electrical actuation of RF MEMS resonator in which the applied force in electrical domain results in beam deflection in mechanical domain. The coupled domain for electrical actuated RF MEMS thus termed as electromechanics. In this coupled equation the force term introduced in Navier–Cauchy equation must be calculated from the electrostatic field. The energy stored in the electrostatic field is given by

$$W_e = \frac{1}{2}\int_{\Omega} D \cdot \xi d\Omega \tag{5.19}$$

The force is the gradient of the stored electrostatic energy as

$$f = \nabla W_e \tag{5.20}$$

Similarly, for the analysis of thermoelastic damping, solid mechanics needs to be solved along with the thermal transport equation. The temperature-induced force term in Navier–Cauchy equation and heat source in the heat transfer equation are given as [9,12]

$$f = \frac{\alpha E}{(1-2\nu)} \nabla \cdot T \tag{5.21}$$

$$\dot{q} = \frac{\alpha E T_0}{(1-2\nu)} \frac{\partial}{\partial t} (\nabla \cdot \mathbf{u}) \tag{5.22}$$

where α is coefficient of thermal expansion and T_0 is initial temperature. Thus, the two physical domains are coupled through the linear thermal expansion coefficient of the material. The constitutive relationship, in this case is modified to $\sigma = D(\varepsilon - \varepsilon_t)$ including the effect of thermal strain ε_t.

Apart from thermoelastic damping, the loss of elastic phonon also occurs through the anchor that attaches the resonating structure with the substrate. The analysis of the leakage of elastic phonon can be done by solving the coupled beam-substrate-electrostatics model [13].

In such type of coupled-physical domain, the physics of all the concerned physical domains need to be solved self-consistently. There are two approaches using which the coupled equations are solved and are depicted in Figure 5.7:

1. Segregated solution: In this approach, each PDE is solved separately and the convergence of the whole system is checked at the end. The algorithm starts with initial guess for all the unknowns. Overall convergence of the coupled

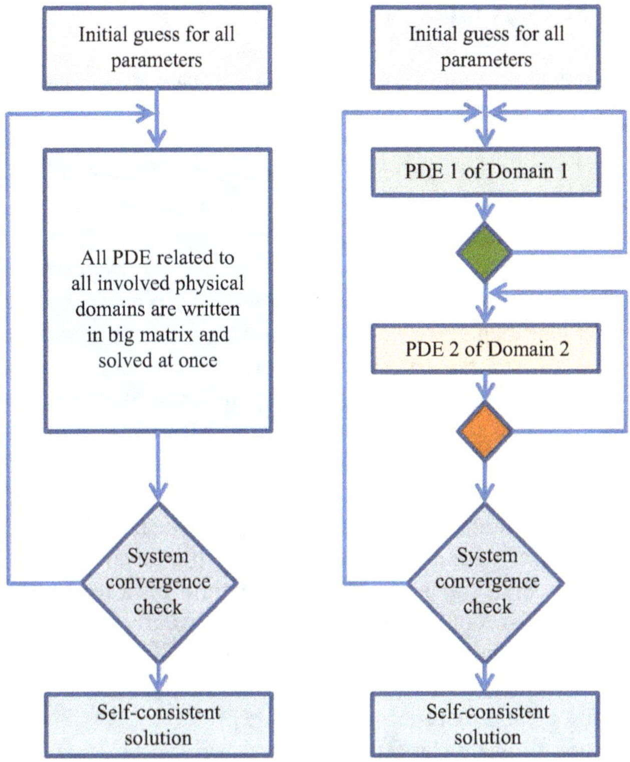

Figure 5.7 Algorithm for the segregated solution and coupled solution approach for solving coupled-domain analysis

equation depends on the degree of coupling. For strong coupling and nonlinear problem, convergence is difficult to obtain. A fast FEM algorithm for electro-actuated RF MEMS resonators is presented in [14] based on this approach.

2. Coupled solution: In the coupled solution approach, all the PDEs are solved simultaneously using techniques like Newton–Raphson method. The initial guess for all the parameters is defined at the beginning of the algorithm. The Jacobian matrix is formed using the old guess which is used to obtain values for the updated guess. The updated guess is fed again for next loop until convergence is reached.

To exploit the computational power of computers for solving PDEs the numerical techniques are used. Using these numerical techniques, the differential equations can be converted into the set simultaneous linear equations which can be assembled in matrix form and solved using computers. In Section 5.4, the most generic numerical technique widely used for device designing and optimization is discussed.

5.4 Finite element method

FEM is a numerical technique to obtain approximate solutions to differential equations that arise from the modeling of physical systems. FEM is most suitable when the geometry of the physical system is an irregular shape. The main advantage of FEM over finite difference scheme is that FEM can handle arbitrary geometry by dividing the domain into smaller mesh elements that could be triangular, rectangular, and tetragonal depending on the problem. The boundary conditions of the differential equations that are specified on irregular shapes can be easily handled by FEM. The differential equations are expressed as algebraic sets of equations that can then be represented into matrix form and numerical techniques can be used to solve these matrix equations.

MEMS structure can have different shapes and structures, and analytically it is not always possible to solve the governing equations due to complex boundary conditions and shapes. FEM is usually employed in such situations to solve the differential equations for the physical laws on the device domain. The real system of MEMS is multiphysics in nature in the sense that it is described completed by considering the set of equations for the physical laws in different domains and finite element is the bridging method between the multiphysics domains. The description of the laws of physics is usually expressed in terms of PDEs and they can be coupled. In this scenario, FEM modeling provides best possible way to handle the equations from different physical laws together with boundary conditions and material properties [5].

Finite element grids are classified in several ways. A grid can be conforming, nonconforming, single-element type and multielement type, locally and globally refined grids, nested and non-nested grids, bisection type grids, sparse grids, and so on. Also, grids can be represented in various formats: YASP grid, structured grid, ALBERTA grid, ALU grid, and so on. The PDEs are converted into the matrices which are assembled on these grid elements. The matrix assembly requires integration calculations that are performed using numerical techniques. Most of the FEM users prefer Gauss quadrature rule as a numerical technique for integration in almost every matrix assembly and postprocessing procedures. The matrix assembly procedure is discussed in detail in Section 5.4.1.5.

There are several commercial software packages that do FEM computation on MEMS devices. Almost all use a common prescription for handling the given simulation problem. We discuss a few of these tools in Section 5.5. The steps involved in the FEM analysis and simulation can be split into the following: preprocessing, processing, and postprocessing. We explain these steps by considering the case of Poisson's equation, which is most common equation encountered in semiconductor device computations. Since by merely running the commercial tools, one does not really get to know how the FEM machinery operates on a physical problem, we give out detailed mathematical equations of FEM assembly process and try to shed light on the prescription of converting governing differential equations or PDEs defined on arbitrary geometry to matrix equations.

5.4.1 Preprocessing

For a given governing differential equations, matrices are to be assembled after obtaining weak form of the PDEs and meshing of the physical domain on which they are defined. These steps come under preprocessing. The two common procedures that are applied on the governing equation to get the weak form are the Variational method and the Galerkin method.

5.4.1.1 Variational method

For some physical problems, it is possible to find a function in which governing differential equation can be embedded implicitly. The solution of the problem is then a function which makes it stationary for the arbitrary variation of the functional. The Variational method works for some physical problems satisfying the conditions; however, the method cannot be used on some more complicated equations.

Governing differential equations and the boundary conditions together are said to define a given problem in strong form whereas when the problem is expressed as an integral expression such as functional that implicitly contains the governing differential equations, then it is called to be in the weak form.

5.4.1.2 Galerkin method

In contrast, the Galerkin method has become more popular since it is easier to use and gives a compatible approximation to the differential equation even when the Variational principle method cannot be applied. When one knows the differential equation but not the functional or the variational principle, application of the method of weighted residues is preferred. Let u be the unknown function and $\mathscr{F}$ be the differential operator for the governing equation:

$$\mathscr{F}(u) = 0 \tag{5.23}$$

In the FEM, we try to find the approximate solution $\bar{u}$ to u which is expressed in terms of node or vertex values $u(x_i)$ on the FEM grid

$$\bar{u}(x) = \sum_i N_i(x)u(x_i) \tag{5.24}$$

where N_i are called as the interpolation functions or the shape functions. The values of the unknown function u on the nodes are determined by setting the product of the residual function $R = \mathscr{F}(\bar{u})$ with the weight function to zero:

$$\int_\Omega R w_i \; d\Omega = 0 \tag{5.25}$$

When the weight functions w_i are the same as the local functions or shape functions N_i on the FEM element, the method is known as Galerkin method.

5.4.1.3 Meshing

A continuum domain is discretized by dividing system domain into elements and a linear solution is assumed in each element. The process of discretizing domain into elements is called as meshing. A mesh can be defined as union of elements, nodes, and intersection elements. For example, in structure, Figure 5.8, as a whole, it is

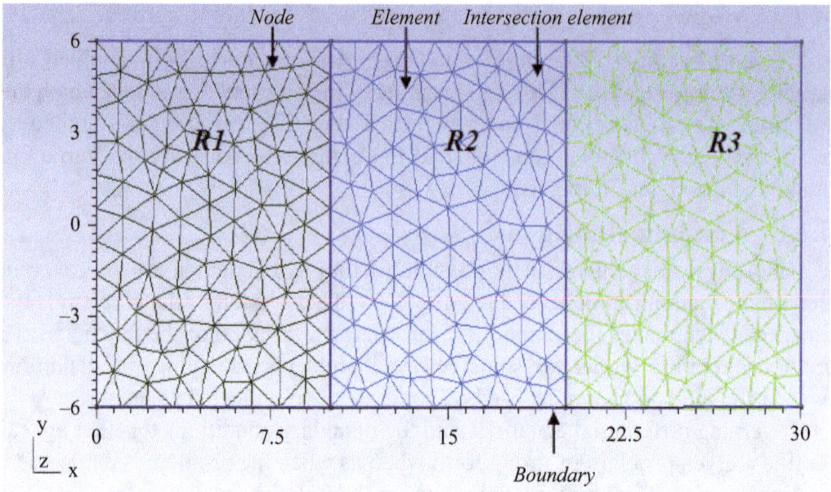

Figure 5.8 Mesh generated using GMSH

called as a domain. Regions R1, R2, and R3 are subdomains and geometric boundaries are also shown. This domain is divided into elements in the meshing process. Elements, nodes, and intersection elements are called as mesh/grid entities. The subdivision of a whole domain into simpler parts has several advantages like accurate representation of complex geometry, inclusion of dissimilar material properties, easy representation of the total solution, and capture of local effects.

There are several algorithms available to do meshing of a given geometry. GMSH is one such structure editor and meshing software which includes MeshAdapt, Delauney, and Frontal algorithms for meshing a given geometry [15].

In the FEM, system domain is divided into subdomains or grid-elements and an interpolation scheme is defined over the subdomains.

Appropriate meshing is one of the crucial steps in FEM prescription. Finer meshing gives accurate results but increases the computation time. Optimal meshing leads to a reliable solution in reasonable computational time.

5.4.1.4 Shape functions

Shape functions are chosen to suit the dimensionality of the problem as well as the type of the elements in FEM. Normally, they are chosen as polynomials functions. Shape functions satisfy a few conditions such as partition of unity and Kronekar-delta unity.

$$\sum_i N_i(x) = 1 \tag{5.26}$$

The shape functions can be linear, for example, in 2D (see Figure 5.9), the linear shape functions can be written as follows:

$$N_0 = 1 - \xi - \eta, \quad N_1 = \xi, \quad N_2 = \eta \quad \text{for } 2D,$$
$$N_0 = 1 - \xi - \eta - \zeta, \quad N_1 = \xi, \quad N_2 = \eta \; N_3 = \zeta \quad \text{for } 3D$$

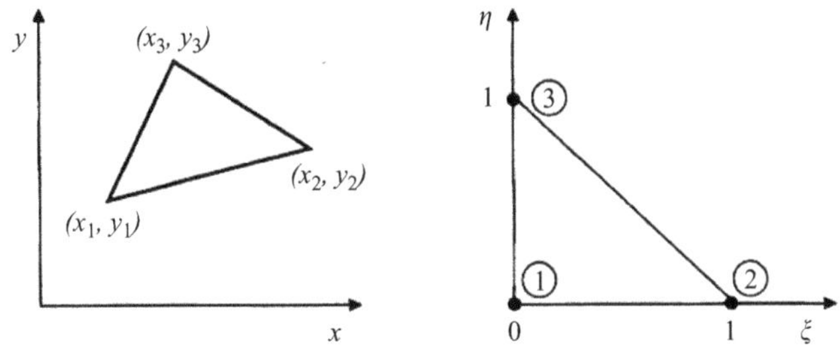

Figure 5.9 Geometry transformation of elements over the reference element

As seen in Figure 5.9, any function $f(x, y)$ can then be linearly approximated in an element as

$$f(x, y) = \sum_{i=0}^{2} f_i N_i(\xi, \eta) \tag{5.27}$$

Other example of shape functions is Hermite polynomials. The 1D, 2D, and 3D nonoverlapping subintervals cover the whole domain where each subinterval is called an "element."

5.4.1.5 Matrix assembly

The differential equations to be solved on the device domain are solved by assembling them in a matrix form. A weak or variational method can be used in which the equation to be solved is multiplied by a well-behaved trial function and integrated over the device domain. Appropriate boundary conditions are then applied to solve the matrix equation. Applying the appropriate boundary condition is essential for obtaining a correct solution for the modes. After applying the boundary conditions, matrix assembly is complete.

5.4.2 Weak formulation of Poisson's equation

The Poisson's equation is given as

$$\nabla(\varepsilon(r)\nabla\phi(r)) = -\rho_q(r) \tag{5.28}$$

where ε is the electric permittivity of material at r, ρ_q is the charge density at r, and ϕ is the potential at r.

Let Ω be the domain over which Poisson's equation is defined.

Charge density ρ_q and electric permittivity ε are specified at all r and the only unknown is potential ϕ. Aim of FEM is to solve this PDE to obtain solution, that is, potential ϕ. Potential ϕ is defined over Ω and satisfies Dirichlet and Neumann boundary conditions. Let position r is defined in Cartesian coordinates (x, y) and boundary conditions are specified as $\phi = V_{ss}$ at Γ_D and $\nabla\phi = Q$ at Γ_N.

If ϕ is the approximate solution of PDE, then taking $\rho_q(r)$ to left-hand side will form a residue R as

$$R = \nabla(\varepsilon(r)\nabla\phi(r)) + \rho_q(r) \tag{5.29}$$

If we define w as the weight function, then approximate solution ϕ can be obtained by minimizing residue in the sense

$$\int_{\Omega} Rw \ d\Omega = 0 \tag{5.30}$$

$$\int_{\Omega} (\nabla(\varepsilon(r)\nabla\phi(r)) + \rho_q(r))w \ d\Omega = 0 \tag{5.31}$$

Weight function w must belong to a class of admissible functions and for this problem it must satisfy the following conditions to being an admissible function:

- w must be zero at boundaries where ϕ is specified,
- w must be unconstrained at boundaries where $\nabla\phi$ is specified, and
- w must be smooth enough for the integral of the weighted residue to be finite.

Integrating (5.31) by parts, we get weighted residual formulation of (5.28) as

$$\int_{\Omega} \varepsilon\nabla\phi.\nabla w \ d\Omega = \int_{\Omega} \rho_q w \ d\Omega - \int_{\Gamma_N} Qw \ d\Omega \tag{5.32}$$

A finite element mesh is generated or defined over the domain Ω such that it is covered with M non-overlapping triangles (called elements) for a two-dimensional (2D) problem or for example, tetrahedra for a 3D problem. The total vertices or nodes of the mesh N decide the size of the assembled matrix to be solved.

As defined earlier the linear shape functions can be defined over every element of the domain Ω. The finite element approximation is done in a finite dimensional subspace consisting of continuous functions ψ_i, $i = 1, 2, \ldots.N$ which are also called as the basis functions. The approximate solution of ϕ is sought which is a linear combination of these basis functions ψ_i with unknown coefficients ϕ_i

$$\phi = \sum_{i=1}^{N} \phi_i \psi_i \tag{5.33}$$

Substituting (5.33) in (5.32) and letting weight function w to be the same as basis function ψ (Petrov–Galerkin formulation), we get:

$$\sum \left(\int_{\Omega} \varepsilon \nabla\psi_i \nabla\psi_j \ d\Omega \right) \phi_j = \int_{\Omega} \rho_q \psi_j \ d\Omega - \int_{\Gamma_l} Q\psi_j \ dS \tag{5.34}$$

Equation (5.34) can be expressed in matrix form as

$$\mathscr{D}x = b \tag{5.35}$$

where

$$\mathscr{D}_{ij} = \int_{\Omega} \varepsilon \nabla \psi_i \nabla \psi_j \; d\Omega,$$

$$b_j = \int_{\Omega} \rho_q \psi_j \; d\Omega - \int_{\Gamma_N} Q \psi_j \; d\Omega,$$

$$x = [\phi_1, \phi_2, \ldots\ldots, \phi_N]^T$$

As discussed earlier, after mesh generation, assembly routines are called which assemble matrix $\mathscr{D}$ and vector b. Matrix $\mathscr{D}$ is sparse and BCRS data type is used to store matrix $\mathscr{D}$. In assembling procedure, integrals are transformed onto the element (shown in Figure 5.9) and computed with an appropriate quadrature rule.

Let $\mathscr{J}$ be the conforming triangulation of the domain Ω and let the transformation map be given by $\gamma : \hat{\Delta} \to \Delta$

$$\begin{aligned} \mathscr{D}_{ij} &= \int_{\Omega} \varepsilon \nabla \psi_i \cdot \nabla \psi_j \; dx \\ &= \sum_{\Delta \in \Upsilon} \int_{\hat{\Delta}} \varepsilon \nabla \psi_i(\gamma(\hat{x})) \cdot \nabla \psi_j(\gamma(\hat{x})) \Theta(\hat{x}) d\hat{x} \\ &= \sum_{\Delta \in \Upsilon} \int_{\hat{\Delta}} \varepsilon (J_\gamma^{-T} \hat{\nabla} \hat{\psi}_i)(\hat{x}) \cdot (J_\gamma^{-T} \hat{\nabla} \hat{\psi}_j)(\hat{x}) \Theta(\hat{x}) d\hat{x} \end{aligned} \tag{5.36}$$

$$\mathscr{D}_{ij} = \sum_{\Delta \in \Upsilon} \sum_{k=1}^{p_1} \varepsilon \omega_k (J_\gamma^{-T} \hat{\nabla} \hat{\psi}_i)(\lambda_k) \cdot (J_\gamma^{-T} \hat{\nabla} \hat{\psi}_j)(\lambda_k) \Theta(\lambda_k) \tag{5.37}$$

where J_γ is the Jacobian of the transformation map γ and $\theta(\hat{x}, \hat{y}) = \sqrt{detJ_\gamma^T J_\gamma}$ is the determinant of the Jacobian. Let λ_k be the quadrature points of the chosen quadrature rule of integration and ω_k the associated weights and p_1 are the total number of quadrature points to evaluate the matrices in (5.37).

Similarly, the right side of (5.32) can be treated in the same manner where we may have to use some other quadrature rule having p_2 quadrature points

$$b_i = \sum_{\Delta \in \Upsilon} \sum_{k=1}^{p_2} \omega_k f(\gamma(\lambda_k)) \psi_i(\gamma(\lambda_k)) \Theta(\lambda_k) \tag{5.38}$$

The matrix entries now can be calculated by running an iteration over all elements of the grid and updating all matrix entries with a nonvanishing contribution on that element.

After assembling, the Dirichlet boundary conditions are inserted by following procedure as can be seen in Figure 5.10. If node j is on Dirichlet boundary then j^{th} row is set to zero except $\mathscr{D}_{jk}$ which is set to 1 and b_j is set to known value at the boundary, in this case V_{ss} which is taken as zero for the example shown in Figure 5.10.

$$\begin{array}{c} \\ \rightarrow j \\ \\ \end{array} \begin{pmatrix} \star & \star & \star & \star & \star & \star & \star \\ 0 & \cdots & 0 & 1 & 0 & \cdots & 0 \\ \star & \star & \star & \star & \star & \star & \star \end{pmatrix} \begin{pmatrix} \star \\ \phi_j \\ \star \end{pmatrix} = \begin{pmatrix} \star \\ 0 \\ \star \end{pmatrix}$$

Figure 5.10 The boundary conditions can be implemented by modifying the lines of $\mathscr{D}$ and b

5.4.3 Processing

Once the matrices are assembled and boundary conditions are applied, the matrices can be supplied to the solvers for obtaining the solutions. The equation to be solved can be of the type $Ax = b$ or generalized eigenvalue equations.

The assembled FEM matrices are largely sparse and to be solved by sparse solvers to save computational time. For solving matrix equations, software packages like LAPACK, ARPACK, PETSc-SLEPc, BLAS, etc. are available that include direct solvers, iterative solvers, and preconditioned solvers.

5.4.3.1 Eigenvalue equation solvers

Generalized eigenvalue problems can be solved using packages SLEPc (Scalable Library for Eigenvalue Problem Computations) and PETSc (Portable, Extensible Toolkit for Scientific Computation). SLEPc is a software library for the solution of large sparse eigenvalue problems on parallel computers [17]. In the case of eigenvalue problems, using libraries is not straightforward and typically the problem is successfully solved only after several cycles of testing and parameter tuning. In the generalized eigenvalue equation such as Schrodinger's equation, both matrices are real and symmetric and eigenvalue E has to be real as it represents energy states. Also, in some situations, entire spectrum of eigenvalues is not desired, only the first few eigenvalues and corresponding eigenvectors are required.

To solve the eigenvalue problems spectral transformations can be used. The spectral transformation (ST) is the SLEPc object that brings in the functionality for acceleration in the time required to obtain the spectrum based on the transformation of the spectrum. Spectral transformations are known as powerful tools for adjusting the way in which eigensolvers behave when dealing with a particular eigenvalue problem. Basically, the original problem is transformed into a new one in which eigenvalues are mapped to new values or position while eigenvectors remain the same. These transformations can be used with the following objectives [17]:

- Compute internal eigenvalues: In some applications, the eigenpairs of interest are not at the extremity (largest magnitude, smallest magnitude, rightmost, and leftmost), but those lie in a certain interval or those closest to a certain value.
- Accelerate convergence: Convergence of eigenvalue problem typically depends on how close the eigenvalues are from each other. Using the spectral transformation technique, difficult eigenvalue distributions can be remapped to a favorable location for the fast convergence.

- Handle some special situations: For instance, in generalized problems $Ax = \lambda Bx$, when the matrix B is singular, it can be necessary to use a spectral transformation.

When there is more than one differential equation in the systems and they are coupled, the system becomes more complex to solve. Newton's scheme can be invoked in this situation to solve the equations together.

5.4.4 Postprocessing

Postprocessing is the last step in FEM which deals with analysis and visualization of the obtained solutions or results. In this step, first the solution is checked thoroughly for possible numerical errors. Then, the behavior of the obtained solution is checked for its consistency with the expected theoretical point of view. Solution of PDE is not the one we require but the derived quantities which, we call as results are to be extracted from the obtained solution. Visualization tools like Paraview can be used to see the simulation result data on the FEM grid. The commercial tools have their own in-built visualization tools.

The details about the FEM implementation of Poisson's equation including preprocessing, matrix assembly, and postprocessing are discussed in next section 5.4.5.

5.4.5 Examples of Poisson's equation solved using FEM

We describe here a straightforward example for solving Poisson's equation in 2D. The finite element matrices are assembled in the Distributed and Unified Numerics Environment (DUNE) which is a modular toolbox for solving differential equations with grid-based methods [16].

5.4.5.1 Triangular charge density in X direction

A 2D grid is generated using GMSH for a square plate. A triangular charge density profile shown in Figure 5.11 is assigned at every node of the grid. Vanishing Dirichlet boundary conditions are inserted on all boundaries and then the potential is obtained by solving Poisson's equation. An expected quadratic Potential profile solution is obtained as shown in Figure 5.12.

5.4.5.2 Laplace equation

Laplace equation is a special case of Poisson's equation in which the charge density is zero. Solution of Laplace equation is a linear potential profile. Figure 5.13 shows a profile obtained by solving Laplace equation using FEM with Dirichlet boundary conditions: $\phi = 1$ at $x = 0$ and $\phi = 0$ at $x = 1$.

The similar steps can be followed for FEM implementation of Euler–Bernoullis and Timoshenko beam equations [18,19]. For solving Navier–Cauchy equation of solid mechanics, the details of FEM matrix assembly, along with its MATLAB implementation, are given in [19]. The FEM implementation of the thermal transport equation for accounting thermoelastic damping, the implementation details are discussed in [20] for contour-mode resonators and the beam resonators in [21]. For the case of gas-damping for perforated microstructure, the

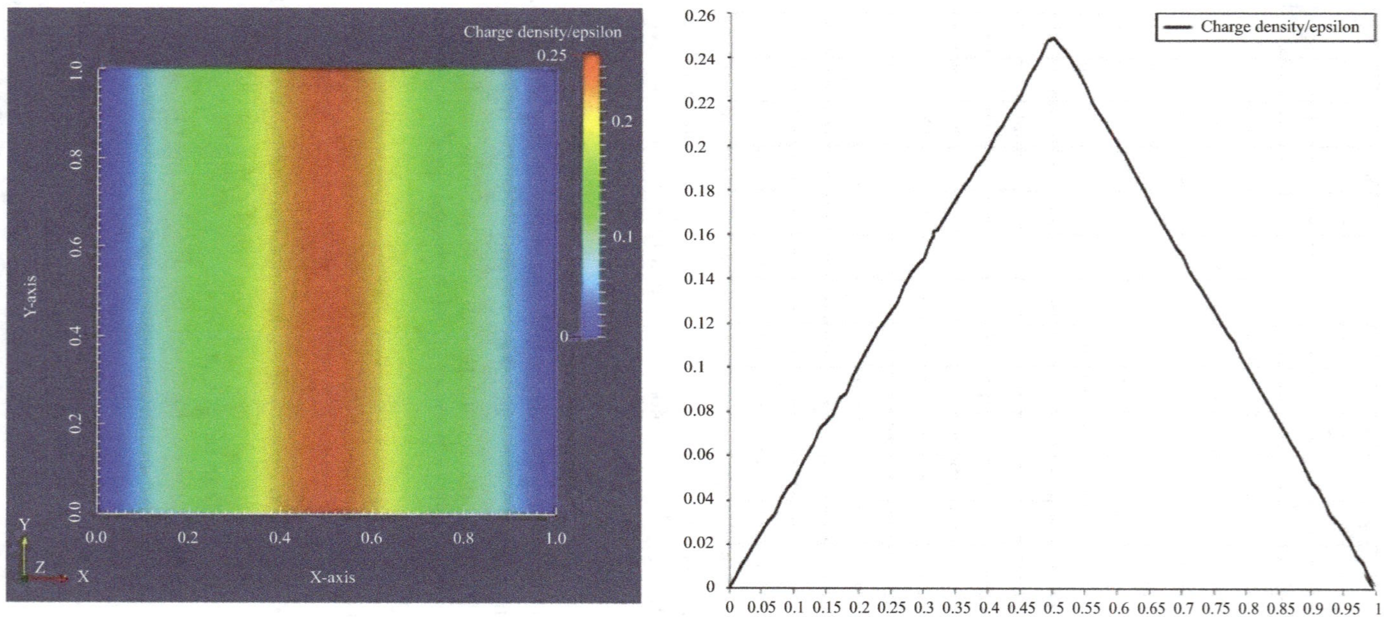

Figure 5.11 Triangular charge density profile

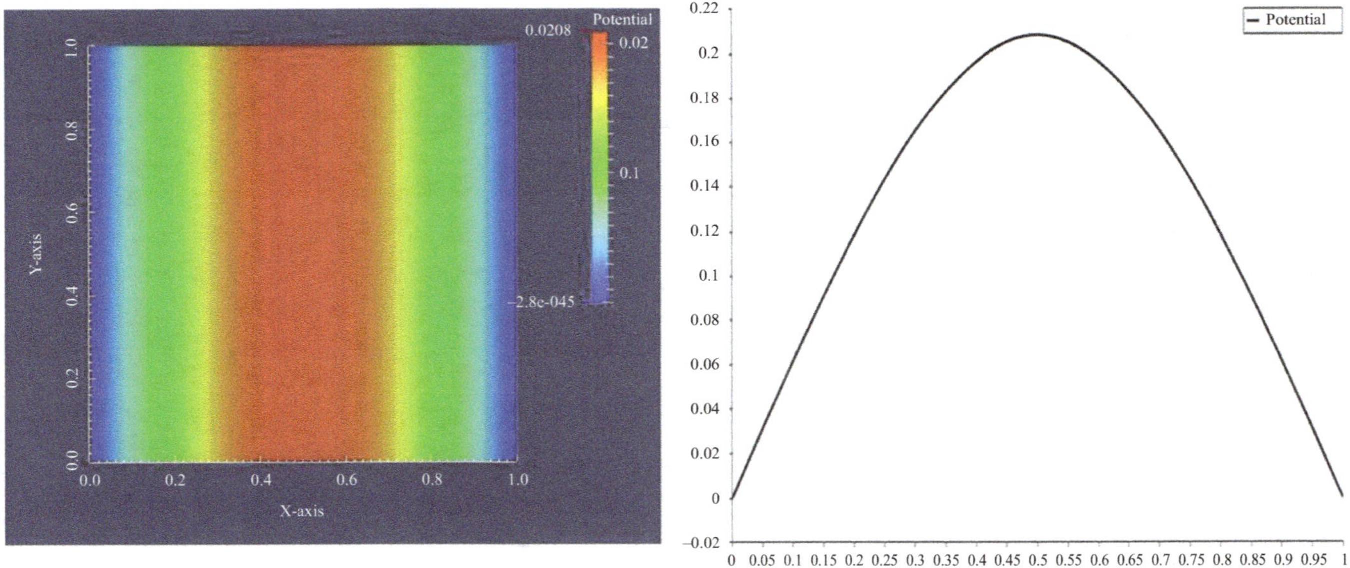

Figure 5.12 Potential obtained by solving Poisson's equation for triangular charge density profile in Figure 5.11

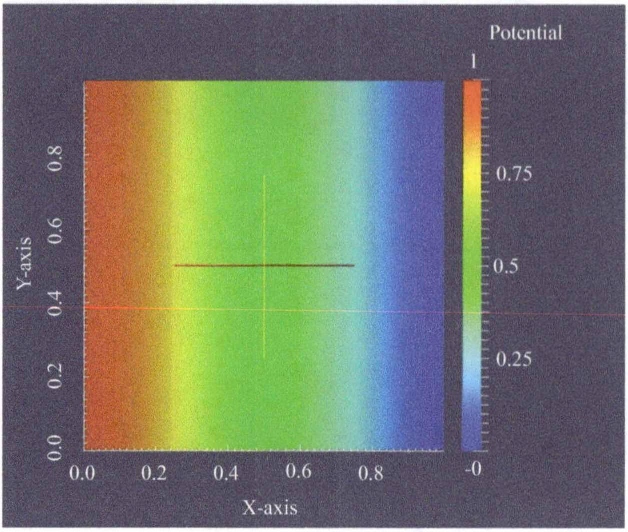

Figure 5.13 Solution to Laplace equation

FEM implementation of Reynolds equation is given in [11]. The method discussed the efficient 2D FEM solver for the analysis of gas-damping.

In the analysis of piezoelectric resonators and filters, the material constitutional relation needs to be solved along with the electromechanics. The details including weak formulation, shape functions, and finite-element matrix assembly for these types of resonators are provided in [22].

5.5 Commercial MEMS design tools

Commercial tools based on physical modeling and FEM are available which are used for designing and optimizing RF MEMS devices. The important tools are CoventorWare, Intellisuite, and COMSOL Multiphysics. In this section, the details and related studies performed using these tools are presented.

5.5.1 CoventorWare

The CoventorWare software is a design and simulation tool for modeling a wide range of devices with accuracy, capacity, and speed that are required for designing real-time MEMS devices. It consists of system-level and physical design approaches. In the system-level approach, behavioral model libraries with a high-speed system simulator are used to generate a 2D layout. The 2D layout is then built into a 3D model in the physical approach [23]. Once the 3D model is created, a mesh will be generated and simulation will be carried out using FEM solvers. This process leads to a 2D layout to be moved to a model ready for fabrication. This software has various design libraries and 3D physics solvers. It also has options for importing and exporting files from third-party software.

CoventorWare has three modules: architect, designer, and analyzer as seen from its basic interface screen in Figure 5.14 [24].

Architect: This allows system-level simulation by allowing CoventorWare simulation tools, Saber generic models, as well as user-generated models and incorporates plotting tools to view and analyze the results in various graphical formats.

Designer: Designer creates the 2D model in the layout editor into a 3D model. In the designer tab, there is access to select material properties database, process editor, and layout editor (Figure 5.15).

Materials properties database is the database that stores properties of all the materials used for the MEMS design as MPD files. Material chosen for each deposition step in the 2D layout needs to be defined in terms of strength, thermal, density, and conductivity. The analysis results depend upon the properties of the materials. CoventorWare has its own database of materials stored as an mpd1.mpd file in a folder named MPD. Once this material is selected, their physical properties get automatically displayed which can be edited to enter different values if required. CoventorWare supports import of new materials to its database. The user needs to enter the name of the material and its corresponding material properties. The user can copy, edit, delete, and import materials to the database.

In the process editor as seen in Figure 5.15, a simulation of the fabrication process flow of MEMS device will be performed. Materials, layers, masks, action

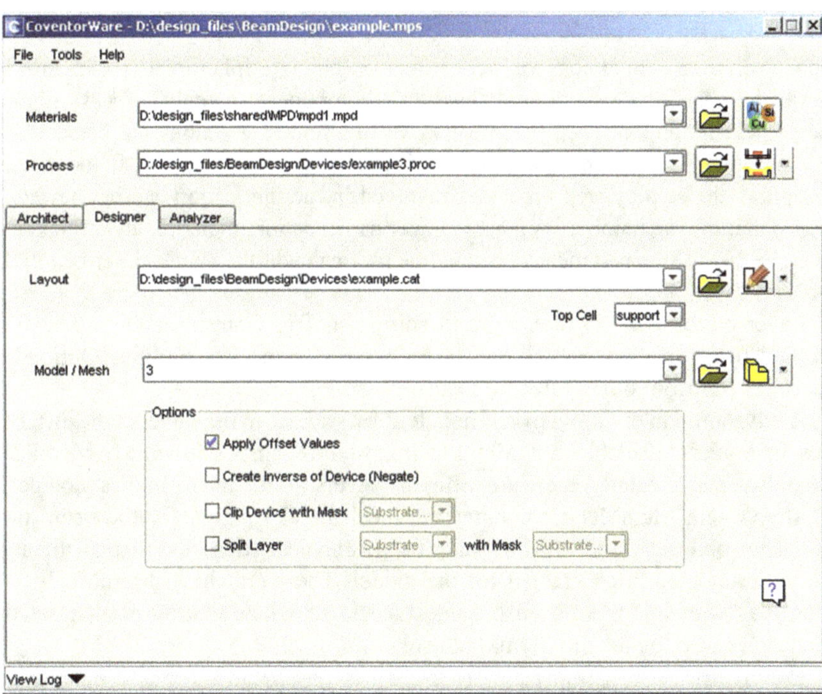

Figure 5.14 Basic interface screen of CoventorWare software [25]

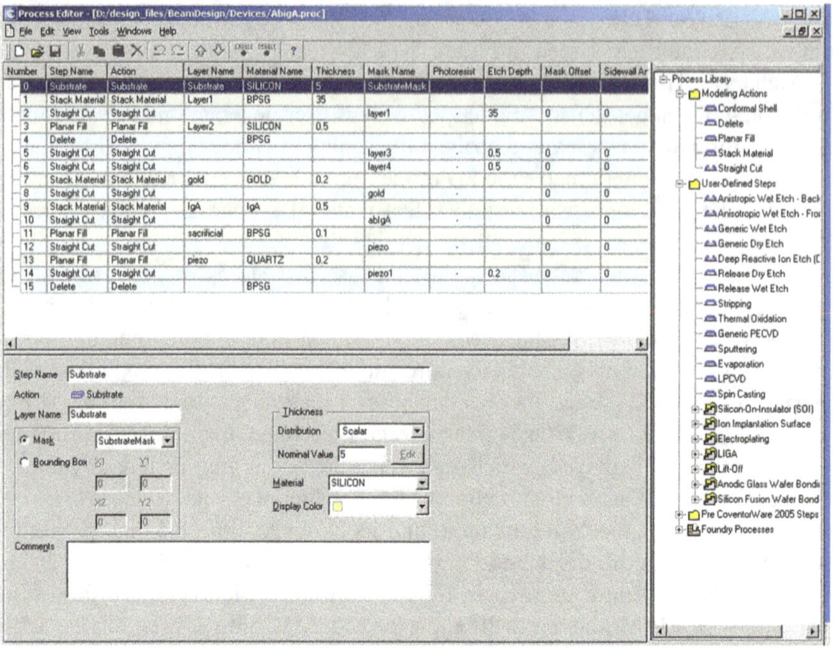

Figure 5.15 Process editor of the CoventorWare software [25]

(deposit or etch), thickness of deposition or etching, photoresist, etc. are all described in the process editor. It has three components namely, process library, process description, and step parameters. All the modeling actions are listed in the process library from which a particular fabrication step can be selected. In the process description, the step by step processes involved in the fabrication can be viewed. It shows detail of step number, step name, layer name, action, material, layer thickness, and more. In the step parameters section, the user can edit the values described in the process description.

Layer editor is a 2D drawing and editing tool in ConventorWare. The files specified using the process editor can be opened in the layer editor. Options to import 2D file from other software is also possible.

Analyzer module: The analyzer module allows access to the solvers. The different types of solvers available are MemElectro, MemMech, CoSolveEM, MemPZR, Memoptics, etc. Similarly, there are different solvers for the microfluidics module.

MemMech: MemMech computes the mechanical solution for the beam problem. The solver uses the FEM to solve for mechanical stress and displacement at each node on each brick created for the model. The hierarchical MemMech BCs window as seen in Figure 5.18 provides access to window sections that set the boundary conditions for the MemMech solver.

CoSolveEM: The CoSolveEM tool couples the electrostatic and mechanical solvers. In an iterative process, the electrostatic results are input to the mechanical solver, and the results are fed back until convergence is reached.

MemElectro: The MemElectro solver produces an electrostatic solution by solving for the charge and capacitance interaction between the beam and ground components of the created model. MemElectro uses the boundary element method (BEM). During the calculation, MemElectro computes the charge on each surface panel and presents a final solution with charge distribution calculated for all the panels in the model.

5.5.1.1 Meshing

Once the 3D model is built, the next step is the mesh generation. This is an important step in finite element analysis. The whole structure is discretized to portions called finite elements. These finite elements are joined to each other using nodes. The mesh consists of the finite elements and the nodes. When an analysis is run, the values from each node is calculated and output from each finite element is found. The different types of meshes are extruded bricks, Manhattan bricks, tetrahedrons, mapped bricks, and surface mesh [24]. The user will decide on the type, size, and how many nodes to be present as per the analysis.

5.5.1.2 Solver and visualizer

The next step after meshing and setting up volume and surface names, the model is ready for inputting in the 3D solver. The different dialog boxes can be accessed and the boundary conditions (BCs) are set for the volume and surfaces. The visualizer reads the 3D results by providing views, axis orientation control, translucency, data probing, and animation. Mechanical stresses, electrostatic charges, heat distribution, etc. are some of the solver results that can be viewed using visualizer.

5.5.1.3 Integrator

The CoventorWare modules can be accessed by other software such as Saber and Cadence. This integration helps in higher level system simulation for the electronic devices which will be connected to the MEMS device [24].

Design of cantilever beam immunosensor based on mass detection using CoventorWare is discussed. A perforated cantilever beam was designed as an immunosensor with the dimensions of Length 200 μm, breadth 10 μm, and thickness 0.5 μm as can be seen in Figure 5.16 [25]. The cantilever beam deflection and frequency change depend upon the change in the mass of the cantilever. The basic interface screen shows the three modules: (a) architect, (b) designer, and (c) analyzer of the CoventorWare software (see Figure 5.14).

The substrate is the fundamental layer for all the process flow. Therefore, this step cannot be edited or deleted or moved to a different position. The borophosphosilicate glass (BPSG) deposit acts as a sacrificial layer for constructing a cantilever beam and is completely etched away at the final sequence. Etching happens by exposing the resist through the mask. The final structure of the cantilever beam after meshing can be seen in Figure 5.17 [25].

In this step, the boundary conditions will be fixed for the cantilever beam by fixing one end and applying an input load to the free end of the cantilever.

The red area in Figure 5.19 shows the maximum deflection while the area in blue represents minimum or zero deflection of the cantilever beam [25].

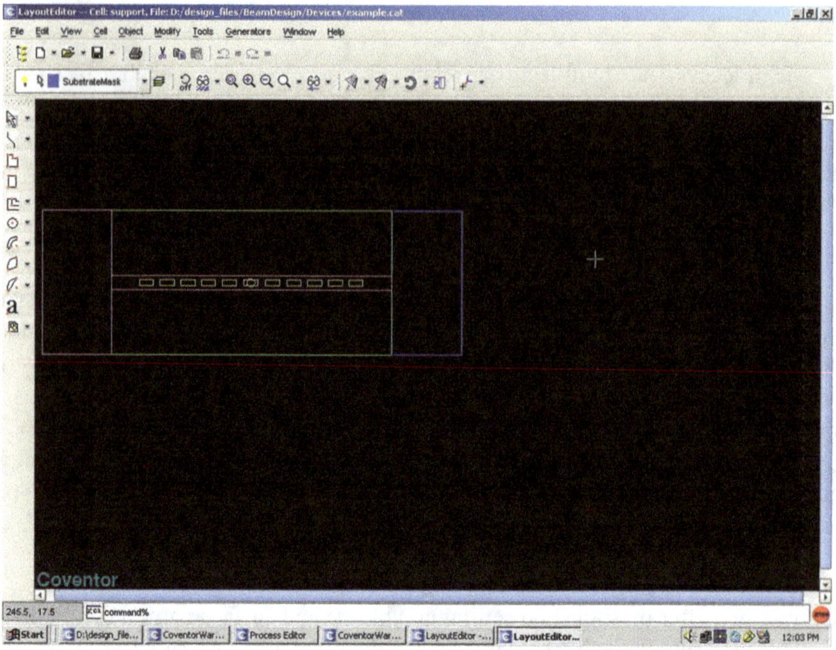

Figure 5.16 Layout of cantilever beam and dimension [25]

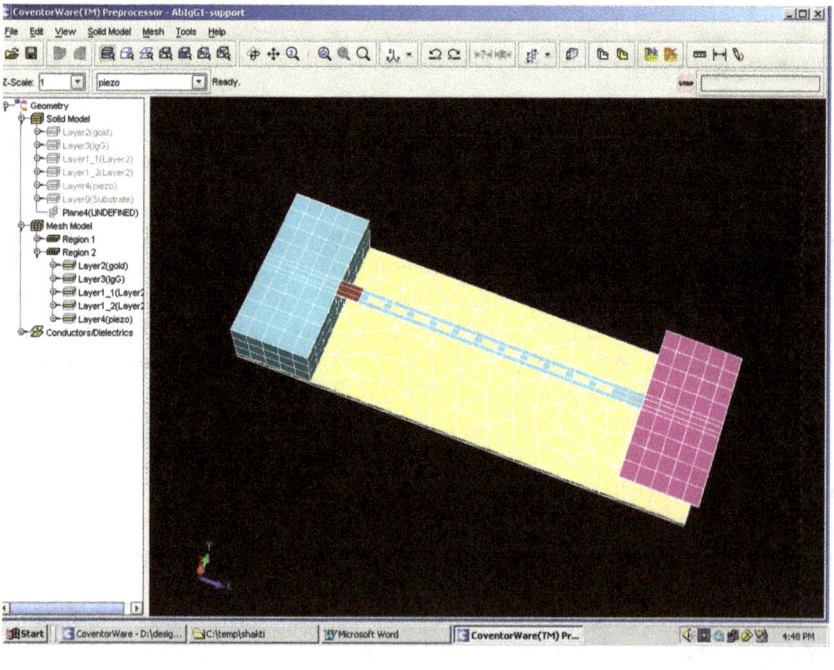

Figure 5.17 Structure and mesh of the cantilever beam designed in CoventerWare software [25]

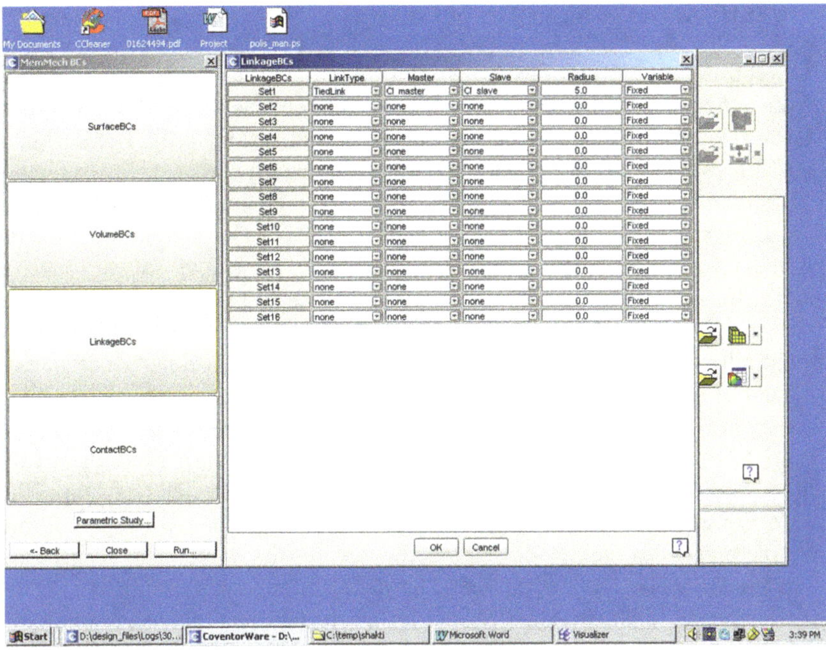

Figure 5.18 Setting the boundary conditions for the cantilever beam [33]

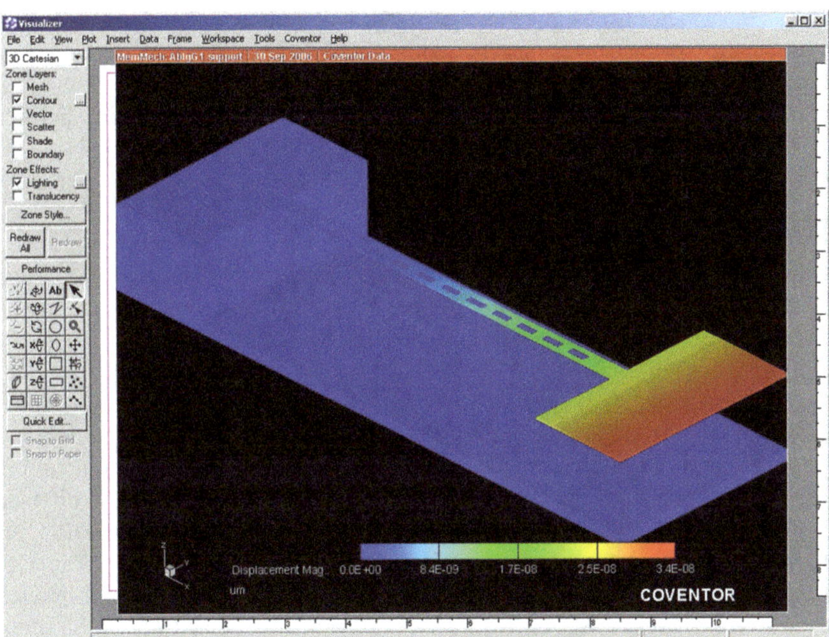

Figure 5.19 Analysis of the cantilever beam: visualiser showing cantilever beam deflection [33]

CoventorWare simulations are performed for many device designs. Lin *et al.* designed a MEMS resonators that could replace the traditional quartz crystal resonators in communications and other applications [26]. The design of the MEMS resonators should match the important parameters such as frequency and temperature stability like the quartz crystal. Therefore, to optimize the design, CoventorWare-based FEA was adapted [27]. The simulation was run using the electromechanical library coupled with statistical solvers such as Monte Carlo simulation. To analyze the mode shapes and frequency of the beam, the CoventorWare module was linked to MATLAB and the simulation was carried out by running the MATLAB code. Thus, adding third party software to work hand in hand with CoventorWare reduces the computation time and many new designs [27]. Similarly, Rahimi *et al.* designed a double air gap parallel plate MEMS capacitor using CoventorWare simulation [28]. Electrostatically actuated and capacitive sensed CMOS-MEMS was modeled using 2008 CoventorWare software [29]. Depending upon the acetone flow, a polymer membrane present in the device absorbs the gas, results in increase of mass and shift in the frequency of the device. The design steps such as initializing the device geometry, importing the materials properties, process flow for creation of 2D layout, construction of a 3D model, meshing the 3D model, fixing the boundary conditions, and analysis using the analyzer tool was carr out. The FEA results were mathematically verified and only 5.3% error was found thus confirming that CoventorWare-based simulation is promising. A high-resolution microaccelerometer is designed using CoventorWare coupled with SABER [30]. A schematic design was built using SABER which was imported to CoventorWare for 3D model building and analysis. Designing with SABER initially helps to reduce the FEA time [30]. CoventorWare simulation helps to understand the MEMS packaging and its effect on the performance [31].

5.5.2 Intellisuite

Intellisuite is a tool used to prototype and manufacture MEMS devices through its design, process, packaging, and system tools [32]. The Intellisuite software consists of process modeling, design tools, and multiphysics modules. The different domains in the process modeling involve mask design, FAB steps, importing materials, and viewing the FAB processes. Designer tool is mainly used for building the 3D model from the process model while the multiphysics module analyzes the device. Some of the modules are similar in functioning to the CoventorWare software. In the process modeling, there are different domains such as AnisE, IntelliEtch, IntelliFAB, MEMaterial, RECIPE 3D, FabViewer, FabSim, and Exposure. In design, Blueprint (Layout), TapeOut, 3DBuilder, and Hexpresso are available. The domains in the multiphysics modules are ThermoElectroMechanical, Electromagnetics&RF-MEMS, Piezo MEMS, Microfluidics&Bio-MEMS, Packaging, and Parametric Designer [32].

To construct a cantilever beam, the required modules are IntelliFAB, 3D builder, and Thermomechanical analyzer.

5.5.2.1 IntelliFAB

IntelliFAB module constructs 3D models directly from the process steps (such as material deposition, wet and dry etching, wafer bonding, and mask definition steps) which are present in the process databases in the IntelliFAB module. The databases can be customized as per the fab requirements. Each process step includes machine parameters which contain details of material properties and mask layouts. As the process sequence involves defining masks, a 2D mask layout that is compatible with DXF and GDSII file formats, standard vendor mask formats need to be created. Figure 5.18 explains the setting of boundary conditions of the cantilever beam. MEMaterial is important when the MEMS structure has low dimensional structures. Nanomaterials, for example, have material properties that are size-dependent. Therefore, correct values of material properties have to be entered to obtain accurate results in the simulation.

5.5.2.2 Builder

The 3D builder is an IntelliSuite module in which the 2D model is built into 3D and meshing of that geometry is done. The model that is built in the IntelliFAB or 3D builder will be exported to the analysis module.

5.5.2.3 Thermoelectromechanical analysis—device analysis

This module couples thermal, electrical and electrostatic, and mechanical analysis. There are five main steps to perform finite element analysis on MEMS devices:

1. Set boundary conditions
2. Mesh structure and locally refine mechanical and electrical meshes
3. Apply loads (voltage, pressure, thermal, etc.)
4. Set calculation type
5. Run analysis and view results

MEMS-based cantilever design and analysis using Intellisuite is performed [33]. The fabrication process for a Silicon cantilever was carried out using the IntelliFAB module and masks for selective etching were designed. A piezoresistive patch (p-type Si) for measuring the strain in the cantilever was attached to the fixed end of the cantilever. The static analysis was carried out using the thermoelectromechanical analysis module. Boundary conditions, material properties, and the input electrical current for the piezo patch were applied. The stress distribution of the microcantilever is shown in Figure 5.20. The red area shows the maximum stress while the region in blue shows minimum stress [33].

5.5.3 COMSOL Multiphysics

COMSOL is a commercial FEM-based multiphysics simulation tool. It can be used for predicting the behavior, designing, and optimization of various engineering devices whose working involves multiple physical phenomena. Various physical domains like solid mechanics, electromagnetics, fluid flow, and heat transfer are included in the tool as modules, which can be coupled according to the need of the simulation. The tool has dedicated MEMS module for designing and analysis of MEMS structure. Various coupled equations for considering electrostatics–solid,

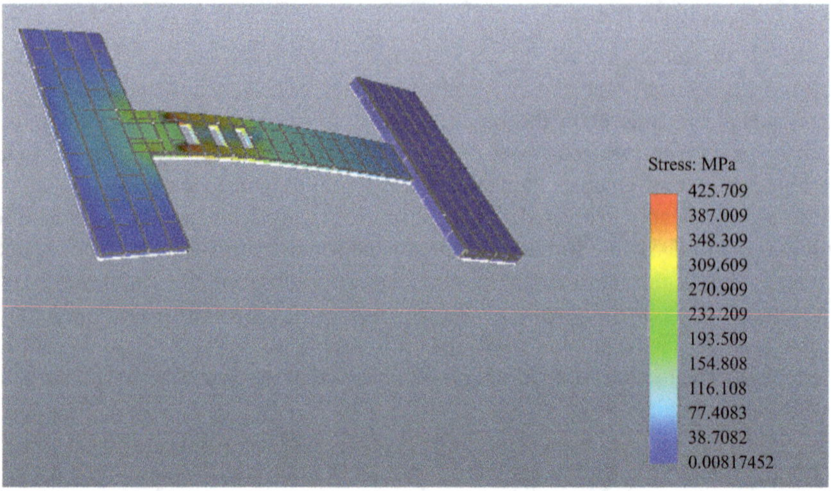

Figure 5.20 Stress distribution on microcantilever [33]

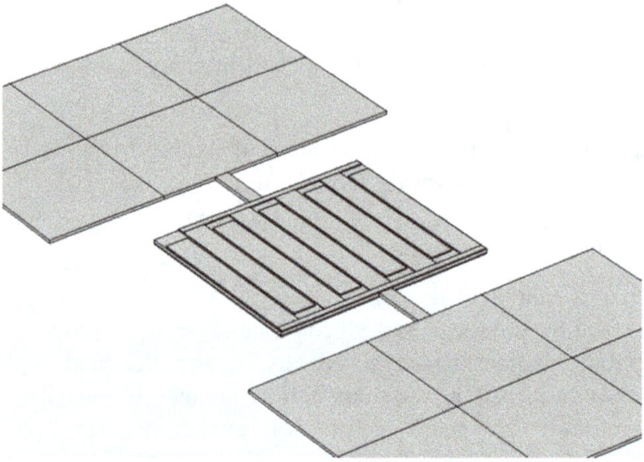

Figure 5.21 The geometry of piezoelectric AlN contour mode resonator designed in COMSOL. The PML is included for modeling anchor loss

thermal–solid, or fluid–solid interactions are present in the MEMS module of the COMSOL tool. In this section, the 3D and 2D simulation of AlN piezoelectric resonator using COMSOL simulation tool is discussed. The damping mechanisms like air damping, thermoelastic damping, and anchor damping are included for the analysis of quality factor for MEMS resonators. For simulating anchor loss, which is the leaking of elastic waves toward substrates, perfectly matched layers (PMLs) that completely absorb the leaking elastic waves from resonator structure are used. These layers will effectively simulate the substrate which is generally very large in comparison to the resonating structure (Figure 5.21).

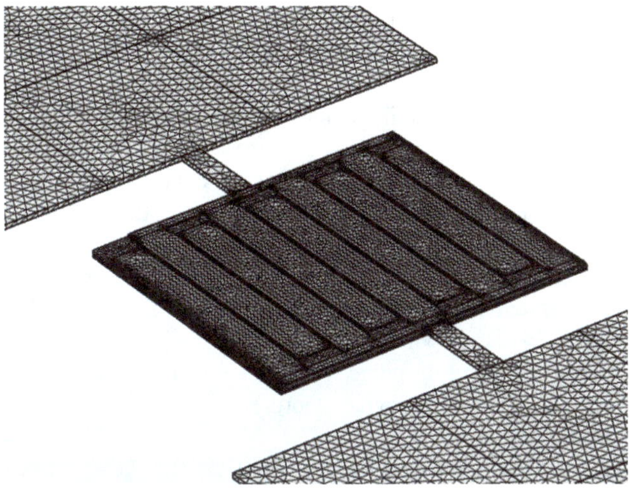

Figure 5.22 The nonuniform mesh for AlN CMR is utilized to reduce the computational resource without affecting accuracy

The steps involved in the simulation of aluminum nitride (AlN), piezoelectric contour mode resonators (CMR) under thickness field excitation (TFE) are presented below. The first step is to define the geometry type, physics involved, and the study required for the device. The geometry type can be defined as 1D/2D/3D or 1D/2D axisymmetric for 2D/3D devices by exploiting the structural symmetry. For the CMR, 2D simulations can be done for finding the approximate location of resonator's eigenfrequency modes while for finding accurate modes detailed 3D simulations with the anchors and resonator length needs to be performed.

Defining physics involves the selection of individual or coupled physical domain required for the device simulation. For example, multiphysics like electromechanics which couples solid mechanics and electrostatics domain needs to be selected for electrically actuated RF MEMS. For AlN CMR, piezoelectric devices module should be selected which combines electromechanics and piezoelectric material constitutional relation. The CMR can work at high frequency (1–10 GHz) for which thermoelastic damping is the major damping mechanism. The thermoelasticity module which couples solid mechanics and heat transport in solids is selected for accounting TED.

The definition of a type of analysis or study namely static, dynamic, or modal analysis which needs to be done is selected in the study section of the COMSOL simulation tool. The important performance parameters for any RF MEMS devices are its natural frequency or modal analysis and its frequency response. For the AlN CMR, modal eigenfrequency and frequency domain analysis is selected for getting the eigenfrequency mode and frequency analysis that include determination of frequency-dependent quality factor and admittance (Figures 5.26 and 5.27).

After the definition of geometry, physics, and study, the next step is to design the device geometry. Any complex geometry can be designed using available simple shapes and their Boolean operations like merging, intersection, extrude, etc. The rich graphical user interface of the tool allows generating any device geometry whose simulation is made possible using FEM for solving PDE on it. While making the geometry, all the dimensions like length and width of the resonator, electrodes, the gap between the electrode, etc. can be defined as the global parameters in the simulator. Doing this will allow the designers to change the geometry dimensions quickly during the device optimization process. Another important damping mechanism which needs to be included in the simulation is anchor loss. To accurately calculate the anchor loss, some part of the silicon substrate is also included with its outer layer as a perfectly matched layer (PML). The CMR geometry designed in COMSOL is shown in Figure 5.21. In AlN CMR, d31 piezoelectric coefficient results in lateral in-plane vibration when a field is applied in the thickness direction. Thus, the device coordinate system needs to be properly selected for getting proper results.

The materials are added next to the geometry. COMSOL has its material database in which popular materials are already included. The material properties required for the selected physical domain are included by default or can also be added in case it is not defined. For example, the essential parameters like Young's modulus E and Poisson's ratio ν for solving solid mechanics and ε for solving electrostatics are defined for the selected material. In the simulation, aluminum of thickness 100 nm is used for the bottom and top electrodes (Figures 5.21 and 5.25).

The physics that was defined in the initial stage needs to be set up which includes setting up the boundary conditions or specifying fixed constraints and the material type, for example, isotropic linear elastic material for metals and anisotropic linear elastic material for silicon. The mechanical Rayleigh damping and dielectric loss for the piezoelectric material are also included under solid mechanics. Under electrostatics physics, input and output electrodes should be defined as terminals and bottom electrode as ground. Voltage is given to the input terminal and the output terminal is taken as floating potential in COMSOL.

As already mentioned, COMSOL relays on solving the physical governing equations using FEM, meshing is the main step for the simulation. COMSOL provides many options for the meshing. By default, free tetrahedral meshing is done for a 3D structure which can be selected from finer to coarser depending on the desired accuracy as can be seen in Figure 5.22 or the physics dependent meshing can be directly selected. Another option for meshing is to define custom meshing by setting maximum/minimum element size.

Finally, the simulation can be run after meshing. First, the eigenfrequency modes are identified which is shown in Figure 5.23 for 3D simulation and in Figure 5.25 for the 2D simulation. The importance of PML for accurately modeling the anchors can be seen from Figure 5.24 in which the leaking of the acoustic vibrational waves can be seen. Following the identification of the resonance mode, frequency analysis around the resonance mode needs to be carried out. The selected physics will be solved using FEM and the results of the selected study can be visualized and analyzed thereafter.

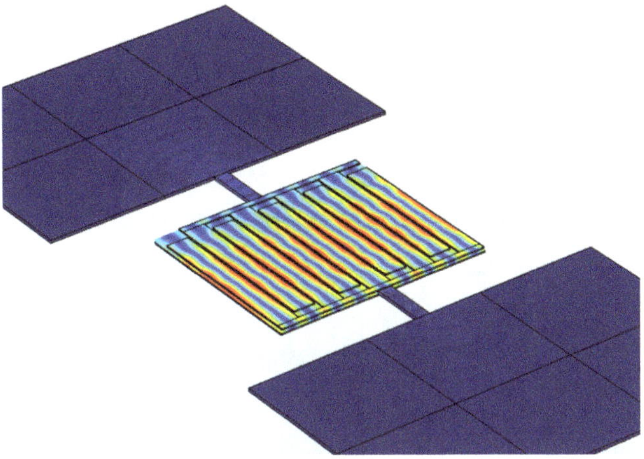

Figure 5.23 The 3D mode shape of AlN CMR with total nine electrodes. The length and width of the AlN film is kept as 15 μm and 18 μm, respectively

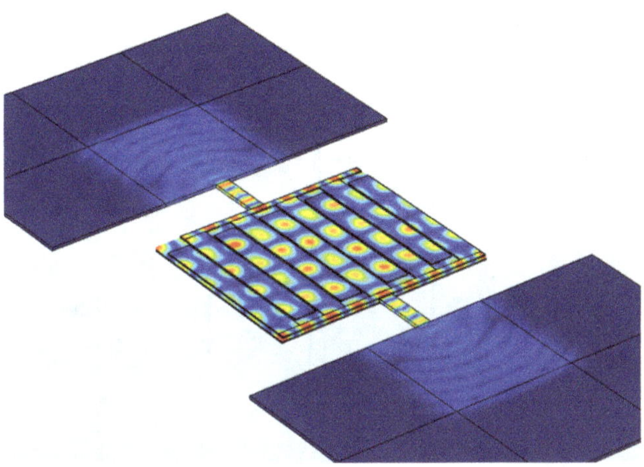

Figure 5.24 The leaking out of the resonator energy through anchors. The leaking out waves through anchors are absorbed by the perfectly matched layers

While calculating eigenfrequency in COMSOL, several options can be set; for example, the number of eigenfrequencies needed, searching of eigenvalues around some specified value, and search methods like largest real, smallest real, etc.

The frequency analysis in the specified frequency range can be performed and the variation of quality factor, admittance in dB and input/output port voltage with the frequency can be plotted as shown in Figure 5.26, Figure 5.27, and Figure 5.28, respectively.

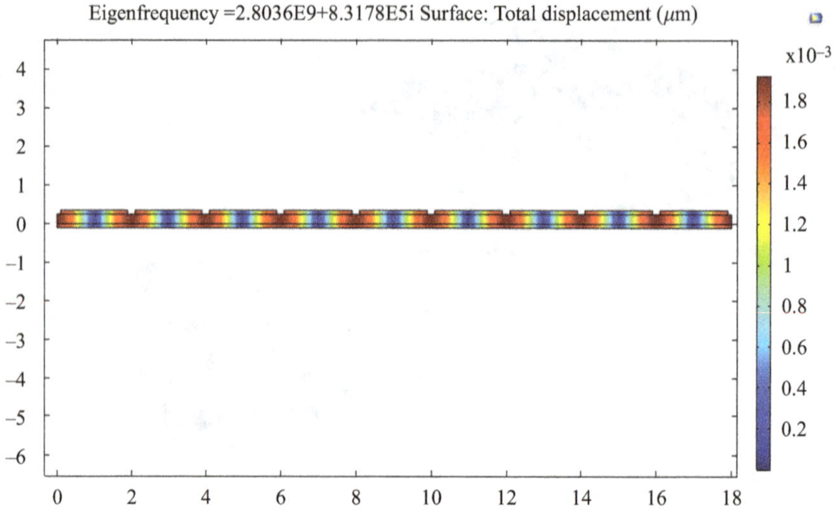

Figure 5.25 The mode shape for the width extensional mode obtained by running 2D simulations

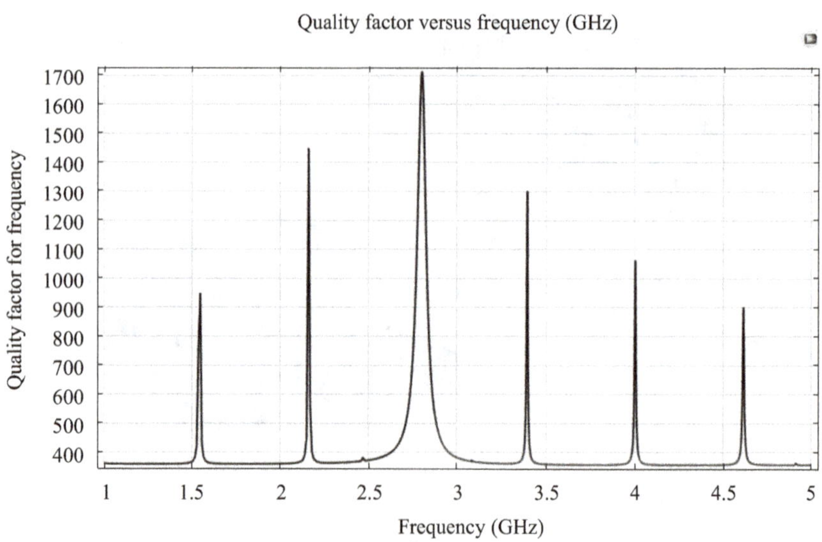

Figure 5.26 The quality factor versus frequency curve for the AlN CMR

Since, for the CMR, the resonance frequency depends on various factors including the electrode width, the gap between electrodes, anchor length/width, and electrode materials, the resonance frequency can be tuned by running accurate FEM simulations by tuning these parameters.

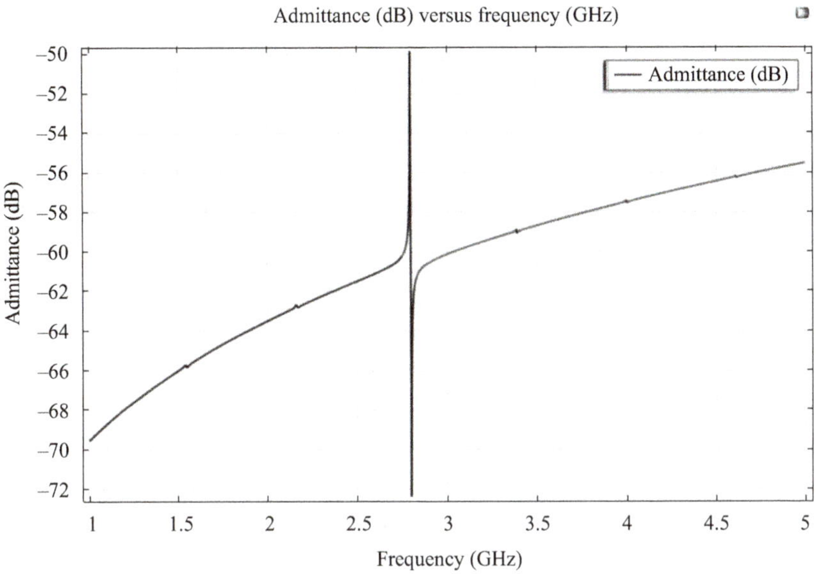

Figure 5.27 The admittance (20log10(abs(Y11))) versus frequency curve for the AlN CMR as obtained from COMSOL

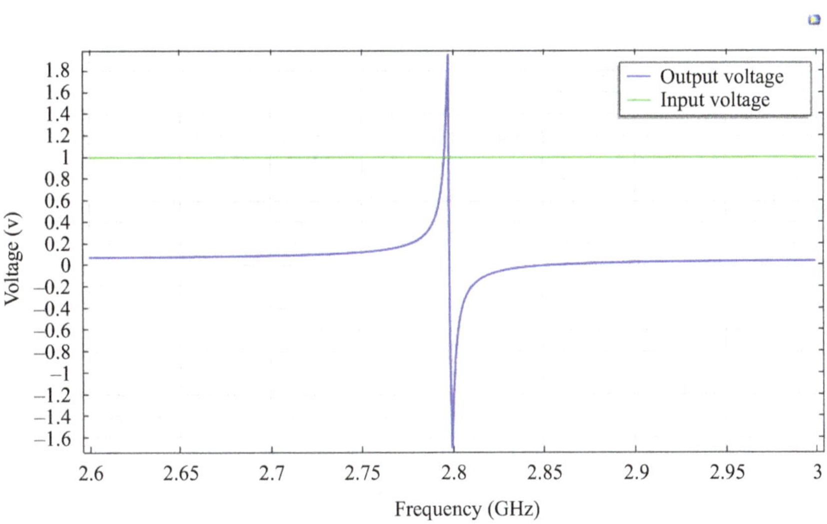

Figure 5.28 The variation of voltage at output port for the constant 1 V applied at the input port. The output voltage sharply peaks at resonance frequency enabling the device to work as filter or resonator

5.6 Summary

RF MEMS technology has shown unmatched potential for super high frequency oscillators and filters used for next generation communication systems. The RF MEMS resonator comprises of submicrometer size resonating structure which can either be suspended in substrate cavity via anchors or can be placed directly on the top of the substrate. The resonator structure can be fabricated using semiconductor or metal having shapes like beam, plates, and ring and can be actuated electrostatically. Another class of RF resonators having piezoelectric layer as a resonating structure in the mechanical vibrations caused by electrostatic actuation is utilized for oscillators. The key performance parameters for the resonators are the resonating frequency, high quality factor, and low motional resistance. To meet the required specification of these parameters, the device dimensions, resonator materials, and proper designing are required which can be achieved using proper device modeling and computer simulations.

Modeling is a process of obtaining a set of mathematical equations which govern resonator's working. The most basic model of mechanical resonator is spring-mass-damper model in which spring represents restoration and damper represents energy loss mechanisms. The series *R-L-C* circuit is the electrical equivalent of the mechanical spring-mass-damper model and can be useful in case supporting circuitry needs to be simulated along with resonator. However, these models are limited to single frequency analysis and are difficult to apply for complex geometries.

As the working of MEMS resonators depends on various material- and structure-dependent mechanisms, physical modeling is the best modeling approach for them. Physical modeling relies on solving the governing equation related to all involved physical mechanisms like electrostatics which governs the applied electrical force, solid-mechanics which governs structure deformation, and fluid flow and thermal transport which govern the damping mechanisms. All these physical mechanisms are coupled with each other during the working of RF MEMS resonators.

Finite-element method is the most popular numerical technique used for solving partial differential equations with the known boundary conditions. This technique is best suited for arbitrary geometry and can be coded easily to make generalized software package. The differential equations are expressed as algebraic sets of equations that can then be represented into matrix form and numerical techniques can be used to solve these matrix equations. The two common procedures that are applied on the governing equation are the variational method and the Galerkin method. A continuum domain is discretized by dividing system domain into elements and a linear solution is assumed in each element. The process of discretizing domain into elements is called as meshing. Once the matrices are assembled and boundary conditions are applied, the matrices can be supplied to the solvers for obtaining the solutions. Postprocessing is the last step in FEM which deals with analysis and visualization of the obtained solutions or results.

There are several commercial software packages that do FEM computation on MEMS devices. The CoventorWare software is a design and simulation tool for modeling wide range of device behavior and simulation of the fabrication process flow of MEMS devices. The designing process of cantilever beam in CoventerWare software is presented. Intellisuite is another software which consists of process modeling, design tools, and multiphysics modules. The simulation of silicon cantilever in Intellisuite is presented. COMSOL is a commercial FEM-based multiphysics simulation tool. The simulation steps and types of analysis that can be performed using COMSOL are discussed and simulation steps for piezoelectric AlN-based contour mode resonators are presented.

References

[1] Nguyen C.T.C. "MEMS technology for timing and frequency control." *IEEE Transactions on Ultrasonics, Ferroelectrics, and Frequency Control*. 2007; **54**(2):251–270.

[2] Yao J.J. "RF MEMS from a device perspective." *Journal of Micromechanics and Microengineering*. 2000;**10**(4):R9.

[3] Yang Y., Lu R., Manzaneque T., *et al.* "Toward Ka band acoustics: lithium niobate asymmetrical mode piezoelectric MEMS resonators." In: *2018 IEEE International Frequency Control Symposium (IFCS)*. IEEE; 2018. pp. 1–5.

[4] Basu J., and Bhattacharyya T.K. "Microelectromechanical resonators for radio frequency communication applications." *Microsystem Technologies*. 2011;**17**(10–11):1557.

[5] Campanella H. *Acoustic Wave and Electromechanical Resonators: Concept to Key Applications*. Artech House Integrated Microsystems Series; 2010.

[6] Tilmans H.A. "Equivalent circuit representation of electromechanical transducers: I. Lumped-parameter systems." *Journal of Micromechanics and Microengineering*. 1996;**6**(1):157.

[7] Heyliger P.R. "When beam theories fail." *Journal of Mechanics of Materials and Structures*. 2013;**8**(1):15–35.

[8] De Silva C.W. *Vibration: Fundamentals and Practice*. CRC Press; 1999.

[9] Lifshitz R., and Roukes M.L. "Thermoelastic damping in micro-and nano-mechanical systems." *Physical Review B*. 2000;**61**(8):5600.

[10] Zener C. "Internal friction in solids. I. Theory of internal friction in reeds." *Physical review*. 1937;**52**(3):230.

[11] Veijola T., and Raback P. "Methods for solving gas damping problems in perforated microstructures using a 2D finite-element solver." *Sensors*. 2007; **7**(7):1069–1090.

[12] Duwel A., Candler R.N., Kenny T.W., *et al.* "Engineering MEMS resonators with low thermoelastic damping." Journal of Microelectromechanical Systems. 2006;**15**(6):1437–1445.

[13] Park Y.H., and Park K. "High-fidelity modeling of MEMS resonators. Part I. Anchor loss mechanisms through substrate." *Journal of Microelectromechanical Systems*. 2004;**13**(2):238–247.

[14] Sumant P.S., Aluru N.R., and Cangellaris A.C. A methodology for fast finite element modeling of electrostatically actuated MEMS. *International Journal for Numerical Methods in Engineering*. 2009;**77**(13):1789–1808.

[15] Geuzaine C., and Remacle J.F. "Gmsh: A 3-D finite element mesh generator with built-in pre-and post-processing facilities." *International Journal for Numerical Methods in Engineering*. 2009;**79**(11):1309–1331.

[16] Bastian P., Blatt M., Dedner A., *et al.* "A generic grid interface for parallel and adaptive scientific computing. Part I: Abstract framework." *Computing*. 2008;**82**(2–3):103–119.

[17] Hernandez V., Roman J.E., and Vidal V. "SLEPc: A scalable and flexible toolkit for the solution of eigenvalue problems." *ACM Transactions on Mathematical Software*. 2005;**31**(3):351–362.

[18] Oñate E. *Structural Analysis with the Finite Element Method. Linear Statics: Volume 2: Beams, Plates and Shells*. Springer Science & Business Media; 2013.

[19] Kwon Y.W., and Bang H. *The Finite Element Method Using MATLAB*. CRC Press; 2018.

[20] Yi Y.B. "Finite element analysis of thermoelastic damping in contour-mode vibrations of micro-and nanoscale ring, disk, and elliptical plate resonators." *Journal of Vibration and Acoustics*. 2010;**132**(4):041015.

[21] Guo X., Yi Y.B., and Pourkamali S. "A finite element analysis of thermoelastic damping in vented MEMS beam resonators." *International Journal of Mechanical Sciences*. 2013;**74**:73–82.

[22] Nguyen W.T., Kumar P., and Leong J. Finite element modelling and simulations of piezoelectric actuators responses with uncertainty quantification. *Computation*. 2018;**6**(4):60.

[23] CoventorWare MEMS Solutions Overview. https://www.coventor.com/mems-solutions/.

[24] CoventorWare User Manual: Doc Ver 2006 Rev A Compatible with CoventorWare Version 2006.

[25] Swarrup J.S., Govardhan K., and Velmurugan V. "Simulation of perforated rectangular cantilever immunosensor for estimation of bacterial pathogens." Health. 2010;**2**(11):1239–1248.

[26] Lin Y.W., Lee S., Li S.S., *et al.* "Series-resonant VHF micromechanical resonator reference oscillators." *IEEE Journal of Solid-State Circuits*. 2004;**39**(12):2477–2491.

[27] MEMS Resonator Design and Simulation. https://www.coventor.com/mems-solutions/resonators/.

[28] Rahimi M., Jamuar S., Hamidon M., *et al.* An optimized 2.4 GHz VCO circuit design and simulation with high-Q MEMS LC-tank. *AEU-International Journal of Electronics and Communications*. 2010;**64**(5):413–424.

[29] Dennis J.O., Rabih A.A.S., Khir M., *et al.* Modeling and finite element analysis simulation of MEMS based acetone vapor sensor for noninvasive screening of diabetes. *Journal of Sensors*. 2016.

[30] Mistry K.K., Swamy K., and Sen S. "Design of an SOI-MEMS high resolution capacitive type single axis accelerometer." *Microsystem Technologies*. 2010; **16**(12):2057–2066.

[31] Lishchynska M., O'Mahony C., Slattery O., *et al.* "Evaluation of packaging effect on MEMS performance: Simulation and experimental study." *IEEE Transactions on Advanced Packaging*. 2007;**30**(4):629–635.

[32] Intellisuite @ Technical Reference Manual, Version 8.2/PC, Part No.30-090-100. 2007.

[33] Swarrup J.S., and Velmurugan V. Modeling of micro-electro mechanical systems based cantilever beam as biosensor for C-reactive protein. *International Journal of Nanoscience*. 2011;**10**(04–05):797–801.

Chapter 6

Fabrication of low-frequency resonators

Amol Morankar[1]

6.1 Introduction

Due to tremendous success in the fabrication of microelectronics components and chips, it has fueled the development in the nonelectronics components such as sensors and systems. It has also resulted in the adaption of VLSI fabrication techniques for the manufacturing of sensors and interfacing with nonelectronics systems. MEMS contain mechanical elements that can be fabricated using VLSI fabrication techniques. It can be merged with on-chip signal processing systems and actuators. MEMS may contain sensors such as pressure sensors, flow sensors, temperature sensors, optical signal handling systems, lab on a chip, etc. Fully integrated MEMS are designed using sophisticated VLSI CAD software and can be batch fabricated using various VLSI-based fabrication tools and techniques. Thus, MEMS components are progressively becoming faster, smaller, efficient, and more functional [1–3].

In this chapter, fabrication of low-frequency MEMS resonator is discussed. The designed device can be fabricated on a 2/4/8-inch silicon wafer as per the requirement. The general specifications of the wafer that can be selected for fabrication of low-frequency MEMS resonator are as follows:

- Size: 2/4/8 inch
- Single-side polished
- Orientation: "100"
- Type: n
- Resistivity: > 1,000 ohm-cm

The performance of the resonator depends on the substrate parasitic and resistivity. The n-type silicon wafer can be selected since it is required to use metals as a structural material and Cr/Au for connecting pads. It would result in low parasitic capacitance and achieve better device performance. The standard surface micromachining process is mostly preferred for the fabrication of low-frequency devices as compared with bulk micromachining and Lithographie, Galvanoformung, Abformung (LIGA). Silicon dioxide can be used as electrical isolation between the

[1]Riva Labs, CIVN, Visvesvaraya National Institute of Technology, Nagpur, India

metal electrodes and the substrate. Silicon dioxide is also preferred as sacrificial material. The standard CMOS fabrication technique employs Cr/Au for connecting pads which can be preferred in resonators as well. Au/Ag/Cu/Al/Ni metal can be used as a structural material in the formation of beams [4–7]. Fabrication of the low-frequency MEMS resonator includes processes such as

- Mask design using Clewin 4 and mask printing
- RCA cleaning of Si wafer
- Wet oxidation for oxide growth
- Spin coating for deposition of photoresist
- Annealing
- Thermal evaporation for Cr/Au deposition
- CVD/LPCVD for deposition
- Lithography: double-sided alignment and exposing to UV light
- Dielectric sputtering for deposition of sacrificial layer/metals
- Electroplating for deposition of thick metal layers
- Liftoff and wet etching for removal of unwanted material

6.2 Fabrication processes

Microfabrication techniques are employed in the microelectronics and MEMS fields, which are based on planar technologies. Various electronic devices and MEMS components are fabricated on the initially flat Si substrate. Wafer-based process is developed for the manufacturing of integrated circuits which can be exploited for the fabrication of MEMS components with or without variations in the process. Bulk micromachining is the term applied to a variety of etching procedures that selectively remove material, usually with help of a chemical etchant whose etching properties are dependent on the crystallographic structure of the bulk material. Wafer-to-wafer bonding is a strategy commonly employed to avoid the restrictions in the type of structures that can be fabricated using bulk micromachining. Since anisotropic etching only removes the desired material exposed after photoresist deposition, bonding of wafers allows for the addition of material to the bulk micromachining. High-aspect-ratio micromachining is a micromachining technique developed (originally in Germany) to allow the fabrication of thick (usually greater than hundreds of microns and up to centimeters thick), high precision, and high-aspect-ratio MEMS structures with approximately vertical sides. Surface micromachining, like bulk micromachining, also starts with a wafer that acts as a surface on which multiple, alternating layers of structural and sacrificial material are deposited and etched. Surface micromachining assists the fabrication of free-form, complex, and multicomponent integrated electromechanical structures [8–11]. It also assists MEMS designer to predict and build devices and systems that are difficult to realize with bulk micromachining or wafer bonded processes. It is very much important to determine a process sequence that creates an accurate desired structure and can be manufactured using available tools at a satisfactory cost. It involves various combinations of technical knowledge and manufacturing experience to achieve the desired device structure [12–13]. The processes involved in the fabrication of the low-frequency MEMS resonator are discussed. Finally, an example to illustrate the steps in fabrication process to get the desired structure is explained.

6.2.1 Wafer cleaning process

The standard set of the wafer cleaning process is termed as RCA cleaning. It involves various steps to remove impurities before any fabrication step. First, it removes all organic coatings using a 7:3 mixture of concentrated sulfuric acid and hydrogen peroxide (piranha). Second, organic residues are removed using 5:1:1 mixture of water, hydrogen peroxide, and ammonium hydroxide. It grows a thin oxide on the wafer which can be removed with the help of dilute HF etch and can be avoided if it is desired to have oxide on the wafer. Finally, ionic contaminants are removed with a 6:1:1 mixture of water, hydrochloric acid, and hydrogen peroxide.

6.2.2 Oxide deposition

High-quality oxide can be deposited on Si wafer using a dry and wet oxidation process. A molecule of oxygen reacts directly with the silicon wafer, forming silicon dioxide. It is processed in tube furnaces, which are operated at temperatures from around 850 °C to 1150 °C. In the dry oxidation process, pure oxygen is used as the oxidant that is made to flow through oxidation furnace with a flow of nitrogen which acts as a diluent. The rate of oxidation depends on the flow rate of oxygen at the silicon–oxide interface. The rate of oxidation decreases as the oxygen diffuses through the silicon surface. Dry oxidation is employed when the highest-quality oxides are required. The diffusion rate of oxygen through oxide can be significantly enhanced through wet oxidation. Water breaks a silicon–oxygen–silicon bond, forming two OH groups, hence the oxidation rate is faster. The water vapor can be provided by oxidizing hydrogen to steam in the furnace. Wet oxidation is used to make thicker oxides, from several hundred nm to μm. If a thick oxide layer is desired, high-pressure steam oxidation or chemical vapor deposition methods can be employed. However, the thickness of the wafer increases due to deposition of oxide since the density of silicon in silicon dioxide is lower than in crystalline silicon. Local oxidation can be achieved with the use of oxygen diffusion barriers such as silicon nitride.

6.2.3 Thin-film deposition

MEMS fabrication process step involves the deposition and subsequent patterning of a thin film. Various methods are developed and employed for performing such depositions, which are mostly referred to as additive processes.

6.2.3.1 Physical vapor deposition

In this deposition method, the metal to be deposited is held in a crucible. The surface of the metal is then heated with an incident electron beam. The flux of vapor atoms from the crucible is guided to reach the wafer under high-vacuum conditions. Evaporation with an e-beam is fairly directional, that is employed to obtain shadowing effects. Shadowing effects through the use of tilted substrate permit one side of elevated features to be coated, while the other side is shadowed. Such a method is mostly employed in the deposition of electrodes used in accelerometers and rate gyroscopes.

6.2.3.2 Sputter deposition

Sputtering is a process in which chemically inert atoms, such as argon, are ionized in a glow discharge also termed as plasma. The ions are accelerated into a target material to be deposited by the electric field in the dark space at the boundary of the plasma. Atoms from the target are knocked out and these atoms are permitted to reach the substrate. Sputtering takes place under a low-pressure gas environment and is comparatively less directional than the e-beam evaporation method. However, it can achieve much higher deposition rates. Sputtering can also be employed with nonmetallic targets such as dielectric silicon dioxide and piezoelectric film zinc oxide and aluminum nitride.

6.2.3.3 Chemical vapor deposition

Chemical vapor deposition is mostly preferred method for deposition in which material to be deposited is introduced into a heated furnace. Various materials can be deposited using this method such as thin-film silicon, silicon dioxide, silicon nitride, etc. Mostly chemical vapor deposition process is performed under low pressure (less than 1 μTorr) termed as LPCVD and involves an inert diluent gas, such as nitrogen, in addition to the reacting species. LPCVD process is performed at high temperatures in the range of 500° to 850 °C. These ranges of temperatures are not suitable to perform deposition on wafer predeposited with metals such as gold, aluminum, chromium, etc. that form eutectics with silicon at 577 °C and 380 °C. Therefore, LPCVD needs to be employed before deposition of metals on the silicon. Deposition rates can be enhanced if the deposition occurs in glow-discharge plasma termed as a plasma-enhanced chemical vapor deposition (PECVD) process. PECVD is mostly employed for the deposition of thin-film dielectrics at low temperatures and used as insulation layers.

6.2.3.4 Electrodeposition

Electrodeposition or electroplating process involves electrochemical solution in which metal ions are deposited onto a substrate. Metals that are suitable for electroplating are gold, copper, chromium, nickel, and magnetic iron–nickel alloys (permalloy). Mostly, the electroplating process involves controlling deposition with an applied electric current. Uniformity in the thickness of the deposited material depends on uniform current density everywhere the plating is done and the quality of the electrochemical solution. Electroplated metals often exhibit rougher surfaces than evaporated or sputtered films. Electroplating is mostly employed to achieve higher thickness and high aspect ratio microstructures in MEMS sensors and systems.

6.2.3.5 Spin coating

Spin coating is the standard method employed for depositing photoresist, a photosensitive polymer layer used in photolithography. It can also be used to deposit polyimide films, which are insulating polymers that can withstand temperatures up to 400 °C. The material to be deposited, typically a polymer or chemical precursor to a polymer, is dissolved in a suitable solvent. The solution is applied on the wafer, and the wafer is spun at high speed. Centrifugal forces in combination with the surface tension of the solution and the viscosity of the solution assist in spreading the film uniformly over the wafer which results in uniform thickness. The postbake (annealing) is used to remove solvent and also converts the amic acid to an imide, due to loss of a water molecule.

6.2.3.6 Photolithography

Photolithography consists of two parts: a photo-process, whereby the desired pattern is photographically transferred from an optical plate to a photosensitive film coated on the wafer, and a chemical or physical process of either removing or adding materials to create the pattern. A photomask contains the pattern to be transferred which is a mixed set of opaque and transparent regions. It is brought into contact with an oxidized silicon wafer coated with photoresist. Ultraviolet light is directed through the mask onto the wafer, exposing the unprotected portions of the resist that changes the chemical properties as a result of the light exposure. Most processes are subtractive that results by removing photoresist material through etching unwanted material chemically. The positive photoresist regions exposed to the UV light become more soluble in the developer than the protected regions. After development and postbake, the protected regions of resist remain on the wafer so that after etching, the opaque regions of the mask remain as oxide and the clear regions are removed. Negative photoresist functions exactly opposite of positive photoresist. Regions that are clear in the negative become dark in the transferred image. The regions exposed to UV light are insoluble in the developer. The final result is a transfer of pattern into the photoresist so that after etching, the opaque regions of the mask become regions cleared of photoresist.

6.2.3.7 Etching

Etching is a process of removing unwanted material through suitable masking layer such as patterned photoresist. The etching is achieved by immersion of the patterned substrate in a suitable chemical that removes the exposed region of the substrate whereas the unexposed regions are protected. The rate of etching and features of the structure depends on the type of substrate, the chemical properties of the etchant, and the masking layer. Wet etching can be used to etch deeply into the substrate and is mostly isotropic. A deep reactive ion etching process is mostly employed to achieve higher etching rate and deep etches into the material or photoresist.

6.2.3.8 Liftoff

Liftoff is used with metals that are difficult to etch with plasmas. A wafer is coated with a resist and is exposed and developed so as to create a slightly re-entrant resist profile. This same effect can be created with a two-part masking layer in which the upper layer is patterned, and the second layer beneath is slightly undercut. When metal is evaporated from a directional source, such as an e-beam heated crucible, the resist profile shadows the side walls. The metal thickness must be only a fraction of the resist thickness. When the resist is stripped, the metal on top of the resist is lifted-off, while the metal deposited directly into the opening of the resist remains.

Microfabrication process steps briefed are mostly employed in the fabrication of low-frequency MEMS resonator devices [14–15]. However, it does provide an overview of many of the steps used in the fabrication of MEMS devices. It demonstrates process sequences and assists in planning by analyzing some of the trade-offs between process and device performance. Process planning by reducing the number of photolithography layers is desired.

6.3 Device fabrication

The steps involved in the fabrication of the device are illustrated with an example. The fabrication of a clamped-clamped (CC) beam array is selected for a better understanding of the process flow. The CC beam array through mixed mechanical and electrical coupling scheme was fabricated using nickel-metal as a structural material that involves four-level lithography processes. It consists of four mechanically coupled CC beams anchored to a substrate using anchors, one input electrode and two output electrodes. Electrically coupling was achieved through the placement of input–output electrodes [16–17]. The structure of the device is depicted in Figure 6.1. Figure 6.2 shows the masks prepared using Clewin software; all the masks were dark field, and mask 2 and 3 are similar.

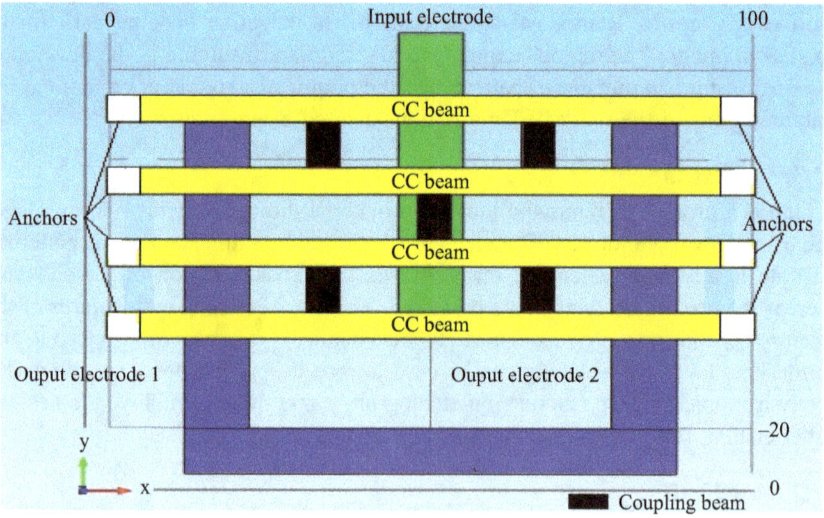

Figure 6.1 CC beam array [16]

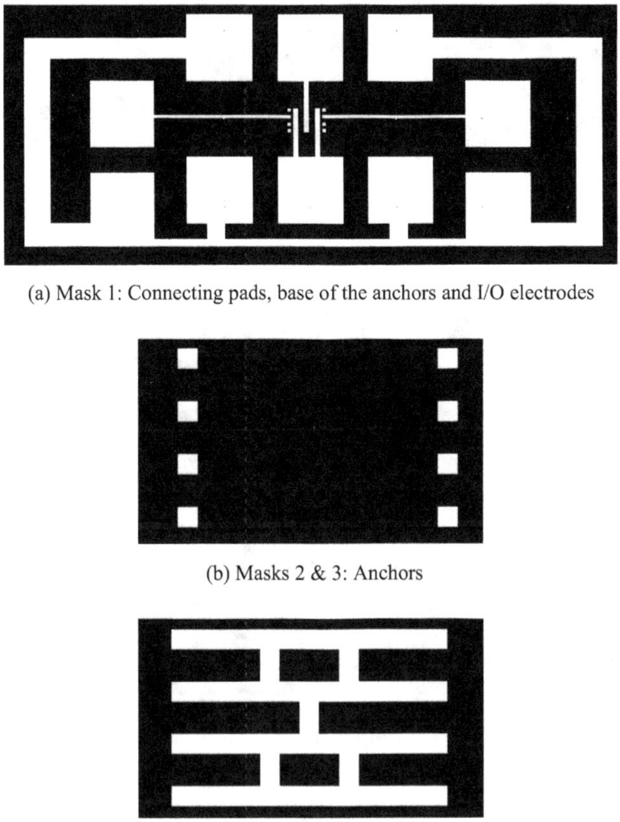

(a) Mask 1: Connecting pads, base of the anchors and I/O electrodes

(b) Masks 2 & 3: Anchors

(c) Mask 4: CC beams and mechanical couplers

Figure 6.2 Mask design using Clewin software: masks 2 and 3 are similar. (a) Mask 1: connecting pads, base of the anchors, and I/O electrodes; (b) masks 2 and 3: anchors; and (c) mask 4: CC beams and mechanical couplers

The steps involved in the fabrication of the device are as follows:

1. RCA cleaning: A two-step process that removes organic contaminants, metal contaminants, and any oxide layer that might have been deposited on the wafer.
2. Oxide deposition: A 1-μm-thick oxide (isolation oxide) was grown on the surface of the n-type silicon wafer using a wet oxidation process refer Figure 6.3. It is an isolation layer between electrodes and substrate.
3. Photoresist deposition 1: The wafers are prebaked at 150 °C for 180 s, then coated with UV-sensitive LOR and photoresist AZ5216 with a thickness of 1 μm using spin coating and then annealing (postbake) at 100 °C for 180 s (see Figure 6.4).

Figure 6.3 Oxide deposition

Figure 6.4 Photoresist deposition 1

Figure 6.5 Lithographically patterned mask 1

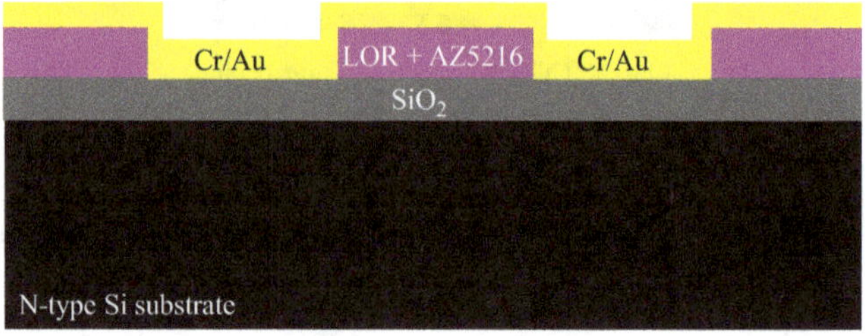

Figure 6.6 Metal deposition 1

4. Expose to UV light: Lithographically patterned by exposing to UV light through the first level mask for 10 s and then developing it in MF319 for 35 s and washed with DI water. The photoresist in exposed areas was removed, leaving behind a patterned photoresist mask for further process (see Figure 6.5).
5. Metal deposition 1: A 120 nm layer of Cr was deposited, followed immediately by the deposition of a 40 nm layer of Au using thermal evaporation (Figure 6.6). The metal was deposited in vacuum at a pressure of about 1 μTorr and the rate of deposition was approximately 0.1 nm/s. This first metal layer forms the base of anchor, input–output electrodes and connecting

pads for RF & DC probes. The thickness was measured using surface profilometer after deposition that varied from 150 nm to 160 nm.

6. Metal patterning 1: Liftoff was performed to remove photoresist from the patterned areas using PG remover. After the liftoff was completed, the remaining photoresist if any was removed using acetone and IPL (see Figure 6.7). Figure 6.8 shows the image of the wafer obtained using an Olympus microscope after completing this step.
7. Photoresist deposition 2: The wafers are again prebaked at 150 °C for 180 s, then coated with UV-sensitive LOR and photoresist AZ5216 with a thickness of 1.5 μm using spin coating and then annealing (postbake) at 100 °C for 180 s (see Figure 6.9).
8. Expose to UV light: Lithographically patterned by exposing to UV light through the second level mask for 10 s and then developing it in MF319 for 35 s and washed with DI water. The photoresist in exposed areas was removed, leaving behind a patterned photoresist mask for further process (see Figure 6.10).

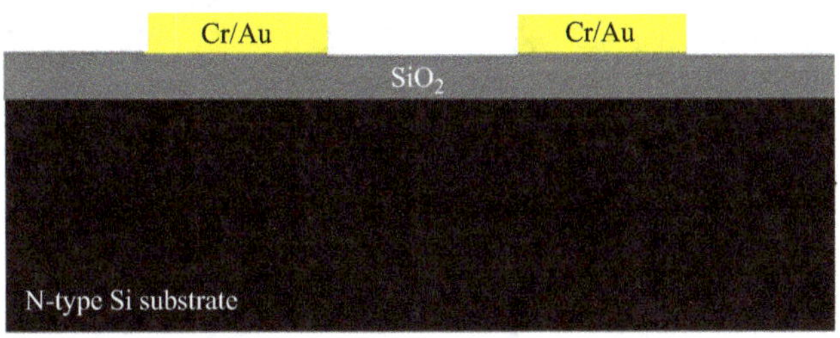

Figure 6.7 Metal patterning 1

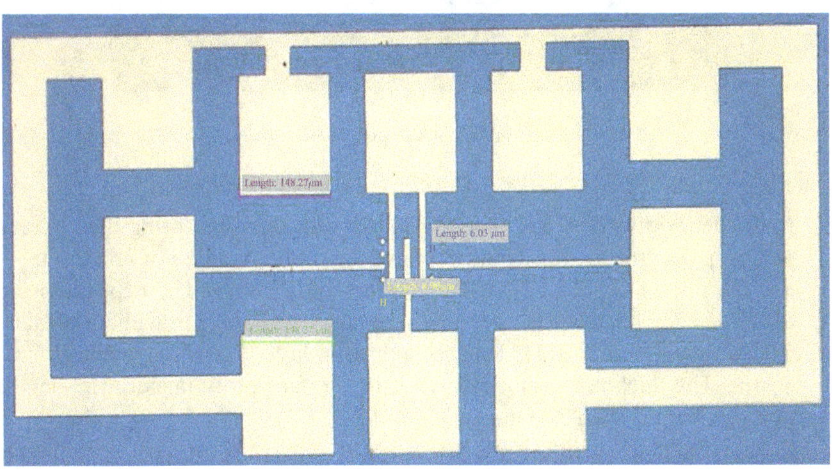

Figure 6.8 Image of the wafer obtained after first metal deposition

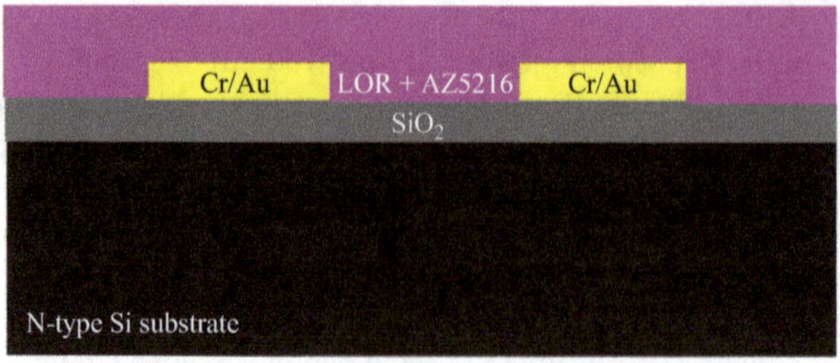

Figure 6.9 Photoresist deposition 2

Figure 6.10 Lithographically patterned mask 2

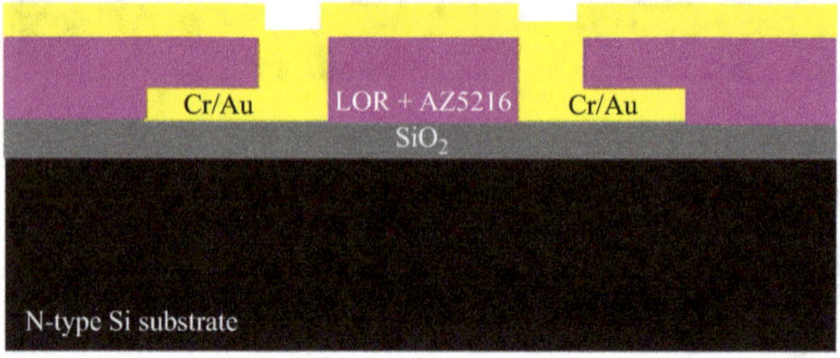

Figure 6.11 Metal deposition 2

9. Metal deposition 2: A 360 nm layer of Cr was deposited using thermal evaporation. The metal was deposited in vacuum at a pressure of about 1 μTorr and the rate of deposition was approximately 0.1 nm/s. The deposition process was completed in two consecutive slots of 180 nm each without breaking vacuum (see Figure 6.11). This second metal layer forms the anchor for CC beams. The thickness was measured using surface profilometer after deposition that varied from 340 nm to 360 nm.
10. Metal patterning 2: Liftoff was performed to remove photoresist from the patterned areas using PG remover. After the liftoff was completed, the remaining photoresist if any was removed using acetone and IPL (see Figure 6.12).

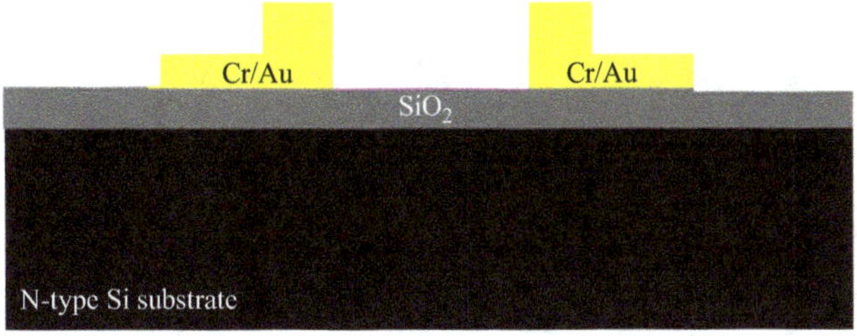

Figure 6.12 Metal patterning 2

Figure 6.13 Misalignment of the anchors

The images and measurements were obtained through an Olympus microscope after depositing the second metal layer. Misalignment of the anchors in first and second layers of metals may be observed for few/all devices as shown in Figure 6.13.

11. Deposition of sacrificial layer: A 360 nm sacrificial layer of silicon dioxide was deposited using dielectric sputtering. It was deposited in vacuum at a pressure of about 0.01 Torr and the rate of deposition was approximately 1 nm/min. The process was completed in two consecutive slots of 180 nm each without breaking vacuum (see Figure 6.14). The deposited silicon dioxide (387 nm thickness) was measured using a surface profilometer.
12. Photoresist deposition 3: The wafers are prebaked at 150 °C for 180 s then coated with UV-sensitive photoresist AZ5216 with a thickness of 1.5 μm using spin coating and then annealing (postbake) at 100 °C for 180 s (see Figure 6.15).
13. Expose to UV light: Lithographically patterned by exposing to UV light through the third level mask for 10 s and then developing it in MF319 for 35 s and washed with DI water. The photoresist in exposed areas was removed, leaving behind a patterned photoresist mask for further process (see Figure 6.16).

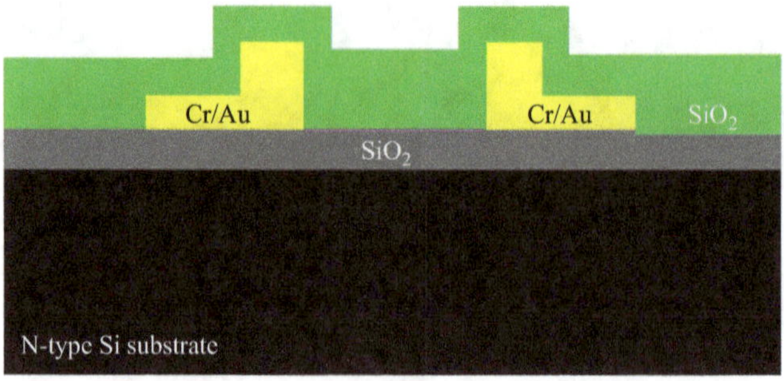

Figure 6.14 Deposition of sacrificial layer (SiO_2)

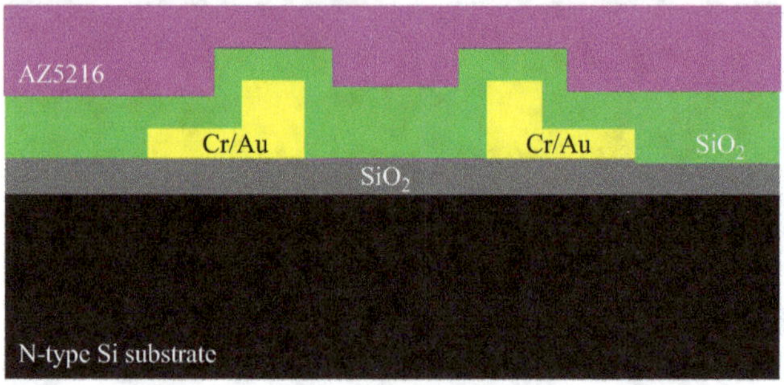

Figure 6.15 Photoresist deposition 3

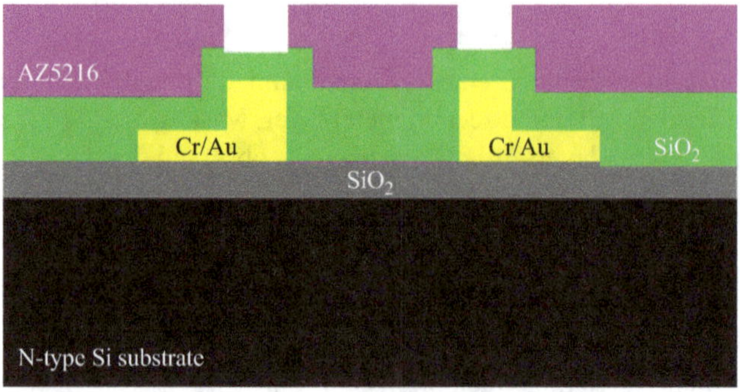

Figure 6.16 Lithographically patterned mask 3

14. Patterning: Wet etching was performed to remove SiO_2 from the patterned areas using 5:1 BHF for 40 s with etch rate 10 nm/s. After the etching was completed, the remaining photoresist if any was removed using acetone and IPL (see Figure 6.17).
15. Photoresist deposition 4: The wafers are prebaked at 150 °C for 180 s then coated with UV-sensitive LOR and photoresist AZ5216 with a thickness of

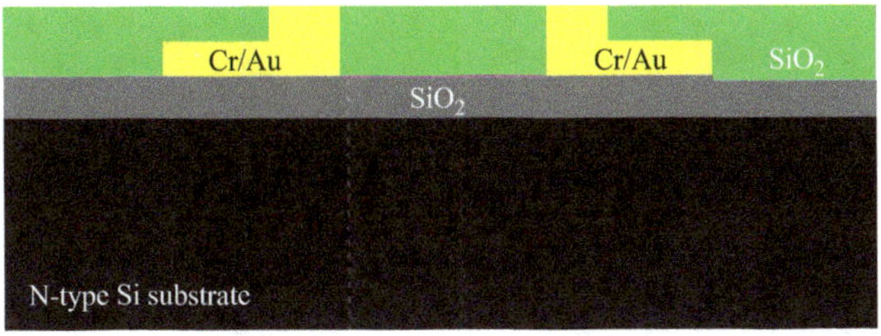

Figure 6.17 Etching SiO_2 and removal of photoresist

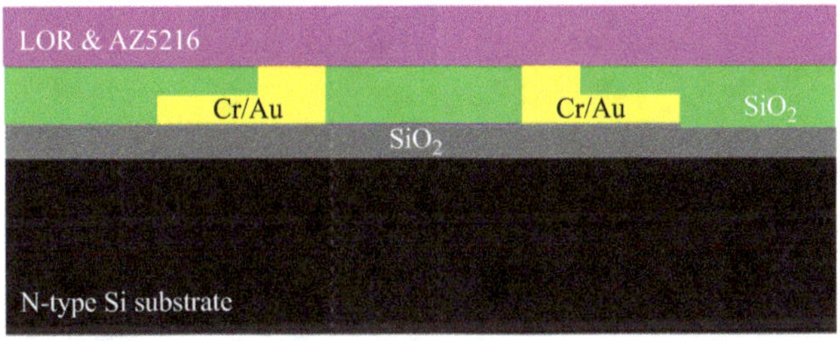

Figure 6.18 Photoresist deposition 4

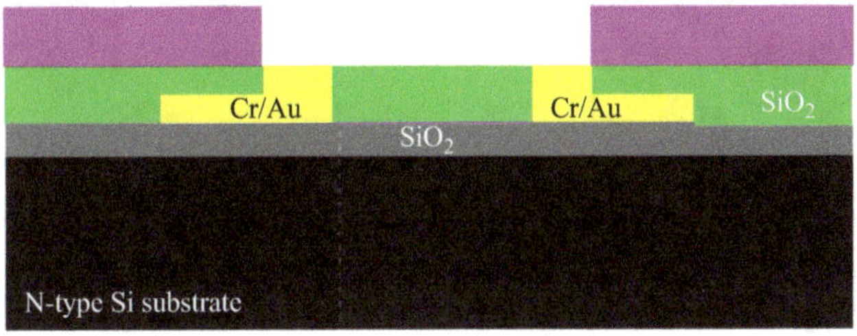

Figure 6.19 Lithographically patterned mask 4

1.5 μm using spin coating and then annealing (postbake) at 100 °C for 180 s (see Figure 6.18).

16. Expose to UV light: Lithographically patterned by exposing to UV light through the fourth level mask for 10 s and then developing it in MF319 for 35 s and washed with DI water. The photoresist in exposed areas was removed, leaving behind a patterned photoresist mask for further process (see Figure 6.19).
17. Metal deposition 3: A 120 nm layer of Cr was deposited using thermal evaporation. The metal was deposited in vacuum at a pressure of about 1 μTorr and the rate of deposition was approximately 0.1 nm/s. This third metal layer forms the CC beams and seed layer for Ni electroplating (see Figure 6.20). The deposited Cr (120 nm thickness) was measured using a surface profilometer.

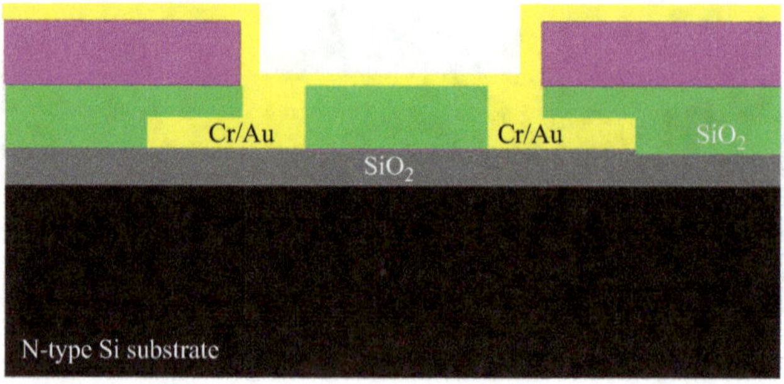

Figure 6.20 Metal deposition 3

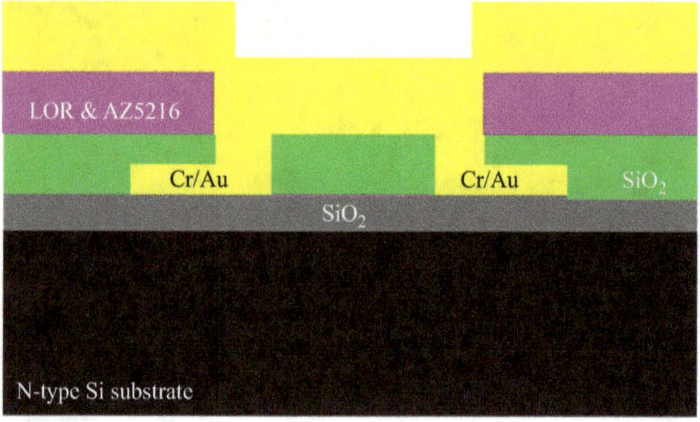

Figure 6.21 Ni electroplating

18. Electroplating: A 380 nm Ni was deposited in Ni electroplating solution for 180 min at a current density of 0.001 A using magnetic stir (see Figure 6.21). The process was included to achieve a higher thickness of the beam around 500 nm. The thickness of 513 nm of the beam was measured using FESEM.
19. Metal patterning: Liftoff was performed to remove photoresist from the patterned areas using PG remover. After the liftoff was completed, the remaining photoresist if any was removed using acetone and IPL (see Figure 6.22).
20. Removal of sacrificial layer: SiO_2 sacrificial layer was removed using a wet etching process in 5:1 BHF solution of 40 s with an etch rate of 10 nm/s (see Figure 6.23).

Figure 6.24 shows the FESEM image of the device obtained after completing all the fabrication steps. The figure clearly shows the four CC beams and input–output electrodes underneath the beam. The transduction gap of 387 nm and a beam thickness of 513 nm was measured using FESEM. Table 6.1 illustrates the complete steps involved in the fabrication of device.

The fabrication of shunt capacitive RF MEMS switch using surface micromachining is illustrated and the fabrication steps involved are depicted in

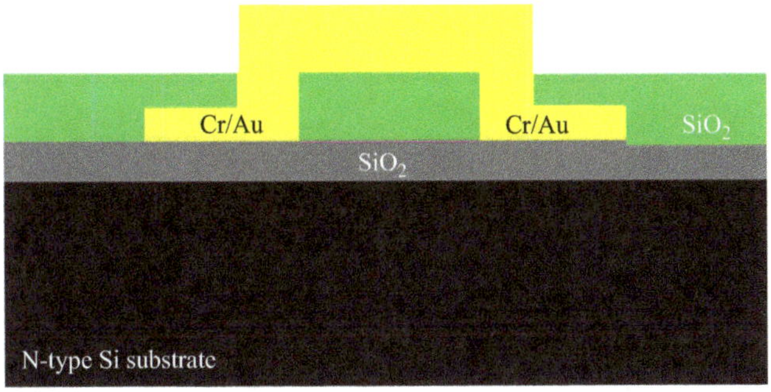

Figure 6.22 Metal patterning 3

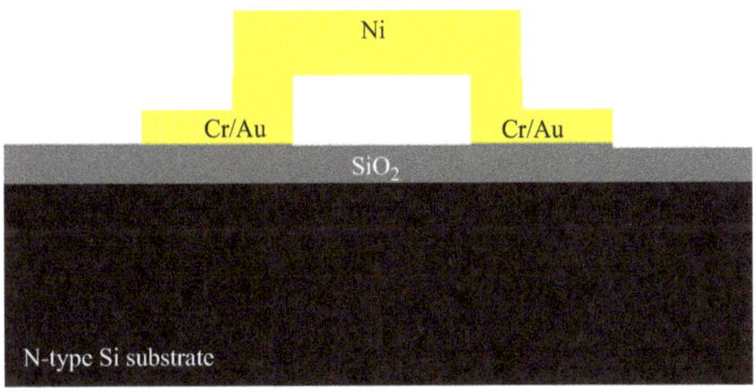

Figure 6.23 Release device after removal of sacrificial layer

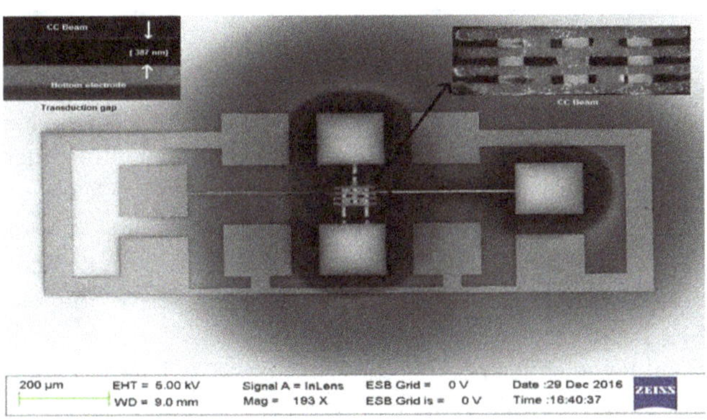

Figure 6.24 FESEM image of the designed device

Figure 6.25 [18]. The example was intentionally selected since it can be exploited in the design of low-frequency resonators. The switch can be fabricated p-type Si <100> single-side polished (SSP) substrate. An insulating SiO_2 layer of 1 µm can be deposited using thermal oxidation. With the use of LOR and S1813 positive

Table 6.1 Process steps

Sr. no.	Particulars
1.	2 inch, n-type Si wafer with <100> and resistivity > 1,000 ohm-cm
2.	RCA cleaning
3.	1-μm-thick oxide deposition using wet oxidation furnace
4.	1-μm-thick photoresist LOR and AZ5216 deposition using a spin coating
5.	Lithographically patterned by exposing to UV light through the mask 1 using DSA and then developing it
6.	Cr/Au deposition of 160 nm (120 nm Cr and 40 nm Au) using thermal evaporator
7.	Metal patterning using liftoff in PG remover
8.	1.5-μm-thick photoresist LOR and AZ5216 deposition using a spin coating
9.	Lithographically patterned by exposing to UV light through the mask 2 using DSA and then developing it
10.	Cr deposition of 360 nm using thermal evaporator
11.	Metal patterning using liftoff in PG remover
12.	SiO_2 sacrificial layer deposition of 360 nm using dielectric sputter
13.	1-μm-thick photoresist AZ5216 deposition using a spin coating
14.	Lithographically patterned by exposing to UV light through the mask 3 using DSA and then developing it
15.	Wet etching was performed to remove SiO_2 from the patterned areas using 5:1 BHF for 40 s
16.	1.5-μm-thick photoresist LOR and AZ5216 deposition using a spin coating
17.	Lithographically patterned by exposing to UV light through the mask 4 using DSA and then developing it
18.	Cr deposition of 120 nm using thermal evaporator
19.	380 nm Ni deposition in Ni electroplating solution
20.	Metal patterning using liftoff in PG remover
21.	Removal of sacrificial layer SiO_2 in 5:1 BHF for 40 s

photoresist materials, CPW and actuation lines can be patterned using the liftoff process. Au metal is mostly preferred for CPW and actuation lines. Cr/Au/Cr metal layers can be deposited using DC sputtering. Mostly Cr is used for better adhesion to silicon and Au that assist in protecting CPW and actuation lines. AlN can be used as a dielectric material, and the AlN material can be deposited using DC sputtering. The AlN can be patterned using reactive ion etching (RIE) method. S1813 positive photoresist material can be used as a sacrificial layer. S1813 sacrificial layer of desired thickness can be coated using a spin coater. The membrane using Au material can be deposited using DC sputtering that helps the membrane to develop the strain. The membrane is patterned using wet etching with developer $KI:I_2:H_2O$ in 4:01:40 ratio, after patterning of the membrane the switch is Figure 2.26 [18].

The low-frequency MEMS resonator sensor is illustrated which is composed of an electrostatically actuated clamped-clamped microbeam applied in the detection of gas sensors [19]. The sensor was fabricated on a 4 silicon wafer coated with a silicon dioxide (SiO_2) layer. The SiO_2 layer was deposited using the plasma-enhanced

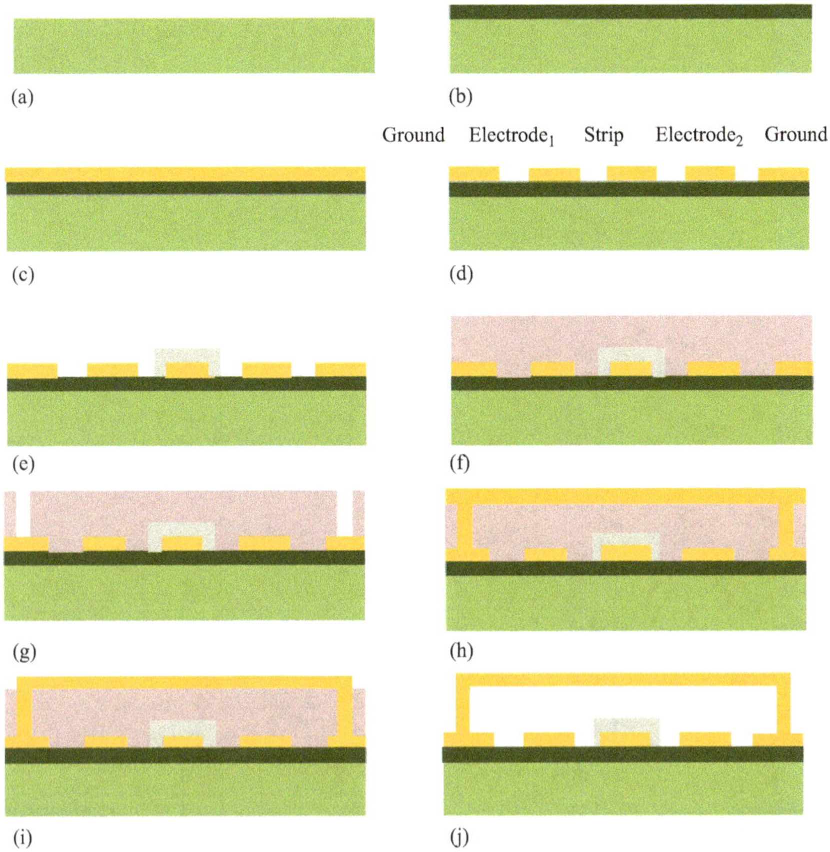

Figure 6.25 Fabrication flow: (a) Si substrate, (b) SiO_2 insulator deposition using thermal oxidation, (c) Cr/Au deposition using DC sputtering PVD process, (d) patterning of CPW and actuation lines using wet etching, (e) deposition and patterning of AlN using DC sputtering and RIE, respectively, (f) deposition of sacrificial layer, (g) patterning of trenches, (h) deposition of membrane metal (Au), (i) patterning of membrane, and (j) membrane release using wet process [18]

physical–chemical vapor deposition (PECVD) to provide the electrical insulation between the device and the wafer. Using the PECVD tool, amorphous silicon was deposited and then polished to the desired thickness. The amorphous silicon was used as sacrificial layer that dictates the air gap between the two electrodes. The sacrificial layer needs to be etched during the final release step. The microbeam was composed of a polyimide coated from the bottom with a Cr/Au/Cr layer that forms the upper electrode of the resonator. The microbeam was also coated from the top with Cr/Au layer to protect the microbeam during the etching process of the polyimide. A uniform MOF layer is grown by dipping the chip in copper acetate

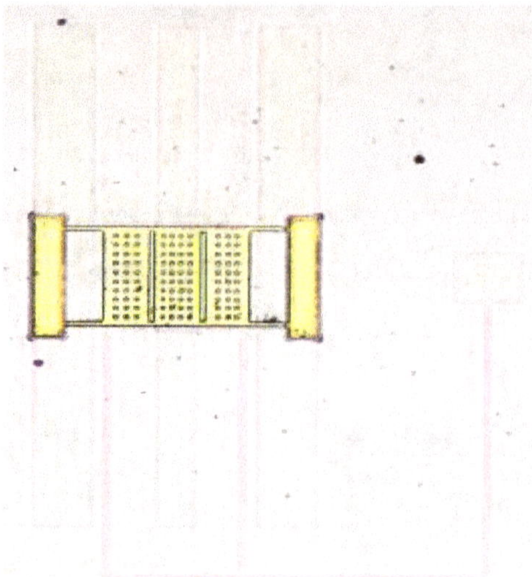

Figure 6.26 RF switch [18]

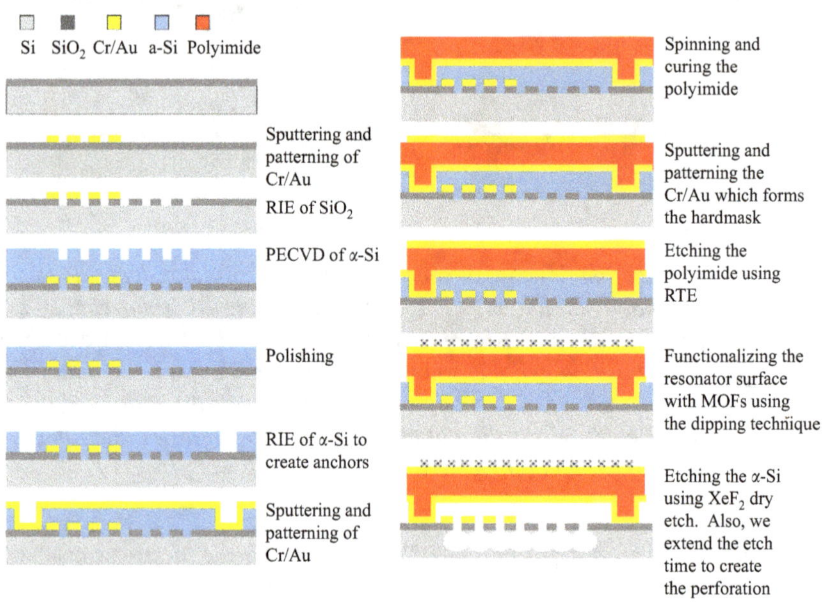

Figure 6.27 Fabrication of microbeam MEMS resonator [19]

metal precursor ethanolic solution and organic ligand (terephthalic acid (bcd)) ethanolic solution for 5 and 10 min, respectively, to grow the Cu(bcd)·xH_2O MOF thin film [19]. Figure 6.27 shows the fabrication steps and Figure 6.28 shows an SEM image and a schematic of the clamped-clamped resonator. To release the

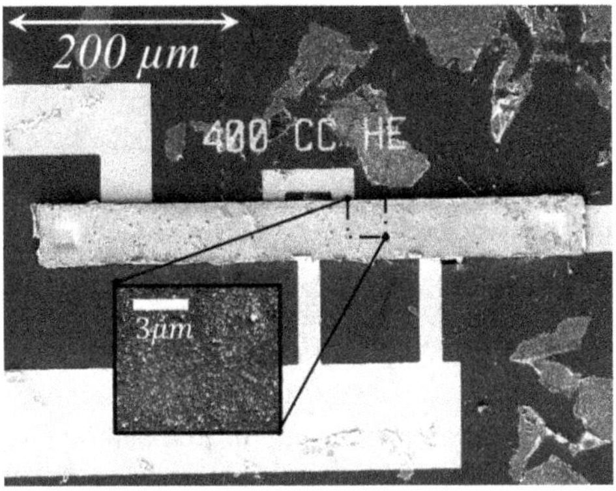

Figure 6.28 SEM image of microbeam [19]

microbeam, the sacrificial layer is fully etched using the XeF_2 dry etchant. Extra etching cycles were conducted to etch the silicon from the opening in the SiO_2 layer and form the perforations in the lower electrode.

6.4 Challenges in fabrication

During the fabrication of the device, certain problems are encountered. Experiments need to be performed to solve those problems before deciding on the parameters of the process. Some of these problems and their solutions are discussed for illustration purposes.

1. The first major problem that is encountered during the writing of the mask. Mask may have small square shape patterns with smaller dimensions such as 6 μm × 6 μm and large size cross-shaped alignment mask with dimensions such as 500 μm × 500 μm. It is observed that the laser writer is unable to print small shapes in such a scenario of varied dimensions. Although the mask may get printed in several attempts, it is necessary to insert variable size alignment mask that would solve the problem at the first instant, for example, 10 μm × 10 μm, 50 μm × 50 μm, 100 μm × 100 μm, and finally 500 μm × 500 μm.
2. Second, misalignment of the mask during lithography is a major problem due to the minimum resolution of the DSA system. Due care has to be taken during the alignment of the mask especially when smaller dimensions need to be aligned. A variable size alignment mask can be used for better alignment.
3. Third, it is essential to determine the etching rate and the etching solution required to etch metal such as Cr/Au/Ni, etc. In case of unsuccessful attempts to achieve at exact etching rate that would be incorporated during the final etching, liftoff process can be preferred instead of etching. Liftoff typically has a disadvantage of generating spikes whenever the metal on top of the resist is lifted off. Hence it is desired to deposit resist with a thickness greater than two times the desired metal thickness.

4. Fourth, selective etching process can be used to remove the sacrificial layer on top of the anchor. It is important to select photoresist that would remain unaffected in 5:1 BHF and other etchants. Various experiments with photoresist need to be performed to remain unaffected in the 5:1 BHF and other etchants.
5. The deposition rate of metal in electroplating solutions depends upon the type of metal and current density. Therefore, it is desired to determine the deposition rate of metal in electroplating solution at varied current density. It is observed that metals in electroplating solutions have deposition rates approximately a few nm/min. However, the rate of deposition in electroplating at room temperature is very less, it can be increased by increasing current density and temperature of electroplating solution. Also increasing the temperature of the electroplating solution may elevate the thermal stability of the deposited metal.

6.5 Summary

The device fabrication process for low-frequency MEMS resonator is explained in this chapter. Challenges faced during the fabrication process and experiments performed to solve those issues are also discussed. All the processes such as metal deposition, dielectric deposition, electroplating, and wet etching must be studied by performing experiments to determine its deposition and etching rate, respectively. The rate of deposition in electroplating at room temperature is very less; it can be increased by increasing the current density and temperature of the electroplating solution. Observations and measurements of dimensions and thickness need to be regularly performed after every fabrication step. It helps to confirm the correctness of the process. Physical characterization and electrical characterization need to be performed on the fabricated device to validate the proposed concept.

References

[1] Basu J., and Bhattacharyya T.K. "Microelectromechanical resonators for radio frequency communication applications." *Microsystem Technologies.* 2011;**17**(10–11):1557–80.

[2] Nguyen C.T.C., Katehi L.P.B., and Rebeiz G.M. "Micromachined devices for wireless communications." *Proceedings of IEEE*. 1998;**86**(8):1756–68.

[3] Héctor J., and Santos D.L. *RF MEMS Circuit Design for Wireless Communications*. Artech House; 2002.

[4] Vardhan V. *et al. RF MEMS and Their Application*. John Wiley; 2003.

[5] Senturia S. *Microsystem Design*. Kluwer Academic Publishers; 2002.

[6] Ananthasuresh G. K. *et al. Micro and Smart Systems*. Wiley India; 2010.

[7] Randy J.R., Héctor J., and Santos D.L. "MEMS for RF/microwave wireless applications: The next wave." *Microwave Journal.* 2001.

[8] Pourkamali S., Ho G.K., and Ayazi F. "Low-Impedance VHF and UHF capacitive silicon bulk acoustic wave resonators—Part I: Concept and fabrication." *IEEE Transactions on Electron Devices*. 2007; **54**(8); 2017–23.

[9] Srikar V.T., and Spearing S.M. "Materials selection in micromechanical design: An application of the Ashby approach." *Journal of Microelectromechanical System*. 2003; **12**(1): 3–10.

[10] Gad-el-Hak M. *The MEMS Handbook MEMS Applications*. Boca Raton, FL: Taylor & Francis; 2006.

[11] Carter J., Cowen A., Hardy B., Mahadevan R., Stonefield M., and Wilcenski S. *PolyMUMPs Design Handbook*. MEMSCAP Inc.

[12] Johnson R. A. *Mechanical Filters in Electronics*. Wiley Series on Filters. Canada: John Willey & Sons;1983.

[13] Lobotniu N., and Garcia E. *Mechanics of Microelectromechanical Systems*. Boston, MA: Kluwer Academic Publishers; 2005.

[14] Bannon F.D., Clark J.R., and Nguyen C.T.C. "High-Q HF microelectromechanical filters." *IEEE Journal of Solid-State Circuits*. 2000;**35**(4): 512–26.

[15] Poddar A. *et al*. "MEMS enabled RF signal source." *IEEE Journal of Microelectromechanical Systems*. 2007:16–19.

[16] Morankar A., and Patrikar R.M. "Dual frequency MEMS resonator through mixed electrical and mechanical coupling scheme." *IET Journal of Circuits, Devices & Systems*. 2018;**12**(1); 88–93.

[17] Morankar A., and Patrikar R.M. "Effective method for temperature compensation in dual band metal MEMS resonator." *22nd International Symposium on VLSI Design and Test (VDAT - 2018)*. Thiagarajar College of Engineering, Madurai, India, June 28–30, 2018. *Springer Communication in Computer & Information Science*. 2019; **892**:233–41.

[18] Rao K.S., Thalluri L.N., Guha K., and Sravani K.G. "Fabrication and characterization of capacitive RF MEMS perforated switch." *IEEE Access*. 2018;**6**(7): 7519–28.

[19] Jaber N., Ilyas S., Shekhah O., Eddaoudi M., and Younis M.I. "Multimode MEMS resonator for simultaneous sensing of vapor concentration and temperature." *IEEE Sensors Journal*. 2018; **18**(24): 10145–53.

Chapter 7

Fabrication of high-frequency resonators

Prasanna Deshpande[1] and Rajesh Pande[2]

7.1 Introduction

Micromechanical resonators are the most popular RF microelectromechanical system (MEMS) devices as their potential advantage is easy integration with CMOS circuitry. The piezoelectrically transduced micromechanical resonators are difficult to fabricate as the piezoelectric layer introduces contamination, residual stress, and process compatibility issues. MEMS are fabricated using some standard process steps including photolithography, wet/dry etching, oxidation, diffusion, low-pressure chemical vapor deposition (LPCVD), and sputter deposition [1]. The block schematic of MEMS fabrication cycle is shown in Figure 7.1.

There are various fabrication techniques involved in MEMS. The fundamentals of some of those techniques were already discussed in Chapter 4. Here, we consider the particular case of laterally vibrating one port contour mode resonator (CMR) devices with 21 interdigitated fingers to show a complete microfabrication process. Although CMRs are just one example of microresonators, the techniques presented in this chapter are also applicable to the general case of MEMS/NEMS resonators [3].

Piezoelectric layer deposition and interdigitated electrode's deposition and patterning are important processes involved in the fabrication of CMR. Releasing of the device is also a challenge.

This chapter focuses on fabrication laterally vibrating contour mode one port ZnO MEMS resonator for global system for mobile communication (GSM) frequency range. The bottom electrode is not used, thus saving the fabrication step. The resonators have been fabricated on Si/SiO_2 by using three mask processes. The device area is small ($W \times L = 57.12\ \mu m \times 18.52\ \mu m$) compared to the previous works reported. The thickness of ZnO piezoelectric film was 250 nm in this fabrication. The piezoelectric ZnO thin film has been used due to its better coupling coefficient, low deposition temperature, excellent bonding, and unique semiconducting and optical properties. Among all the piezoelectric materials, zinc

[1]Department of Electronics and Communication Engineering, Shri Ramdeobaba College of Engineering and Management, Nagpur, India
[2]Department of Electronics Engineering, Shri Ramdeobaba College of Engineering and Management, Nagpur, India

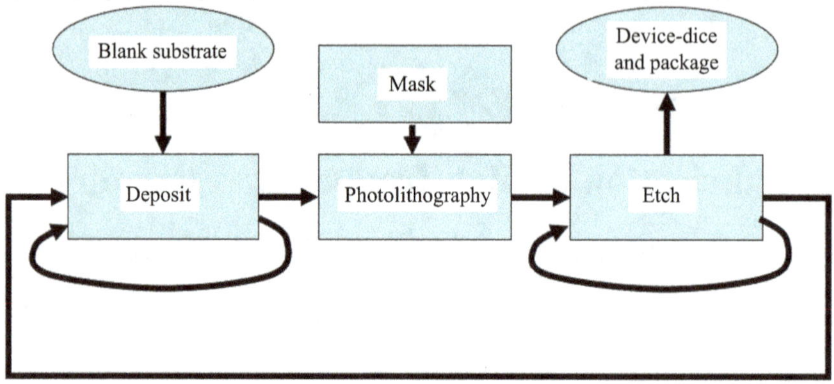

Figure 7.1 A general block diagram of MEMS fabrication cycle [2]

oxide is considered as a pollution-free green piezoelectric material. The ZnO is highly tensile and may undergo huge mechanical deformation for a long duration unaffected by the temperature variation. The synthesis of ZnO thin films or nanoparticles has been investigated in the past. The ZnO thin film can be deposited at room temperature and a variety of acidic etchants are also available [4].

In the fabrication process, a simple and inexpensive wet etching method for ZnO has been incorporated over costly deep reactive ion etching (DRIE). As ZnO is a low toxic, an excellent biocompatible/biodegradable and the stable material, hence, this device may be the best convenient detector of GSM range for mass/pressure sensing for medical application [4].

The chapter explains the fabrication issues while fabricating the CMR with interdigitated electrodes and the actions to solve them. The steps to solve the issues and then fabricates the fully released device efficiently are also discussed. Using scanning electron microscope (SEM), ellipsometry, and interferometer, the CMR's structural analysis is carried out.

In the following sections, the concepts, physical principles, fabrication equipment, materials, and chemical products involved in microfabrication will be studied. The study comprises fundamental on oxidation, metallization, sputtering deposition, and micromachining techniques regarding the device fabrication [3].

7.2 Main steps for fabrication

7.2.1 Oxidation of silicon

Oxidation is used to passivate and interface the substrate and the device through a dielectric SiO_2 film. SiO_2 grown at high temperatures is also known as thermal oxide. However, SiO_2 can also be grown on other materials (different from Si). In this case, chemical vapor deposition (CVD) is usually employed to obtain a high-quality oxide. In the case of CMR implementations, either the thermal or the deposited SiO_2 may have thicknesses of hundreds of nanometers.

7.2.2 Metallization and piezoelectric layer deposition

Metallization is the formation of metal films for interconnections, ohmic contacts, rectifying metal-semiconductor contacts, and protections. Metallic thin films can be deposited on the surface of dielectric, conductor, or semiconductor materials. Vacuum evaporation, sputtering, CVD, platting, and electroplating are some of the employed techniques. In CMR processes, Pt, Al, Ti, Cr, and Mo, etc. may be deposited on top of the SiO_2 passivation layer by RF sputtering. Sputtering is the deposition of compound materials and refractive metals through the removal of the surface atoms or molecular fragments from a solid cathode (target). By bombarding it with positive ions from an inert gas (argon), removed atoms or molecular fragments deposit on the substrate forming the thin film [3].

7.2.3 Micromachining techniques

The fabrication technologies in MEMS are categorized into surface micromachining and bulk micromachining and molding techniques.

7.2.3.1 Surface micromachining process

The fabrication of micromechanical structures by deposition and subsequent wet etching is surface micromachining. Surface micromachining is one of the most common technologies used to manufacture MEMS devices. In surface micromachining, films are deposited on a substrate and patterned, using photolithography, to create micromechanical devices. The films normally alternate between structural and sacrificial layers, with the MEMS parts being made from the structural layers. The sacrificial layers serve to support the structural components during the fabrication. After the structural layers are patterned, the sacrificial material is removed, often using wet chemical etching. The result is freestanding MEMS parts that can move relative to the fixed substrate.

Most early surface micromachining used polycrystalline silicon (polysilicon) as the structural layers and an oxide of silicon as the sacrificial material. Depending on the desired application, MEMS developers have used metals, oxides, and nitrides of silicon, and even polymers for both structural and sacrificial films [1,5–8].

7.2.3.2 Bulk micromachining process

Bulk micromachining, the oldest of the micromachining technologies, is accomplished by removing the material from a substrate to create holes, cavities, channels, or other desired shapes. Early bulk micromachining was accomplished using isotropic or anisotropic wet etching of silicon or glass substrates. In particular, several chemicals such as KOH (potassium hydroxide) or TMAH (tetramethylammonium hydroxide) etch a silicon substrate preferentially depending on the crystalline planes in the direction of etching. The etch rate for these chemicals is tens to hundreds of times faster in the [100] crystalline plane compared to the [111] plane. This effect has been used to create a wide variety of features using simple wet etching. Another common technique of bulk micromachining uses a DRIE plasma etcher. Using this technique, silicon, as well as some other materials,

(a) (b) (c)

Figure 7.2 (a) Surface micromachined, (b) bulk micromachined, and (c) molding [5]

can be etched very quickly and very anisotropically, making possible very thick structures with small widths [1,5–8].

The simple, as well as the most complex, resonator structures can be fabricated on a silicon substrate. In contrast to surface micromachining, bulk machining etches into a silicon wafer, either isotropically or anisotropically, to form micro-structures. The advantage of bulk micromachining over surface micromachining is that it yields high aspect ratio structures.

7.2.3.3 Molding process

Finally, MEMS parts are often made by creating a mold, which may then be filled to create the desired part. Molds have been made from a variety of polymers, including some types of photoresist, as well as metal and deep-etched silicon wafers. Photolithography is normally used to define the mold pattern. If metal parts are desired, the mold may be filled by electroplating. Polymer parts may be created by pouring or pressing the precursor into the mold. After the part has been molded, it may be removed from the mold by either etching the mold away, or, if the mold is to be used again, by peeling away the mold.

Micro-molding was first performed in Germany, where it was called LIGA, an acronym for the German words "lithography", "electroplating", and "molding". The original LIGA process required an X-ray source to fully expose thick layers of photosensitive material, but many molding techniques have since been developed that use visible or ultraviolet (UV) light sources. However, because of this history, many molding processes are still referred to as LIGA or LIGA-like processes [1,5–8].

The SEM in Figure 7.2 represents the above three micromachine technologies.

7.3 Sputter-deposited ZnO and its XRD pattern

Figure 7.3(a–d) shows the SEM images of sputtered zinc oxide at different magnifying levels. The grain size, surface roughness, and c-orientation indicate the quality of zinc oxide film. Besides, high resistance is also desirable to obtain higher breakdown voltage. Typical resistance value ranges from 10^{15} to 10^{16} Ω-cm [2]. The crystalline structure of ZnO can be observed by X-ray diffraction (XRD) patterns recorded on diffractometer. All the XRD measurements were carried out in

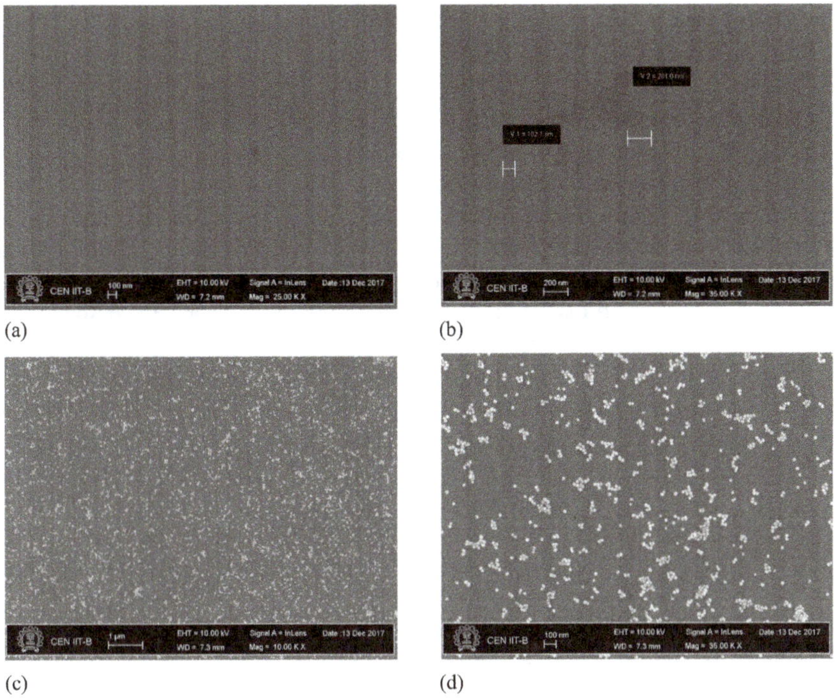

Figure 7.3 SEM images (at different magnifying level) of the sputter-deposited ZnO [7,10] (We acknowledge the INUP, IITB, India)

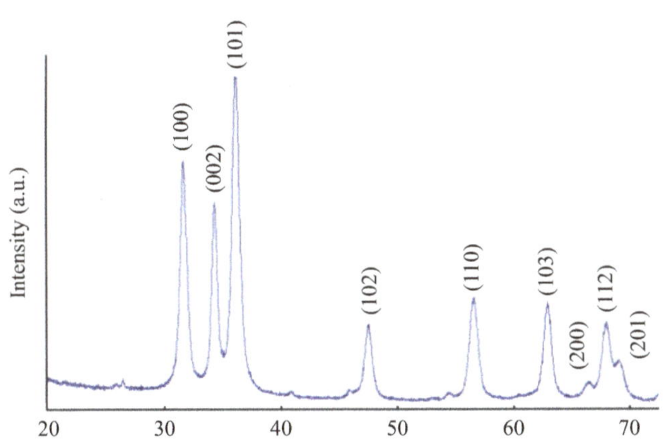

Figure 7.4 ZnO XRD (X-ray diffraction) pattern [9,10]

the range between 20° and 70°. All the measurements were carried out at room temperature. All the diffraction peaks have indices (100), (002), (101), (102), (110), (103), (200), (112), and (201) matching with pure hexagonal crystalline structure of zincite zinc oxide. XRD graph shows that there are no additional XRD peaks corresponding to any impure phases. Figure 7.4 presents ZnO XRD pattern [9,10].

7.4 Selection of materials for the fabrication of micromechanical resonator

MEMS devices comprise many materials. Silicon (Si) with <100>/<110> orientation and silicon dioxide (SiO_2) have proven the best combination for the substrate and the sacrificial layer respectively enabling CMOS compatibility furthermore [1]. Apart from that, silicon/silicon nitride is also being researched in most of the applications such as MEMS planer and suspended inductor for low-pass filtering. For this fabrication, 2″ silicon wafer of <100> orientation with 4–7 Ω-cm resistivity was being selected over a SiO_2 layer of 1,000 nm thick have been deposited by the thermal wet oxidation method.

ZnO thin film has been employed due to its better coupling coefficient, low deposition temperature (maybe at room temperature), excellent bonding, unique semiconducting, and optical properties (Figure 7.5). The pollution-free green piezoelectric ZnO material is highly tensile and may undergo huge mechanical deformation for a long duration unaffected by temperature variation. Similarly, a variety of acidic etchants are also available to etch ZnO effectively. In this fabrication process, a simple and inexpensive wet etching method for ZnO has been incorporated over costly DRIE [11–16]. The low toxic and biocompatible feature of ZnO can make it comfortable in medical applications. The highly conductive aluminum (Al) metal of 100 nm to pattern 21 IDT fingers (electrodes) has been deposited by thermal evaporation after ZnO deposition step [4]. Table 7.1 lists the parameters used in the deposition of zinc oxide by dielectric sputter.

7.5 Lithography—hard mask

Lithography is the transfer of geometric shapes on a mask to smooth surface. Figure 7.6 illustrates schematically the lithographic process used to fabricate circuit chips. The exposing radiation is transmitted through the "clear" parts of the mask. The chromium/glass mask is used with UV light. Other types of exposing radiation are electron, X-rays, or ions. Shadow printing may be employed where the gap between mask and wafer is small [17].

7.5.1 Optical lithography

The optical lithography uses optical radiations on a silicon wafer using photoresist layer. Optical lithography comprises the formation of images with visible or UV radiation in a photoresist using contact, proximity, or projection printing. The photoresists used for optical lithography are of two types. A negative resist on exposure to light becomes less soluble in a developer solution, while a positive resist becomes more soluble. Commercial negative resists are Kodak Microneg 747 and positive resists are PMMA (poly-methyl methacrylate), MP-2400, and HR-206. The developer solvent dissolves the unexposed resist. Photoresists are being developed for exposure at shorter wavelengths where higher resolution is possible. The exposed resist swells as the uncrosslinked molecules are dissolved. Optical lithography lies near 0.4 μm, although 0.2 μm features may eventually be printed under carefully controlled conditions [17].

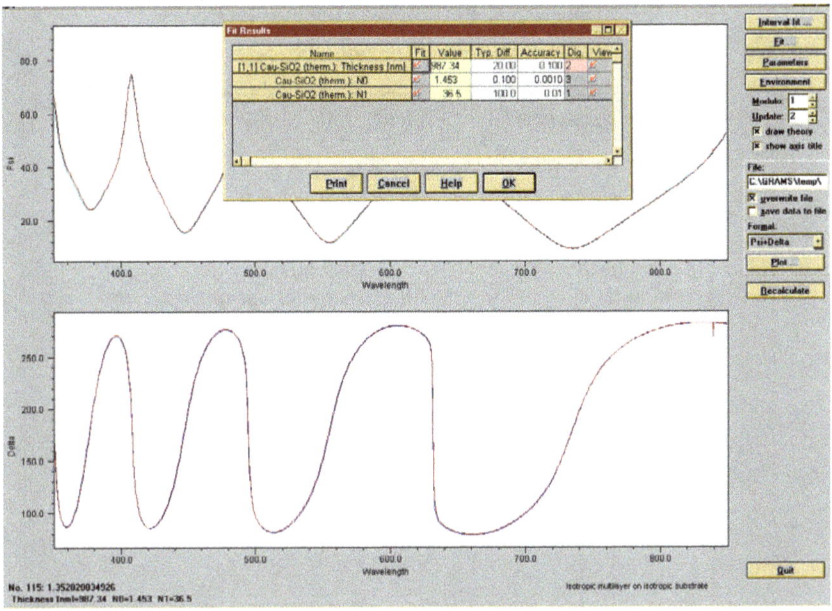

(a)

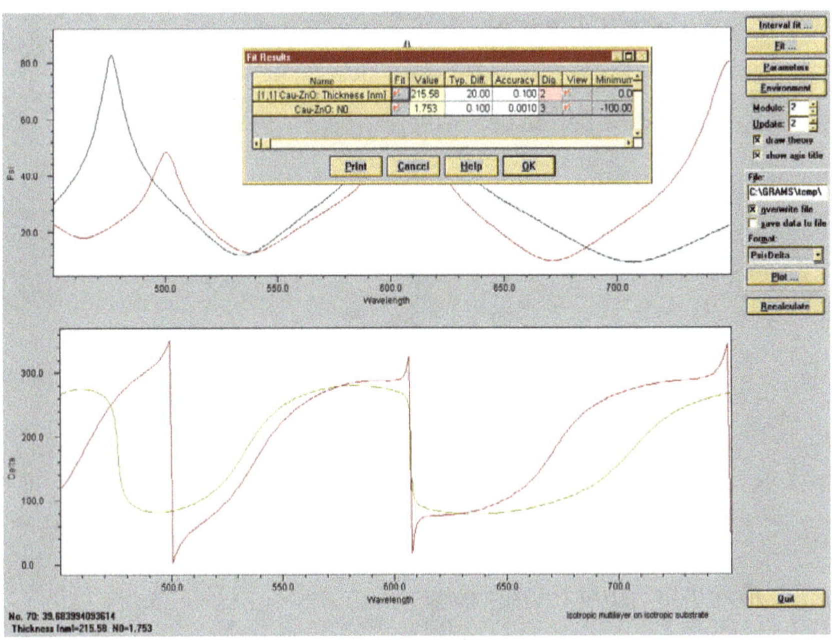

(b)

Figure 7.5 Thickness observation on ellipsometry (a) SiO_2 and (b) ZnO [7,10] (We acknowledge the INUP, IITB, India)

Table 7.1 The parameters used during zinc oxide deposition (dielectric sputter) [4,10]

Sputtering parameter	Value
Zinc oxide thickness	250 nm
RF power	100 W
Gas O_2:Ar	1:1
Process pressure	2.2×10^{-6} mbar
Base pressure	5.4×10^{-6} mbar
Deposition time	30–35 min after vacuum creation

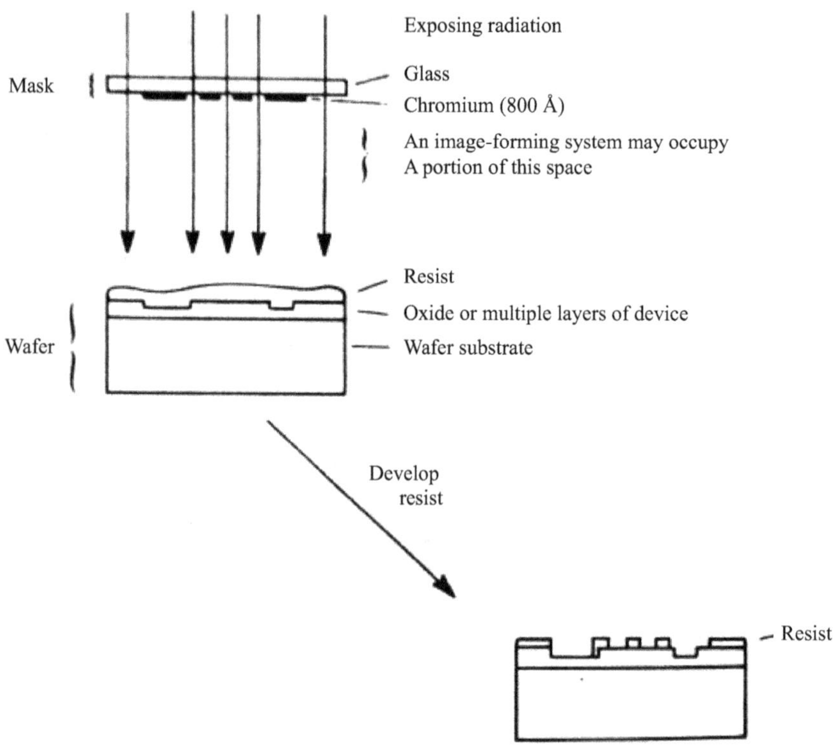

Figure 7.6 Lithography generation [17]

7.5.2 Electron beam lithography

Electron lithography offers higher resolution than optical lithography because of the small wavelength of the 10–50 keV electrons. The resolution of electron lithography systems is not limited by diffraction, but by the electron scattering in the resist and by the various aberrations of the electron optics. Scanning electron-beam systems have been under development for several decades, and commercial systems are available. The electron beam exposure system machine has proved to be the best photomask pattern generator. It is widely used in mask shops. Because of the serial nature of the pattern writing, throughput is less much than for optical system. However, some special products such as microwave transistors have for many years been manufactured by direct wafer patterning. In the first application to low-volume integrated circuits, some levels were pattern optically and some by electron beam [17].

Electron exposure of resists occurs through bond breaking (positive resists) or the formation of bonds or crosslinks between polymer chains (negative resists). The incident electrons have energies far greater than bond energies in the resist molecules, and so all these energies are effective. Both bond scission and bond formation occur simultaneously. Which predominates determines whether the resists are positive or negative. In a negative resist, electron-beam-induced crosslinks between molecules make the polymer less soluble in the developer solution. One crosslink per molecule is sufficient to make the polymer insoluble. Resists sensitivity increases with increasing molecular weight. If molecules are larger, then fewer crosslinks are required per unit volume for insolubility. The polymer molecules in the unexposed resists will have a distribution of lengths or molecular weights and thus a distribution of sensitivities to radiation. The narrower the distribution, the higher will be the contrast, γ. In a positive resist, the scission process predominates, the exposure leading to lower molecular weights and greater solubility. Again, high molecular weight and narrow distribution are advantageous [17].

When electrons are incident on resists or other material, they enter the material and lose energy by scattering, thus producing secondary electrons and X-rays. The fundamental process limits the resolution of electron resists to an extent that depends on resists thickness, beam energy, and substrate composition. More electrons are scattered back into the resists from a GaAs substrate than from a silicon substrate.

Table 7.2 lists a few readily available electrons resist. Many other resists are under development. PMMA is the highest resolution resists known. MP-2400 is an

Table 7.2 Some electron resists [17]

Resist	Polarity	Sensitivity (C/cm^2) @ 20 kV	Resolution (μm)	γ
PBS (Mead Tech.)	+	1.8×10^{-6}	0.5	1.7
PMMA (KTI Chem.)	+	1×10^{-4}	<0.1	2
EBR-9 (Toray Ind.)	+	1.2×10^{-6}	0.5	3
FBM-110 (Daikam Ind.)	+	1.5×10^{-6}	1.5	5
AZ 2400 (Shipley Co.)	+	2×10^{-4}	0.5	2
COP (Mead Tech.)	−	5×10^{-7}	1.5	0.8
OEBR-100 (Tokyo Okha)	−	5×10^{-7}	1.5	0.8
SEL-N (Somar Ind.)	−	1×10^{-6}	1	0.6
GMCIA (AT & T)	−	7×10^{-6}	0.5	1.7
CMS (Toyo Soda)	−	2×10^{-6}	0.7	1.5
RE-4000 N (Hitachi Chem.)	−	3.5×10^{-6}	1	1.3

optical resist that is also electron sensitive. Values for sensitivity and resolution are approximate. Because faster electrons penetrate more deeply, more current is required at higher voltages. A resist is about one-half as sensitive for 20 keV electrons as it for 10 keV electrons [17].

7.6 Preparation of hard mask for interdigitated laterally vibrating CMR

The mask layout used for optical lithography is prepared using CleWin, printed and stuck on a glass plate to be used during optical lithography UV exposure. Laser writer was employed for writing with a minimum feature size of 100 nm. Figure 7.7 depicts the illustration of hard mask prepared using the software CleWin (www.wieweb.com).

7.7 Challenges in fabricating laterally vibrating CMR and issues related to lithography and its solution

There were certain issues faced while fabricating microresonators and it seemed very important to include discussion on these issues along with the modifications made on the fabrication process to solve these issues.

1. The first issue faced with respect to the window openings for cavity formation in the SiO_2 layer. The electron beam lithography (EBL) on Si/SiO_2/ZnO, shown in Figure 7.8(a) has been done to observe the time required to get undercut below the device. But while etching in buffered hydrofluoric acid (BHF) 5:1 the PMMA started peeled off within 5 min damaging the whole structure, as shown in Figure 7.10(b) and (c).

   Some experimentation has been performed and optimized to get undercut using optical lithography as photoresist S1813 sustain in BHF. One mask plate was taken and optical lithography has been performed followed by the sample etched in BHF 5:1 for 18 min. The undercut was observed as shown in Figure 7.9 [18].
2. The second important issue was the deposition of Cr/Au (10 nm/80 nm) for the alignment marks. In the first level of EBL, the sample was kept for lift-off into AZ100 initially for 24 h but no lift-off has been processed. Some optimization has been done and thereafter decided to keep the sample into AZ100 and heat it at 90 °C for 2 h; this results in a proper lift-off (Figure 7.10(a) and (b)) [18].
3. The major issue was the mask. After the lift-off, the writing of the mask plate has been done for window openings by the laser writer. There are six rows and six columns shown in Figure 7.11(a). In each column, the spacing between windows is different than that of the neighboring column because the interdigitated electrode must be accommodated between these spacing even if there is misalignment issue or reduction in spacing during the BHF etch. Considering this issue, the spacing was increased [18].

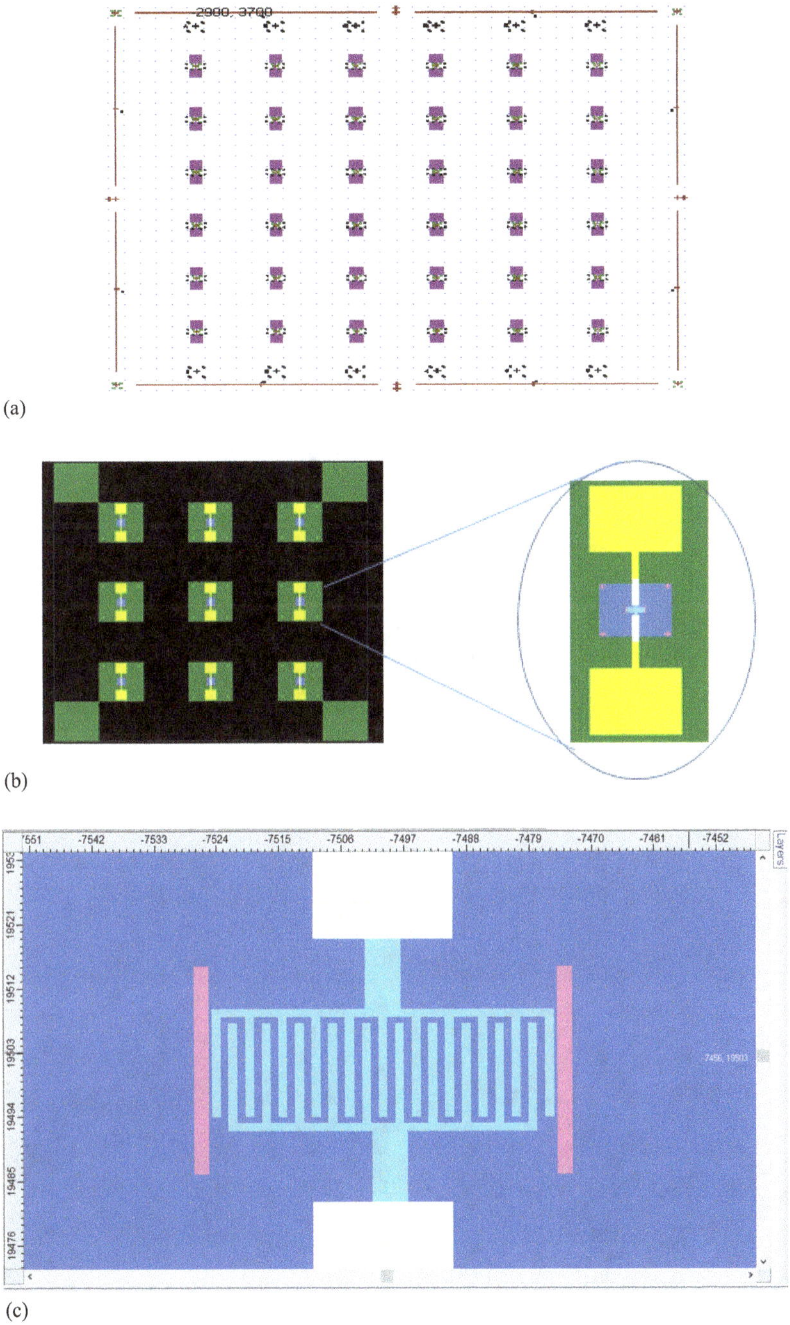

Figure 7.7 Illustration of the hard mask in CleWin (a) 6 × 6 devices, (b) device under magnification, and (c) single device [18]

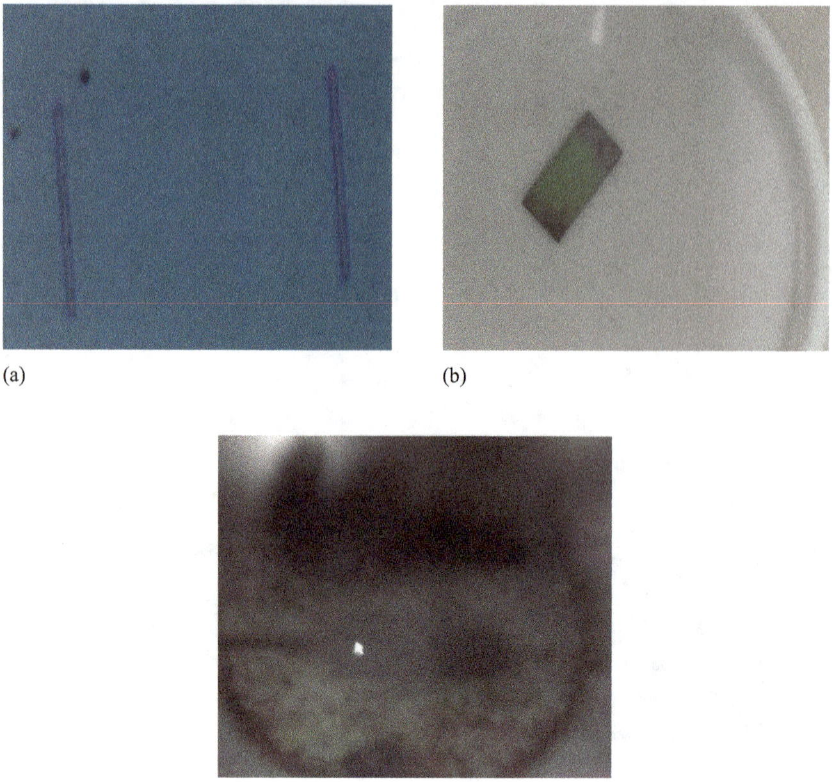

(a) (b) (c)

Figure 7.8 (a) EBL for window opening, (b) sample in BHF PMMA started peeling, and (c) structure got damaged [18]

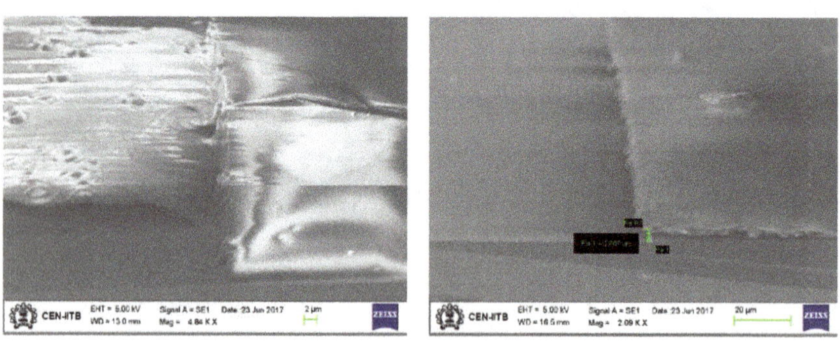

(a) (b)

Figure 7.9 SEM images of (a) ZnO etch and (b) undercut [18] (We acknowledge the INUP, IITB)

(a) (b)

Figure 7.10 (a) Image after a 24-h dip in AZ100 (no lift-off observed) and (b) lift-off started after 1 h of heat at 90 °C

4. It was observed during second-level lithography that the masks layer for EBL and optical lithography should be a mirror image of one another this is because while doing optical lithography we flip the mask plate and then UV to be exposed. The window openings were observed and the ZnO etching has been performed. BHF etching for 20 min has also been performed to confirm undercut (BHF 5:1 is 10 nm/min) in SiO_2. The Olympus microscope images are shown in Figure 7.12(a) and (b). The spacing required for interdigitated electrode was 50.4 μm but it was observed that the first column was of no use as the spacing left between the windows was found less. The other columns were observed fine to use for the further process [18].

The SEM image of the undercut as illustrated in Figure 7.13(a) has been observed to check lateral etch but after etching for 20 min, the whole pit or undercut was not obtained. So, further 15 min etching was done and it was seen that the whole device got etched out, Figure 7.13(b). So, it was noted that the BHF etching period must be between 20 and 35 min for a proper device [18].

Considering all these issues faced during the fabrication process, some experimentations have been carried out and the solutions have been incorporated and accordingly, we have developed an optimized recipe for our device fabrication, as shown in Figures 7.14 and 7.15. The optimized process time for each step, optimized dimensions of the device, proper selection of the etchant, SEM/Olympus images after each step were helpful to us to get our resonator fabricated properly. The exact design and the dimensions of laterally vibrating CMR with 21 interdigitated fingers and zinc oxide as piezoelectric material are explained in Chapter 4.

7.8 Laterally vibrating CMR fabrication process

The fabrication process used for this fabrication is standard surface micromachining type, where films are deposited on the substrate and patterned using lithography to create micromechanical devices. The sacrificial layers serve to support the

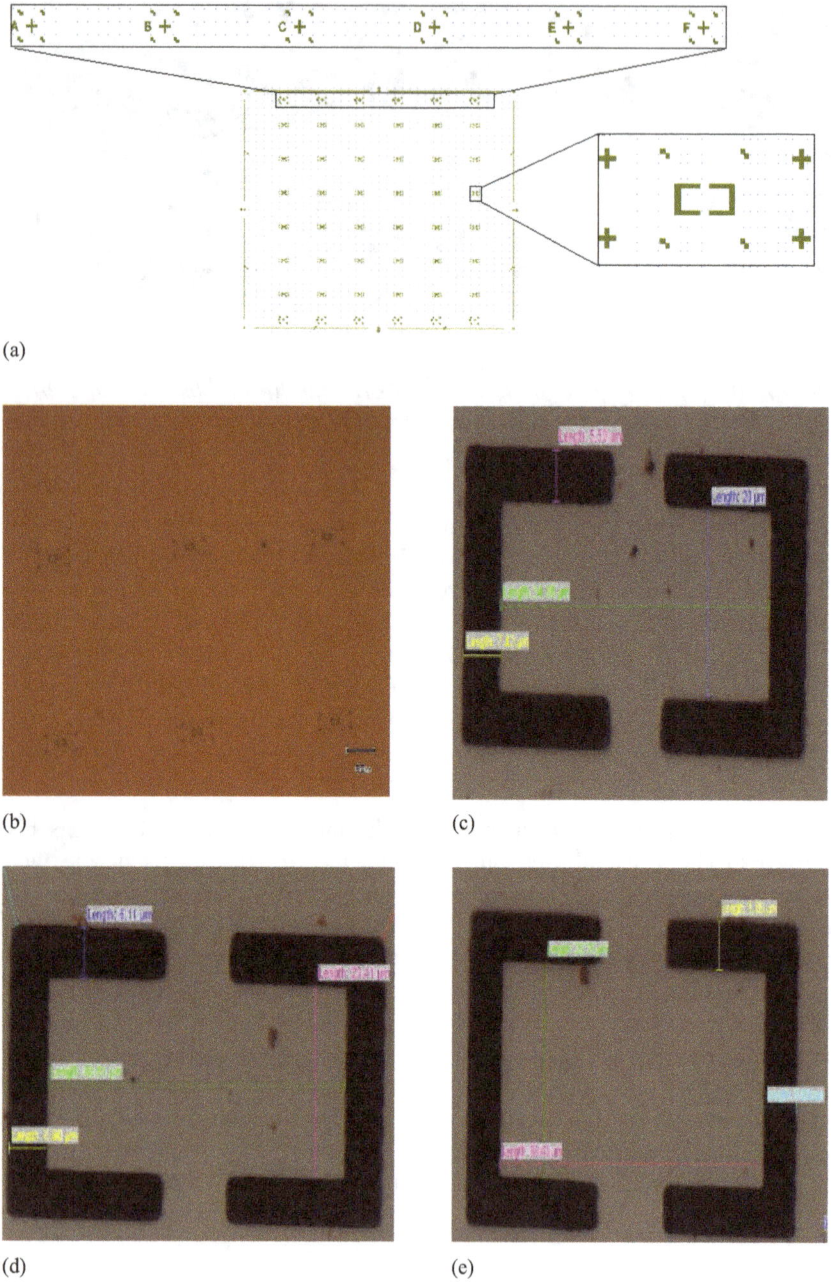

Figure 7.11 (a) Mask plate in CleWin for second-level lithography, (b) mask plate with six devices, (c) device of the second column, (d) device of the third column, and (e) device of the fourth column [18]

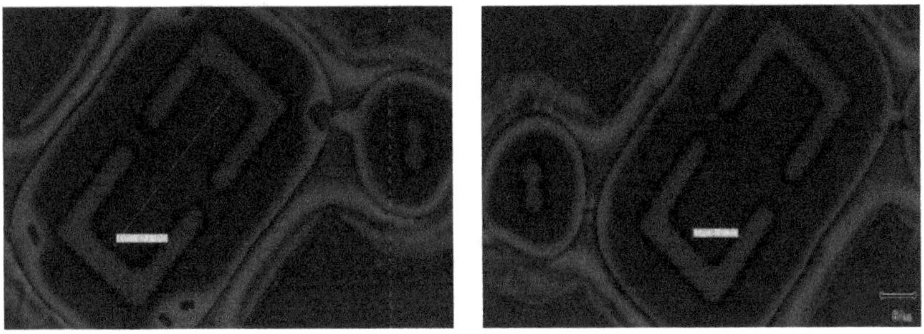

(a) (b)

Figure 7.12 (a) Window in the first column with spacing approximately 47.22 μm and (b) window in the second column with spacing approximately 50.85 μm [18]

structural components during fabrication. The structural layers are patterned and the sacrificial material is removed, using wet chemical etching. In this work, the fabrication process incorporates two EBL and two optical lithography steps, as shown in Figure 7.14(a)–(i). These steps involve metal deposition and etching. The fabrication begins with the deposition of isolation layer silicon dioxide (SiO_2) of 1,000 nm on the top of p-type\100 orientation 200 silicon wafer (4–7 Ω-cm resistivity) by thermal wet oxidation at a furnace temperature of 1,000 °C for 3 h. The 250 nm ZnO film was deposited on SiO_2 using dielectric sputter with radio frequency power of 100 W. The thickness of SiO_2 and ZnO layer has been confirmed on the ellipsometer (Figure 7.5(a) and (b)). The first step of EBL has been used to locate alignment marks for which El19 resist bilayer PMMA 4% is spun for 30 s at 3,000 rpm. The writing takes 3 h followed by rinsing in MIBK (methyl isobutyl ketone) developer for 30 s, thereafter rinsing in IPA (isopropanol) for 90 s. The next step is Cr/Au deposition (10 nm/80 nm), then lift-off using AZ-100 photoresist. The optical lithography has been performed next by mask aligner (SUSS MJB4) for the definition of cavity opening for which PPR (positive photoresist) is deposited at 3,000 rpm for 30 s. The expose period was 10 min followed by rinsing in MF319 (Microposit developer) for 30 s followed by rinsing in DI water for 10 s. ZnO thin film is etched by wet etching in phosphoric acid (H_3PO_4) + acetic acid (CH_3COOH) + DI water (1:1:30) followed by SiO_2 etching in BHF (5:1). The next EBL has been performed to write interdigitated (IDT) fingers for which again El19 resist bilayer PMMA 4% is spun for 30 s at 3,000 rpm. The writing takes 3 h followed by MIBK for 30 s, thereafter IPA for 90 s. The aluminum (Al) of 100 nm thin is evaporated for IDT fingers. These fingers are patterned by lift-off (Microposit^{TM} 1165 remover). The next optical lithography (process as described above) has been performed by using mask aligner (SUSS MJB4) for the definition of contact pads. The contact pads have been developed by Cr/Au (10 nm/80 nm) deposition, followed by the lift-off process [4].

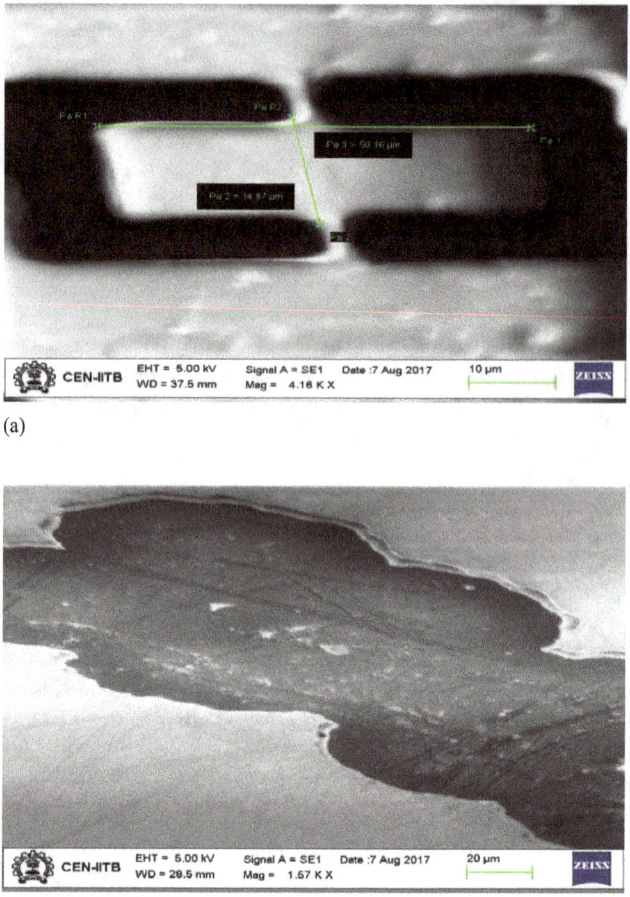

(a)

(b)

Figure 7.13 (a) Undercut, BHF etching for 20 min and (b) BHF etching for 35 min, the device got damaged [18] (We acknowledge the INUP, IITB, India)

The fabrication flow of the above process is also determined by its flowchart representation, as shown in Figure 7.15.

The step-wise SEM images of the fabricated resonator are shown in Figure 7.16. The cavity/undercut formation after ZnO and SiO_2 etching is shown in (a), whereas Raith SEM after third level EBL for interdigitated structure (vertically flip mask) is depicted in (b). A complete device with a 21 IDT pattern of 100 nm thick aluminum metal deposited with alignment marks of Cr/Au (10 nm/80 nm) is shown in (c). Part (d) of Figure 7.16 shows the magnifying cross-section image of only IDT structure followed by the formation of Cr/Au metal contact pads in (e). Totally 36 (6 × 6) devices have been accommodated on a single sample of 1 cm × 1 cm as shown in (f).

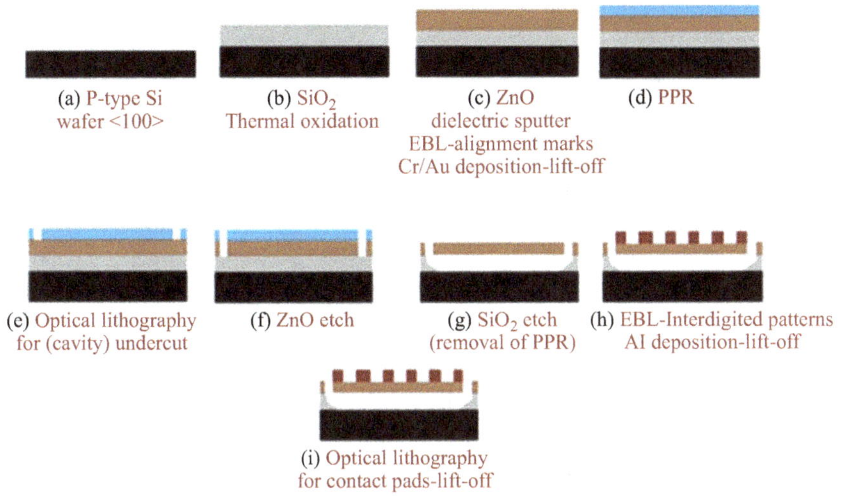

Figure 7.14 Fabrication process flow implemented [4] (Fabrication of MEMS resonator is carried out at the IITB, India-under INUP program)

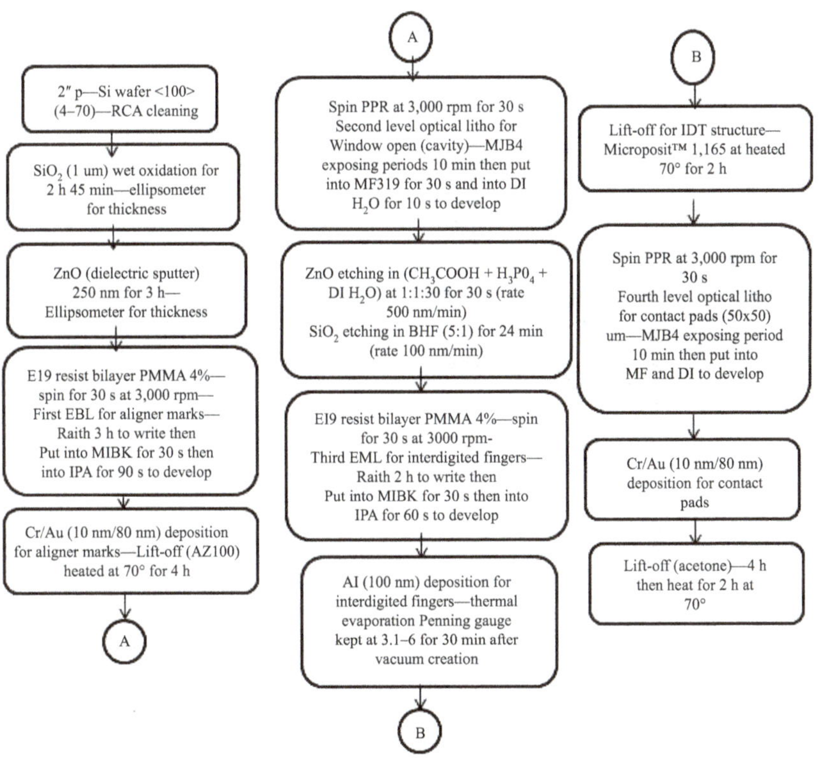

Figure 7.15 The device fabrication flowchart—a recipe (Fabrication of MEMS resonator is carried out at the IITB under INUP program)

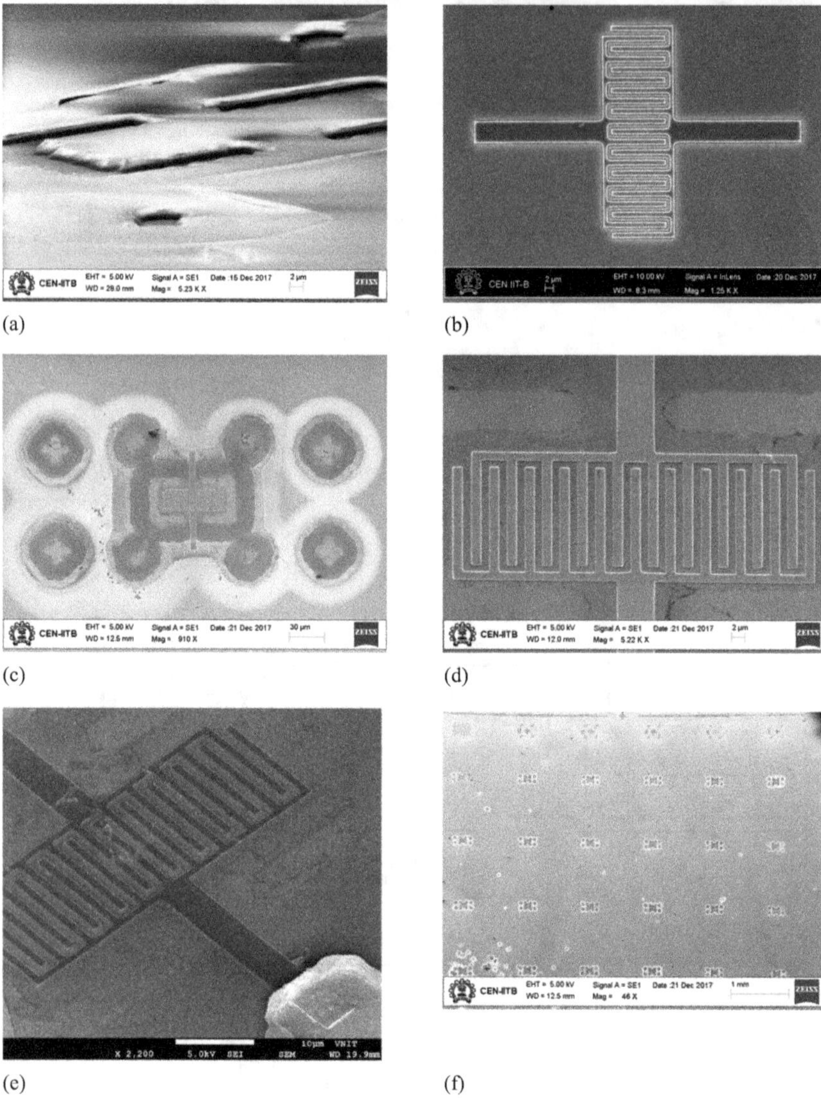

Figure 7.16 *SEM images of fabricated resonator: (a) Cavity/undercut formation, (b) IDT pattern before Al deposition, (c) IDT pattern after Al deposition and with alignment marks, (d) magnifying image of only IDT pattern after Al deposition [4], (e) device with contact pads [4], and (f) number of devices with alignment marks on one quarter (We acknowledge the INUP, IITB, India) (We acknowledge INUP, IITB, VNIT, Nagpur, India)*

7.9 Tunability in resonance frequency

From device fabrication, we may observe that the actual targeted resonance frequency has deviated from its originally designed structure. This may be due to the electromechanical properties of the materials used for the piezo-resonator during fabrication. This deviated resonance frequency of the fabricated device must be retained to its originally designed vales for its better performance. Hence, to retain its actual frequency, some tuning procedures must be incorporated in MEMS manufacturing processes. Some of these procedures implement post-fabrication techniques for changing/modifying the resonance frequency, whereas others are intended for tuning of the device during its operation [3].

To correct the deviations of operating frequency, post-process trimming is usually necessary for commercialized products. Depending upon the standards, the channel-select architecture may entail the simultaneous fabrication of tens to hundreds of narrowband filters with 0.05% to 0.1% bandwidth and 0.05% to 0.1% center frequency spacing. Mass loading of MEMS resonator and laser trimming are two commonly used methods for post-process fine frequency tuning. By adding materials on the surface of the structure, the desired resonant frequency could be achieved. Here, the mass loading-based tuning performance of microelectromechanical resonators may be described by one or more metrics, through the mass responsivity is the main parameter being considered. The inverse mass responsivity is defined as the number of grams deposited on the device needed to produce a change of 1 Hz in its resonance frequency. However, they are expensive and impractical for a filter bank with more than ten filters. It would, therefore, be advantageous to develop fine frequency selection techniques or active trimming techniques to define the relative center frequency of each resonator in an array [19].

Another tuning methodology is a voltage-controlled tuning (DC tuning). It is more convenient than laser trimming or other methods, and it can be done in real-time, anytime. The tuning process should be fast, efficient, and if possible, automated [3,19–21].

7.10 Summary

CMR fabrication technology comprises many steps and has to be developed to achieve full-process compatibility. Among these steps, piezoelectric layer deposition and a variety of micromachining technologies, and the formation of cavity are the key processes. Also, different characterization techniques have to be implemented in order to evaluate the main steps of the process. Structural, crystallographic, thickness and etching analysis can be studied by current-art instrumentation like XRD, ellipsometry, AFM, SEM, microscopy, etc. Although a variety of process parameters should be considered, advantages and challenges of each technology should be sought on considering the possibilities and limitations of the technology available to the designer.

Facilities at the INUP, IITB—fabrication lab

(Source of information & Images: http://www.iitbnf.iitb.ac.in/iitbnf)

RCA cleaning bench

- RCA chemicals used: DI water, NH_4OH, H_2O_2, HCl, HF
- Piranha chemicals used: H_2SO_4, H_2O_2
- HF cleaning: 2% HF
- Substrates used: Si, glass
- Substrate size: Max 4″ diameter wafer

Silicon and metal etching bench

- Used for Si, polysilicon, nitride, metal etching
- Si etchant: TMAH, HNA
- SiO_2 etchants: 5:1 BHF and 7:1 BHF
- Silicon nitride: HNA
- Al etchant:
- Cr-Au etchant: CEN for Cr and KI + I_2 + Di water for Au

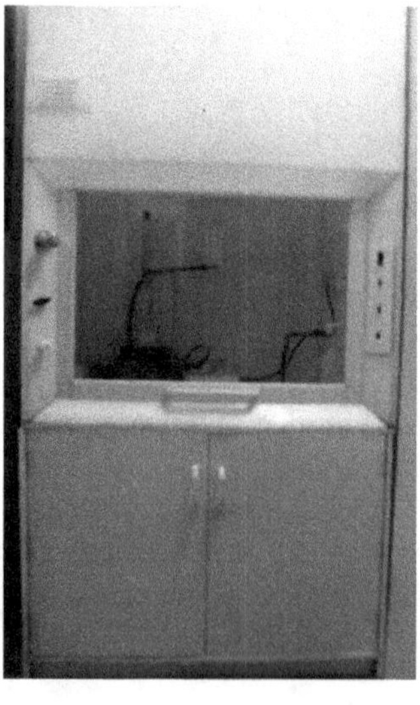

Furnaces 2″

Thermal oxidation-wet
All tubes can handle 2″ wafers.
Pyrogenic oxidation furnace
Temperature up to 1,150 °C—gases H_2 and O_2
Double-walled furnaces
Uniform growth of oxide—only dry oxidation of silicon wafers
Compound semiconductor furnace
This furnace is used for various compound semiconductor (GaN, GaAs, etc.) processes
Post metallization anneal (PMA) furnace
This furnace is dedicated to post aluminum annealing
Annealing furnace (Argon ambient)
Max. temp.: 900 °C, ambient: argon
Boron diffusion furnace
Dedicated for making junctions in Si with solid source dopant and spin on dopant. Max. temp. is 1,100 °C, gases used: O_2, H_2, and N_2.
Phosphorus diffusion furnace:
Dedicated for making junctions in Si with solid source dopant and spin on the dopant.
Max. temp. is 1,100 °C, gases used: O_2, H_2, and N_2

Spectroscopic ellipsometry (SE 800)

Specifications

- Spectral range: 240 nm–930 nm
- Incident angles: 40°–90°, 5° steps, generally kept at 70° fixed
- Parameters can be extracted: Thin-film thickness, refractive index, uniformity of films and layer stacks

Capabilities

- Substrate used: Si substrate or transparent substrate like glass
- Sample size: minimum 1 cm × 1 cm, maximum 6″ wafer
- Thickness: minimum 1–2 nm, maximum 100 μ

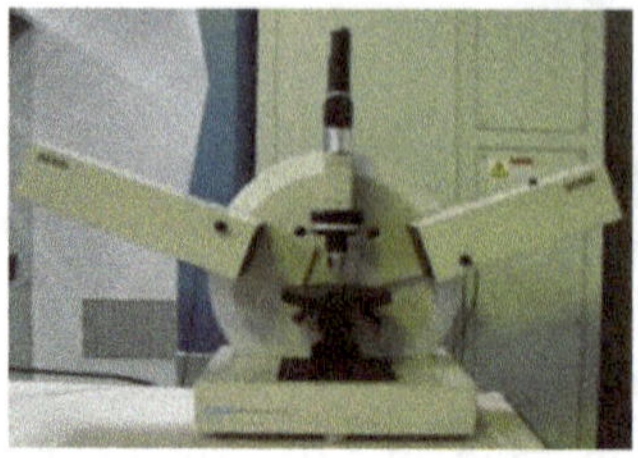

PMMA spinner

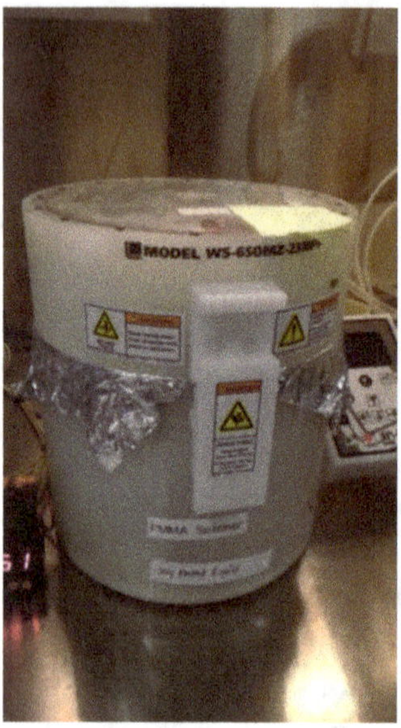

Olympus microscope

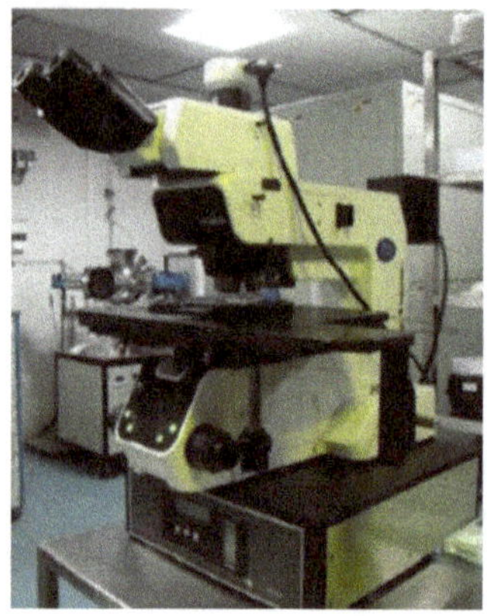

Thermal evaporator (Al)	Thermal evaporator (Cr/Au)
Purposes: To perform thin-film deposition of Al by thermal evaporation **Process capabilities** **Wafersize:** 2″ at a time 2–3 wafers can be processed if required **Number of current sources to head the filament:** Two **Number of filaments that can be fitted in the system:** Two **Material:** Al **Base pressure:** 8×10^{-5} mbar **Substrate temperature:** Room temperature	**Specifications** Hind HiVac vacuum-coating unit (Model12A4D) Resistive heating evaporation system, which has a diffusion pump backed by rotary pump to achieve chamber base vacuum of 1×10^{-6} mbar Digital thickness monitor: Manual quartz crystal thickness monitor with a resolution of 0.1 A/s rate of evaporation Sequential or simultaneous evaporation of metals

(Continues)

(*Continued*)

Thermal evaporator (Al)	Thermal evaporator (Cr/Au)
Coolant required: Liquid nitrogen **Additional facility:** Shadow masks of 1 mm diameter available for making MOS capacitors 	**Process capabilities** Substrates used: Silicon, glass, polymers Substrate size: Two 2″ substrates can be used simultaneously Substrate temperature: Room temperature to 250 °C Materials deposited: Chromium and gold Chamber base vacuum: 1×10^{-6} mbar 

RAITH150—Lithography

RAITH150 is used for ultra-high-resolution patterning, pattern inspection, and dimensional meteorology. It has the ability to handle a wide range of samples including up to 8″ wafers.

Specifications

- TFE filament with beam size < 2 nm at 20 kV
- Beam current range 5 pA–20 nA
- Beam energy 100 eV–30 kEV
- Stage travel range 150 × 150 × 20 mm
- Current density: 20,000 A/m^2
- Current stability: 0.5%/8 h
- Minimum line width < 20 nm
- Stitching accuracy < 40 nm (means + 3 sigma)
- Overlay accuracy < 40 nm (means + 3 sigma)

Process capabilities:

- Imaging, lithography, EBID

Imaging:

- The substrate used: Si, sapphire, Pt–Si, glass plate
- Substrate history: Powdered sample cannot be used; nonconductive sample needs 10–20 nr Au/Al-coated on them to avoid charging
- Substrate size: For surface imaging, small pieces up to 100 mm or 4″ wafer for cross-section imaging at 45° (20 mm *W* × 10 mm *H*) at max 90° (20 mm *W* × 7 mm *H*)

Deposition

- The substrate used: Si (till now)
- Substrate size: Small pieces (up to 10 mm to 2″) wafer material that can be deposited: platinum, tungsten, SiO_2

Specifications of MJB-4 Standard Mask Aligner

Configured for topside alignment (TSA).
Primary exposure wavelength: 365 nm–405 nm (350 W mercury lamp)
Support substrates: 2″–4″ diameter
UV lamp: 350 W Hg arc lamp and 500 W Hg arc lamp
Support mask: 3″–5″
Line/Space resolution: 0.6 μm
Alignment accuracy: 0.2 μm
Lithography type: Soft contact mode, vacuum mode, low-vacuum mode, gap mode, hard contact mode
Substrate size: Max. size 4″ and min. size—quarter or less 2″ wafers

Microtech laser writer

Equipment compatible with 3″,4″ and 5″ mask plates
Gallium nitride diode laser (405 nm)
The illumination by a red source (640 nm) to which both resist and photo-emulsions are not sensitive.
Writing area up to 4″ × 4″
5,000 Å iron oxide mask plate
3,100 Å photoresist
Resolution: 4 μm for lens 5, 10 μm for lens 4
Field of view after development: up to100 μm
Software package: Laser Draw-2D
Design software: CleWin

Dielectric sputter system (advanced processing technology)

Purpose: To perform thin-film deposition of dielectric materials
Process capabilities
Wafer size: 2″
Dielectric targets: hafnium oxide, silicon oxide, tantalum oxide
Gases involved: argon, oxygen, nitrogen
Base pressure: 2×10^{-6} mbar
Operating pressure: 0.015–0.023 mbar substrate temperature: room temperature
Target size: 125 mm (diameter) sputtering distance: 13 cm
Two RF power supplies with matching networks RF matching can be manual or auto
Sputter target details
Target shape: circular disk
Diameter: 125 mm
Thickness of the target: 1 mm–5 mm (as per requirement)
Backing plate: copper backing plate bonding of backing plate with a target: indium metallic bonding (indium metallic bonding is must for copper)

EVO 18 SEM facility

EVO 18 SEM is the scanning electron microscope from Carl Zeiss, Germany system.
Specification filament: Tungsten secondary e-image resolution: 50 nm (depends on a sample)
BSD detector: Available
Tilt: 0°–60°
Rotation: 360°
EHT: 200 V–30 kV
Magnification: up to 50 K~100 K (depends on the sample)
Process capabilities
Imaging modes: Surface and cross-sectional
Sample holder: maximum 9 stubs (~1 cm dia.) can be mounted
Substrates used: Si, glass, sapphire, Ge

Substrate size: $(2 \times 2 \times Z)$ mm to $(10 \times 10 \times Z)$ mm [for surface imaging] $(4 \times 4 \times Z)$ mm to $(8 \times 8 \times Z)$ mm [for cross-sectional imaging], where Z is the variable substrate thickness (it can vary from 200 μ to 2 mm), depending on the substrate type like Si/glass substrates

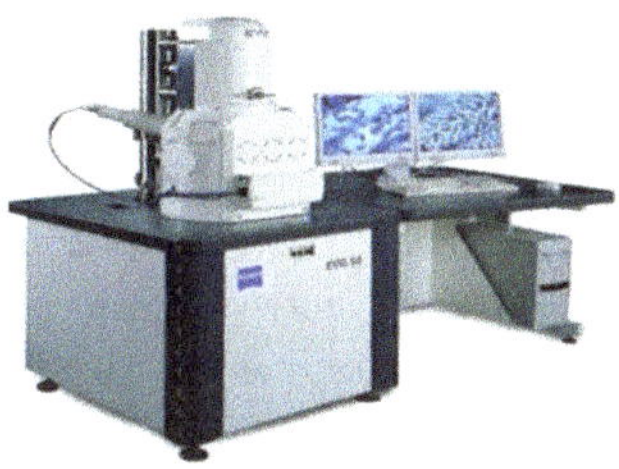

Acknowledgements

We acknowledge Indian Nanoelectronics User Program (INUP), IITB, Mumbai, India and Centre for VLSI & Nanotechnology, VNIT, Nagpur, India. We are thankful to Dr. S. Balpande, Dr. D. Khushalani, and Ms. P. Talekar of RCOEM, Nagpur, India.

References

[1] Piazza G., "MEMS Resonators for Frequency Control and Sensing Applications," *Penn Micro and Nano Systems Laboratory* (PmaNS Lab), Department of Electrical and Systems Engineering, University of Pennsylvania, Philadelphia, PA 19104, USA, Power point presentation slides, http://pmans.ese.upenn.edu/, 2012.

[2] Yan L., "Piezoelectrically Transduced Low Impedance MEMS Resonators," *Ph.D. Thesis*, University of California, Irvine, 2005.

[3] Campanella H., "Acoustic wave and electromechanical resonators-Concepts to key applications", *Artech House*, Norwood, MA 02062, ISBN-13: 978-1-60783-977-4.

[4] Deshpande P.P., Pande R.S., and Patrikar R.M., "Fabrication and characterization of zinc oxide piezoelectric MEMS resonator", *Microsystem technologies* (2019). https://doi.org/10.1007/s00542-019-04509-w.

[5] https://compliantmechanisms.byu.edu/content/introduction-microelectro mechanicalsystems-mems.

[6] Khushalani D.G., Dubey V.R., Bheley P.P., Kalambe J.P., Pande R.S., and Patrikar R.M., "Design optimization and fabrication of microcantilever for switching application," *Sensors and Actuators A: Physical*, 2015, 225, 1–7.

[7] Gupta A., Denton J.P., McNally H., and Bashir R., "Novel Fabrication Method for Surface Micromachined Thin Single-Crystal Silicon Cantilever Beams," *IEEE Journal of Micromechanical Systems*, 2003, Volume 12, Issue 2, pp. 185–192.

[8] Rinaldi M., Zuniga C., Zuo C., and Piazza G., "Super-High-Frequency Two-Port AlN Contour-Mode Resonators for RF Applications," *IEEE Transactions on Ultrasonics, Ferroelectrics, and Frequency Control*, 2010, Volume 57, Issue 1, pp. 38–45.

[9] Talam S., Kurumuri S.R., and Gunnam N., "Synthesis, Characterization, and Spectroscopic Properties of ZnO Nanoparticles", *ISRN Nanotechnology*, vol. 2012, Article ID 372505, 6 pages, 2012. https://doi.org/10.5402/2012/372505.

[10] Deshpande P.P, and Pande R.S., "Low motional resistance lateral field extensional UHF MEMS resonator", *International Journal of Engineering and Advanced Technology (IJEAT)*, Volume-8, Issue-5, June 2019, pp. 2194–2198, ISSN:2249-8958(Online), Blue Eyes Intelligence Engineering & Science Publications.

[11] Balpande S.S., Pande R.S., and Patrikar R.M., "Design and low cost fabrication of green vibration energy harvester," *Sensors and Actuators A: Physical*, 2016, 251, 134–141.

[12] Singh S., and Chakrabarti P., "Simulation, Fabrication, and Characterization of ZnO Based Thin Film Transistors Grown by Radio Frequency Magnetron Sputtering," *Journal of Nanoscience and Nanotechnology*, 2012, 12, 1880–85.

[13] Zhang Y., Nayak T.R., Hong H., and Cai W., "Biomedical Applications of Zinc Oxide Nanomaterials," *Current Molecular Medicine*, 2013, 13, 1633–45.

[14] Kołodziejczak-Radzimska A. and Jesionowski T., "Zinc Oxide- From Synthesis to Application: A Review," *Materials*, 2014, 7, 2833–81 DOI:10.3390/ma7042833.

[15] Gryba T., Carlier J., Wang S., Zhao X., Guo S., and Lefebvre J-E., "One port contour-mode ZnO piezoelectric MEMS resonator", *Microelectronic Engineering* 2011, 88, 3003–10.

[16] Kwon J.W., and Kim E.S., "Fine ZnO patterning with controlled sidewall-etch front slope," In *HH 2004 Proceeding Solid-State Sensor, Actuator and Microsystems Workshop*, 2002, 223–26.

[17] Sze S.M., "VLSI Technology", *Tata McGraw-Hill Publications*, Eleventh reprint 2008, Reprinted in India by arrangement with the Tata McGraw-Hill Companies, Inc., New York.

[18] Deshpande P.P., Talekar P.M., Khushalani D.G., and Pande R.S., "Design Optimization and fabrication issues of MEMS resonator", *International Journal of Technical Innovation in Modern Engineering & Science (IJTIMES)*, e-ISSN: 2455-2585, Volume 4, Issue 5, 2018, pp. 1372–1379.

[19] Yen T-T., "High-Q Aluminum Nitride RF MEMS Lamb Wave Resonators and Narrowband Filters," *Ph.D. Thesis*, University of California, Berkley, 2012.

[20] Borrow H., "Frequency tunable MEMS-based timing oscillators and narrowband filters", Electrical engineering and computer science, University of California at Berkeley *Technical report* no. UCB/EECS-2015-255, December 18, 2015.

[21] Wang K., and Nguyen C.T.C., "High-order medium frequency micromechanical electronic filters," *Journal of Microelectromechanical Systems*, 1999, 8, 534–57.

Chapter 8
Filter and oscillator design using SAW/BAW resonators

Motoaki Hara[1]

8.1 Introduction

In frequency control above the GHz band, surface acoustic waves (SAWs) and bulk acoustic waves (BAWs) are used, rather than the bending-mode or contour-mode vibration of the elastic body generally used in silicon-based resonators. Piezoelectric materials are employed as a resonating body in SAW/BAW devices to obtain a sufficient electromechanical coupling coefficient since the amplitude of vibration decreases with increasing operating frequency. In this chapter, we will describe SAW/BAW devices using piezoelectric materials and their applications.

Historically, SAWs were an object of interest to seismologists since their derivation by Lord Rayleigh [1], rather than for their application to electronic devices via the excitation of vibration on a piezoelectric substrate [2]. However, since the report of the efficient excitation of SAWs on a piezoelectric substrate using an interdigitated transducer (IDT), industrial applications of SAWs have expanded with the establishment of processing technologies in the semiconductor industry. In particular, the progress of thin-film technologies and the photolithographic patterning of metal films with high resolution have been key factors in the development of SAW devices.

SAW devices were initially developed as communication devices for military use in Europe and the United States and were applied to delay lines, transversal filters, pulse compression filters and the like [3–9]. However, with the progress of digital processors, the need for these technologies gradually declined. On the other hand, in the 1970s, applications such as intermediate frequency (IF) filters for television tuners became commonly used with the development of consumer communication devices [10,11], then the market for SAW devices for mobile communications, typified by mobile phones, expanded rapidly from the 1990s. In their application to mobile phones, the problems of insertion loss and power durability inherent in SAW devices have also been reduced by numerous technical innovations, and SAW devices have become indispensable devices in the RF front-end circuitry [12–18].

[1]National Institute of Information and Communications Technology, Japan

Quartz and ceramic resonators are classified as BAW devices, and the history of BAW devices is never short in comparison with that of SAW devices. However, the use of quartz and ceramic devices as MEMS resonators is not discussed in this text, and acoustic devices using a piezoelectric thin film will be described as BAW devices here. BAW devices were initially in the form of high-overtone acoustic resonators (HBARs), as shown in Figure 8.1 [19]. In an HBAR, a piezoelectric film is employed as a transducer to strongly excite a high-overtone mode in a low-elastic-loss material such as quartz, yttrium–aluminium garnet (YAG), spinel or sapphire. The industrial application of HBARs is, however, limited to acousto-optic modulators because the electromechanical coupling coefficient is small, and many harmonics are adjacently owing to the overtones. To avoid such unnecessary adjacent harmonics and achieve a strong resonance with a large coupling coefficient, it is necessary to trap the fundamental-mode vibration in the piezoelectric film. A composite resonator, whose diaphragm structure is fabricated beneath the piezoelectric transducer by thinning the substrate with MEMS technologies, as shown in Figure 8.2, was proposed by three research groups at almost the same time [20–22]. In recent years, such resonators have been called thin-film bulk acoustic resonators (FBARs) and have been widely employed as filter devices

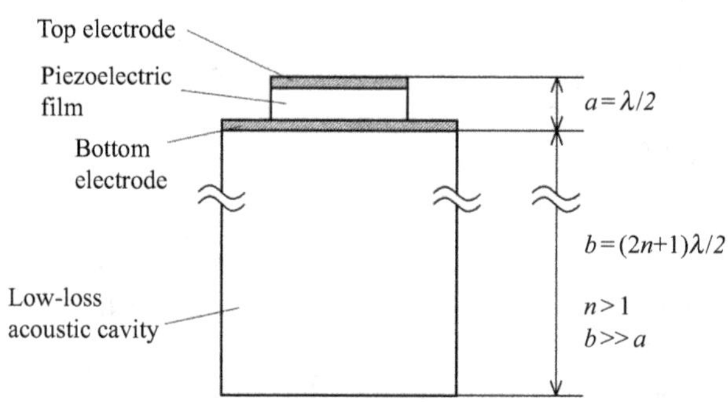

Figure 8.1 Configuration of HBAR

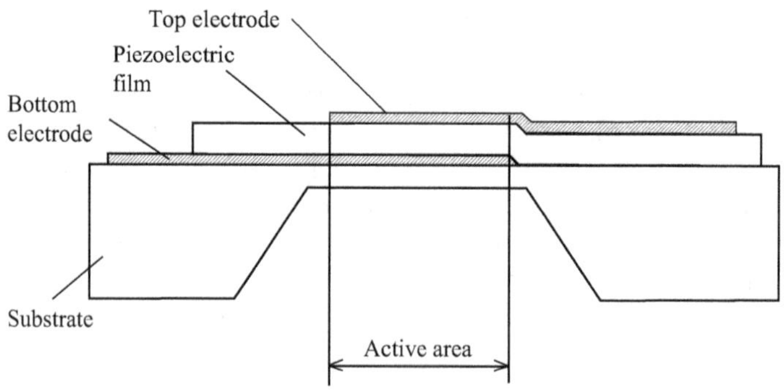

Figure 8.2 Configuration of FBAR

placed closest to the antenna in mobile communication devices in the GHz band in a way similar to that in the case of SAW devices since around the 2000s [23–25]. An interesting feature of FBARs is that the resonator is fabricated on the substrate in a process based on film deposition. This feature in principle enables the acoustic device to be integrated with the active circuits. Indeed, monolithic oscillators integrating an FBAR and an amplifier were first reported in the early 1980s and are still frequently discussed at present [26,27].

In this chapter, SAW and FBAR resonators are regarded as MEMS devices and their structures are classified. In addition, we introduce RF front-end circuits, which are the main application of SAW/FBAR resonators, and review the topology of the filters and duplexers used in SAWs/FBARs. Also, as a technique of improving the performance of these acoustic devices, doping and bonding techniques are explained, referring to recent studies. Nowadays, the demand for linearity and isolation is heightening in mobile phones with advanced architectures, and these trends are mentioned briefly in this chapter. Finally, GHz-band oscillators are also becoming a new research field of FBARs, which we briefly discuss.

8.2 Basis of the RF front-end circuit

Acoustic wave devices have established a solid position in communication systems, especially in high-frequency circuits typified by the front ends of mobile phones, owing to the sustained efforts of many researchers and engineers. To help the reader's understanding, here we will describe the configuration of RF front-end circuits, which are the main application of acoustic wave devices, and the key parameters in acoustic wave devices. There are many excellent textbooks providing the details of communication technologies and integrated circuit design, so here we will only briefly outline the essential parts of acoustic wave devices.

8.2.1 Multiple access systems

8.2.1.1 Duplexing: uplinks and downlinks

When subscribers use a communication system, the two kinds of systems in Figure 8.3 must be occupied. Figure 8.3(a) shows the uplink or transmitter (Tx) channel and Figure 8.3(b) shows the downlink or receiver (Rx) channel. Since these two kinds of channels are used at the same time, their signals must be separated into orthogonal dimensions. This separation is called duplexing, which is realised using the frequency or time domain, called frequency domain duplexing (FDD) and time domain duplexing (TDD), respectively.

8.2.1.2 Multiple access

The frequency bands, which are separated into the Tx and Rx channels, must then be allocated efficiently to multiple users as dedicated channels. For the allocation, time division, frequency division, code division or a combination thereof is used. Dividing the dimension of a signal among multiple users and performing allocation is called multiple access.

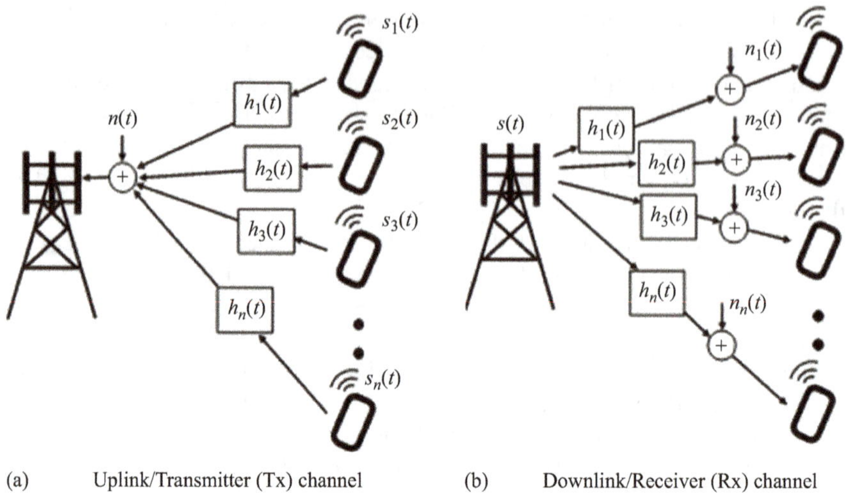

Figure 8.3 Uplink/down channels: radiation characteristic of antenna, h: propagation characteristics, n: AWGN

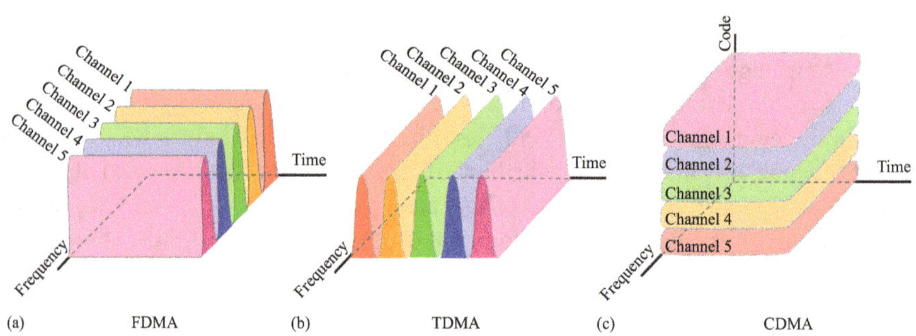

Figure 8.4 Classification of multiple access

Frequency division multiple access: FDMA

In an FDMA system, the dimension of the signal is separated on the frequency axis for multiple users, as shown in Figure 8.4(a). The channel is occupied by one user, but when the user has finished taking the channel, it is released to other users. Since the allocated channel is always changed in every call, it is necessary to switch the frequency of the carrier and implement a system capable of quickly capturing the carrier frequencies on each terminal. In an FDMA system, the channel allocated to each user becomes a narrowband, allowing the influence of frequency-selective fading to be suppressed. On the other hand, the overhead, which is indispensable in digital communication such as for channel estimation, becomes longer owing to the limited bandwidth. However, channel separation in the frequency domain has a long history of use in wireless communication tools with a fixed frequency, such as radio and TV, and FDMA was used extensively in early cellular systems.

Time division multiple access: TDMA

In a TDMA system, a periodic time frame is set, slotting this frame, and sharing among users, as shown in Figure 8.4(b). Since signals cannot be continuously handled along the time, all data including audio data are accumulated (buffered), compressed and transmitted as a burst signal. The advantage of TDMA is that transceiver activation is limited to a specific time slot within the communication frame. As a result, power consumption can be greatly reduced. Furthermore, it is possible to compress the bandwidth by digital compression. As a result, an expensive frequency band can be effectively allocated.

The implementation of A/D conversion and digital compression in a TDMA system increases the complexity of the transceiver system. Also, the use of time slots and frames makes the synchronisation accuracy between the base station and the terminal more severe. However, these problems have been gradually mitigated owing to the progress in signal processing using large-scale integration (LSI). TDMA is effectively utilised in combination with FDMA to alleviate the upper limit of users that can be accommodated in a limited frequency band.

However, the above-mentioned problem of synchronisation accuracy between the base station and the terminal becomes greater in the uplink (Tx channel) of TDMA. If signals from multiple users are aggregated via different delay paths, the time domain orthogonality among users may be impaired. The adjustment of the synchronisation at the base station or access point is indispensable among the users who share the same frame. However, this increases the length of the overhead in the transmission data.

Code division multiple access: CDMA

In CDMA, signals of different users are spread and orthogonalised by modulation using a spreading code while occupying the same time frame and frequency band (Figure 8.4(c)). It is also possible to assign multiple codes to a single user, which makes it possible to flexibly assign the number of channels required by the user. The most common spreading technique in CDMA is the use of a multiuser spread spectrum with direct spreading and frequency hopping.

In CDMA, the spectra of users are packed in the same time frame and frequency band since they can be diffused to suppress interference with each other. However, this feature induces a new risk called the perspective problem. This problem is caused by the fact that the intensity of the disturbance wave superimposed on the desired signal increases when approaching other user terminals or access points. If the intensity of the disturbance wave being spread reaches the same level as that of the demodulated desired signal due to the proximity, the receiving sensitivity is seriously deteriorated owing to the reduction in S/N ratio. To solve this problem, the transmission power of each terminal is controlled so that the power received by all users is almost equal regardless of the distance of the terminal. This technique is called transmission power control (TPC), which is indispensable in CDMA systems.

8.2.2 Communication architecture

As described in the previous section, each user is assigned a dedicated channel at the time of use. In general, the bandwidth of the allocated channel is extremely narrow and no more than several hundred kHz (<0.02% of the fractional bandwidth). Therefore, in the design of transceiver circuits, coding, compression and modulation with high efficiency are utilised in the baseband circuitry. To alleviate the design requirements for these processes in the baseband, the RF front-end circuit as the front stage of the baseband process or must have the following functions:

1. Band selection: reduction in the intensities of unwanted signals from other communication standards in use.
2. Downconversion of the desired signal to the low-frequency band (baseband) to alleviate the design requirement for the channel selection.
3. The above two steps must be carried out with sufficient linearity to reduce gain suppression and intermodulation.

In this section, we explain the importance of the figure of merit in RF circuits for the acoustic wave filter design. We will deal with the design of acoustic wave filters in the next chapter. In addition, we describe the architecture utilised in actual systems by focusing on the receiver system. Advanced digital processing in the baseband is covered in many excellent textbooks and will not be discussed here. Also, regarding the transmitter architecture, its important points are similar to those of the receiver architecture, so it is not discussed in this section.

8.2.2.1 Basic concept

Channel selection and downconversion

It is difficult to pick up the allocated channel in the RF band. This is because the channel allocated to each user is narrow and the channels of the other users are very close. For example, when attempting to pick an adjacent channel up using a filter in the GHz band, a high Q-value of at least 10,000, which is impractical, is required for the filter. Also, since the channel allocated to each user is updated in every use, the centre frequency of the filter must be quickly changed following this update, which is also not possible.

In an actual system, the received signal is downconverted to the intermediate-frequency band (IF band) by multiplication (mixing) with a local oscillator employing quartz to relax the requirement of channel selection. Regarding changing the allocated channel, the desired signal is always downconverted to the same frequency within the IF band, regardless of the channel allocation, by incorporating a frequency synthesis with the local oscillator to adjust the mixing frequency.

Band selection

Channel selection is carried out to suppress the interference of adjacent channels. Band selection is different from channel selection and is performed to prevent the mixing of unnecessary signals from different communication standards. To sufficiently attenuate the unnecessary signals from other standards, a filter device with a sharp band edge is needed. To achieve characteristics with a steep cut-off, it is effective to increase the number of stages of filters, although this intrinsically has a trade-off with the loss. The loss of the filter not only decreases the sensitivity of the receiver but also severely affects power consumption. For a transmitter system requiring a high transmission power of about 1 W, even for losses of 1 to 2 dB, 200 to 300 mW of power is wasted as heat.

Increasing the performance of band selection is key to improving the linearity. A principal source of unnecessary signals in the out band is the transmitter paired with the receiver system in the FDMA receiver. The leakage of signals from the transmitter generates an unwanted offset and compresses the dynamic range in which sufficient linearity is obtained in the receiver system. The isolation characteristic between Rx and Tx is a crucial figure of merit for the duplexer used in FDMA-based systems.

8.2.2.2 Receiver architecture

Heterodyne architecture

It was previously mentioned that downconversion using a multiplier is implemented at the receiving system to mitigate the severe design requirements for channel selection. Figure 8.5 schematically shows a block configuration of downconversion and the signal spectrum. The input signal ω_{in} moves to $\pm(\omega_{\text{in}} \pm \omega_{\text{LO}})$ by multiplying by the sinusoidal wave $A_0\cos\omega_{\text{LO}}t$ generated by the local oscillator and is extracted as the signal $\omega_{\text{in}} - \omega_{\text{LO}}$ by using a low-pass filter. This process is called downconversion, and the centre frequency of the converted signal is called the IF. A receiver having such a configuration is called a heterodyne receiver.

The use of a heterodyne receiver is an excellent technique for facilitating channel selection, but it invokes a problem of the image frequency. The image is, as shown in Figure 8.6, a disturbance signal that is separated from the input signal by ω_{IF} in the frequency spectrum and is on the opposite side of the impulse of the local oscillator ω_{LO} from the desired signal ω_{in}. This image signal is overlapped with the desired input signal through the downconversion process. To eliminate this undesired image, an image rejection filter is essential in heterodyne receivers, and SAW devices have been widely used as rejection filters. However, the need for image rejection filters has been decreasing with the establishment of zero-IF techniques. These new architectures will be described later.

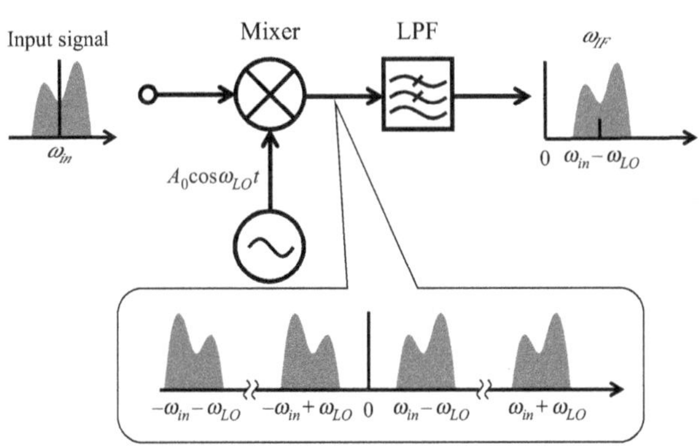

Figure 8.5 Downconversion process in heterodyne receiver

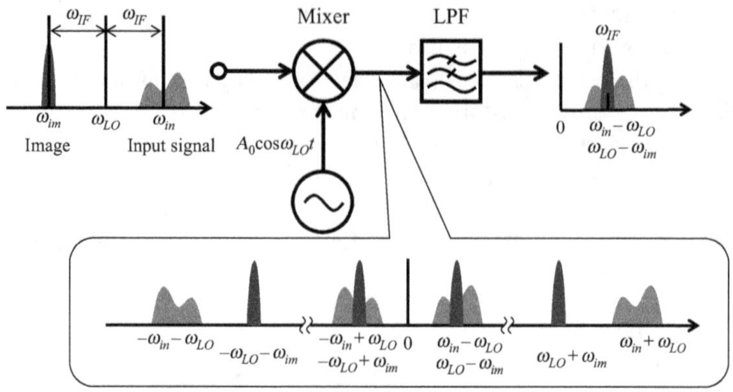

Figure 8.6 Problem of image frequency

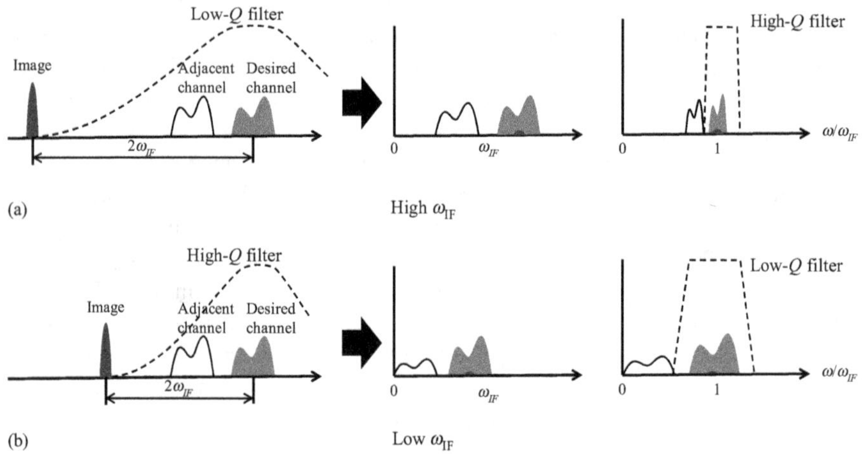

Figure 8.7 Performance trade-off between image rejection and channel selection filters

Dual conversion architecture

In a heterodyne receiver, there is an apparent trade-off between image removal and channel selection in the positioning of the IF band. If ω_{IF} is set to a higher value, as shown in Figure 8.7(a), image rejection becomes easy. However, a higher Q-value is required for the filter used in channel selection, making channel selection difficult. On the other hand, by setting ω_{IF} to a lower value, the Q-value required for the channel selection filter is relaxed, but the attenuation of the unnecessary image is deteriorated (Figure 8.7(b)).

A solution to this dilemma is downconversion with multiple stages. Dual downconversion (two-stage downconversion) is usually chosen to avoid an increase in the spurious caused by the mixing (Figure 8.8). By using this architecture, while obtaining a sufficient image rejection rate, it is possible to improve the signal-to-noise (SN) ratio by including an amplifier and to enhance the effective Q-value of the channel selection filter for the subsequent stage.

Architecture with zero second IF

We can recognise that it is difficult to reduce the cost, size and power consumption in the configuration of Figure 8.8, where there are multiple interstage filters and quartz oscillators. To overcome this difficulty, the second IF band should be set to zero, meaning that the second frequency ω_{LO} coincides with the centre frequency of the desired signal. As a result, the image signal becomes the desired signal itself and an image rejection technique is no longer necessary.

By setting the IF to zero, the spectrum on one side becomes negative and is cut down. In FM modulation, quadrature phase shift keying (QPSK), Gaussian-filtered minimum shift keying (GMSK), quadrature amplitude modulation (QAM) and the like, the signal quality is critically degraded since the frequency spectrum has asymmetry with respect to the carrier frequency. As a countermeasure against this deterioration, orthogonal downconversion is adopted. In orthogonal downconversion, downconversion is performed using two local oscillators whose phases are orthogonal to each other (Figure 8.9). The two output signals, $x_{BB,I}(t)$ and $x_{BB,Q}(t)$ in Figure 8.9, are called orthogonal baseband signals.

Using multiple quartz oscillators not only is unfavourable from the point of view of cost and size reduction but also generates spontaneous frequency pulling by mutual coupling among the oscillators. In many cases, the configuration in Figure 8.10 using a frequency divider and a quartz oscillator is used in the architecture with zero second IF.

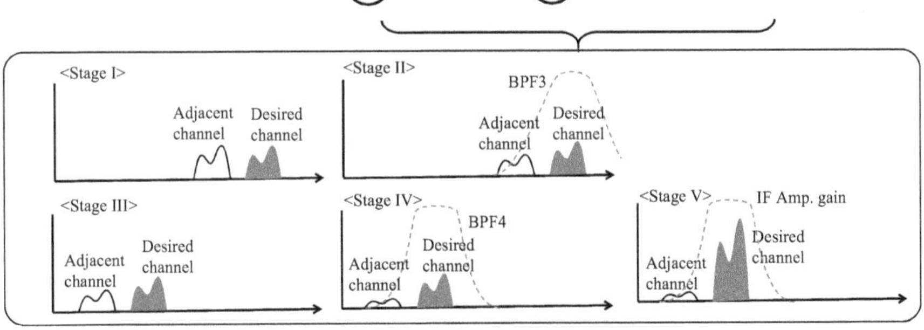

Figure 8.8 Architecture of dual-IF receiver

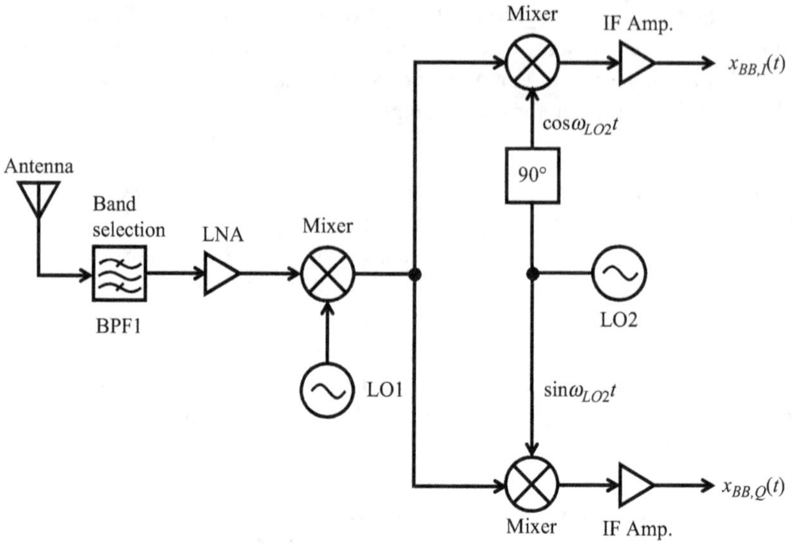

Figure 8.9 Architecture of receiver with zero second IF

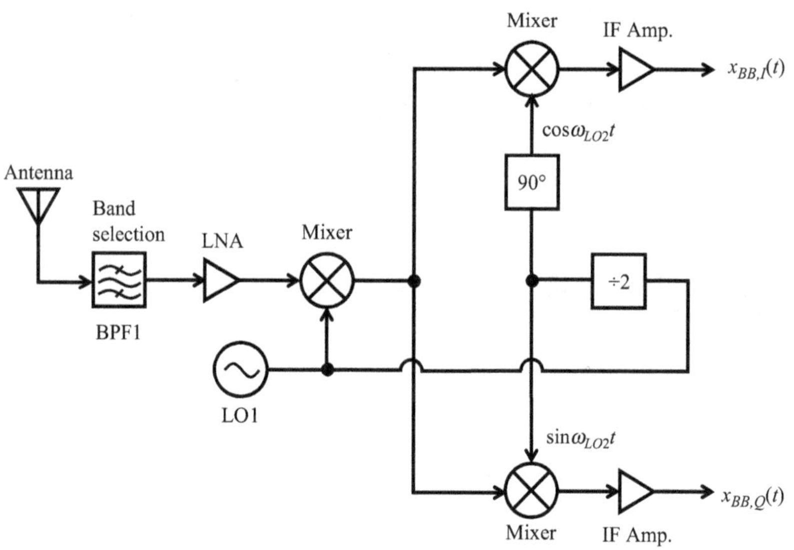

Figure 8.10 Practical architecture of receiver with zero second IF

Direct-conversion architecture

Considering Figures 8.9 and 8.10, a simpler architecture can be readily conceived, as shown in Figure 8.11, which is called a direct conversion, zero-IF or homodyne architecture. However, this architecture still has many problems and is a relatively

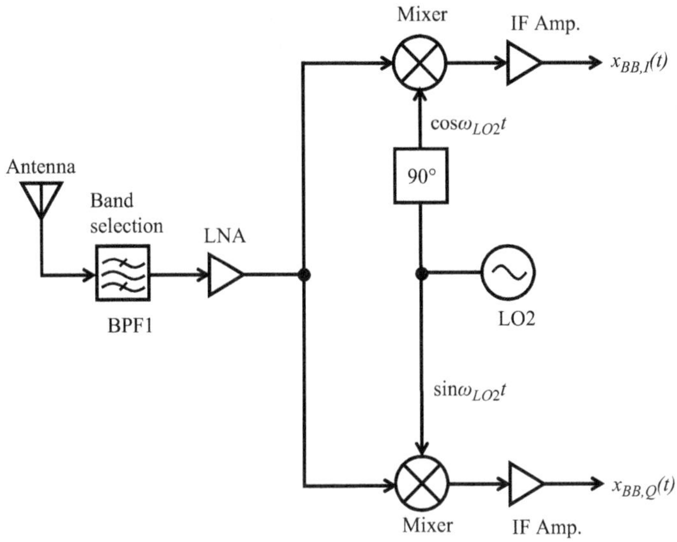

Figure 8.11 Architecture of direct-conversion receiver

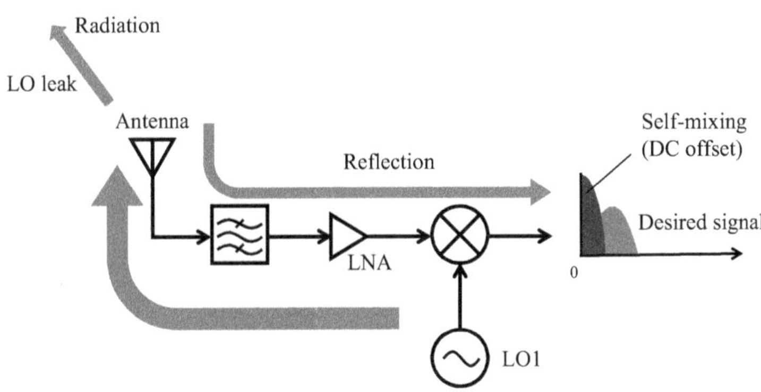

Figure 8.12 Problems in direct-conversion architecture

new technology that finally entered the market in the 1990s to 2000s. In the direct-conversion architecture, since the frequency of the LO is the same as the carrier frequency of the desired signal, we encounter the following severe technical problems (see Figure 8.12):

(a) **Problem of LO leakage:** The impulse of the LO that radiates from the antenna to the outside is called LO leakage. In direct conversion, since the frequency of the LO coincides with the communication band, there is a possibility that the sensitivity of other receivers operating in the same band will deteriorate. So, the LO leakage must be sufficiently suppressed in the RF design.

(b) **Problem of DC offset:** Part of the LO leakage is reflected at the antenna end and introduced into its own low-noise amplifier (LNA). The amplified signal generates a large DC offset in the baseband by self-mixing. This large offset squeezes the dynamic range of the signal processing in the baseband and causes the saturation of the system in the worst case.

Moreover, when simplifying the system configuration, we must additionally take countermeasures against the problems of even-order distortion and flicker noise. We must also carefully consider the mismatches of the two orthogonal signals (I and Q signals) used in the downconversion. Many of these problems have already been overcome by designing elaborate circuits including the system layout. However, since the discussion of active circuits including the circuit layout is beyond the scope of this book, readers are referred to some of the outstanding textbooks already written on this topic.

8.2.3 Remarks

As some readers will have noticed, the evolution of the RF frontend has focused on how to remove the interstage filters and crystal oscillators, which are off-chip components, and how to integrate the RF system. Considering this trend, will antenna duplexers become unnecessary in the future? The answer is no. In fact, the demand for the duplexers used for band selection is still strong and shows no signs of declining. This is caused by the number of communication standards continuing to increase, mainly in the sub-6 GHz band, and the importance and complexity of band selection are also increasing owing to the high demand for wireless communication.

The growing need for wireless communication is relevant to the emergence of smartphones. Such new devices have many channels to connect the network, explosively increasing the amount of data handled by individuals. The situation is considerably different from the earlier phones that transmitted analogue voices with occupied lines and exchanged simple text. Owing to the increase in communication capacity, carrier aggregation (CA) using multiple bands simultaneously has advanced, and the increased complication and parallelisation of the RF front-end circuitry are being promoted. Consequently, the number of communication standards incorporated into a terminal has increased, and the current high demand for duplexers is being maintained.

8.3 Resonators and filters

8.3.1 Classification of resonators

8.3.1.1 SAW resonators

What kind of configuration comes to mind when hearing 'a SAW resonator'? A SAW resonator has the structure shown in Figure 8.13. Some readers may recognise such a configuration since interdigital transducers (IDTs) used to efficiently excite SAWs have already been touched on in the introduction of this chapter. The basic configuration of the resonator shown in Figure 8.13 does not vary so much, although there

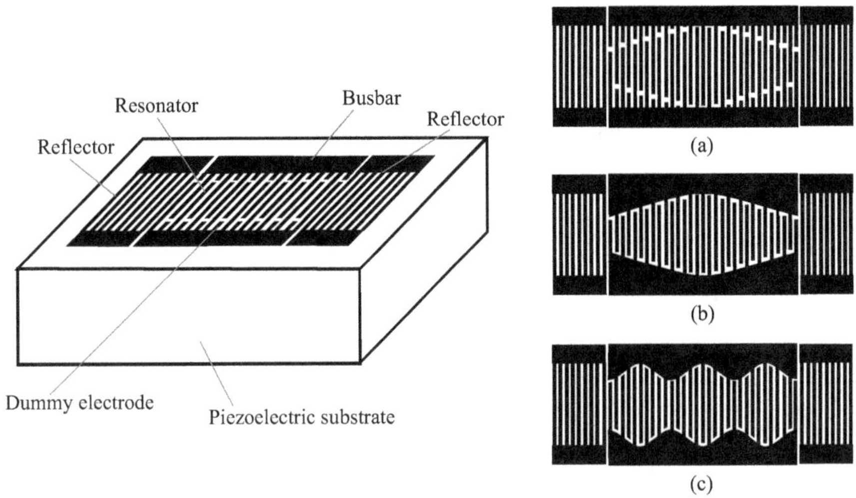

Figure 8.13 SAW resonator configuration and variation of layout for apodisation

are some variations in the structure used for apodisation to suppress the transverse mode orthogonal to the travelling direction of the SAW.

Among SAWs, Rayleigh-type SAWs are well known. However, from a practical viewpoint, Rayleigh-type SAWs, which have a relatively low propagation velocity, are rarely used in communication devices. As the desired frequency increases, the pitch of the IDT decreases, which is limited by the photolithography technology and ohmic loss. Thus, a sufficiently high propagation speed is required. Since the amplitude of the wave decreases with increasing frequency, the coupling coefficient must also be kept high.

Here, we will expand the discussion from Rayleigh-type SAWs to SH-type SAWs, Love-type SAWs and boundary wave devices in a stepwise manner. Before discussing each type of SAW resonator, let us classify waves propagating through a substrate. Unlike bulk waves propagating in an infinite medium, the concept of an interface must be included when modelling SAW propagation, and waves are classified into longitudinal (L), shear vertical (SV) and shear horizontal (SH) waves as follows from the relationship between the interface, amplitude and propagation direction (Figure 8.14):

L wave: A wave having an amplitude in the direction parallel to the propagation direction. An L wave cannot propagate alone along with an interface.

SV wave: A wave having an amplitude in the direction perpendicular to the propagation direction and interface. An SV wave cannot propagate alone along with the interface.

SH wave: A wave orthogonal to the propagation direction and having an amplitude in the direction parallel to the interface. It can propagate along with the interface without coupling with another wave.

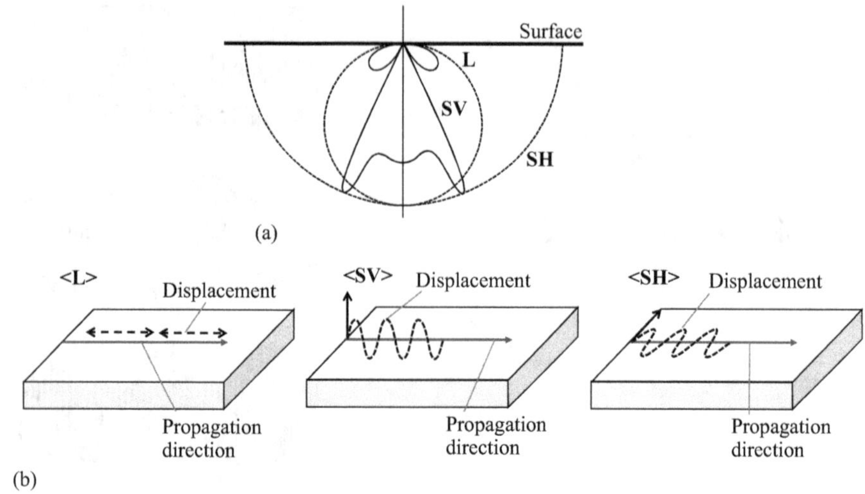

Figure 8.14 Classification of propagating waves: radiation pattern into a semi-infinite substrate (a), relationships between displacement, propagation direction and surface (b)

Rayleigh-type SAWs

Rayleigh-type SAWs consists of L and SV waves that cannot propagate alone on a surface, but a couple and propagate while exchanging vibration energy with each other. Since Rayleigh-type SAWs are generated by the coupling of pure mechanical vibrations and can propagate in all media including anisotropic substrates, they are not affected by electrical boundary conditions such as the piezoelectricity of the substrate. Therefore, Rayleigh-type SAWs are unsuitable for application in practical electronic devices. Compared with SH-type SAWs, their low propagation speed is one of the reasons why Rayleigh-type SAWs are not commonly used in electronic devices. However, the Rayleigh-type SAW has an advantage that there is little leakage into the substrate and the propagation loss is extremely small since its propagation velocity is slower than the SV bulk wave into the substrate.

SH-type SAWs

SH waves can exist on a surface, as described above. In other words, since there is no coupling of waves, such as that between an L wave and an SV wave in a Rayleigh-type SAW, SH waves spread isotropically and attenuate conventionally (Figure 8.14(a)). However, by selecting a suitable piezoelectric substrate and its crystal orientation, SAWs can be generated using SH waves. These are Bleustein–Gulyaev–Shimizu (BGS) waves [28–30]. In a BGS wave, piezoelectricity and an SH wave are coupled and propagate on a surface while exchanging their energy. The substrate and crystal orientations used to actually obtain BGS waves (SH-type SAWs) are typically 41° YX $LiNbO_3$, 64° YX $LiNbO_3$ and 36° YX $LiTaO_3$ [7,14].

For these particular orientations and materials, there is almost no leakage of energy into the bulk substrate. Indeed, leakage into the substrate occurs when slightly changing the orientations. Such SH-type SAWs including leakage into the substrate are called leaky SAWs (LSAWs) or pseudo-SAWs [31]. BGS waves are special cases of LSAWs.

LSAWs are generated by the coupling between piezoelectricity and SH waves and the piezoelectric effect depends on the properties of the piezoelectric material. Therefore, SH-type SAWs using LSAWs can easily realise high electromechanical coupling coefficients and excellent temperature characteristics compared with Rayleigh-type SAWs. For this reason, more than 90% of SAW devices currently in practical application use LSAWs.

Love-type SAW

Whether the leakage of an LSAW occurs is determined by the magnitude relation between the velocity of the surface wave, in which the piezoelectricity and SH wave undergo coupling, and the velocity of the bulk wave propagating into the substrate. That is, if the velocity of the LSAW is sufficiently lower than the wave velocity in the bulk, which indicates energy diffusion into the substrate, the LSAW is trapped as a surface wave on the substrate without leakage loss. The LSAW velocity can be adjusted by selecting the substrate orientation as described above, but it is not possible to give an optimum solution for all performance characteristics such as the temperature characteristic, coupling coefficient and mechanical loss. The LSAW velocity can also be adjusted by depositing a heavy film that suppresses the propagation velocity of the SAW. An SH-type SAW whose leakage has been suppressed by the deposition of an additional film is called a Love-type SAW [32,33]. By the way, in seismology, the Love wave can be observed in the area, where the soft ground is accumulated on the hard ground, as a side-sliding-vibration only in the soft ground.

An example of the use of Love-type SAWs is in a SAW device using a Cu grating on a $LiNbO_3$ substrate. The 0° YX $LiNbO_3$ has a high electromechanical coupling coefficient, but the bulk wave velocity is low and the loss due to the energy leakage into the substrate cannot be ignored. In this case, Shimizu and Mizunuma showed that the SH mode can be confined to the surface by using an Au or Ag electrode with a low wave velocity in the bulk [34]. A Cu grating was developed following this proposal. In actual developments, 15° YX $LiNbO_3$ has been used rather than 0° YX $LiNbO_3$ to suppress spurious responses [35,36].

SiO_2 is also an attractive deposition material with a low bulk wave velocity and a specific negative temperature characteristic, and SiO_2 films are widely used as temperature compensation (TC) films for resonators. By depositing SiO_2 on a resonator, a Love-type SAW is obtained. In other words, SiO_2 not only expands the choice of substrates with a high coupling coefficient but also improves the temperature characteristic. However, specific techniques are required to deposit the film on an IDT. If the film is conformally deposited, the surface becomes bumpy, reflecting the shape of the electrode. These bumps introduce unwanted spurious responses into the frequency spectrum. As simple methods of obtaining a flat surface, the deposition and firing of a spin-on-glass and polish planarisation are commonly used. However, with these methods, it is difficult to precisely control the SiO_2 thickness. Thus, they are not commonly used in communication applications that require the matching of frequency bands. Depositing the oxide between electrodes in advance to make the surface flat, then depositing the oxide again to the desired thickness is one of the techniques used to eliminate the unwanted bumps and spurious responses (Figure 8.15(a)) [37,38]. Dry etching after deposition is also an effective method of reducing the bump size to a value smaller than the wavelength of the Love wave (Figure 8.15(b)) [39,40].

Boundary wave resonators

When rigid plates are bonded with glue, vibrations are concentrated in the soft glue and the bonding may break. Such vibrations in the soft layer are called Stoneley or boundary waves [41,42]. This phenomenon has provided many suggestions for the development of acoustic devices. SAW devices have vibrations on their surface, so careful packaging is required. If a boundary wave can be utilised, the vibrating part can be confined inside the substrate, enabling an ultra-compact acoustic device that does not require a package to be accomplished.

Considering the breakdown of glue from the viewpoint of acoustics, it is understood that the vibration energy is concentrated in the boundary layer with low acoustic impedance when a layer with low acoustic impedance is sandwiched between layers with high acoustic impedance. Shimizu's report [34], which mentions Love-type SAWs, provides an excellent insight into forming electronic devices with layers of low acoustic impedance. That is, a boundary wave device is obtained by fabricating an Au electrode with a low acoustic impedance on a piezoelectric substrate ($LiNbO_3$) and depositing SiO_2 with a sufficient thickness, which has a higher acoustic impedance than the Au electrode layer (Figure 8.16) [43–45].

As described later in the filter design section, a package-free configuration is extremely effective for realising ultra-miniaturised filters. However, the package

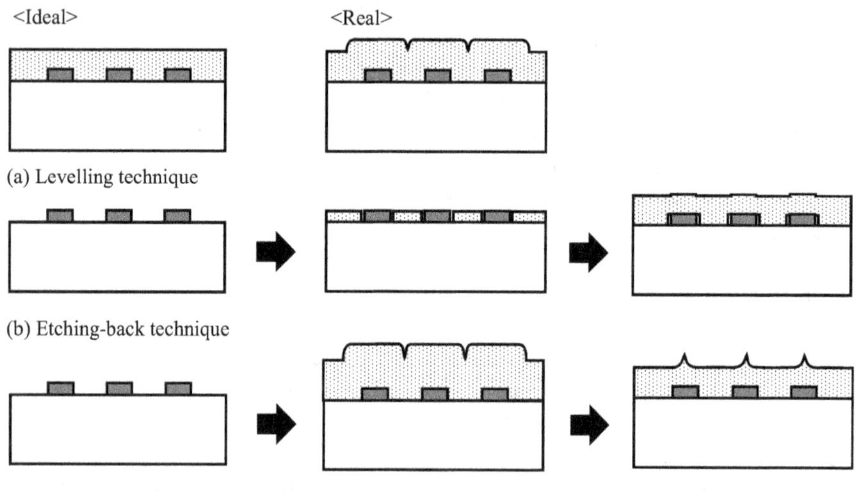

Figure 8.15 Microfabrication techniques used to form a flat SiO_2 surface

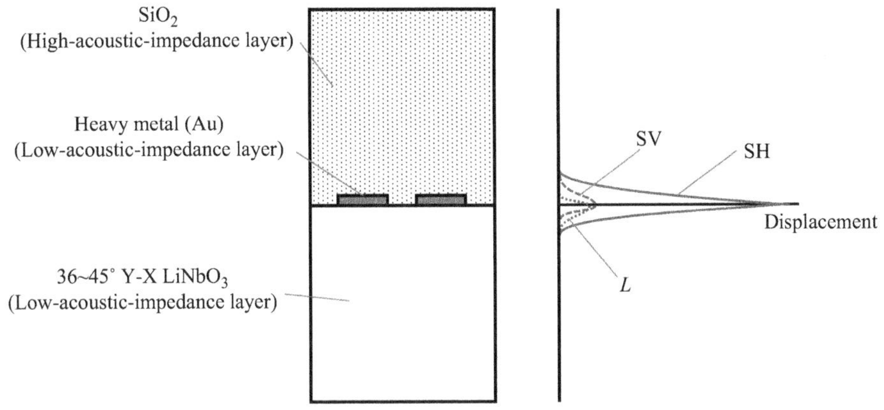

Figure 8.16 Boundary wave device

not only protects the device but also acts as a passive component for adjusting the impedance matching and the arrangement of the attenuation poles in the frequency spectrum. The application of boundary wave devices should be carefully considered.

8.3.1.2 BAW resonators

Assuming that a piezoelectric thin film is deposited on an elastic substrate and vibrated, the vibration energy leaks and dissipates to the substrate as shown in Figure 8.14. To suppress such dissipation, an acoustic insulation structure is required between the piezoelectric transducer and the substrate. Note that such an acoustic insulation structure is used not only to suppress energy dissipation but also to confine the suitable fundamental mode in a piezoelectric transducer for signal processing applications. To apply the BAW resonators to signal processing such as filtering, it is desirable to sufficiently separate the main resonance from the harmonics and spurious response and to be given a high coupling coefficient. Thus, micromachined structures used to confine the fundamental mode, as shown in Figure 8.2, were reported in the 1980s by Nakamura *et al.*, Lakin and Wang and Grudkowski *et al.* [20–22]. In these structures, abutting overtone harmonics were purged out and the coupling coefficient was enhanced by thinning the support substrate as shown in Figure 8.17. We discuss BAW resonators by classifying them into thin-film bulk acoustic resonators (FBARs) and solidly mounted resonators (SMRs), which are typical acoustic isolation structures, in the following section.

FBAR-type BAW resonators

The simplest way to confine the fundamental mode in a piezoelectric thin film is to thin and remove the substrate beneath the resonator as shown in Figure 8.17 [20–22]. Such a structure with a free-standing piezoelectric resonator is called a thin-film bulk acoustic resonator (FBAR). There are several names for this structure, but this textbook uses FBARs.

In the early FBARs, the anisotropic etching of the silicon substrate by an alkaline etchant was used (see Figure 8.2). However, from the viewpoint of

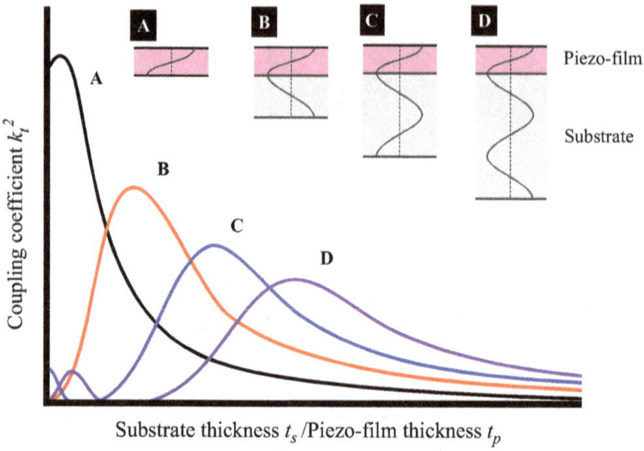

Figure 8.17 Fundamental mode confinement

efficient utilisation of the chip area, a structure fabricated by vertical plasma etching (Bosch process) was proposed as shown in Figure 8.18(a). This technology has been commercialised as a filter for use in 5-GHz-band wireless LANs [23].

The most popular structure of FBARs is shown in Figure 8.18(b) [25]. This was the first FBAR structure to be applied to a duplexer (US PCS). In this structure, a buried sacrificial layer is utilised, which is fabricated by the oxide deposition into the shallowly etched substrate (swimming-pool-like structure) and chemical mechanical polishing (CMP). By using the buried sacrificial layer, a double-sided process is no longer necessary and a wide air gap is obtained, which is effective for avoiding sticking and reducing the time required for sacrificial layer etching.

The integration of active circuits and BAW devices was first reported in the 1980s [26]. The structure of these BAW resonators had a narrow airgap fabricated by etching of a sacrificial layer patterned on the substrate. However, since the stress control of piezoelectric films had not been established at that time, it was too difficult to obtain a high yield without a special technique as shown in Figure 8.18(b). In recent years, a piezoelectric film deposition technique in which the film quality and stress can be independently controlled has become available [46,47]. So, an air-gap-type FBAR with a reasonable yield has been proposed, as shown in Figure 8.18(c), in which the piezoelectric resonator is deformed into a dome shape by the stress applied to the film to create and maintain the air gap [48].

SMR-type BAW resonator

Bragg reflection films are widely used as reflection films for electromagnetic wave (light). The mechanism of Bragg reflector also can be effective for acoustic wave, and the structure in Figure 8.19(a) was proposed as a BAW resonator [49]. That is, the acoustic insulation between the resonator and the substrate is achieved by alternately depositing layers with high and low acoustic impedances, each with a thickness of $\lambda/4$. This structure is called a solidly mounted resonator (SMR) because it does not require a difficult MEMS process to fabricate the free-standing film.

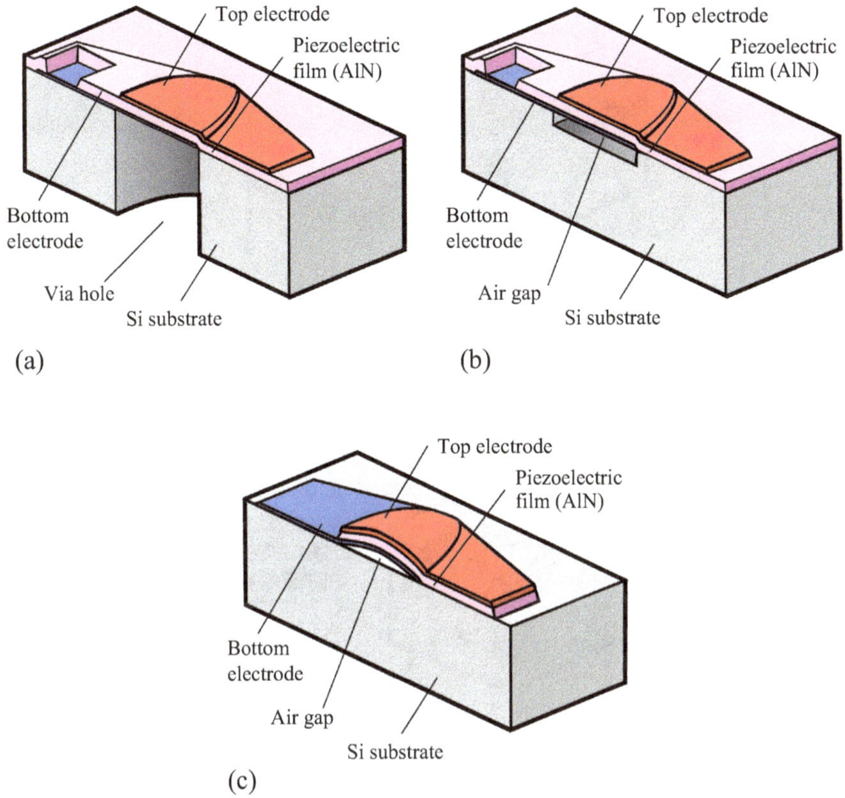

Figure 8.18 FBAR-type BAW resonators

In an application where multiple resonators must be laid out with a high density, such as a ladder filter, the resonators suffer from capacitive coupling with each other when using the structure shown in Figure 8.19(a) [24]. Thus, the structure shown in Figure 8.19(b) is often practically used. In this structure, Bragg acoustic reflectors are built in a swimming-pool-like structure fabricated on a silicon substrate. The swimming-pool-like structure is the same as that in Figure 8.18(b) and is fabricated by shallow etching and CMP.

As shown in Figure 8.14, the leak of vibrational energy into the substrate is classified into three types. In the FBAR type, the air layer insulates the vibration regardless of the type of vibration. On the other hand, in the SMR type, the leakage of SV and SH waves cannot be ignored even if only the longitudinal vibration in the thickness direction is used for the main resonance. The design of Bragg acoustic reflection films requires not only the repeated deposition of layers with a thickness of $\lambda/4$. They are specialised to suppress the longitudinal waves but also optimisation to suppress the shear-type vibrations [50].

Since an air layer is ideal for acoustic isolation, SMR-type BAW resonators are inferior to the FBAR type in terms of the *Q*-value. However, in the SMR type, a

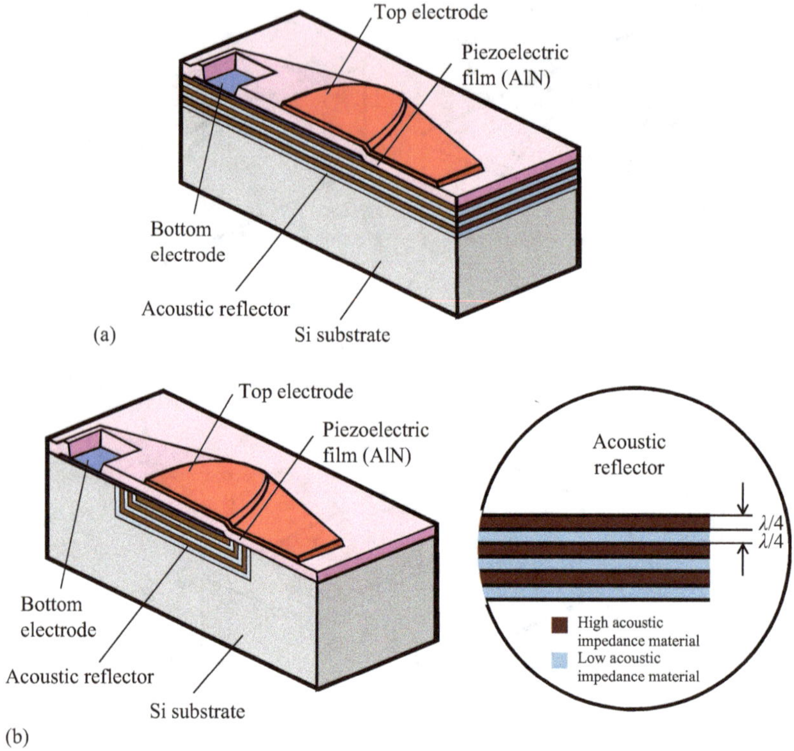

Figure 8.19 SMR-type BAW resonators

multilayer film can be utilised to provide a high degree of design freedom that is not available in the FBAR type. These attractive features will be discussed later in this chapter.

8.3.2 Filter design

8.3.2.1 Characterisation of resonators

SAW and BAW resonators have a double-resonance characteristic, as shown in Figure 8.20, comprising a resonance of mechanical vibration and an anti-resonance generated by mutual interference between the mechanical vibration and the electrostatic capacitance. It is known that the double resonance in Figure 8.20 is approximated by a Butterworth–Van Dyke (BVD) equivalent circuit. In particular, the modified BVD equivalent circuit shown in Figure 8.21 is often used for BAW resonators [51]. For such an equivalent circuit, the Q-values (loaded resonance Q ($Q_{\rm rl}$) and loaded anti-resonance Q ($Q_{\rm al}$)) are defined as follows:

$$Q_{\rm rl} = \frac{\omega_r L_1}{R_s + R_1} = \frac{1}{(R_s + R_1)\omega_r C_1} \tag{8.1}$$

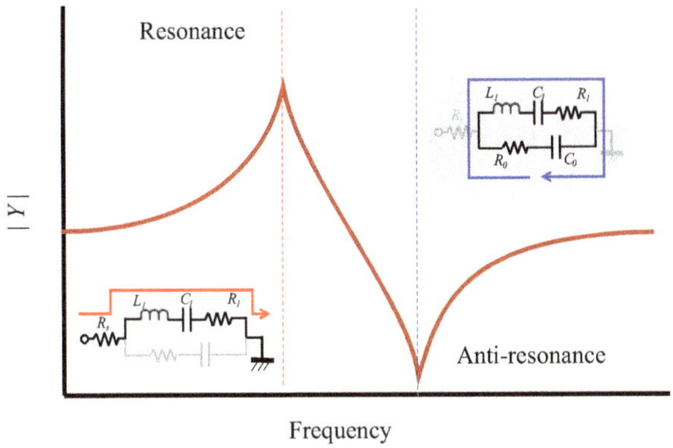

Figure 8.20 Frequency spectrum of acoustic resonator

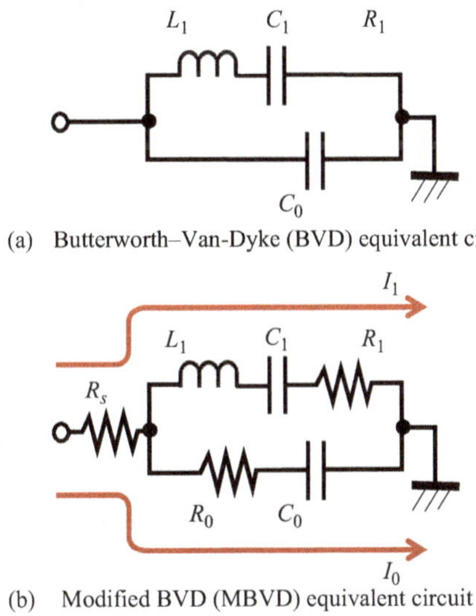

Figure 8.21 Lumped equivalent circuit of acoustic resonator

$$Q_{\mathrm{al}} = \frac{\omega_a L_1}{R_1 + R_0} = \frac{1}{(R_1 + R_0)\omega_a}\left(\frac{1}{C_1} + \frac{1}{C_0}\right) \tag{8.2}$$

where ω_r and ω_a are the following angular resonance and anti-resonance frequencies, respectively.

$$\omega_r^2 = \frac{1}{L_1 C_1} \tag{8.3}$$

$$\omega_a^2 = \frac{1}{L_1}\left(\frac{1}{C_1} + \frac{1}{C_0}\right) \tag{8.4}$$

The frequency band where mechanical resonance can act is also of interest. The electromechanical coupling coefficient k_t^2 is widely used as an evaluation parameter for the width of such a frequency band. The effective electromechanical coupling coefficient k_{eff}^2 is more directly related to the passband width of filter applications and is also used as an evaluation parameter. k_t^2 and k_{eff}^2 are defined as follows:

$$\begin{aligned} k_t^2 &= \frac{\pi f_r}{2 f_a} \tan \frac{\pi}{2} \left(\frac{f_a - f_r}{f_a} \right) \\ &\approx \frac{\pi^2}{4} \frac{f_r}{f_a} \frac{f_a - f_r}{f_a} = \frac{\pi^2}{8} \frac{C_1}{C_0} \frac{C_0 - C_1}{C_0} \end{aligned} \tag{8.5}$$

$$k_{\text{eff}}^2 = \frac{f_a^2 - f_r^2}{f_a^2} \tag{8.6}$$

where f_r and f_a are the resonance and anti-resonance frequencies, respectively.

Here, we mention a problem when using the above coupling coefficient as an evaluation parameter for the double-resonance characteristic. Upon decreasing the electromechanical coupling coefficient, the resonance and anti-resonance frequencies become close to each other. In this case, the resonance and anti-resonance peaks become sharper, and the *Q*-value of each resonance increases. Conversely, if the coupling coefficient is increased, the *Q*-value is degraded. This is called the *Q*-*k* trade-off. The electromechanical coupling coefficient is suppressed when a parasitic capacitance is connected to the resonator in parallel or when the vibration of the resonator is mechanically constrained. The improvement of the *Q*-factor due to the degradation of the coupling coefficient is not relevant to the reduction of the loss element of the resonator. To improve the resonator performance, it is necessary to pay attention to the superficial improvement resulting from the *Q*-*k* trade-off.

Also, the *Q*-*f* trade-off is a bothersome problem in the characterisation of resonators. It is a factor of this trade-off that increasing loss cycles per unit time with increasing frequency. Although this is not superficial, there is a cautionary point when discussing the *Q*-*f* trade-off. Even with resonators of the same dimensions, an increase in the operation frequency causes the input impedance to decrease. In such a case, the resonance and anti-resonance *Q*-values are estimated, respectively, larger and smaller than the actual loss. Therefore, the relationship between the resonance frequency and the *Q*-value is not a simple inversely proportional one and has a strong design dependence.

Bode-*Q* is useful for evaluating the mechanical *Q* of a resonator separately from its electrical characteristics. Although Bode-*Q* is defined by (8.7), which is based on the MBVD equivalent circuit, (8.9) is now more generally employed since it can be used to calculate *Q* regardless of the spurious response [52,53].

$$Q_m(f) = \frac{\omega}{2} \frac{C_0 V(f)^2 + C_1 V(f)^2 + L_1 I_1(f)^2}{R_1 I_1(f)^2 + R_0 I_0(f)^2 + R_s (I_1(f) + I_0(f))^2} \tag{8.7}$$

$$Q_m(f) = -\omega \frac{\delta \phi}{\delta \omega} \frac{|\Gamma(f)|}{1 - |\Gamma(f)|^2} \tag{8.8}$$

Here, ϕ and Γ are the phase and reflection coefficients, respectively.

Filter design

In the early 1970s, SAW transversal filters were widely used as IF filters for TV tuners. A transversal filter is a combination of two IDTs, as shown in Figure 8.22 [54–56]. By weighting the input IDTs, it is possible to design a frequency spectrum and phase characteristics with a high degree of freedom. The drawback of transversal filters is a high insertion loss. When two IDTs are moved close to each other to reduce the insertion loss, as shown in Figure 8.23, mode coupling is induced, and

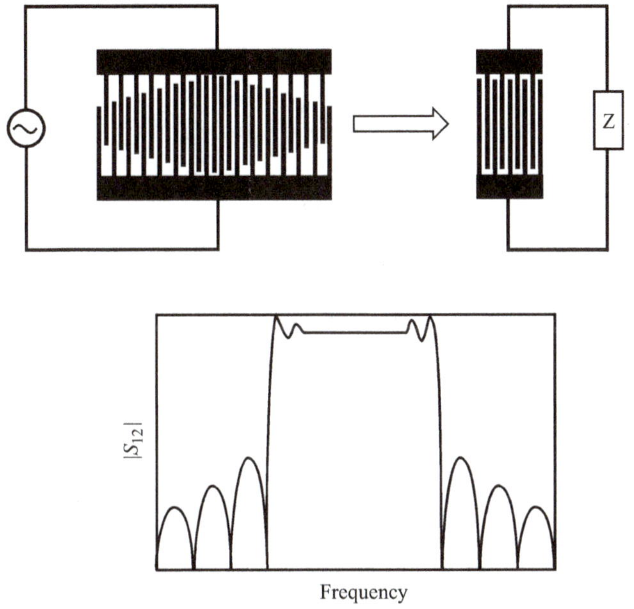

Figure 8.22 SAW transversal filter

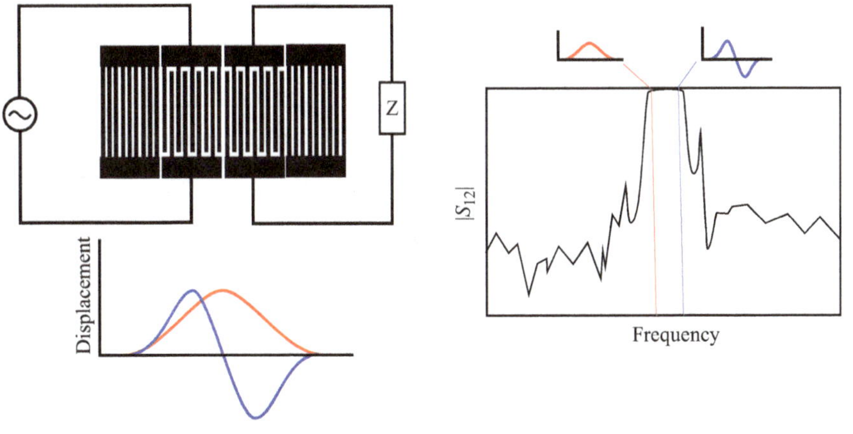

Figure 8.23 Double-mode SAW (DMS) filter

it is possible to obtain a filter characteristic with high sharpness while also reducing the insertion loss [57,58]. Such a filter is called a double-mode SAW (DMS) filter because it is based on a principle different from the design of transversal filters using propagation characteristics. As shown in Figure 8.24, there are several variations of the DMS filter depending on the coupled modes. The concept of DMS filters can be applied to BAW resonators as shown in Figure 8.25. Such BAW filters are called stacked crystal filters (SCFs) [59–61].

A ladder filter topology has been proposed to utilise the double-resonance characteristics of SAW/BAW resonators and is widely used for wireless application nowadays [17,18]. The ladder filter uses two types of resonators having different resonance frequencies, and high-frequency and low-frequency resonators are connected to series and parallel branches, respectively (Figure 8.26). Thus, the filter characteristics are obtained by the electrical coupling of two types of resonators.

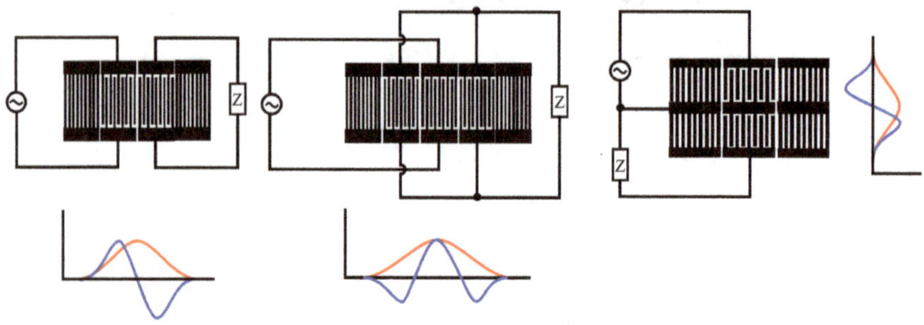

Figure 8.24 Variations of DMS filter

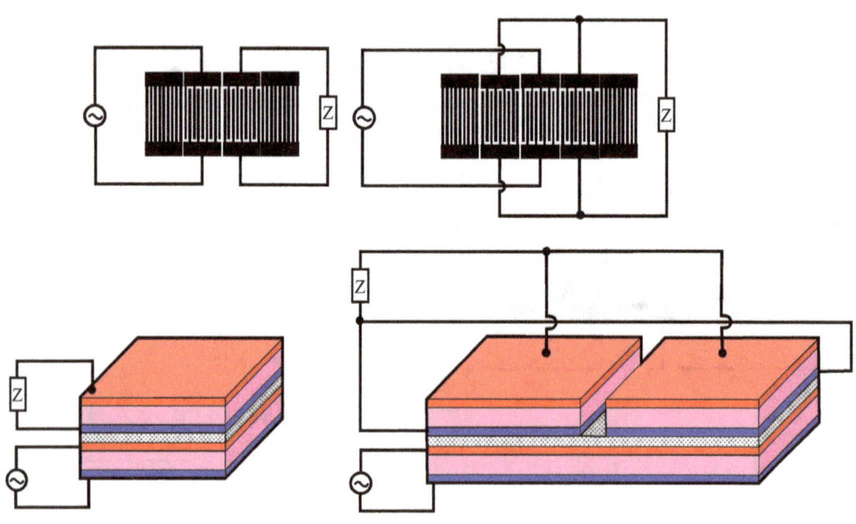

Figure 8.25 Stacked crystal filter (SCF) and its similarity to the DMS filter

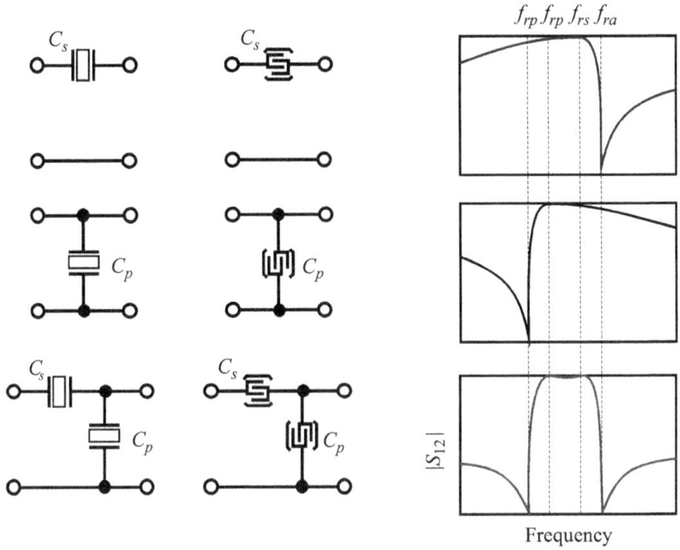

Figure 8.26 Ladder filter topology

The advantages of ladder filters area low insertion loss and a steep roll-off characteristic due to the high Q of the acoustic resonators. The acoustic resonators are electrically coupled in the ladder filter, and the design theory is based on the conventional constant-K-type filter. Setting the nominal impedance to R, the capacitances of the series and parallel resonators, C_s and C_p, have the following relationship:

$$R = \frac{1}{\omega_c^2 C_s C_p} \tag{8.9}$$

where ω_c is the centre angular frequency of the filter.

The ratio of C_s to C_p is determined by the trade-off between the insertion loss of the passband and the out-of-band suppression level. Also, the stage number of filters is determined by the trade-off between the insertion loss and the steepness at the band edge. One of the drawbacks of ladder filters is flyback due to the capacitive pass in the wideband characteristics. Thus, as shown in Figure 8.27, an attenuation pole must be deployed in the frequency spectrum by connecting the inductance to the parallel resonator in series or the series inductor in parallel. These inductors are often fabricated in a ceramic package.

8.4 Performance improvement

We will introduce some examples of techniques for improving the performance of resonators. Here, only the outlines of the improvement technologies are given, so please read the reference papers to obtain tips and details. The characteristics to be improved will be described in terms of the temperature coefficient, which is a key

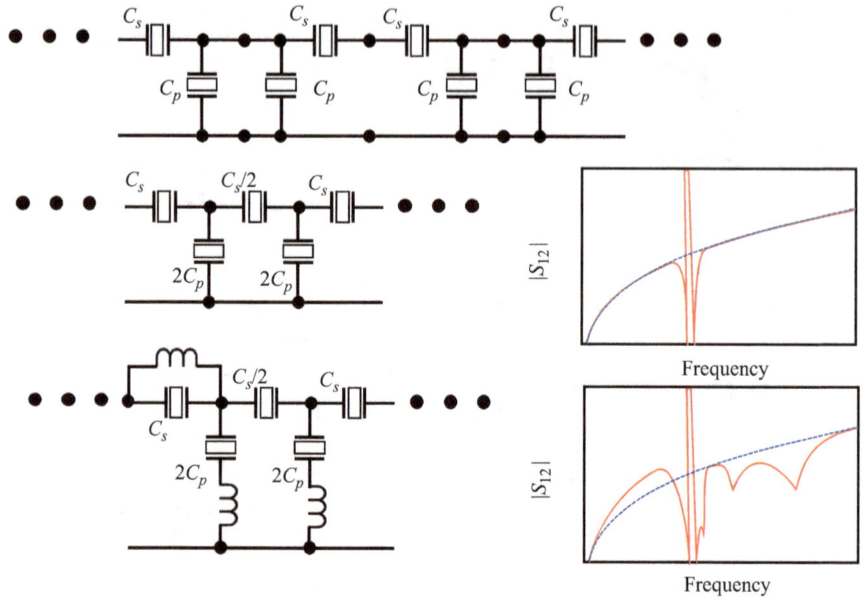

Figure 8.27 Design of attenuation poles

for commercialisation, In addition to the coupling coefficient or Q factor discussed in the previous section. We will also discuss the suppression of spurious responses due to the Lamb mode, which was considerably improved in the 2000s. Note that the spurious suppression technology was developed for BAW resonators and has spread to SAW resonators and filters.

The main application of acoustic devices is duplexing in FDD systems. When a filter is configured using the ladder topology, different characteristics are required for each resonator (Figure 8.28). Therefore, the requirement for the Q factor or the spurious characteristic may be mitigated depending on the resonator. Most of the characteristic improvement techniques have a trade-off relation with the evaluation parameters. Improvement technologies applied to each resonator must be carefully selected on the basis of Figure 8.28.

8.4.1 Improvement of coupling coefficient

The improvement of the coupling coefficient leads to the broadening of the filter bandwidth. In view of the remarkable increase in the amount of communication in recent years, the expansion of the filter bandwidth can be considered a key current requirement. Also, as described in the previous section, the increase in k_t^2 can also be replaced with the improvement of the Q factor on the basis of the Q-k trade-off.

8.4.1.1 Improvement of *k* with electrode material

Electrodes with high acoustic impedance (high Z_a) can enhance the electromechanical coupling coefficient k_t^2 [62]. In SAW devices, where the displacement

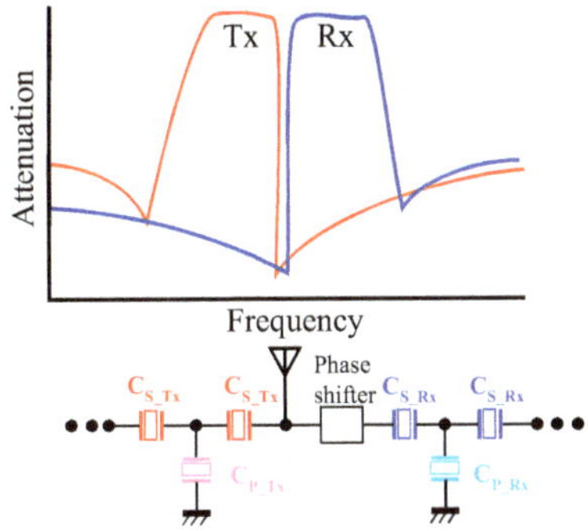

		Q_r	Q_a	spurious
Tx	P	Not important	Important	Not important
	S	Important	Important	Crucial
Rx	P	Important	Important	Not important
	S	Important	Not important	Crucial

Figure 8.28 Importance of parameters for each resonator

is concentrated in the piezoelectric crystal, there is not much benefit of high-Z_a electrodes, and there are also disadvantages due to the high electrical resistance of such electrodes. In contrast, in BAW devices, where the electrodes are excited as part of the resonator, the use of a high-Z_a material is essential [63–65]. Since the electrode does not have piezoelectricity, charge regeneration from displacement is not expected even if the electrode generates a thermal loss. Therefore, high-Z_a electrodes are preferred in BAW devices to suppress the deformation in the electrode.

Furthermore, when changing the thickness ratio of an electrode, the coupling coefficient is varied and has the highest point (Figure 8.29). The mechanism of this effect is described in detail in [66]. It is shown in Figure 8.30 why the coupling coefficient has an optimum point that depends on the thickness of the electrode. The electric field is uniform in the piezoelectric layer and zero in the electrode. On the other hand, the strain is successively distributed from the piezoelectric layer to the electrode and becomes zero at the surface. From the viewpoint of the coupling coefficient, it is desirable that the distributions of the electric field and strain overlap as much as possible in the piezoelectric layer as shown in Figure 8.30(b). If the electrode is thicker than the optimum value (Figure 8.30(c)), the strain energy in the electrode increases and cannot return to the electrical energy, and the electromechanical coupling is weakened. In the case where the electrode is thinner than the optimum value (a), the displacement distribution of the piezoelectric layer is dragged to the surface, where there is zero strain, resulting in a large difference from the electrical field distribution.

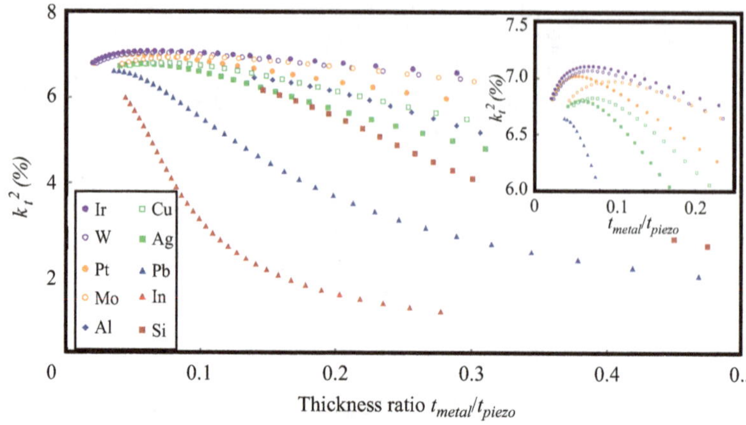

Figure 8.29 Design of attenuation poles

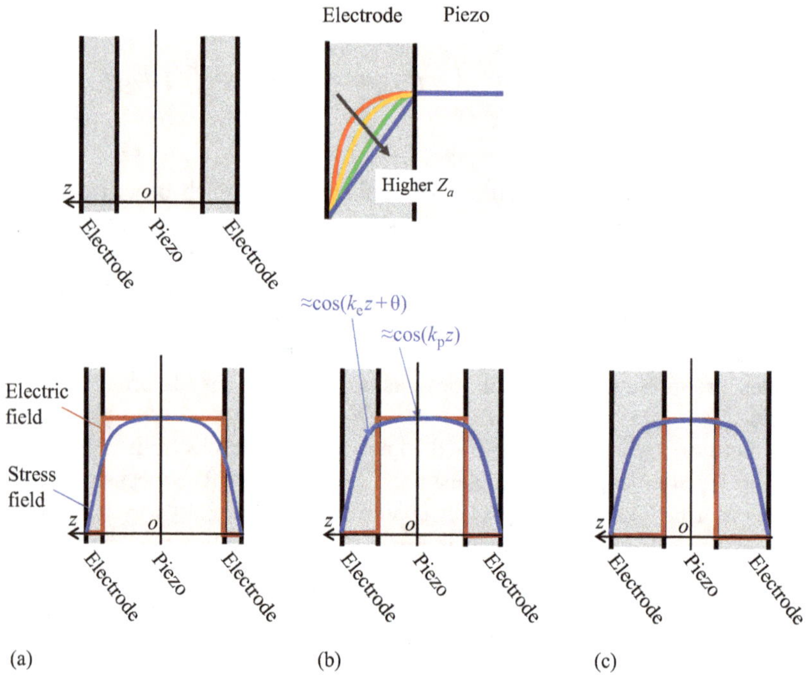

Figure 8.30 Design of attenuation poles

8.4.1.2 Enhancement of piezoelectricity

A material-based approach to enhancing the piezoelectricity itself is also being developed. In BAW devices, AlN is generally employed because of its low dielectric constant, low leakage current and process compatibility for active circuitry.

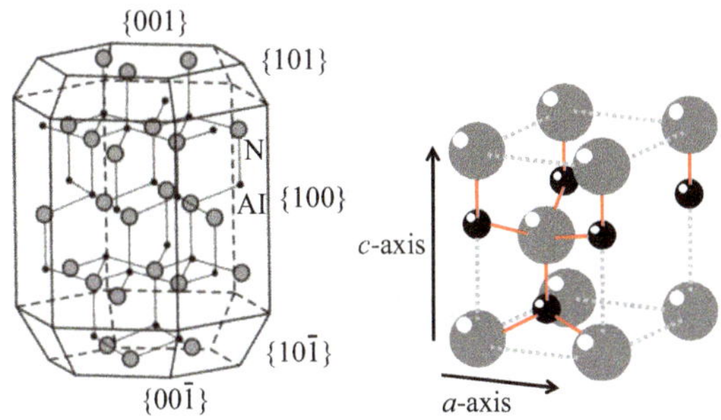

Figure 8.31 Crystal structure of AlN (wurtzite structure)

Therefore, the bandwidth of the filter has thus far been limited by the coupling constant of AlN [67]. AlN has a wurtzite structure, as shown in Figure 8.31. When substituting a Sc atom into an Al site and deforming the crystal structure, the *a*- and *c*-axes are expanded and compressed, respectively. The *c*-axis is a piezoelectric axis and the proximity of Al and N atoms resulting from the doping contributes to the increase in piezoelectricity [68].

If the ease of substitution at the Al site is estimated from the formation of energy by first-principles calculation, the extremely low formation energy of Sc suggests its superiority as a dopant. In fact, the experimental search for dopants as alternatives to Sc has not achieved significant results. As a possible alternative to the doping of Sc, there is a method of co-doping in consideration of the charge neutralisation conditions. Specifically, the co-doping of Mg–Hf or Mg–Zr is performed. Using this co-doping technology, the piezoelectricity can be enhanced as much as by Sc doping [69,70].

8.4.2 Improvement of Q

The mechanical loss can be reduced in various ways, but the suppression of leakage and the unwanted vibration mode are the main approaches. Regarding the suppression of leakage, the first step is to trace the radiation pattern of the vibration from the resonator by laser-based optical observation. For example, in a BAW device, the radiation pattern is observed around the anti-resonance frequency, and the electrode edge is often thickened to confine the vibration energy within the resonator (Figure 8.32(b)) [71,72]. The vibration energy can also be confined by removing the piezoelectric material at the periphery of the resonator (Figure 8.32(c)) [73,74].

The suppression of unwanted vibration can be analysed by carrying out a two-dimensional Fourier transformation of optically observed images. Vibration with a displacement perpendicular to the polarisation axis of the piezoelectric material results in a loss of vibration energy. In the case of BAW devices, since the polarisation axis is generally perpendicular to the wafer surface, it is not easy to clearly observe the vibration mode that becomes a loss. In this case, the analysis of BAW

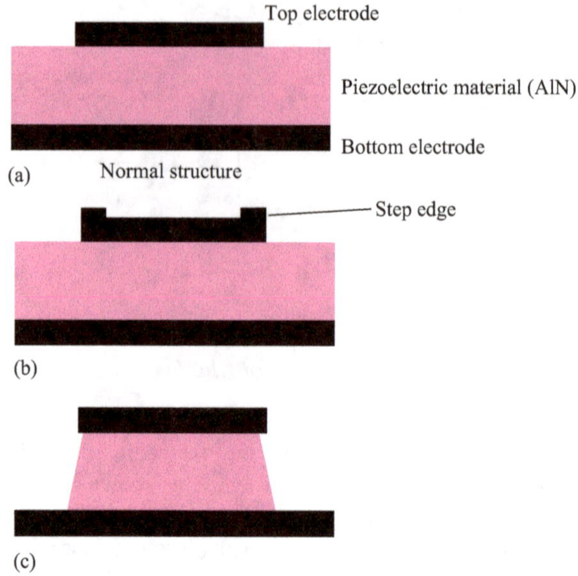

Figure 8.32 Vibration energy confinement structure

devices is assisted by a contour image calculated by the finite element method (FEM) [74].

In recent years, energy leakage to the substrate in SAW devices has been revealed from the radiation pattern in the substrate obtained from the contour image acquired by FEM, and the technology used to suppress energy radiation in the direction of the substrate thickness by multilayering the substrate is attracting interest [75–78].

8.4.3 *Low-*T_{cf} *technologies*

Temperature characteristics are important in frequency control devices such as filters. RF filters are generally designed in consideration of the frequency variation with the temperature, with a margin added to the required specification. In other words, the suppression of the variation with the temperature leads to the reduced insertion loss and improved bandwidth of the product.

As temperature compensation techniques, a substrate having a small temperature coefficient of elasticity can be selected or a temperature compensation film can be used. The former is realised, for example, by bonding a piezoelectric substrate such as $LiTaO_3$ or $LiNbO_3$ to a sapphire substrate having a very small temperature coefficient [79]. The latter is realised by depositing a temperature compensation film having a negative temperature coefficient on the SAW/BAW device [37–40,80].

SiO_2 has a positive temperature characteristic for elasticity, unlike common materials, and is widely used as a temperature compensation film. However, SiO_2

has poor adhesion to metal films and the optimisation of the manufacturing flow is necessary. In addition, since the acoustic loss of SiO_2 is not very small, when it is vibrated as part of a resonator, particularly when it is adopted in a BAW device, it is essential to reduce the acoustic loss by plasma reforming or by removing the impurities and voids.

To reduce the acoustic loss of a temperature compensation film, doping into the SiO_2 film and increasing the temperature coefficient allow the thickness of the temperature compensation film to be reduced. A typical dopant is fluorine (SiOF) [81–83].

8.4.4 Suppression of spurious responses

Analysis using a dispersion diagram is essential for understanding the generation principle of spurious responses and their suppression techniques. In this section, guided waves propagating in a piezoelectric plate are first introduced, and the derivation and interpretation of the dispersion curve are described. Next, the dispersion diagram is used to explain the generation principle and the method of suppressing the spurious responses in BAW devices based on some successful cases. Finally, we mention an instance where the spurious suppression technique for BAW devices was successfully applied to SAW devices.

8.4.4.1 Dispersion in a piezoelectric plate

When considering a wave propagating through an infinite medium, the frequency and wavenumber have a linear relationship, with the acoustic velocity as the proportionality constant, and the acoustic velocity only depends on the crystal orientation. However, a wave propagating through a plate becomes a two-dimensional guided wave and cannot be represented by a simple linear relation. By setting a single excitation frequency, the wavenumber is discretised to some specific values that satisfy the boundary conditions. Here, when the excitation frequency is swept, the discrete wavenumbers can be used to draw continuous curves on the plane of frequency and wavenumber. These curves are called dispersion curves and the diagram is called a dispersion diagram.

In the following, we derive the dispersion curves for a simple BAW device consisting of an electrode–piezoelectric film–electrode structure as shown in Figure 8.33.

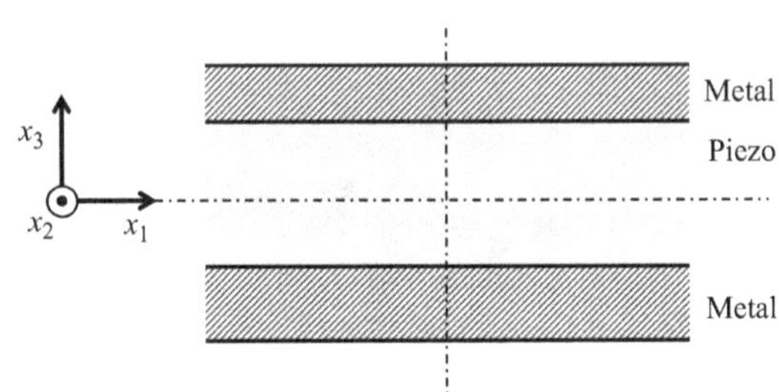

Figure 8.33 Vibration energy confinement structure

Governing equations – The propagation of an acoustic wave in a piezoelectric plate is governed by the following five equations (in the metal plate, the electrical terms cancel):

- **equation of motion**

$$T_{ij,i} = \rho \ddot{u}_j \tag{8.10}$$

- **charge equation (Poisson's equation)**

$$D_{i,i} = 0 \tag{8.11}$$

- **relation between strain and mechanical displacement**

$$S_{ij} = \frac{1}{2}\left(u_{i,j} + u_{j,i}\right) \tag{8.12}$$

- **relation between the electric field and potential**

$$E_i = -\phi_{,i} \tag{8.13}$$

- **linear piezoelectric constitutive equations**

$$T_{ij} = c_{ijkl}{}^{E} S_{kl} - e_{kij} E_k \tag{8.14}$$

$$D_i = e_{ikl} S_{kl} + \varepsilon_{ik}{}^{S} E_k \tag{8.15}$$

where T_{ij}, D_i, S_{ij} and E_i are the components of stress, electrical displacement, strain and electric field, respectively. ρ is the mass density, u is the mechanical displacement, φ is the potential, and $c_{ijkl}{}^{E}$, e_{kij} and $\varepsilon_{ik}{}^{S}$ are the elastic, piezoelectric and dielectric constants, respectively.

Material constants – The arrays of elastic, piezoelectric and dielectric constants for a piezoelectric material poled in the x_3 direction are

$$c_{pq}^{E} = \begin{pmatrix} c_{11} & c_{12} & c_{13} & 0 & 0 & 0 \\ c_{12} & c_{11} & c_{13} & 0 & 0 & 0 \\ c_{13} & c_{13} & c_{33} & 0 & 0 & 0 \\ 0 & 0 & 0 & c_{44} & 0 & 0 \\ 0 & 0 & 0 & 0 & c_{44} & 0 \\ 0 & 0 & 0 & 0 & 0 & c_{66} \end{pmatrix}$$

$$c_{66} = \frac{1}{2}(c_{11} - c_{12}) \tag{8.16}$$

$$e_{ip} = \begin{pmatrix} 0 & 0 & 0 & 0 & e_{15} & 0 \\ 0 & 0 & 0 & e_{15} & 0 & 0 \\ e_{31} & e_{31} & e_{33} & 0 & 0 & 0 \end{pmatrix}$$

$$\varepsilon_{ij}^{S} = \begin{pmatrix} \varepsilon_{11} & 0 & 0 \\ 0 & \varepsilon_{11} & 0 \\ 0 & 0 & \varepsilon_{33} \end{pmatrix},$$

where p or $q = 1, 2, 3, 4, 5, 6$ is replaced by 11, 22, 33, 23 or 32, 31 or 13, 12 or 21, respectively (engineering notation).

Boundary conditions – At the boundary between the electrode and the piezoelectric material, displacement and strain are in successive and the potential is 0. The surface of the electrode is traction-free.

Both symmetric and asymmetric wave components should be assumed because the model is asymmetric in the thickness direction. Under the rectangular Cartesian coordinate system x_i (i = 1, 2, 3), each component of the displacement u_i and electric potential ϕ in the piezoelectric film can be expressed as

$$\begin{aligned}
u_1 &= jA_{p1}e^{j(\omega t-\xi x_1)}(C_p \sin \eta x_3 - D_p cos\ \eta x_3)\\
u_2 &= 0\\
u_3 &= A_{p3}e^{j(\omega t-\xi x_1)}(C_p \sin \eta x_3 + D_p \cos \eta x_3)\\
\varphi &= B_p e^{j(\omega t-\xi x_1)}(C_p \sin \eta x_3 + D_p \cos \eta x_3)
\end{aligned} \tag{8.17}$$

where ξ and η are the propagation constants in the lateral and thickness directions, respectively, and ω is the angular frequency.

Substituting (8.17) into (8.10) and (8.11) using (8.12–8.15), the following simultaneous equations are obtained:

$$\begin{vmatrix}
c_{11}^E\xi^2 + c_{44}\eta^2 - \rho\omega^2 & (c_{13}^E + c_{44}^E)\xi\eta & (e_{31}+e_{15})\xi\eta\\
(c_{44}^E + c_{13}^E)\xi\eta & c_{44}^E\xi^2 + c_{33}^E\eta^2 - \rho\omega^2 & e_{15}\xi^2 + e_{33}\eta^2\\
(e_{15}+e_{31})\xi\eta & e_{15}\xi^2 + e_{33}\eta^2 & -\varepsilon_{11}\xi^2 - \varepsilon_{33}\eta^2
\end{vmatrix} = 0 \tag{8.18}$$

For given ξ and ω, three basic solutions, $\eta^{(i)}$ (i = 1, 2, 3), can be obtained by solving (8.18). To satisfy the boundary conditions, the general solutions are expressed as a sum of these solutions as follows:

$$\begin{aligned}
u_1 &= je^{j(\omega t-\xi x_1)}\sum_{i=1}^{3} A_{p1}^{(i)}(C_p^{(i)} \sin \eta^{(i)}x_3 - D_p^{(i)} \cos \eta^{(i)}x_3)\\
u_2 &= 0\\
u_3 &= e^{j(\omega t-\xi x_1)}\sum_{i=1}^{3} A_{p3}^{(i)}(C_p^{(i)} \sin \eta^{(i)}x_3 + D_p^{(i)} \cos \eta^{(i)}x_3)\\
\varphi &= e^{j(\omega t-\xi x_1)}\sum_{i=1}^{3} B_p^{(i)}(C_p^{(i)} \sin \eta^{(i)}x_3 + D_p^{(i)} \cos \eta^{(i)}x_3)
\end{aligned} \tag{8.19}$$

The following simultaneous equations are also obtained for the top and bottom electrodes, similarly to the piezoelectric film:

$$\begin{vmatrix}
c_{11}\xi^2 + c_{44}\eta'2 - \rho\omega^2 & (c_{13} + c_{44})\xi\eta'\\
(c_{44} + c_{13})\xi\eta' & c_{44}\xi^2 + c_{33}\eta'2 - \rho\omega^2
\end{vmatrix} = 0 \tag{8.20}$$

where η' and c_{pq} is the propagation constant in the thickness direction and the elastic constant in the electrode, respectively.

Solving (8.20), we can obtain the following general solutions for the bottom and top electrodes:

$$u_{t,b_1} = je^{j(\omega t - \xi x_1)} \sum_{i=1}^{2} A_{t,b1}^{(i)} (C_{t,b}^{(i)} \sin\eta'(i)x_3 - D_{t,b}^{(i)} \cos\eta'(i)x_3)$$

$$u_{t,b_2} = 0$$

$$u_{t,b_3} = e^{j(\omega t - \xi x_1)} \sum_{i=1}^{2} A_{t,b3}^{(i)} (C_{t,b}^{(i)} \sin\eta'(i)x_3 + D_{t,b}^{(i)} \cos\eta'(i)x_3) \tag{8.21}$$

Subscripts t and b denote variables in the top and bottom electrodes, respectively. Substituting (8.19) and (8.21) into the boundary conditions, we obtain the linear homogeneous algebraic equations of $C_p^{(i)}$, $D_p^{(i)}$, $C_{t,d}^{(j)}$ and $D_{t,d}^{(j)}$ ($i = 1,2,3$, $j = 1,2$). The roots of the determinant of this matrix give the relationship between ω and ξ. This relationship is used to draw a dispersion diagram. However, the determinant is a transcendental equation; thus, a numerical method is required to obtain the dispersion diagram.

Figure 8.34 shows the dispersion curve when Mo and AlN are respectively used as the electrodes and piezoelectric film in the configuration shown in Figure 8.33. Here, the propagation constant and frequency are normalised as

$$\Omega = \frac{\omega}{\frac{\pi}{2H}\sqrt{\frac{c_{44}^E}{\rho}}} \tag{8.22}$$

$$\Gamma = \frac{\xi}{\pi/2H} \tag{8.23}$$

where $2H$ and ρ are the thickness and density of the AlN film, respectively.

In this figure, the curve labelled by S indicates the symmetric mode, and the curve labelled by A indicates the asymmetric mode. The subscript shows the order of the mode. Also, TE indicates the thickness extension vibration mode and TS indicates the thickness shear mode. The propagation constant is a complex number, and TS_1 and TE_1, and A_1 and A_0 are connected by a branch on the imaginary-number plane. AlN-based BAW devices use the fundamental mode of TE, so point P on the TE branch corresponds to the resonance frequency.

The positional relationship of the TS_1 and TE_1 branches on the dispersion diagram can be changed by changing Poisson's ratio σ. If σ is less than 1/3, the TE branch becomes lower in frequency than the TS branch, and if it is more than 1/3, the TE branch becomes higher in frequency than the TS branch. σ is 0.39 ($>$1/3) for ZnO and 0.25 ($<$1/3) for AlN. For example, when ZnO and AlN are used as the piezoelectric film of a BAW resonator, the dispersion curves are those shown in (a) and (b) of Figure 8.35, respectively.

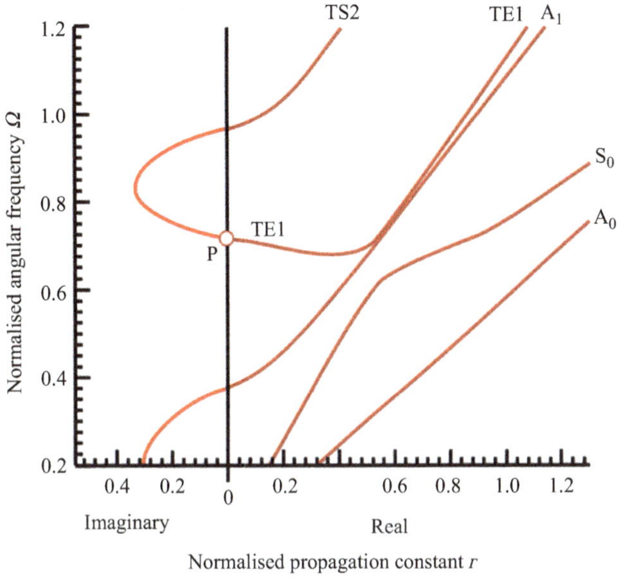

Figure 8.34 Typical dispersion diagram of Mo/AlN/Mo FBAR

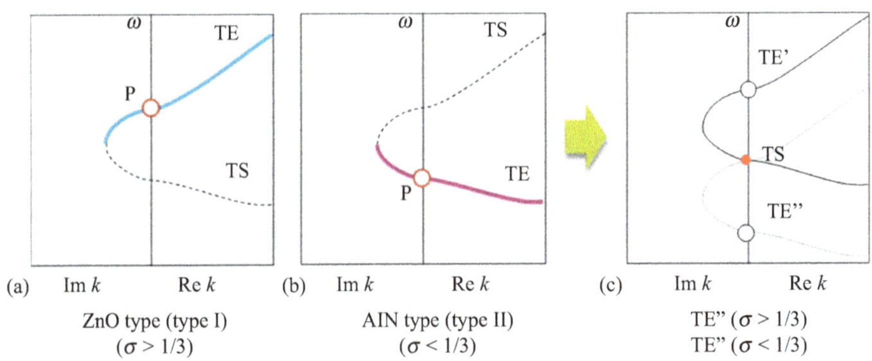

Figure 8.35 Variation of dispersion diagram with Poisson ratio

In the case of a ZnO (type I) piezoelectric film, the propagation constant becomes a real number at frequencies higher than the resonance point P and an imaginary number at lower frequencies. In contrast, in the case of an AlN (type II) piezoelectric film, the propagation constant becomes a real number at frequencies lower than the resonance point P and an imaginary number at higher frequencies. When a propagation constant is a real number, the mechanical vibration propagates in the resonator plane in the guided wave mode (Lamb wave mode). On the other hand, when it is an imaginary number, the mechanical vibration cannot propagate and becomes an evanescent mode.

8.4.4.2 Energy-trapping mode

The theory of energy trapping using the electrode effect is one of the most important concepts in BAW devices. In a type-I BAW device, the propagation constant becomes an imaginary number below the resonance frequency. Here, considering the peripheral region without electrodes (Figure 8.36), the dispersion curve shifts to a higher frequency because there is no surface shortening effect. When viewed from the electrode region, the propagation constant becomes an imaginary number around the resonator. As a result, vibrational energy is confined in the electrode region, making it possible to obtain resonance with a high Q-value. In addition, since vibrational energy does not propagate to the periphery, it is possible to separate the main resonance from the spurious response due to the reflection at the substrate edge, such as contour-mode vibration. On the other hand, in a type-II BAW resonator, when the dispersion curve of the peripheral region shifts to a higher frequency, the propagation constant in the peripheral region becomes a real number at the resonance point, and vibration energy leaks to the peripheral region.

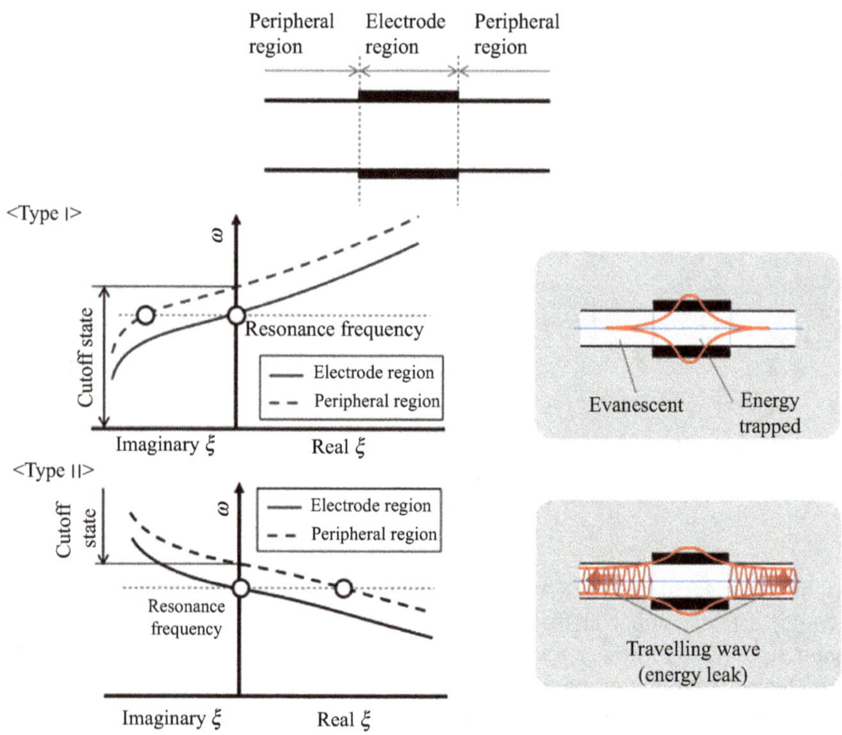

Figure 8.36 Concept of energy trapping

The above concept is applied to quartz and ceramic oscillators in which the electrode is extremely thin relative to the piezoelectric layer. In thin-film BAW devices, the electrode is relatively thick, and the electrode effect should be considered as not only a surface shortening effect but also a mass loading effect. Therefore, the relationship between the dispersion curves in the electrode and peripheral regions is that shown in Figure 8.36, and the leakage of vibrational energy is caused by coupling to other vibrational modes. The degree of coupling is described in detail in, for example, [84,85].

In a thin-film BAW resonator, to realise an energy-trapping effect similar to that in a quartz crystal or ceramic resonator, it is necessary to shift the frequency of the dispersion curve in the electrode region without using the peripheral region. For example, as shown in Figure 8.37, in a type-I device, the electrode edge is thinned, and in a type-II device, the electrode edge is thickened, as already partly illustrated in Figure 8.32. However, this method causes a slight decrease in the coupling coefficient because the edge regions act as parallel capacitance.

8.4.4.3 Spurious responses and piston modes

When a BAW device is evaluated as a one-port resonator, as shown in Figure 8.38, spurious responses due to the Lamb mode are generated. The spurious responses are concentrated from the resonance frequency to the anti-resonance frequency in a type-I resonator and below the resonance frequency in a type-II resonator. If the energy-trapping structure shown in Figure 8.37 is employed, these spurious responses become more emphasised.

In quartz or ceramic resonator, the electrode area is huge and there is sufficient attenuation of propagating transverse Lamb waves for them not to be observed strongly. However, in a thin-film BAW, the electrode area is small and the spurious responses due to Lamb waves are not sufficiently attenuated. Therefore, the

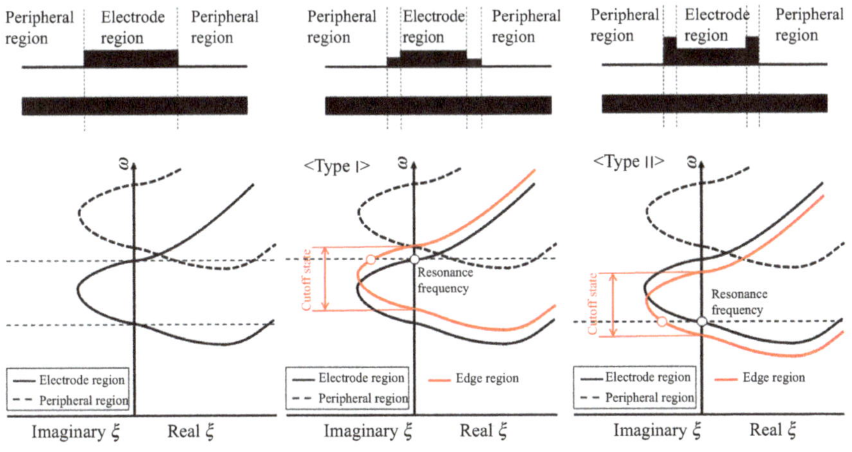

Figure 8.37 Energy-trapping structure in a thin-film BAW device

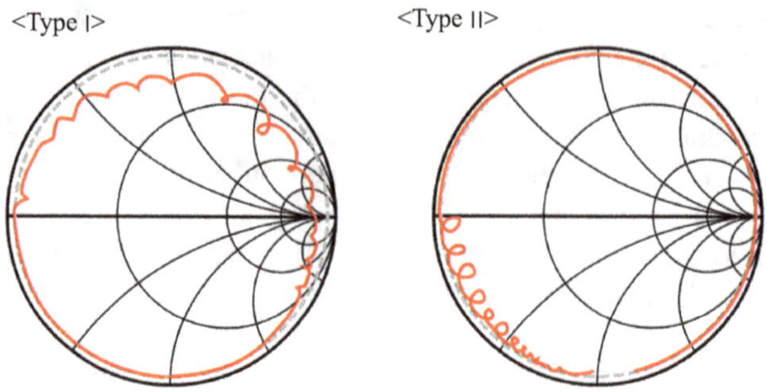

Figure 8.38 Spurious responses on the Q-circle

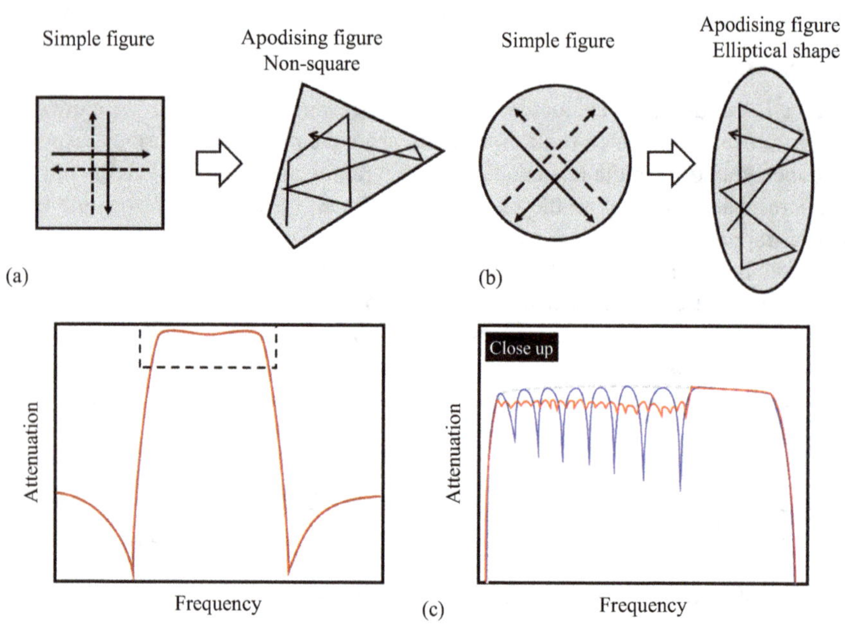

Figure 8.39 Effect of apodisation in the BAW device

technology for the suppression of spurious responses is essential for the application to filters and duplexers.

As shown in Figure 8.39, spurious responses can be apodised by a non-square or elliptical resonator. However, since these apodisations do not eliminate the spurious response, the insertion loss is increased even if the spurious responses can be made small, as shown in Figure 8.39(c), when applying a filter. Additionally, from the viewpoint of chip area efficiency, the irregular shapes shown in Figure 8.39 are not desirable.

In the case of an infinite flat plate without edges, of course, no spurious responses are generated. Also, if it is possible to ideally cut out and pick up the modes, as shown in Figure 8.40, which are called piston modes, naturally, no spurious response should be generated. However, in an actual BAW device, there is an influence of the peripheral region, and the piston mode is subject to distortion from the support. This distortion is the cause of the spurious responses [86]. As shown in Figure 8.32(c), it may be considered that the piston mode can be realised by removing the peripheral region. However, in a BAW device, electrodes, which have a very different hardness from the piezoelectric film, are deposited at the top and bottom, and the reflection of Lamb waves from the edge is not uniform [74]. Therefore, distortion is generated at the electrode edge even in the configuration in Figure 8.32(c) and spurious modes cannot be eliminated.

To obtain the piston modes, the velocity of Lamb waves in the distortion region should be reduced. By reducing the velocity, as shown in Figure 8.41, the distortion region can be spatially compressed, and a pseudo piston mode is generated. How to realise the pseudo piston mode in an AlN-based BAW resonator (type II) is described below [86,87]. Between the thick electrode region and the resonator region, a thin electrode region is deployed as shown in Figure 8.42. In the thin electrode region, the dispersion curve shifts to a lower frequency and the propagation velocity of the Lamb waves can be suppressed. Thus, we obtain a vibration mode similar to the piston mode, in which the width of the area of distortion is greatly reduced. In type-I BAW resonators, it is possible to suppress the spurious responses by making the electrode thick, in contrast to type-II resonators (AlN-based resonator), and spurious responses can be suppressed with a simpler edge structure than that in type-II resonators.

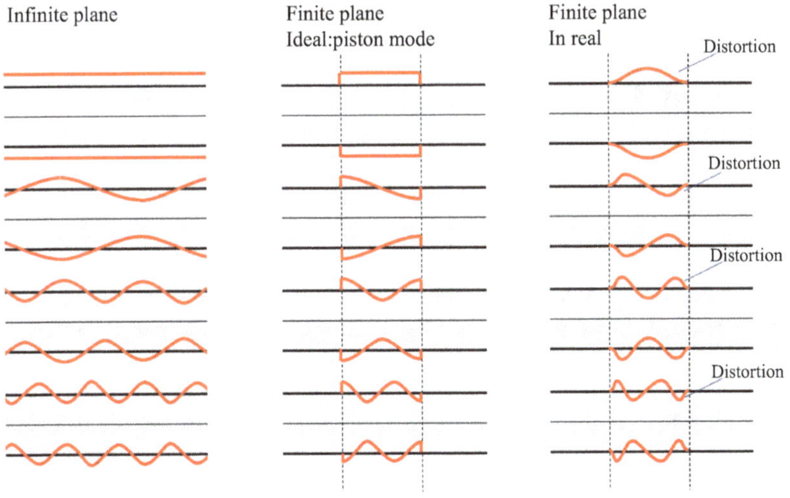

Figure 8.40 Piston modes

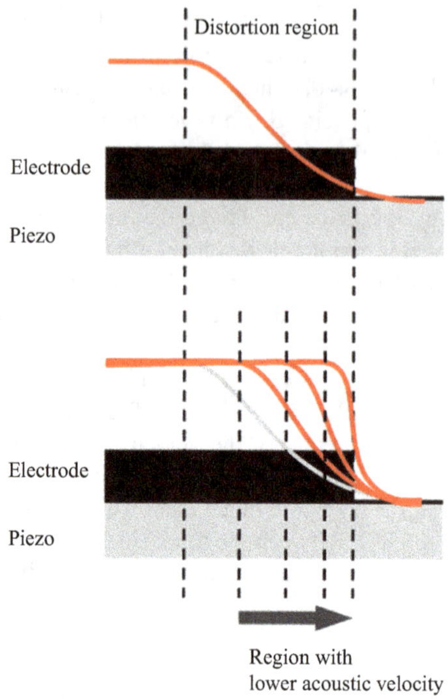

Figure 8.41 Pseudo piston mode

8.4.4.4 Design of dispersion diagram

We have mainly studied the frequency shift of the dispersion curve due to the electrode effect. Although it has been explained that only the frequency of the TE branch can be shifted by controlling Poisson's ratio, it is not easy to use Poisson's ratio as a design parameter because of the limited number of piezoelectric films suitable for BAW devices. There is also a method of thickly depositing an electrode material with a high Poisson's ratio such as Al. However, such a material has a high material loss in general, so this method is not practically used.

Although a type-1 resonator is preferable from the viewpoint of the manufacturing cost, is there any other way apart from selecting ZnO as a piezoelectric film? One solution has been proposed for SMRs, where materials with high and low acoustic impedances are alternately deposited similarly to that in a Bragg reflector to provide acoustic isolation (see Figure 8.19). SiO_2 is commonly used as a material with low acoustic impedance in SMRs. It is possible to shift only the frequency of the TS branch by changing the thickness of the topmost oxide layer, as shown in Figure 8.43, which is the layer closest to the resonator and subjected to vibration with a large amplitude [88]. By increasing the SiO_2 thickness, it becomes possible to use type-II AlN resonators as type-1 resonators.

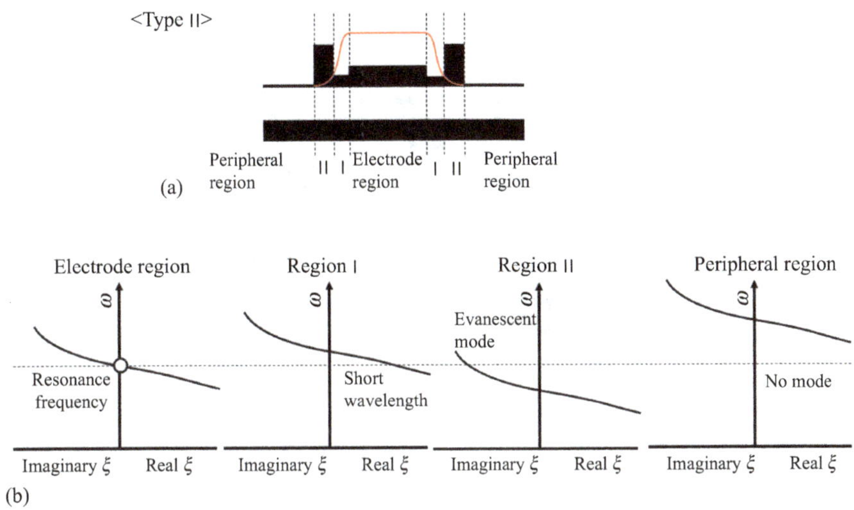

Figure 8.42 Stepped edge structure

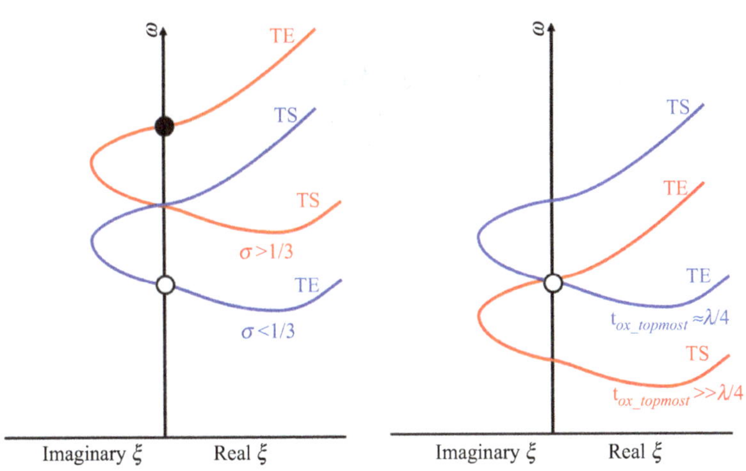

(a) σ changing (b) Changing the thickness of topmost oxide

Figure 8.43 Design of dispersion diagram

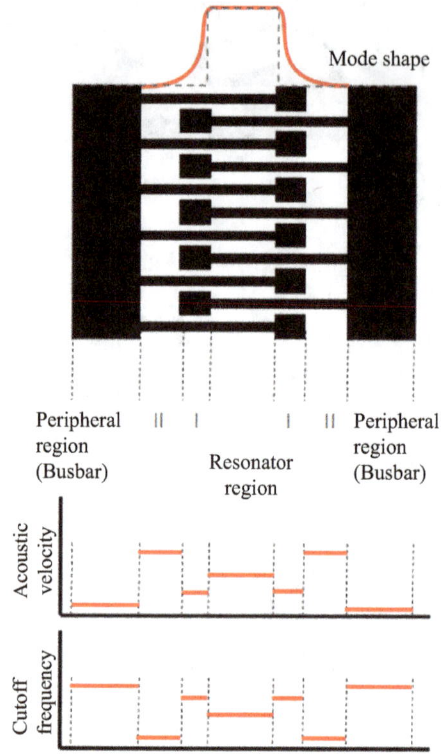

Figure 8.44 Utilisation of the piston mode in the SAW device

8.4.4.5 Feedback from BAWs to SAWs

In the field of SAW devices, spurious responses in the filter passband have been suppressed using apodisation techniques as shown in Figure 8.13. However, owing to the recent progress of technology for suppressing spurious responses in BAW devices, a method of controlling the acoustic velocity at the edge of devices has also been proposed for SAW devices (Figure 8.44) [89]. In addition, similarly to that in SMR BAW devices, a method of suppressing energy loss in the thickness direction of the substrate by using multiple acoustic reflectors is also being developed for SAW devices, which will be very interesting [75].

8.5 Summary

We have introduced the operation principle and methods for improving the characteristics of SAW and BAW devices in this chapter. Among the various MEMS resonators, SAW and BAW devices have been among the most successfully commercialised fields, particularly for wireless RF applications. Therefore, we have described the communication architecture, which is the common ground in device development. Also, the improvement of the characteristics has been described in terms of the requirements of such applications. We hope that your learning will encompass these ideas. With regard to techniques for improving performance, a very wide range of knowledge of engineering from materials to numerical dispersive analysis is required. Therefore, we have only given an outline of these fields. Details can be found in the books and papers cited in this chapter.

References

[1] Rayleigh L. '*On waves propagating along the plane surface of an elastic solid*'. Proceedings of London Mathematical Society. 1885;**s1–17**(1):4–11

[2] White R.M., and Voltmer F.W. '*Direct piezoelectric coupling to surface elastic waves*'. Applied Physics Letters. 1965;**7**(12):314–16

[3] Martin T.A. '*The IMCON pulse compression filter and its applications*'. IEEE Transactions on Microwave Theory and Techniques. 1973;**21**(4):186–94

[4] Williamson R.C., and Smith H.I. '*The use of surface-elastic-wave reflection gratings in large time-bandwidth pulse-compression filters*'. IEEE Transactions on Microwave Theory and Techniques. 1973;**21**(4):195–205

[5] Hartmann C.S., Bell Jr. D.T., and Rosenfeld R.C. '*Impulse model design of acoustic-wave filters*'. IEEE Transactions on Microwave Theory and Techniques. 1973;**21**(4):162–75

[6] Kino G.S., Lufvik S., Shaw H.J., Shereve W.R., White J.M., and Winslow D.K. '*Signal processing by parametric interactions in delay-line devices*'. IEEE Transactions on Microwave Theory and Techniques. 1973;**21**(4):244–55

[7] Gerard H.M., Smith W.R., Jones W.R., and Harrington J.B. '*The design and applications of highly dispersive acoustic surface-wave filters*'. IEEE Transactions on Microwave Theory and Techniques. 1973;**21**(4):176–86

[8] Bell Jr. D.T., and Li R.C.M. '*Surface-acoustic-wave resonators*'. Proceedings of the IEEE. 1976;**64**(5):711–21

[9] Hays R.M., and Hartmann C.S. '*Surface-acoustic-wave devices for communications*'. Proceedings of the IEEE. 1976;**64**(5):652–71

[10] Devries A.J., and Adler R. '*Case history of a surface-wave TV IF filter for color television receivers*'. Proceedings of the IEEE. 1976;**64**(5):671–6

[11] Takahashi S., Hirano H., Kodama T., *et al.* '*SAW IF filter on $LiTaO_3$ for color TV receivers*'. IEEE Transactions on Consumer Electronics. 1978;**24**(3):337–48

[12] Shibayama K., Yamanouchi K., Sato H., and Meguro T. '*Optimum cut for rotated Y-cut $LiNbO_3$ crystal used as the substrate of acoustic-surface-wave filters*'. Proceedings of the IEEE. 1976;**64**(5):595–7

[13] Yamanouchi K., and Shibayama K. '*Propagation and amplification of Rayleigh waves and piezoelectric leaky surface waves in $LiNbO_3$*'. Journal of Applied Physics. 1972;**43**(3):856–62

[14] Nakamura K., Kazumi M., and Shimizu H. '*SH-type and Rayleigh-type surface waves on rotated Y-cut $LiTaO_3$*'. Proceedings of Ultrasonics Symposium; Phoenix, USA, Oct. 1977, pp. 819–22

[15] Hashimoto K., and Yamaguchi M. '*Non-leaky, piezoelectric, quasi-shear-horizontal type SAW on X-cut $LiTaO_3$*'. Proceedings of IEEE Ultrasonics Symposium; Chicago, USA, Oct. 1988, pp. 97–101

[16] Kadota M., Yoneda T., Fujimoto K., Nakano T., and Takata E. '*Resonator filters using shear horizontal-type leaky surface acoustic wave consisting of heavy-metal electrode and quartz substrate*'. IEEE Transactions on Ultrasonics, Ferroelectrics, and Frequency Control. 2004;**51**(2):202–10

[17] Ikata O., Miyashita T., Matsuda T., Nishihara T., and Satoh Y. '*Development of low-loss band-pass filters using SAW resonators for portable telephones*'. Proceedings of IEEE Ultrasonics Symposium; Tucson, USA, Oct. 1992, pp. 111–15

[18] Ikata O., Satoh Y., Uchishiba H., *et al.* '*Development of small antenna duplexer using SAW filters for handheld phones*'. Proceedings of IEEE Ultrasonics Symposium; Baltimore, USA, Oct. 1993, pp. 111–14

[19] Sliker T.R., and Roberts D.A. '*A thin-film CdS-quartz composite resonator*'. Journal of Applied Physics. 1967;**38**(5):2350–8

[20] Nakamura K., Sasaki H., and Shimizu H. 'A piezoelectric composite resonator consisting of a ZnO film on an anisotropically etched silicon substrate'. Japanese Journal of Applied Physics. 1980;**20**(Supplement 3):111–14

[21] Lakin K.M., and Wang J.S. '*Acoustic bulk wave composite resonators*'. Applied Physics Letters. 1981;**38**(3):125–7

[22] Grudkowski T.W., Black J.F., Reeder T.M., Cullen D.E., and Wagner R.A. '*Fundamental-mode VHF/UHF miniature acoustic resonators and filters on silicon*'. Applied Physics Letters. 1980;**37**(11):993–5

[23] Nishihara T., Yokoyama T., Miyashita T., and Satoh Y. '*High performance and miniature thin film bulk acoustic wave filters for 5 GHz*'. Proceedings of IEEE Ultrasonics Symposium; Munich, Germany, Oct. 2002, pp. 969–72

[24] Aigner R., Ella J., Timme H.J., Elbrecht L., Nessler W., and Marksteiner S. 'Advancement of MEMS into RF-filters applications'. Digest of International Electron Devices Meeting (IEDM); San Francisco, USA, Dec. 2002, pp. 897–900

[25] Ruby R., Bradley P., Larson III J.D., and Oshmyansky Y. '*PCS 1900 MHz duplexer using thin film acoustic resonators (FBARs)*'. Electronics Letters. 1999;**35**(10):794–5

[26] Satoh H., Suzuki H., Takahashi C., Narahara C., and Ebata Y. '*A 400 MHz one-chip oscillator using an air-gap type thin film resonator*'. Proceedings of IEEE Ultrasonics Symposium; Denver, USA, Oct. 1987, pp. 363–7

[27] Dubois M.A., Carpentier J.F., Vincent P., *et al.* '*Monolithic above-IC resonator technology for integrated architectures in mobile and wireless communication*'. IEEE Journal of Solid-State Circuits. 2006;**41**(1):7–16

[28] Bleustein J.L. '*A new surface wave in piezoelectric materials*'. Applied Physics Letters. 1968;**13**(12):412–13

[29] Gulyaev Y.V. '*Electroacoustic surface waves in solids*', Soviet Physics Journal of Experimental and Theoretical Physics Letters. 1969;**9**:37–8

[30] Ohta Y., Nakamura K., and Shimizu H. '*Piezoelectric surface shear waves*'. The Institute of Electronics, Information, and Communication Engineering (IEICE) Technical Report. 1969; US69-3 (in Japanese)

[31] Engan H., Ingebrigsten K.A., and Tonning A. '*Elastic surface waves in a-quartz: observation of leaky surface waves*'. Applied Physics Letters, 1967;**10**(11): 312–13

[32] Auld B.A. *Aco*ustic waves and fields in solids Vol.2. New York: Wiley and Sons; 1973. pp. 135–61

[33] Love A.E.H. Some problems of geodynamics. Cambridge: Cambridge University Press; 2015. pp. 144–78

[34] Shimizu H., and Mizunuma Y. 'Piezoelectric substrate for SAW devices with very high electromechanical coupling coefficient ($K^2 \approx 0.3$)'. The Institute of Electronics, Information, and Communication Engineering (IEICE) Technical Report. 1982; US82-35 (in Japanese)

[35] Hashimoto K., Asano H., Omori T., and Yamaguchi M. '*Ultra-wideband surface acoustic wave devices using Cu-grating/rotated-YX-$LiNbO_3$-substrate structure*'. Japanese Journal of Applied Physics. 2004;**43**(5B):3063–6

[36] Hashimoto K., Asano H., Matsuda K., Yokoyama N., Omori T., and Yamaguchi M. '*Wideband Love wave filters operating in GHz range on Cu-grating/rotated-YX-$LiNbO_3$-substrate structure*'. Proceedings of IEEE Ultrasonics Symposium; Montreal, Canada, Aug. 2004, pp. 1330–4

[37] Kadota M., Nakao T., Taniguchi N., *et al.* '*SAW substrate with coupling factor and excellent temperature stability suitable for duplexer of PCS in US*'. Proceedings of IEEE Ultrasonics Symposium; Montreal, Canada, Aug. 2004, pp. 1970–5

[38] Kadota M., Nakao T., Taniguchi N., *et al.* '*Surface acoustic wave duplexer for US personal communication service with good temperature coefficient*'. Japanese Journal of Applied Physics. 2005;**44**(6B):4527–31

[39] Takayama R., Nakanishi H., Iwasaki Y., Sakuragawa T., and Fujii K. '*US-PCS SAW duplexer using high-Q SAW resonator with SiO_2 coat for stabilizing temperature characteristics*'. Proceedings of IEEE Ultrasonics Symposium; Montreal, Canada, Aug. 2004, pp. 959–62

[40] Nakamura H., Nakanishi H., Tsurunari T., Matsunami K., and Iwasaki Y. '*A small-sized SAW duplexer on a SiO_2/IDT/$LiNbO_3$ structure for wideband CDMA application*'. Proceedings of IEEE Ultrasonics Symposium; New York, USA, Oct. 2007, pp. 488–91

[41] Stoneley R. '*Elastic waves at the surface of separation of two solids*'. Proceedings of the Royal Society of London. 1924;A106(738):416–28

[42] Maerfeld C., and Tournois P. '*Pure shear elastic surface wave guided by the interface of two semi-infinite media*'. Applied Physics Letters. 1971;**19**(4): 117–18

[43] Yamashita T., Hashimoto K., and Yamaguchi M. '*Highly piezoelectric shear-horizontal-type boundary waves*'. Japanese Journal of Applied Physics. 1997;**36**(5B):3057–9

[44] Yamaguchi M., Yamashita T., Hashimoto K., and Omori T. '*Highly piezoelectric boundary waves in Si/SiO$_2$/LiNbO$_3$ structure*'. Proceedings of IEEE International Frequency Control Symposium; Pasadena, USA, May 1998, pp. 484–8

[45] Kando H., Yamamoto D., Mimura M., *et al.* '*FR filter using boundary acoustic wave*'. Proceedings of IEEE Ultrasonics Symposium; Vancouver, Canada, Oct. 2006, pp. 188–91

[46] Mishin S., Marx D.R., Sylvia B., Lughi V., Turner K.L., and Clarke D.R. '*Sputtered AlN thin films on Si and electrodes for MEMS resonators: relationship between surface quality microstructure and film properties*'. Proceedings of IEEE Ultrasonics Symposium; Honolulu, USA, Oct. 2003, pp. 2028–32

[47] Mishin S., Sylvia B., and Marx R. '*Improving manufacturability of AlN deposition used in making bulk acoustic wave devices*'. Proceedings of IEEE Ultrasonics Symposium; Rotterdam, the Netherlands, Sept. 2005, pp. 215–18

[48] Taniguchi S., Yokoyama T., Iwaki M., Nishihara T., Ueda M., and Satoh Y. '*An air-gap type FBAR filter fabricated using a thin sacrificial layer on a flat substrate*'. Proceedings of IEEE Ultrasonics Symposium; New York, USA, Oct. 2007, pp. 600–3

[49] Lakin K.M., McCarron K.T., and Rose R.E. '*Solidly mounted resonators and filters*'. Proceedings of IEEE Ultrasonics Symposium; Seattle, USA, Nov. 1995, pp. 905–8

[50] Marksteiner S., Kaitila J., Fattinger G.G., and Aigner R. '*Optimisation of acoustic mirrors for solidly mounted BAW resonators*'. Proceedings of IEEE Ultrasonics Symposium; Rotterdam, the Netherlands, Sept. 2005, pp. 329–32

[51] Larson III J.D., Bradley P.D., Wartenberg S., and Ruby R.C. '*Modified Butterworth–Van Dyke circuit for FBAR resonators and automated measurement system*'. Proceedings of IEEE Ultrasonics Symposium; San Juan, USA, Oct. 2000, pp. 863–8

[52] Feld D.A., Parker R., Ruby R., Bradley P., and Dong S. '*After 60 years: a new formula for computing quality factor is warranted*'. Proceedings of IEEE Ultrasonics Symposium; Beijing, China, Nov. 2008, pp. 431–6

[53] Ruby R., Parker R., and Feld D.A., '*Method of extracting unloaded Q applied across different resonator technologies*'. Proceedings of IEEE Ultrasonics Symposium; Beijing, China, Nov. 2008, pp. 1815–18

[54] Hartmann C.S. '*Weighting interdigital surface wave transducers by selective withdrawal of electrodes*'. Proceedings of IEEE Ultrasonics Symposium; Monterey, USA, Nov. 1973, pp. 423–6

[55] Tancrell R.H., and Holland M.G. '*Acoustic surface wave filters*'. Proceedings of IEEE. 1971;**59**(3):393–409

[56] McClellan J.H., Parks T.W., and Rabiner L.R. '*A computer program for designing optimum FIR linear phase digital filters*'. IEEE Transactions on Audio and Electroacoustics. 1973;AU-21(6):506–26

[57] Morita T., Watanabe Y., Tanaka M., and Nakazawa Y. '*Wideband low loss double mode SAW filters*'. Proceedings of IEEE Ultrasonics Symposium; Tucson, USA, Oct. 1992, pp. 95–104

[58] Tanaka M., Morita T., Ono K., and Nakazawa Y. '*Narrow bandpass filter using double-mode SAW resonators on quartz*'. Proceedings of the 38th Frequency Control Symposium; Philadelphia, USA, May 1984, pp. 286–93
[59] Ballato A., and Lukaszek T. 'A novel frequency selective device; the stacked-crystal filter'. Proceedings of the 27th Annual Symposium on Frequency Control; Cherry Hill, USA, Jun. 1973, pp. 262–9
[60] Stokes R.B., and Crawford J.D. '*X-band thin film acoustic filters on GaAs*'. IEEE Transactions on Microwave Theory and Techniques. 1993;**42**(6): 1075–80
[61] Lakin K.M., Belsick J., McDonald J.F., and McCarron K.T. '*High performance stacked crystal filters for GPS and wide bandwidth applications*'. Proceedings of IEEE Ultrasonics Symposium; Atlanta, USA, Oct. 2001, pp. 833–8
[62] Lakin K.M., Belsick J., McDonald J.F., and McCarron K.T. '*Improved bulk wave resonator coupling coefficient for wide bandwidth filters*'. Proceedings of IEEE Ultrasonics Symposium; Atlanta, USA, Oct. 2001, pp. 827–31
[63] Yokoyama T., Nishihara T., Taniguchi S., *et al*. '*New electrode material for low-loss and high-Q FBAR filters*'. Proceedings of IEEE Ultrasonics Symposium; Montreal, Canada, Aug. 2004, pp. 429–32
[64] Bradley P., Kim J., Ye S., Nikkel P., Bader S., and Feng C. '*2X size and cost reduction of film bulk acoustic resonator (FBAR) chips with tungsten electrodes for PCS/GPS/800 MHz multiplexers*'. Proceedings of IEEE Ultrasonics Symposium; New York, USA, Oct. 2007, pp. 1144–7
[65] Devos A., Olivares J., Clement M., Iborra E., Rimmer N., and Rastogi A. '*Picosecond ultrasonics as a helpful technique for introducing a new electrode material in BAW technology: the iridium case*'. Proceedings of IEEE Ultrasonics Symposium; New York, USA, Oct. 2007, pp. 1443–6
[66] Kaitila J. '*Review of wave propagation in BAW thin film devices - progress and prospects*'. Proceedings of IEEE Ultrasonics Symposium; New York, USA, Oct. 2007, pp. 120–9
[67] Muralt P., Antifakos J., Cantoni M., Lanz R., and Martin F. '*Is there a better material for thin film BAW applications than AlN?*'. Proceedings of IEEE Ultrasonics Symposium; Rotterdam, the Netherlands, Sept. 2005, pp. 315–20
[68] Akiyama M., Kamohara T., Kano K., Teshigawara A., Takeuchi Y., and Kawahara N. '*Enhancement of piezoelectric response in scandium aluminium nitride alloy thin films prepared by dual reactive cosputtering*'. Advanced Materials. 2009;**21**(5):593–6
[69] Iwazaki Y., Yokoyama T., Nishihara T., and Ueda M. '*Highly enhanced piezoelectric property of co-doped AlN*'. Applied Physics Express. 2015;**8**(6): 061501
[70] Yokoyama T., Iwazaki Y., Onda Y., Nishihara T., Sasajima Y., and Ueda M. '*Effect of Mg and Zr co-doping on piezoelectric AlN thin films for bulk acoustic wave resonators*'. IEEE Transactions on Ultrasonics, Ferroelectrics, and Frequency Control. 2014;**61**(8):1322–8

[71] Feng H., Fazzio S., Ruby R., and Radley P. Thin film bulk acoustic resonator with a mass loaded perimeter, US7280007B2, Oct. 9, 2007

[72] Nishihara T., Hara M., Taniguchi S., Iwaki M., Yokoyama T., and Ueda M. '*Piezoelectric thin film resonant element and circuit component using the same*'. US8125123B2, Feb. 28, 2012

[73] Tsutsumi J., Iwaki M., Iwamoto Y., *et al.* '*A miniaturized FBAR duplexer with reduced acoustic loss for the W-CDMA application'*. Proceedings of IEEE Ultrasonics Symposium; Rotterdam, the Netherlands, Sept. 2005, pp. 93–6

[74] Hara M., Ueda M., and Satoh Y. '*A thin-film bulk acoustic resonator and filter with optimal edge shapes for mass production*'. Ultrasonics. 2013;**53**(1):90–6

[75] Takai T., Iwamoto H., Takamine Y., *et al.* '*High-performance SAW resonator on new multi-layered substrate using $LiTaO_3$ crystal*'. IEEE Transactions on Ultrasonics, Ferroelectrics, and Frequency Control. 2017;**64**(9):1382–9

[76] Kadota M., and Tanaka S. '*Wideband acoustic wave resonators composed of hetero acoustic layer structure*'. Japanese Journal of Applied Physics. 2018;**57**(7s1):07LD12

[77] Gomi M., Kataoka T., Hayashi J., and Kakio S. '*High-coupling leaky surface acoustic waves on $LiNbO_3$ or $LiTaO_3$ thin plate bonded to high-velocity substrate*'. Japanese Journal of Applied Physics. 2017;**56**(7s1):07JD13

[78] Inoue S., and Solal M. '*Spurious free SAW resonators on layered substrate with ultra-high Q, high coupling and small TCF*'. Proceedings of IEEE Ultrasonics Symposium; Kobe, Japan, Oct. 2018

[79] Miura M., Matsuda T., Ueda M., *et al.* '*Temperature compensated $LiTaO_3$/sapphire SAW substrate for high power applications*'. Proceedings of IEEE Ultrasonics Symposium; Rotterdam, the Netherlands, Sept. 2005, pp. 573–6

[80] Parker T.E., and Wichansky H. '*Temperature-compensated surface-acoustic-wave devices with SiO_2 film overlays*'. Journal of Applied Physics. 1979;**50**(3): 1360–9

[81] Lee S., and Park J. '*Effect of fluorine on dielectric properties of SiOF films*'. Journal of Applied Physics. 1996;**80**(9):5260–3

[82] Yoshimaru M., Koizumi S., and Shimokawa K. '*Structure of fluorine-doped silicon oxide films deposited by plasma-enhanced chemical vapor deposition*'. Journal of Vacuum Science and Technology A. 1997;**15**(6):2908–14

[83] Matsuda S., Hara M., Miura M., *et al.* '*Use of fluorine-doped silicon oxide for temperature compensation of radio frequency surface acoustic wave devices*'. IEEE Transactions on Ultrasonics, Ferroelectrics, and Frequency Control. 2012;**59**(1):135–8

[84] Thalmayr F., Hashimoto K., Omori T., and Yamaguchi M. '*Frequency domain analysis of Lamb wave scattering and application to film bulk acoustic resonators*'. IEEE Transactions on Ultrasonics, Ferroelectrics, and Frequency Control. 2010;**57**(7):1641–7

[85] Thalmayr F., Hashimoto K., Ueda M., Omori T., and Yamaguchi M. '*Quantitative analysis of power leakage in a film bulk acoustic resonator device at the antiresonance frequency*'. Japanese Journal of Applied Physics. 2010; **49**(7s):07HD11

[86] Kaitila J., Ylilammi M., Ella J., and Aigner R. '*Spurious resonance free bulk acoustic wave resonator*'. Proceedings of IEEE Ultrasonics Symposium; Honolulu, USA, Oct. 2003, pp. 84–7

[87] Thalhammer R., Kaitila J., Zieglmeier S., and Elbrecht L. '*Spurious mode suppression in BAW resonators*'. Proceedings of IEEE Ultrasonics Symposium; Vancouver, Canada, Oct. 2006, pp. 456–9

[88] Fattinger G.G., Marksteiner S., Kaitila J., and Aigner R. '*Optimisation of acoustic dispersion for high performance thin film BAW resonators*'. Proceedings of IEEE Ultrasonics Symposium; Rotterdam, the Netherlands, Sept. 2005, pp. 1175–8

[89] Solal M., Gratier J., Aigner R., *et al.* '*Transverse modes suppression and loss reduction for buried electrodes SAW devices*'. Proceedings of IEEE Ultrasonics Symposium; San Diego, USA, Sept. 2010, pp. 624–8

Chapter 9

Testing and verification of MEMS resonator filters

Vasu Pulijala[1]

9.1 Introduction

Before designing the layout of the MEMS resonator filter, it must be decided how testing has to be performed. Once the appropriate test setup has been chosen, provisions have to be made in the design to accommodate the test structures. For example, MEMS resonator filters could be tested directly on-wafer or on-package, depending on its end usage. In case MEMS resonators are integrated along with circuits for monolithic solutions, on-wafer testing would be preferred. If the MEMS resonators are used as a standalone product, such as in a PCB environment, then on-package testing is preferred. In both the cases for accurate test performance of the device, the parasitic effects have to be eliminated. The popular equipment employed for both the test cases is vector network analyzer (VNA) which is shown in Figure 9.1 and spectrum analyzer. But the interface between the device to be tested and the equipment is different. On-wafer testing requires probes and corresponding calibration standards which could be designed on-wafer or a standard substrate could be used. On-package testing requires connectors and discrete calibration standards. Test structures for removing the parasitic effects have to be designed appropriately in both cases.

While testing any device, it can be treated as a black box and the output obtained for a given input is processed further to understand its performance. Typically at lower frequency ranges, the input and output variables are either voltages, currents, or a combination of them. In the millimeter-wave regime measuring voltages and currents are difficult. So scattering parameters (S-parameters) are employed to characterize the black-box in the high-frequency ranges.

To measure the MEMS resonator filter's performance S-parameters are often employed. Measurements at high frequencies involve large parasitics in the equipment, cables, connectors, testing structures, and so on. All the parasitic effects have to be eliminated to precisely characterize the device under test (DUT). The process of removing these unwanted effects is called "calibration".

[1]Centre for VLSI and Nanotechnology, Visvesvaraya National Institute of Technology, Nagpur, India

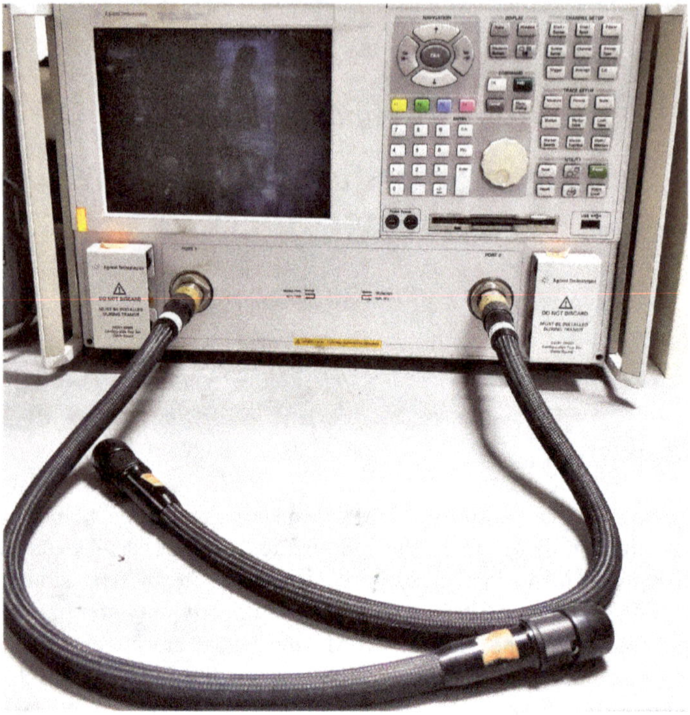

Figure 9.1 Agilent vector network analyzer

This chapter briefly revises the basics of scattering parameters, followed by an overview of the network analyzer and, various calibration techniques are described.

9.2 *S*-parameters

Assume an electrical circuit with two pairs of terminals connected to external circuits and let two terminals constitute a port. Such a circuit is called a two-port network. The model of the two-port network is used in the analysis of an electrical circuit. A two-port network can be considered as a "black box" which helps us to calculate the response of the network to signals applied to the ports, without solving for all the internal network voltages and currents.

The two-port mathematical models: impedance-parameters (Z), admittance-parameters (Y), hybrid-parameters (H), inverse hybrid-parameters (G), and ABCD-parameters are usually employed at low frequencies. If we consider a circuit that operates at high frequency, greater than 1 GHz, it is very difficult to create a short circuit or open circuit. Therefore, a new representation of the two-port network at microwave frequencies is needed. The appropriate representation is called scattering matrix and scattering parameters are defined in terms of traveling waves, which do not vary with position, unlike voltage and current waves.

S-parameters differ from the other two-port parameters, in the sense that they do not use open or short circuit conditions to characterize a linear electrical network; instead, matched terminations are used. These terminations are much easier to use at high signal frequencies than open-circuit and short-circuit terminations. Moreover, the quantities are measured in terms of power.

9.2.1 Derivation of scattering parameter for two-port network

Figure 9.2 shows the symbol of the two-port network. The impedance parameters for this two-port network can be expressed as

$$V_1 = Z_{11}I_1 + Z_{12}I_2 \tag{9.1}$$

$$V_2 = Z_{21}I_1 + Z_{22}I_2 \tag{9.2}$$

As can be seen from the above equations, port voltages are expressed in terms of port currents. Experimentally it is easy to open or short circuit the port to measure various Z-parameters at low frequency. For example, Z_{11} can be determined by open circuiting port 2. If port 2 is open-circuited then there will be no current flowing through port 2, hence I_2 becomes zero so from z_1, Z_{11} is obtained. Similarly, we can get all other two-port parameters by either open circuiting or short-circuiting a port. At high frequencies, the metal interconnects employed for short or open lead to inductances and capacitances, so we rely on scattering parameters to characterize the two-port network.

For a two-port network as shown in Figure 9.3 with incident wave a_1 and reflected wave b_1 at port 1, and incident wave a_2 and reflected wave b_2 at port 2, S-parameters are given as

$$b_1 = S_{11}a_1 + S_{12}a_2 \tag{9.3}$$

$$b_2 = S_{21}a_1 + S_{22}a_2 \tag{9.4}$$

or in the matrix form, we can write:

$$\begin{pmatrix} b_1 \\ b_2 \end{pmatrix} = \begin{pmatrix} S_{11} & S_{12} \\ S_{21} & S_{22} \end{pmatrix} \begin{pmatrix} a_1 \\ a_2 \end{pmatrix} \tag{9.5}$$

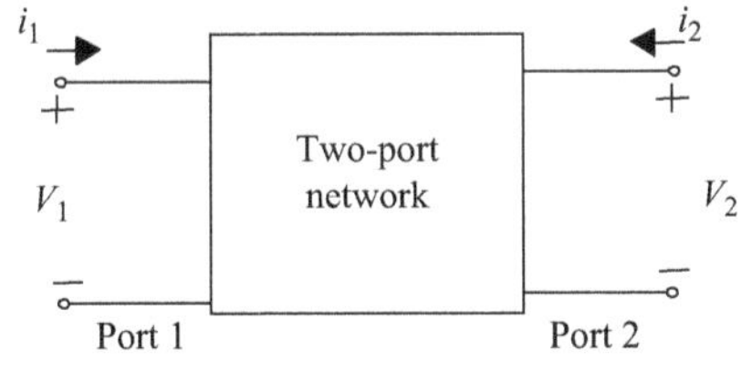

Figure 9.2 Two-port network: V_1 is the voltage across port 1, V_2 is the voltage across port 2, I_1 is the current through port 1, I_2 is the current through port 2

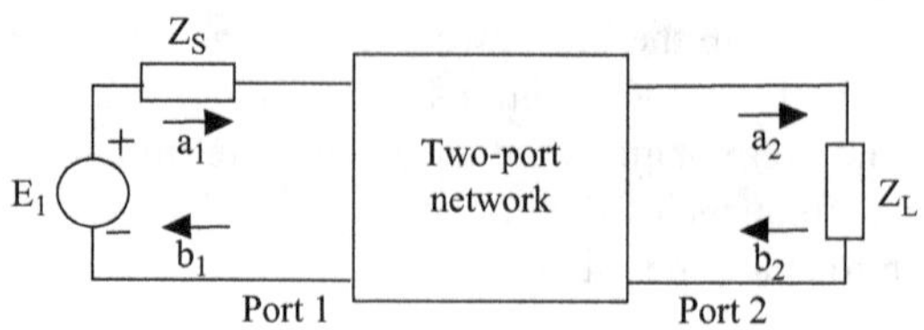

Figure 9.3 Two-port network for S-parameter measurement

These traveling waves can be defined in terms of V_1, I_1 and V_2, I_2 and a impedance Z_0:

$$a_1 = \frac{V_1 + Z_0 I_1}{2\sqrt{Z_0}} \tag{9.6}$$

$$b_1 = \frac{V_1 - Z_0 I_1}{2\sqrt{Z_0}} \tag{9.7}$$

$$a_2 = \frac{V_2 + Z_0 I_2}{2\sqrt{Z_0}} \tag{9.8}$$

$$b_2 = \frac{V_2 - Z_0 I_2}{2\sqrt{Z_0}} \tag{9.9}$$

Adding and subtracting (9.6) and (9.7), we get voltage and current at port 1:

$$V_1 = \sqrt{Z_0}(a_1 + b_1) \tag{9.10}$$

$$I_1 = \frac{1}{\sqrt{Z_0}}(a_1 - b_1) \tag{9.11}$$

Adding and subtracting (9.8) and (9.9), we get voltage and current at port 2:

$$V_2 = \sqrt{Z_0}(a_2 + b_2) \tag{9.12}$$

$$I_2 = \frac{1}{\sqrt{Z_0}}(a_2 - b_2) \tag{9.13}$$

The parameters S_{11}, S_{12}, S_{21}, and S_{22} represent reflection and transmission coefficients of the two-port network measured at port 1 and port 2. Assuming port 1 as input and port 2 as the output of the two network scattering parameters are explained as follows:

1. Input reflection coefficient (S_{11}):

$$S_{11} = \left.\frac{b_1}{a_1}\right|_{a_2=0} \tag{9.14}$$

Input reflection coefficient is given as the ratio of the reflected and incident wave at the input port when the reflection at port 2 is zero ($a_2 = 0$), which means output port is terminated with the equal value of the characteristic impedance, due to which wave which is traveling to the load will get totally absorbed and no energy is reflected. S_{11} tells us about the accuracy of the input matching.

2. Reverse transmission coefficient (S_{12}):

$$S_{12} = \left.\frac{b_1}{a_2}\right|_{a_1=0} \tag{9.15}$$

The reverse transmission coefficient is given as the ratio of the reflected wave at port 1 to the incident wave at port 2 when there is proper matching at port 1 ($a_1 = 0$). For this, the input port should be connected with a load with an equal value of the characteristic impedance. S_{12} tell us about the reverse isolation of the two-port network which means how much signal at port 2 is coupled to the signal at port 1.

3. Forward transmission coefficient (S_{21}):

$$S_{21} = \left.\frac{b_2}{a_1}\right|_{a_2=0} \tag{9.16}$$

Forward transmission coefficient is given as the ratio of the wave incident on the load attached to port 2 to the wave at the input of port 1 when the reflection from the load attached to port 2 is zero ($a_2 = 0$). S_{21} tells us about the gain of the two-port network.

4. Output reflection coefficient (S_{22}):

$$S_{22} = \left.\frac{b_2}{a_2}\right|_{a_1=0} \tag{9.17}$$

Output reflection coefficient is given as the ratio of the reflected and incident wave at the output port (port 2) when the reflection at port 1 ($a_1 = 0$) is zero. S_{21} tells us about the accuracy of the output matching.

Often it is necessary to convert the *S*-parameters to *Z*- or *Y*-parameters to analyze a two-port network. The following tables help us in quick conversion of *S*- to *Z*- or *Y*-parameters and vice versa.

9.2.2 Conversion of S*-parameter to Z-parameter*

For *Z*-parameter, we know that

$$[V] = [Z][I] \tag{9.18}$$

where

$$[V] = \begin{pmatrix} V_1 \\ V_2 \end{pmatrix} \quad [I] = \begin{pmatrix} I_1 \\ I_2 \end{pmatrix} \quad [Z] = \begin{pmatrix} Z_{11} & Z_{12} \\ Z_{21} & Z_{22} \end{pmatrix}$$

Equation (9.18) in terms of the incident and reflected wave becomes

$$[V^+] + [V^-] = [Z]([I^+] - [I^-]) \tag{9.19}$$

or

$$([Z] + [Z_0])[I^-] = ([Z] - [Z_0])[I^+] \tag{9.20}$$

where Z_0 is assumed to be real and

$$[V^+] = [Z_0]([I^+] \tag{9.21}$$

$$[V^-] = [Z_0]([I^-] \tag{9.22}$$

$$[Z_0] = \begin{pmatrix} Z_0 & 0 \\ 0 & Z_0 \end{pmatrix} \tag{9.23}$$

Therefore, scattering matrix in terms of Z-parameter is given by

$$[S] = \frac{[b]}{[a]} = \frac{[I^-]}{[I^+]} = ([Z] + [Z_0])^{-1}([Z] - [Z_0]) \tag{9.24}$$

and solving for $[Z]$, we can obtain

$$[Z] = [Z_0]([1] + [S])([1] - [S])^{-1} \tag{9.25}$$

Equations (9.24) and (9.25) give us the conversion from S to Z and Z to S. The below table shows the conversion from S-parameter to the Z- and Z- to S-parameter [3].

	S-parameter	**Z-parameter**
S_{11}	S_{11}	$\frac{(Z_{11} - Z_0)(Z_{22} + Z_0) - Z_{12}Z_{21}}{\Delta Z}$
S_{12}	S_{12}	$\frac{2Z_{12}Z_0}{\Delta Z}$
S_{21}	S_{21}	$\frac{2Z_{21}Z_0}{\Delta Z}$
S_{22}	S_{22}	$\frac{(Z_{11} + Z_0)(Z_{22} - Z_0) - Z_{12}Z_{21}}{\Delta Z}$
Z_{11}	$Z_0 \frac{(1 + S_{11})(1 - S_{22}) + S_{12}S_{21}}{(1 - S_{11})(1 - S_{22}) - S_{12}S_{21}}$	Z_{11}
Z_{12}	$Z_0 \frac{2S_{12}}{(1 - S_{11})(1 - S_{22}) - S_{12}S_{21}}$	Z_{12}
Z_{21}	$Z_0 \frac{2S_{21}}{(1 - S_{11})(1 - S_{22}) - S_{12}S_{21}}$	Z_{21}
Z_{22}	$Z_0 \frac{(1 + S_{22})(1 - S_{11}) + S_{12}S_{21}}{(1 - S_{11})(1 - S_{22}) - S_{12}S_{21}}$	Z_{22}

9.3 Conversion of *S*-parameter to *Y*-parameter

For *Y*-parameter we know that

$$[I] = [Y][V] \tag{9.26}$$

where

$$[V] = \begin{pmatrix} V_1 \\ V_2 \end{pmatrix} \quad [I] = \begin{pmatrix} I_1 \\ I_2 \end{pmatrix} \quad [Y] = \begin{pmatrix} Y_{11} & Y_{12} \\ Y_{21} & Y_{22} \end{pmatrix}$$

In terms of the incident and reflected wave, we have

$$[I^+] - [I^-] = [Y]([V^+] + [V^-]) \tag{9.27}$$

$$-([Y] + [Y_0])[V^-] = ([Y] - [Y_0])[V^+] \tag{9.28}$$

where

$$[Y_0] = \begin{pmatrix} \dfrac{1}{Z_0} & 0 \\ 0 & \dfrac{1}{Z_0} \end{pmatrix} \tag{9.29}$$

$$[S] = \frac{[b]}{[a]} = \frac{[V^-]}{[V^+]} = ([Y] + [Y_0])^{-1}([Y_0] - [Y]) \tag{9.30}$$

and solving for y we can obtain

$$[Y] = [Y_0]([1] - [S])([1] + [S])^{-1} \tag{9.31}$$

Equations (9.30) and (9.31) give us the conversion from S to Y and Y to S. The below table shows the conversion from S-parameter to the Y- and Y- to S-parameter.

	***S*-Parameter**	***Y*-Parameter**
S_{11}	S_{11}	$\dfrac{(Y_0 - Y_{11})(Y_{22} + Y_0) + Y_{12}Y_{21}}{\Delta Y}$
S_{12}	S_{12}	$\dfrac{-2Y_{12}Y_0}{\Delta Y}$
S_{21}	S_{21}	$\dfrac{-2Y_{21}Y_0}{\Delta Y}$
S_{22}	S_{22}	$\dfrac{(Y_{11} + Y_0)(Y_0 - Y_{22}) + Y_{12}Y_{21}}{\Delta Y}$
Y_{11}	$Y_0\dfrac{(1 - S_{11})(1 + S_{22}) + S_{12}S_{21}}{(1 - S_{11})(1 - S_{22}) - S_{12}S_{21}}$	Y_{11}

(Continues)

(*Continued*)

	S-Parameter	*Y*-Parameter
Y_{12}	$Y_0\dfrac{2S_{12}}{(1-S_{11})(1-S_{22})-S_{12}S_{21}}$	Y_{12}
Y_{21}	$Y_0\dfrac{2S_{21}}{(1-S_{11})(1-S_{22})-S_{12}S_{21}}$	Y_{21}
Y_{22}	$Y_0\dfrac{(1+S_{22})(1-S_{11})+S_{12}S_{21}}{(1-S_{11})(1-S_{22})-S_{12}S_{21}}$	Y_{22}

9.4 Network analyzers

Network analyzers are RF test instruments that can measure both active as well as passive devices. It can perform the measurements for inductors, filters, mixers, transistors, switches, amplifiers, and many more. The network analyzers are of two types: scalar, and vector. A brief history of network analyzers could be found in [1].

1. Scalar network analyzer (SNA): A scalar network analyzer is a form of RF network analyzer that only measures the amplitude properties of the device under test. Spectrum analyzer is an example of an SNA. Spectrum analyzer is useful for spectrum analysis of an electrical signal. It is employed to measure modulation, noise, and distortion.
2. Vector network analyzer (VNA): The VNA measures not only the amplitude response but also the phase. A VNA can measure *S*-parameters, VSWR, loss, gain, isolation, and group delay of an *n*-port network. The capability of the network analyzer depends on its specifications such as frequency range, number of test ports, output power, and IF bandwidth.

9.4.1 Vector network analyzer

To understand how a VNA operates, it is useful to see a basic block diagram of the test instrument which is shown in Figure 9.4. The diagram shows the very basic blocks of the VNA including the signal ports, directional couplers, the reference signal receiver, measured signal receiver, and finally the processor and display.

The RF signal is generated from an RF source which then goes through a switch. The switch is either connected to port 1 or port 2. Suppose S_{11} is to be measured then RF source switch will route the signal towards port 1. The signal first goes through the directional coupler 1. The directional coupler has two outputs and one input. One of the outputs goes towards reference receiver 1. Receiver down-converts this reference signal and then it is processed. Another output signal goes towards the device under test (DUT) through directional coupler 2. From DUT, the signal gets reflected and goes towards measurement receiver 1. Here, this signal is down-converted and processed. The VNA can then compare the data

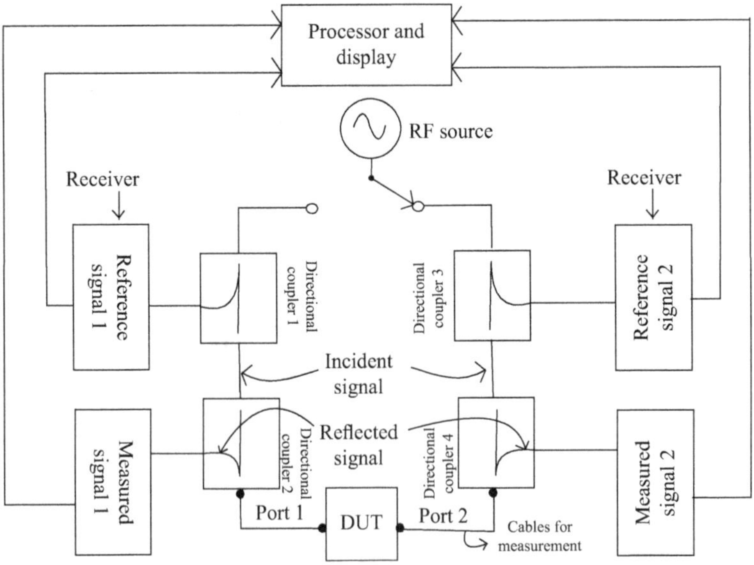

Figure 9.4 Basic block diagram representation of a vector network analyzer

obtained from measured signal 1 with the data obtained from reference signal 1 to calculate S_{11}.

For the measurement of S_{21}, the RF source switch again routes the signal towards port 1. This signal goes towards DUT through directional coupler 1. After this, the signal goes towards reference signal receiver 1 and also towards DUT. We get a reflected signal from DUT which is the measured signal 1. For the S_{21} measurement transmitted signal is considered which is the measured signal 2 obtained from directional coupler 4. This signal is downconverted and processed. After this VNA compares the data obtained from measurement signal 2 with the reference signal 1 to calculate S_{21}. Similarly, other S-parameters are measured.

In general, the interface between the VNA and DUT constitutes connectors/probes and cables. The losses associated with connectors/probes and cables are also reflected in the measurements performed for a DUT. The process of removing the errors is described in the next section.

9.4.2 Spectrum analyzer

A spectrum analyzer is used to measure the various frequency components and their corresponding strength in a signal. From the spectrum analyzer measurements, harmonic distortion, intermodulation distortion output power, power bandwidth, signal-to-noise ratio and so on could be figured out. A spectrum analyzer is shown in Figure 9.5. A typical output of the spectrum analyzer is shown in Figure 9.6. Spectrum analyzers normally employ linear frequency scale on the x-axis and logarithmic scale on the y-axis to display the amplitudes of spectrum components.

Figure 9.5 Spectrum analyzer

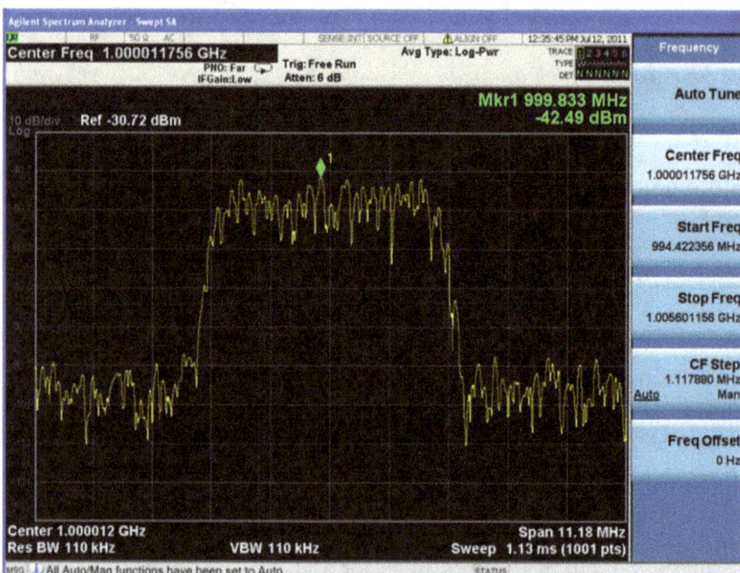

Figure 9.6 Typical display of a spectrum analyzer (Image source: https://en.wikipedia.org/wiki/File:SpectrumAnalyzerDisplay.png)

Spectrum analyzers employ two methods to measure the spectrum, swept-tuned, and Fourier analysis. Swept-tuned receivers are the common type of spectrum analyzers. It works on superheterodyne principles. A simplified block diagram of a superheterodyne spectrum analyzer is shown in Figure 9.7. Heterodyne refers to mixing operation which translates frequency and super means frequencies superior to audio range. Such analyzers perform a frequency sweep in the range of

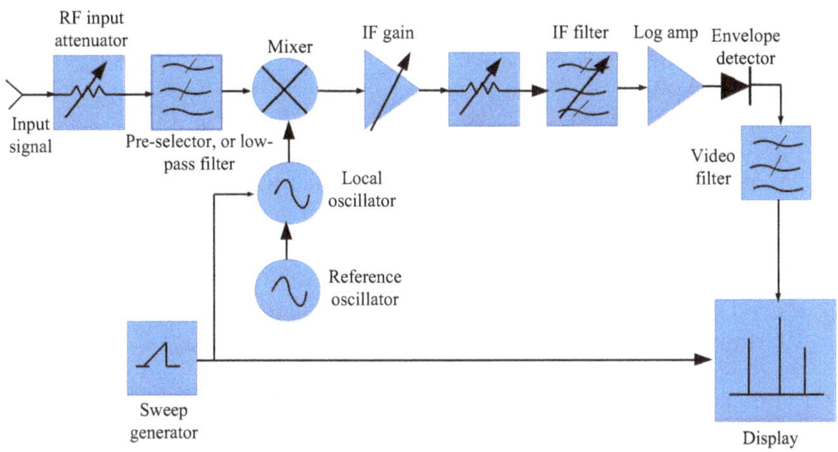

Figure 9.7 A simplified block diagram of superheterodyne spectrum analyzer

interest. RF attenuator, mixer, intermediate frequency (IF) gain stage, an IF filter, envelope detector, video filter, local oscillator, sweep generator, and display are critical components of the analyzer.

The input signal is fed to the RF input attenuator, which is a step attenuator to adjust the level of the signal before the mixing stage to avoid gain compression and distortions to high-level signals. The output of the attenuator is fed to a low-pass filter or preselector which does not allow out-of-band signals to reach the mixer stage. Mixer stage mixes the filtered input signal and the local oscillator signal generating the IF component and other harmonics. The signals which have their frequency in the passband of the IF filter range are further processed. The IF gain stage adjusts the vertical position of the signals on the display without changing the reference level. This stage compensates for the RF input attenuator stage. The output of the IF filter is rectified by the envelope detector, filtered through the low pass filter and displayed on the screen. The sweep generator tunes the local oscillator frequency across the range of frequencies of interest. In this manner, we could view various frequency signals and their amplitude levels.

These days we find both the spectrum analysis and network analysis options in a single handheld device.

9.5 Signal flow graphs and error models

To accurately measure the scattering parameters of the DUT, various errors have to be corrected. In general, errors are classified as drift, random, and systematic [2].

- Drift errors are due to change in the performance over time and they vary slowly. For example, the errors caused by the changes in temperature can be classified as drift errors. To minimize the drift errors, measurements have to be done at stable ambient conditions. Random errors are due to uncontrollable changes and cannot be removed by error correction mechanisms. They include connector wear, noise and so on.

- Random errors like noise can be minimized by averaging a large number of measurement traces.
- Systematic errors are the ones that are corrected using calibration. They are caused by stationary errors in the measurement setup. Systematic errors are considered to be constant over time. Reflection and transmission errors due to imperfections in the instrument and test set up can be classified as systematic errors. Following are the six types of systematic errors:
  - Directivity and crosstalk errors relating to signal leakage by the coupler. Cable and connector mismatch increases these errors.
  - Source and load impedance mismatches relating to reflections
  - Frequency response errors caused by reflection and transmission tracking within the test receivers. These errors are significant because of the cable length mismatch between the measurement ports.

Signal flow graphs aid in analyzing microwave circuits and are employed to describe the systematic error models involved in VNA measurements. Signal flow graphs have two nodes corresponding to each terminal of the port of the *n*-port network [3]. Branches of the graph show the relation between a pair of the nodes and they have a direction associated with them. Figure 9.8 depicts the signal flow graph of the two-port network. The nodes a_1 and a_2 correspond to the incident waves at the input and output of the two networks in Figure 9.8(a) and nodes b_1 and b_2 correspond to the reflected waves. The branches between nodes describe the *S*-parameters of the network.

Let us consider a one-port network connected with an oscillator and reflectometer. The systematic errors of the reflectometer can be lumped into an error adapter and made perfect as shown in Figure 9.9 [4,5]. Though a_0 and b_0 are the incident and reflected waves measured at the perfect reflectometer, a_1 and b_1 are

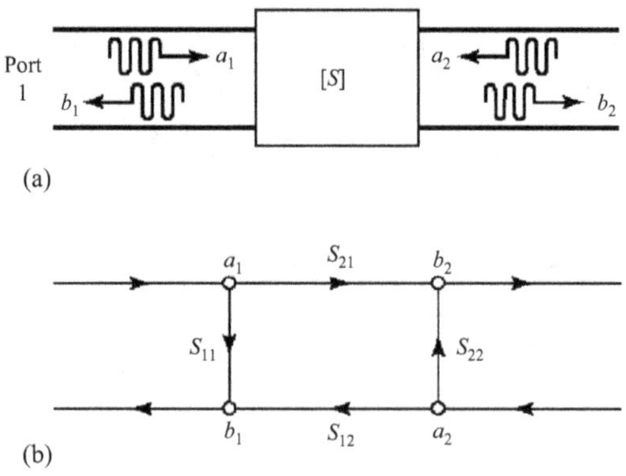

Figure 9.8 Signal flow graph of a two-port network

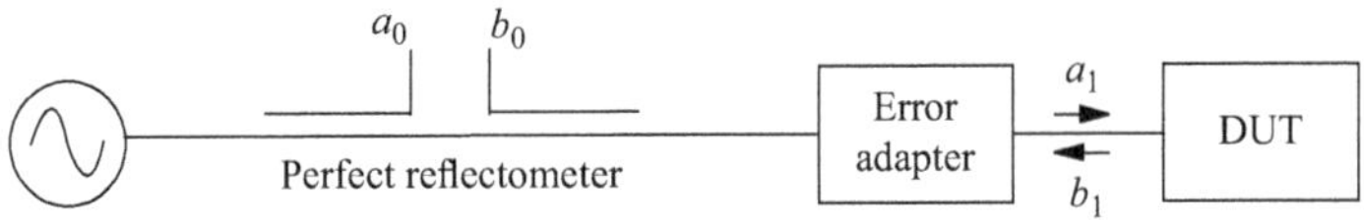

Figure 9.9 One-port network with perfect reflectometer and error adapter

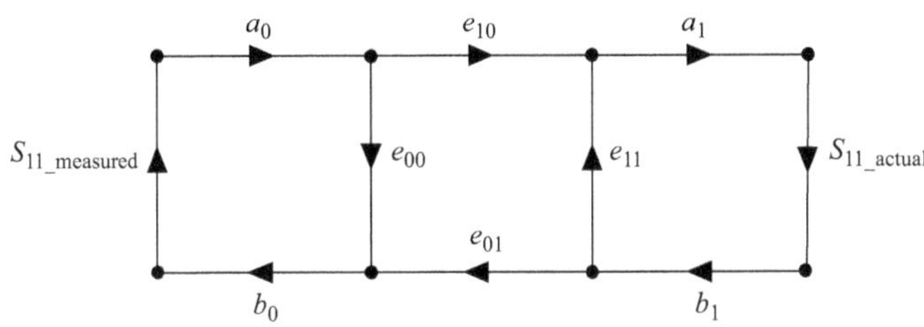

Figure 9.10 Signal flow graph of one-port network

the actual incident and reflected waves at the one-port network. The signal flow graph of this setup is shown in Figure 9.10. The relation between the measured reflection coefficient and the measured coefficient is established in the equations below.

$$a_1 = a_0e_{10} + b_1e_{11} \tag{9.32}$$

$$b_0 = a_0e_{00} + b_1e_{01} \tag{9.33}$$

The actual reflection coefficient of the one-port network is

$$S_{11a} = b_1/a_1 \tag{9.34}$$

while the measured reflection coefficient of the one-port network is

$$S_{11m} = b_0/a_0 = (a_1e_{00} - \Delta_e b_1)/(a_1 - e_{11}b_1) = (e_{00} - \Delta_e S_{11a})/(1 - e_{11}S_{11a}) \tag{9.35}$$

where

$$\Delta_e = e_{00}e_{11} - e_{10}e_{01} \tag{9.36}$$

From (9.35) and (9.36), it could be observed that if we could determine three terms e_{00}, e_{11}, and $e_{10}e_{01}$, the relation between actual and reflected coefficient could be established. So, the signal flow graph could be modified as shown in Figure 9.11. The three error terms are known as directivity (e_{00}), source match (e_{11}), and reflection-tracking errors ($e_{10}e_{01}$) of the one-port network.

To solve for the three error terms, (9.35) is used as three simultaneous equations with three unknowns. The three equations can be established by measuring the three known standards such as short ($S_{11a} = -1$), open ($S_{11a} = 1$), and, a

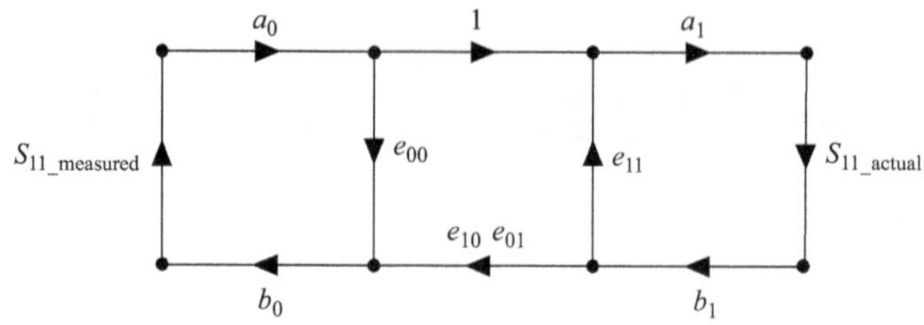

Figure 9.11 Modified signal flow graph of one-port network

load ($S_{11a} = 0$). The solution of these equations gives us the error terms. The error terms once calculated are utilized to give us the actual reflection coefficient of an unknown device under test.

Similarly, when measurements are performed for a two-port network, six of the systematic errors are present for the forward and the reverse directions, giving rise to a total of 12 error terms as shown in Figure 9.12 [6]. Solving the signal flow graphs in the forward and reverse directions will give us four equations containing the four actual *S*-parameters of the DUT and twelve error terms. The calibration methods employed would be used to determine the error terms to subsequently deduct the systematic errors from the actual measurements. Few of the popular calibration methods are described in the next section.

Employing the calibration techniques removes port match, directivity, tracking, and main leakage path errors. Errors related to noise, receiver non-linearity, drift, stability, repeatability are not removed by calibration. Errors due to drift, stability after calibration could be minimized by repeating the calibration at regular intervals. Noise- and linearity-related errors can be minimized by employing a linear calibration procedure.

9.6 Calibration

Calibration takes into account the systematic errors from the equipment and accessories and removes them from actual measurements. This type of calibrations is also called as error correction as it corrects the effects of additional components and also corrects the systematic errors of VNA. Calibration is defined as the "set of operations that establish, under specified conditions, the relationship between values of quantities indicated by a measuring instrument or measuring system, or values represented by a material measure or a reference material, and the corresponding values realized by standards" [8].

VNA measurements can be performed on-wafer or off-chip. Each of these test set-ups has its interface with the device under test. To remove the systematic errors from the VNA, cables, accessories and probes or connectors, on-wafer calibration or off-chip calibration has to be performed. Calibration to be performed for these test set-ups are different and are discussed further. The following discussion is

Forward model

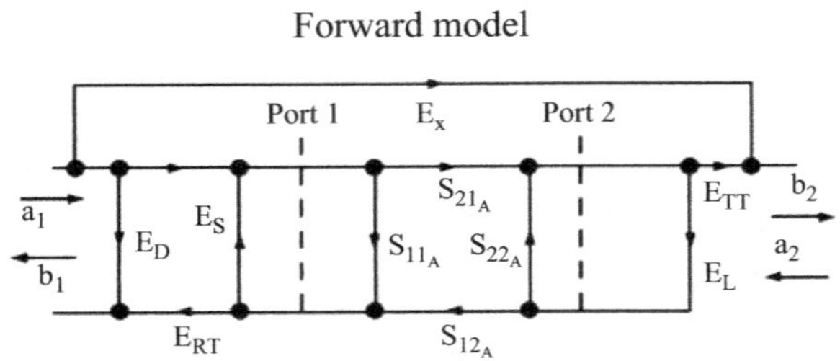

Reverse model

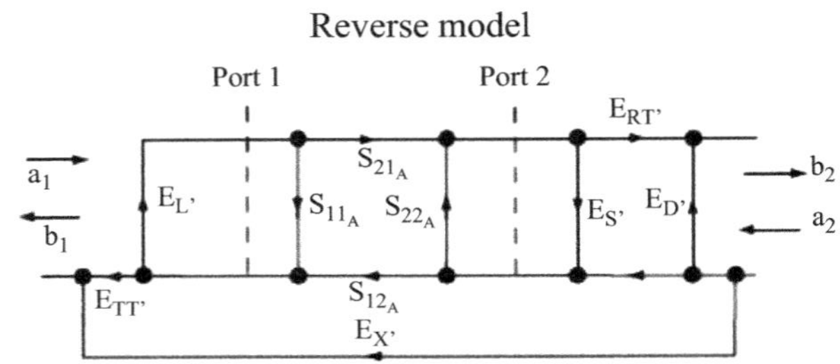

E_D = Fwd directivity

E_S = Fwd source match

E_{RT} = Fwd reflection tracking

E_L = Fwd load match

E_{TT} = Fwd transmission tracking

E_X = Fwd isolation

$E_{D'}$ = Rev directivity

$E_{S'}$ = Rev source match

$E_{RT'}$ = Rev reflection tracking

$E_{L'}$ = Rev load match

$E_{TT'}$ = Rev transmission tracking

$E_{X'}$ = Rev isolation

-Notice that each actual S-parameter is a function of all four measured S-parameters

-Analyzer must make forward and reverse sweep to update any one S-parameter

$$S_{11a} = \frac{\left(\frac{S_{11m}-E_D}{E_{RT}}\right)\left(1+\frac{S_{22m}-E_D'}{E_{RT}'}E_S'\right)-E_L\left(\frac{S_{21m}-E_X}{E_{TT}}\right)\left(\frac{S_{12m}-E_X'}{E_{TT}'}\right)}{\left(1+\frac{S_{11m}-E_D}{E_{RT}}E_S\right)\left(1+\frac{S_{22m}-E_D'}{E_{RT}'}E_S'\right)-E_L'E_L\left(\frac{S_{21m}-E_X}{E_{TT}}\right)\left(\frac{S_{12m}-E_X'}{E_{TT}'}\right)}$$

$$S_{21a} = \frac{\left(\frac{S_{21m}-E_X}{E_{TT}}\right)\left(1+\frac{S_{22m}-E_D'}{E_{RT}'}(E_S'-E_L)\right)}{\left(1+\frac{S_{11m}-E_D}{E_{RT}}E_S\right)\left(1+\frac{S_{22m}-E_D'}{E_{RT}'}E_S'\right)-E_L'E_L\left(\frac{S_{21m}-E_X}{E_{TT}}\right)\left(\frac{S_{12m}-E_X'}{E_{TT}'}\right)}$$

$$S_{12a} = \frac{\left(\frac{S_{12m}-E_X'}{E_{TT}'}\right)\left(1+\frac{S_{11m}-E_D}{E_{RT}}(E_S-E_L')\right)}{\left(1+\frac{S_{11m}-E_D}{E_{RT}}E_S\right)\left(1+\frac{S_{22m}-E_D'}{E_{RT}'}E_S'\right)-E_L'E_L\left(\frac{S_{21m}-E_X}{E_{TT}}\right)\left(\frac{S_{12m}-E_X'}{E_{TT}'}\right)}$$

$$S_{22a} = \frac{\left(\frac{S_{22m}-E_D'}{E_{RT}'}\right)\left(1+\frac{S_{11m}-E_D}{E_{RT}}E_S\right)-E_L'\left(\frac{S_{21m}-E_X}{E_{TT}}\right)\left(\frac{S_{12m}-E_X'}{E_{TT}'}\right)}{\left(1+\frac{S_{11m}-E_D}{E_{RT}}E_S\right)\left(1+\frac{S_{22m}-E_D'}{E_{RT}'}E_S'\right)-E_L'E_L\left(\frac{S_{21m}-E_X}{E_{TT}}\right)\left(\frac{S_{12m}-E_X'}{E_{TT}'}\right)}$$

Figure 9.12 Two-port error correction. Image courtesy: [7]

assuming two-port measurements. This could be easily extended to any *n*-port VNA. A brief overview of calibration could be found in [9].

9.6.1 On-wafer measurement

On-wafer measurement requires a probe station (Figure 9.13), RF probes, DC probes (Figure 9.14), calibration standards, apart from VNA, connectors, and cables. DC probes are optional if passive devices are being tested. The calibration standards for on-wafer measurements can be implemented on-wafer or an Impedance standard substrate can be employed.

Connections between the probes and VNA are made using probe cables and corresponding connectors (Figure 9.15). The torque provided while connecting them has to be appropriate enough as indicated in the data-sheets to establish the necessary coupling. If excessive torque is applied, the threading in the connector might wear out and might also damage the signal line.

Figure 9.13 Probe station

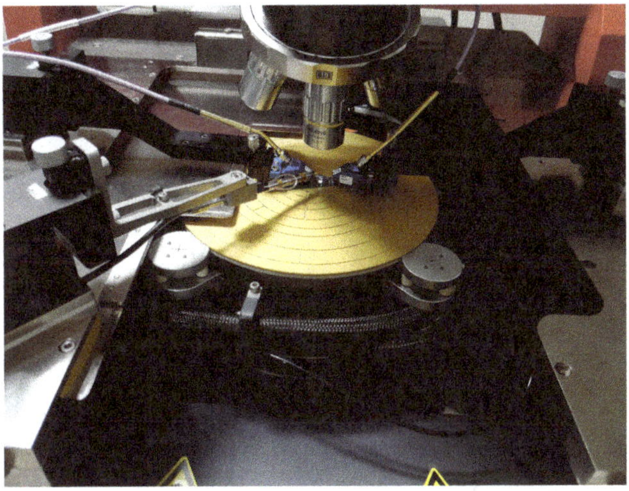

Figure 9.14 RF and DC probes

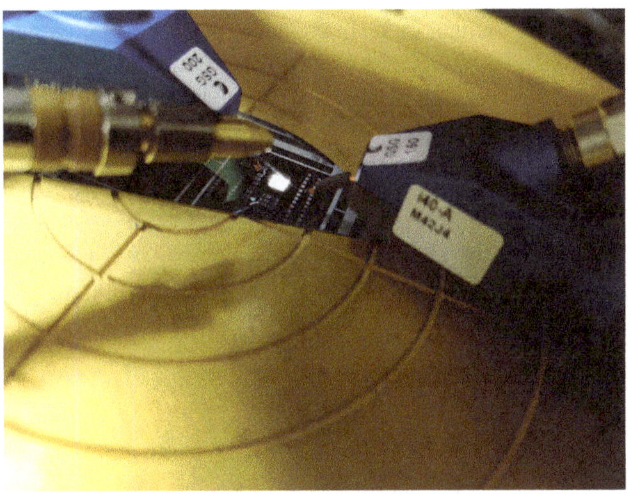

Figure 9.15 Zoomed in view of probes

RF probes come in a variety of footprints: ground-signal (GS), signal-ground-signal (SGS), ground-signal-ground (GSG), ground-signal-signal-ground (GSSG), and ground-signal-ground-signal-ground (GSGSG). GSG and GSGSG are the popular configurations used and they are based on single-ended and differential coplanar waveguides, respectively. Differential probes are useful for devices when differential ports. RF probes with 50 Ω impedance are usually employed for the measurements. DC probes are used to provide required DC voltages and ground to the devices.

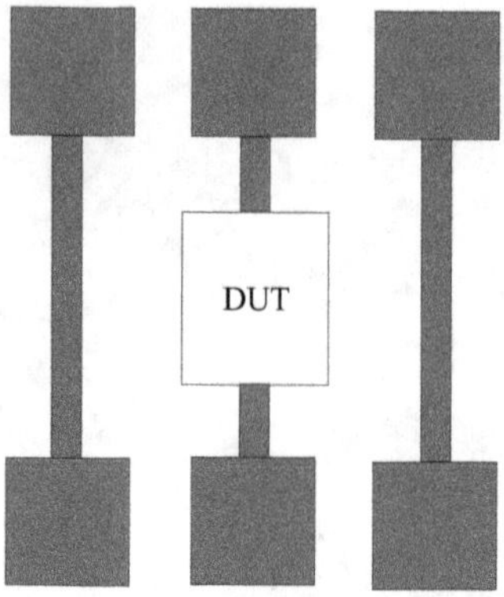

Figure 9.16 GSG RF pads connected to DUT with interconnect lines

For the RF probes to make contact with the device on-wafer RF pads are needed. Pads have to be designed depending on (1) configuration of the RF probes (GS, GSG, GSGSG), (2) diameter of the RF probe tips, and, (3) pitch of the RF probes. (Pitch is the distance between adjacent tips of the RF probe.) Pad dimensions have to be large enough to accommodate the RF probe tip and also should take into account the over-travel as soon they land (after touching down probe tips slide on the pad). But very large pads will offer large parasitics and hence their size has to be optimized to minimize the losses. Any electromagnetic field solver will aid in the optimal design of the pads. Pads designed for integrated circuit measurements require electrostatic discharge (ESD) protection to avoid unwanted high voltages which might breakdown the devices. Incorporation of ESD protection is optional for MEMS devices depending on its sensitivity to unwanted high voltages. RF pads are connected to the device either directly or with the interconnect lines (Figure 9.16).

9.6.2 On wafer calibration standards

For accurate measurements, it is ideal to fabricate the calibration standards on the same wafer where the device exists. The two popular calibration standards are short-open-load-through (SOLT) [4] and through-reflect-line (TRL) [10]. SOLT is calibration is good enough for frequencies <20 GHz. TRL is highly accurate compared to SOLT but its implementation is a challenge at lower frequencies. Unless the devices have a lower range of measurement frequencies (<1 GHz), TRL is the preferred calibration standard. In either of these techniques, VNA measures these standards and compares the values with the values provided in the calibration kit and removes the errors from the subsequent measurements after calibration. This process of calibration also moves the reference planes of measurement until the actual DUT eliminating all the parasitics associated with pads and interconnects.

SOLT: One of the popular and often-used calibration standards is the SOLT technique in the coplanar waveguide environment. We consider here the symmetrical case of the CPW environment.

The short standard of SOLT is implemented by connecting the signal pad to both ground pads with the same metal as shown in Figure 9.17. If the pads are implemented using multiple metal layers then the short connection also has to be implemented employing vias. The short fabricated is not an ideal short at high frequencies. So, the electrical equivalent of the short standard can be obtained from any standard EM simulator and has to be entered in the calibration kit.

The open standard can be implemented either on-wafer by having open unconnected pads or by lifting the RF probes in the air at a certain distance above the substrate (Figure 9.18). At high frequencies (>20 GHz) fringing fields may affect the measurement if the open pads are utilized. Better results are obtained with RF probes in air. An appropriate calibration standard has to be entered in the calibration kit for on-wafer open pad measurement or in-air measurements. The parasitic effects associated with RF probes in the air can be obtained from their datasheet.

Load standards are implemented by connecting standard resistance between the signal and ground pads (Figure 9.19). In most of the on-chip measurements, the

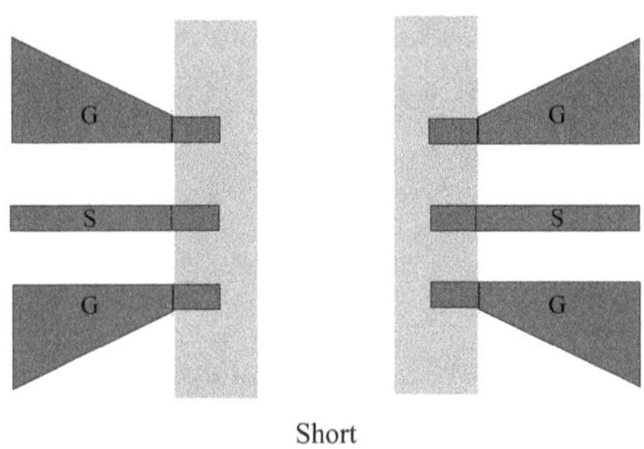

Figure 9.17 Short standard

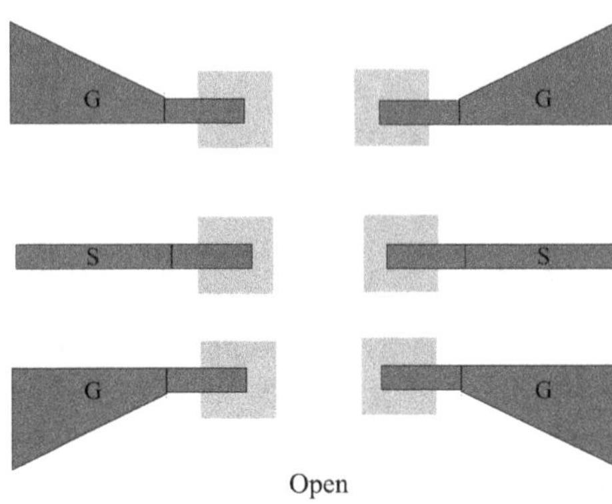

Figure 9.18 Open standard

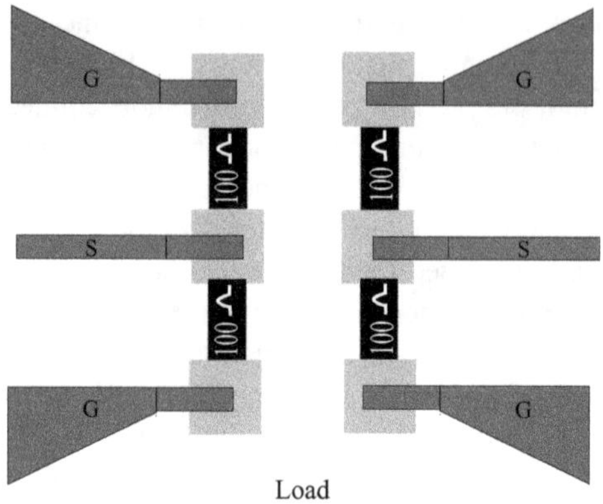

Load

Figure 9.19 Load standard

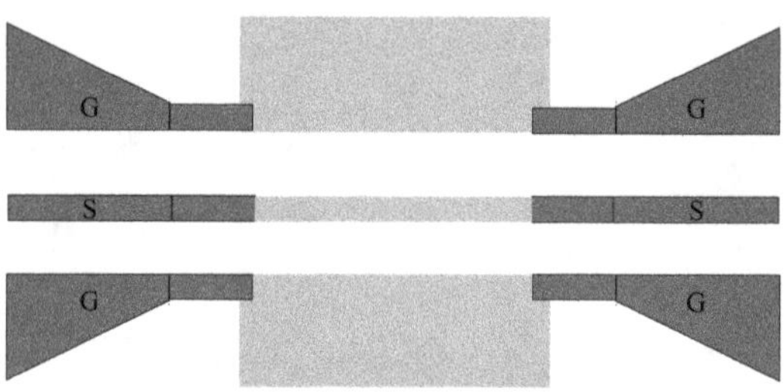

Thru

Figure 9.20 Thru standard

50 Ω standard is employed. Two 100 Ω on-chip resistors can be connected between each of the ground and signal pad. Else this resistance can be realized by using an appropriate length of the transmission line. The combination of these two 100 Ω resistors in parallel will give a 50 Ω resistor. The load standard is not just resistance but depending on the configuration is a series/parallel combination of inductance and capacitance. The simulation of the standard will give the parasitic effects associated with it and the electrical equivalent values should be mentioned in the calibration kit.

Through (thru) standard is a simple transmission line of nonzero length between two sets of GSG pads (Figure 9.20). While short, open and load are individually measured for both the probes in a two-port measurement, thru measurement has to be done with both the probes placed on the pads of the standard. Appropriate delay of the chosen thru line has to be entered in the calibration kit before the calibration. The delay for the through-line is from edge to edge between the pads. Hence, the placement of probes must be done accordingly.

TRL: TRL calibration constitutes of measuring a through (thru) standard (Figure 9.21), reflect standard (open or short) (Figure 9.22) and a line (thru but of

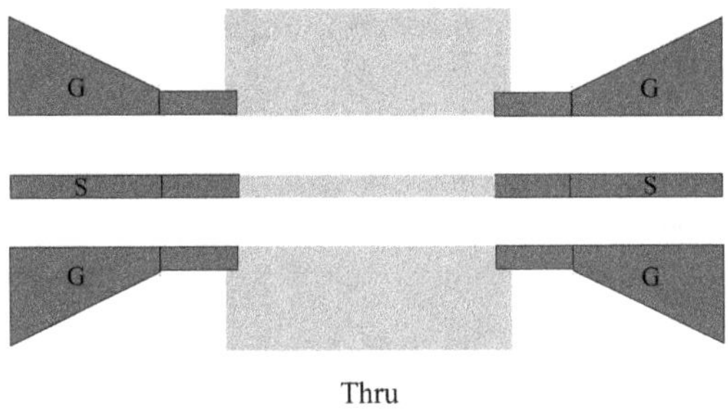

Figure 9.21 Thru standard

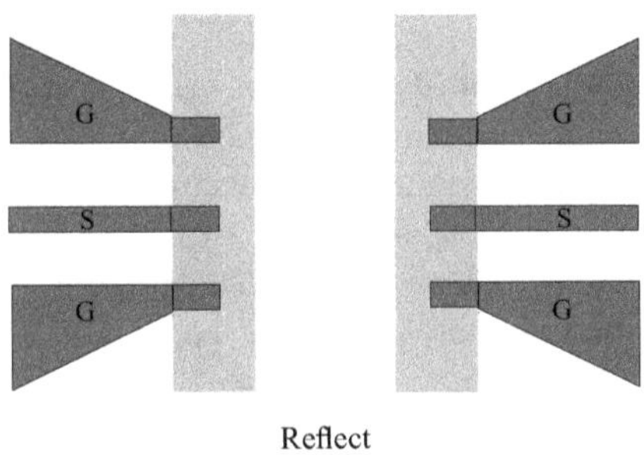

Figure 9.22 Reflect standard

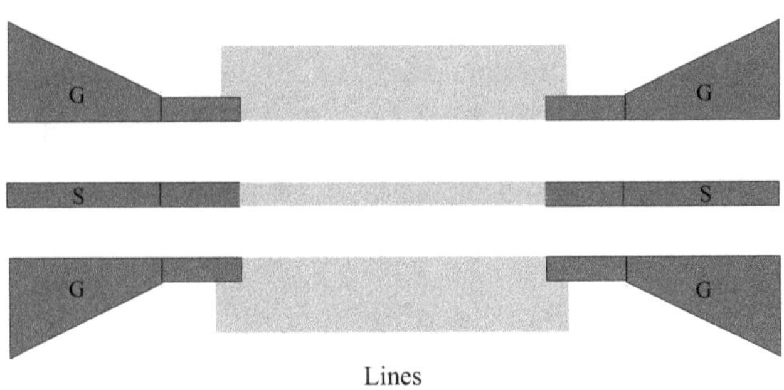

Figure 9.23 Lines standard

different length) (Figure 9.23). TRL gives accurate results compared to the SOLT technique but there is a limitation at lower frequencies (<1 GHz). Through standard is the same as the through in the SOLT calibration technique. Line standard is also a through but of different lengths. The electrical length difference between line and through standards must be different by 20° to 160° range. If wider bandwidths are to be measured, then multiple lines with different lengths are employed. In case of lower frequencies (<1 GHz), the length of the line standard will be very large to be accommodated. Reflect standard can be an open or a short standard.

9.6.3 Impedance standard substrate calibration and on-wafer DUT de-embedding

Implementation of calibration standards along with MEMS devices increases area and may not be cost-effective. In such cases, off-chip calibration standards that are available on an impedance standard substrates (ISS) are utilized (Figure 9.24). Impedance calibration substrate which contains various calibration standards is provided by the vendor of RF probes.

By using a calibration substrate, pad parasitic effects cannot be removed, unlike the case where all the standards are implemented together with the device on the same wafer. Hence, separate de-embedding structures have to be incorporated on the wafer.

If calibration substrate (ISS) is used for the removal of systematic errors, then the reference plane for the measurement shifts to the tip of the probes. When measurements are performed with this kind of calibration, the effects of pads on DUT are also included in the *S*-parameters obtained. To obtain the performance of DUT alone, it is necessary to de-embed the pad parasitic effects. For de-embedding pad effects from DUT, several techniques are available. The simplest is the open-DUT de-embedding [12]. In addition to the DUT with probe pads another structure without the DUT is fabricated, that is, it contains only the open pads corresponding

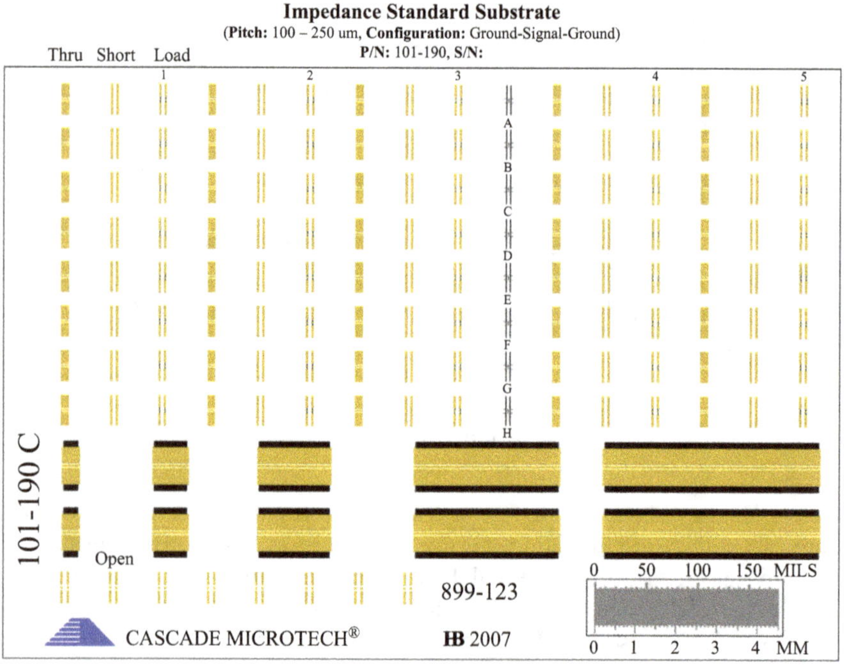

Figure 9.24 Impedance standard substrate. Image courtesy: [11]

to the probe placement. Following procedure is followed to de-embed the pad effects:

- Obtain the S-parameters of the DUT. Convert them to Y-parameters (Y_{DUT})
- Obtain the S-parameters of the open pads structure. Convert them to Y-parameters (Y_{open})
- Subtract the Y_{open} from Y_{DUT} ($Y_{de\text{-}embedded}$)
- Convert $Y_{de\text{-}embedded}$ into S parameters, which correspond to DUT alone

This technique eliminates the parasitic capacitance and resistance of the pads alone. In case there are interconnects of significant length from the pads to DUT, then a cascade-based de-embedding technique can be employed [13]. This involves the fabrication of an open and a thru structure in addition to DUT. Assuming the interconnect lengths at the input and output are l_1 and l_2, respectively, the through structure has an interconnect which is the sum of l_1 and l_2. The following steps need to be followed to obtain the de-embedded S-parameters.

- Obtain the S parameters of DUT (S_{dut}), open (S_{open}) and thru (S_{thru}) structures
- Characteristic impedance and the propagation constant of the interconnect line should be calculated [14]. The open-DUT de-embedding technique has to employed to obtain the S-parameters of the interconnect line alone from S_{thru} and S_{open}

$$\gamma = \ln\left(1 - S_{11}^2 + S_{21}^2/2S_{21} + -K\right) \tag{9.37}$$

where S_{11} and S_{21} are the S-parameters of the interconnect line alone.

- ABCD matrices of the input $[A_{in}]$ and output $[A_{out}]$ interconnect lines are obtained from the matrix below by substituting l_1 and l_2 in place of l, respectively.

$$\begin{bmatrix} A & B \\ C & D \end{bmatrix} = \begin{bmatrix} \cosh y\, l & Z_c \cosh y\, l \\ \sinh y\, l/Zc & \cosh y\, l \end{bmatrix}$$

- Calculated the ABCD matrix of the pad alone A_{pad} by obtaining the Y-parameters from S_{open} as follows

$$\begin{bmatrix} 1 & 0 \\ Y_{11open} + Y_{12open} & 1 \end{bmatrix}$$

- Calculate ABCD matrices of the input port $[A_{in\ port}]$ and output port $[A_{out\ port}]$ which are cascaded with pad and the interconnects utilizing $[A_{in\ port}] = [A_{pad}]$ $[A_{in}]$ and $[A_{out\ port}] = [A_{out}]\ [A_{pad}]$
- Obtain the ABCD matrix of DUT with probe pads $[A_{dut}]$ from its S parameters (S_{dut}). The ABCD matrix of DUT alone $[A_{de\text{-}embed}]$ can be obtained by $[A_{de\text{-}embed}] = [A_{in\ port}]^{-1}\ [A_{dut}]\ [A_{out\ port}]^{-1}$
- Obtain S parameters of the de-embedded structure alone from $[A_{de\text{-}embed}]$.

The above method may not give accurate results when the devices are implemented on lossy substrates. The four-step de-embedding method addresses the issues of substrate coupling and contact effects and is appropriate for MEMS devices implemented on lossy substrates such as CMOS technologies.

9.6.3.1 Four-step de-embedding method

Generally, for performing on-wafer measurements, impedance standard substrates are used to move the reference plane to the tip of the probes. And the above described de-embedding techniques are employed to remove the pad parasitic effects. These calibration methods are accurate until the substrate losses of the chip containing the devices are comparable to that of impedance standard substrate. If the substrates are lossy then the effects of substrate parasitics are ignored. Hence a four-step method will be useful as it considers the contact effects and the substrate coupling [15]. In this method, the standards are fabricated on the same substrate on which the devices are present. This method includes four steps of de-embedding with the test fixture given below.

Following the de-embedding method that accounts for the interaction between test-fixture and underlying substrate and is accurate for gaps large enough to hold most relevant RF CMOS devices and compensates for the dangling open standard. This method includes four steps of de-embedding with the test fixture given below.

The representative impedance model of the test fixture is shown in Figure 9.25. Test fixture includes device under test and interconnects connecting pads and DUT. As shown in Figure 9.25, S1 and S2 represent signal pads whereas G1 and G2 represent ground pads for port 1 and port 2, respectively. Figure 9.25 represents the equivalent circuit of the impedance model. The equivalent circuit is obtained by considering the symmetry on the test fixture as shown in Figure 9.26.

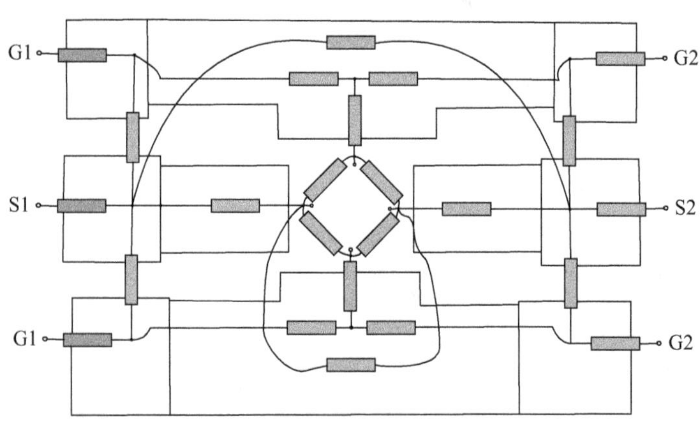

Figure 9.25 Impedance model of the test-fixture

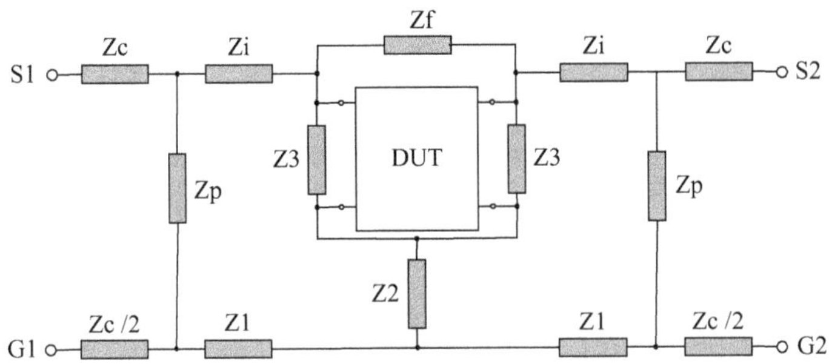

Figure 9.26 Equivalent impedance model of the test-fixture

- **Step 1**
  First step is to measure Z-parameters of the test-fixture including DUT which is given by

$$Z = \begin{pmatrix} Z_{11} & Z_{12} \\ Z_{21} & Z_{22} \end{pmatrix} \tag{9.38}$$

  where Z_c is the contact resistance due to metal pads, Z_p is coupling between the signal pad and ground pad, Z_i, Z_1 is the impedance from pad to DUT boundary as interconnect which is used to connect pad to DUT will have impedance at higher frequencies, Z_2 is the impedance of dangling leg used to connect the DUT and surrounding substrate to ground, Z_3 is used to denote the direct and substrate-carried coupling from the ground leg to the device input/output, Z_f represents the direct and substrate-carried coupling from input to output.
- **Step 2**
  This step involves the de-embedding of Z_c and Z_p (Figures 9.27 and 9.28).

  Z_c is de-embedded as

$$Z' = Z - \begin{pmatrix} \frac{3}{2} & 0 \\ 0 & \frac{3}{2} \end{pmatrix} Z_c \tag{9.39}$$

  Z_c is extracted from the S-parameter measurement of simple short standards (shown in Figure 9.29) by using the following equation.

$$Z_c = \frac{2}{3} Z_{11,ss} \tag{9.40}$$

  Step 2 is completed by converting the modified parameters, Z', into Y-parameters, $Y' = Z'-1 * Z_p$, Z_p is extracted from the S-parameter measurement of simple short and simple open standards shown in Figures 9.29 and 9.30 by using the following equation.

$$Z_p = Z_{11,so} - Z_{11,ss} \tag{9.41}$$

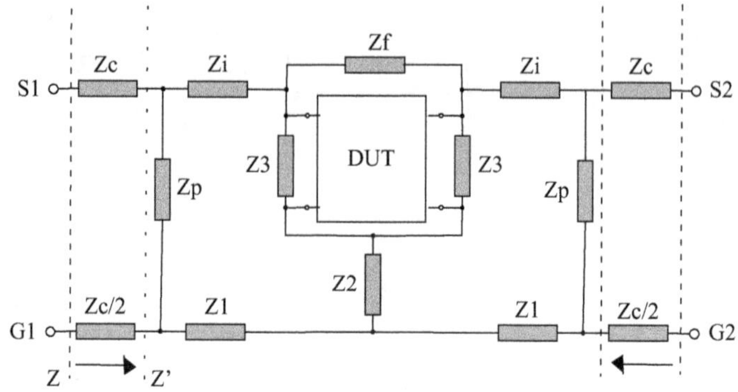

Figure 9.27 De-embedding of Z_c

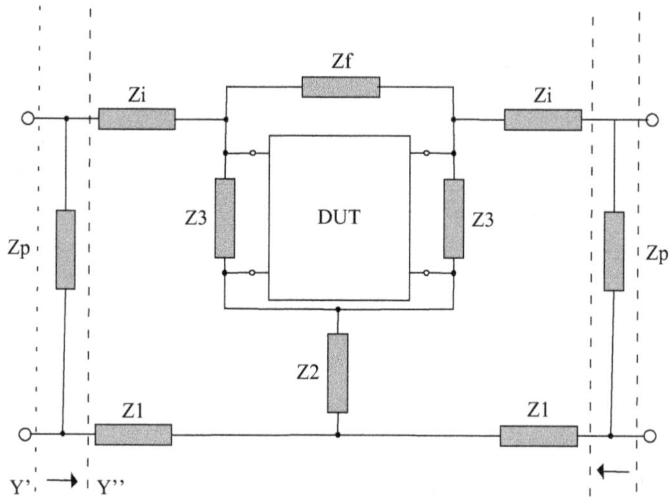

Figure 9.28 De-embedding of Z_p

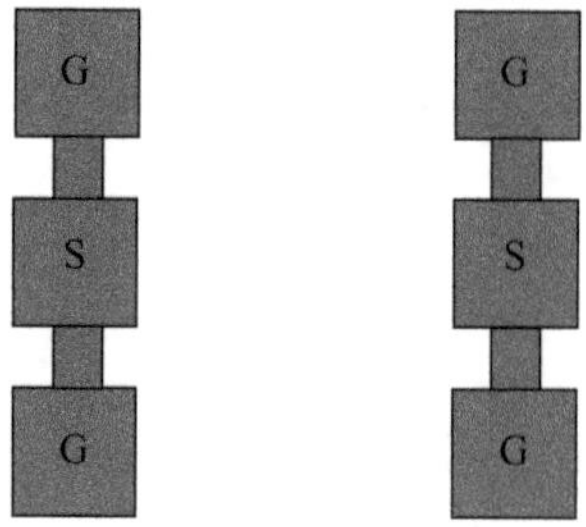

Figure 9.29 Simple short standard

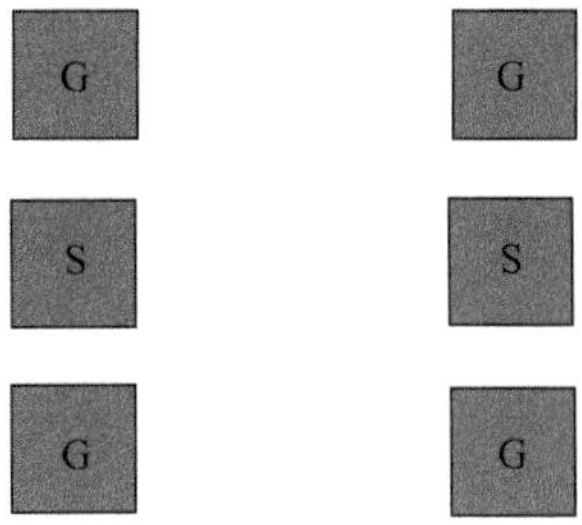

Simple open

Figure 9.30 Simple open standard

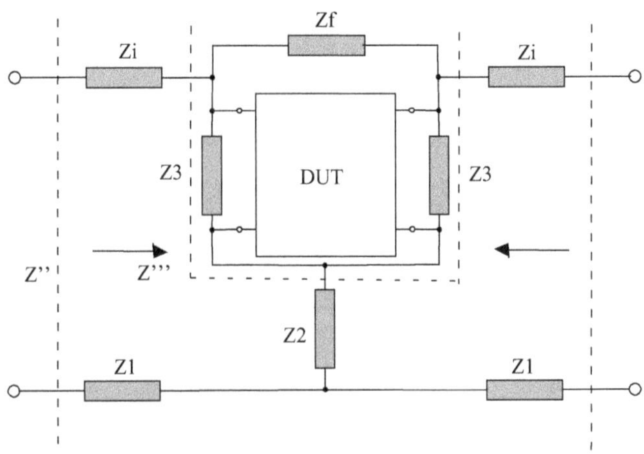

Figure 9.31 De-embedding of impedances $Z_i + Z_1$ and Z_2

Z_p is de-embedded as

$$Y'' = Y' - \begin{pmatrix} 1 & 0 \\ 0 & 1 \end{pmatrix} \frac{1}{Z_p} \tag{9.42}$$

- **Step 3**
  This step includes the de-embedding of impedances $Z_i + Z_1$ and Z_2 (Figure 9.31). The impedances are subtracted from Z'' matrix which is obtained from step 2.

$$Z'r' = Z'r - \begin{pmatrix} Z_i + Z_1 + Z_2 & Z_2 \\ Z_2 & Z_i + Z_1 + Z_2 \end{pmatrix} \tag{9.43}$$

Z_i, Z_1, Z_2, Z_3 are extracted from the following equations by using short and open standards, as shown in Figures 9.32, 9.33, and 9.34. In the following

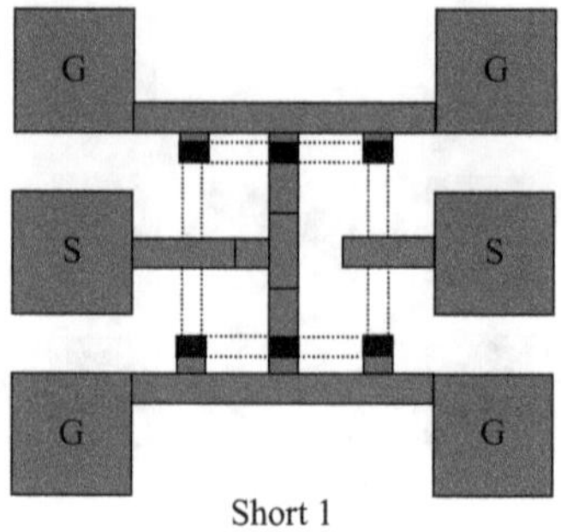

Figure 9.32 Short1 standard

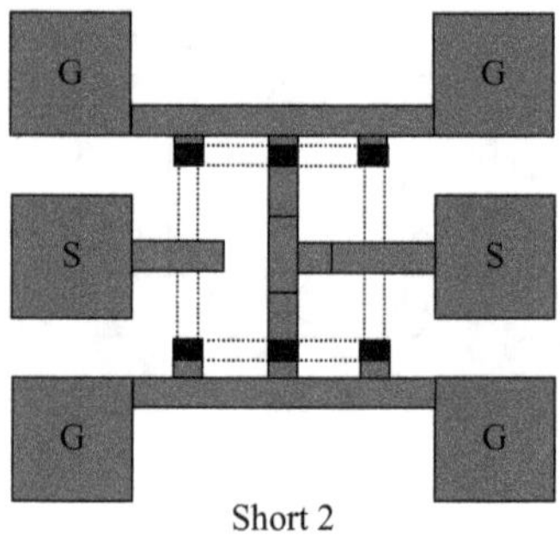

Figure 9.33 Short2 standard

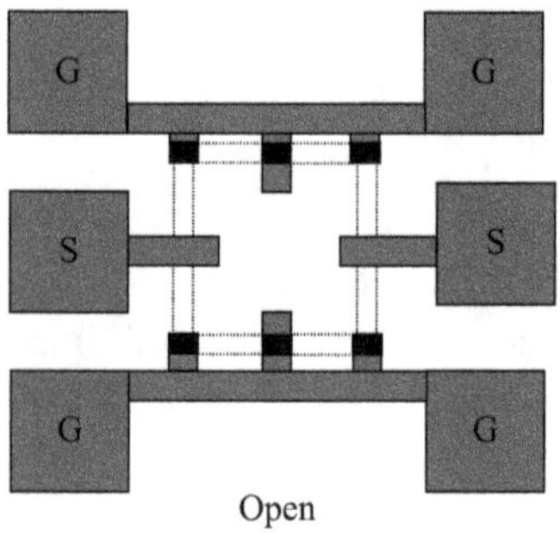

Figure 9.34 Open standard

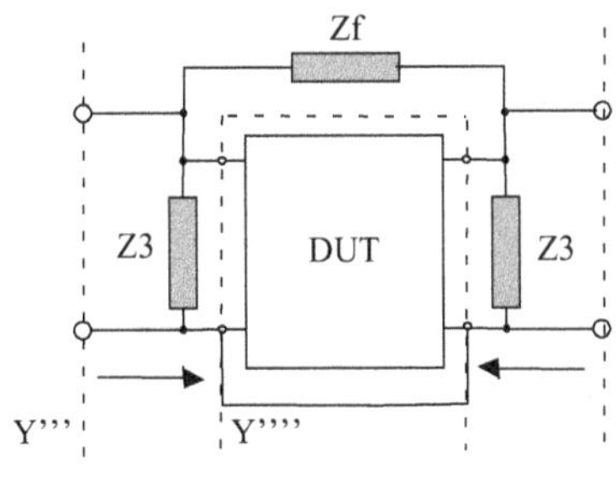

Figure 9.35 Obtaining the parameters for DUT alone

equations, all the parameters on the left-hand side are known from the above step.

$$z''_{11,o} = Z_i + Z_1 + Z_2 + Z_3||(Z_3 + Z_f) \tag{9.44}$$

$$z''_{21,o} = Z_2 + \frac{Z_3^2}{2Z_3 + Z_f} \tag{9.45}$$

$$z''_{11,s} \simeq (1+\alpha)(Z_i + Z_1) + Z_2 \tag{9.46}$$

$$z''_{21,s} \simeq Z_2 \tag{9.47}$$

$$z''_{22,s} \simeq Z_i + Z_1 + Z_2 + Z'_f||Z'_3 \tag{9.48}$$

$$z''_{12,s} \simeq Z_2 \tag{9.49}$$

- **Step 4**
  This is the final step which is shown in Figure 9.35. It consists of the conversion of de-embedded Z-parameters back into corresponding Y-parameters.

$$Y''\prime\prime = Y'\prime' - \begin{pmatrix} \frac{1}{Z_3} + \frac{1}{Z_4} & -\frac{1}{Z_f} \\ -\frac{1}{Z_f} & \frac{1}{Z_3} + \frac{1}{Z_4} \end{pmatrix} \tag{9.50}$$

And finally, the Y-parameters obtained are converted to S-parameters. This method accurately predicts the series losses and coupling parasitic of a fabricated test fixture suitable for two-port measurements on RF devices.

9.6.4 Verification of calibration

It is necessary to check if the calibration performed is correct or not. The S parameters obtained before and after SOLT calibration using standard impedance substrate are shown in Figures 9.36 and 9.37. Before calibration S_{11} and S_{22} are nonzero with value increasing with increasing frequency while the reflection coefficients S_{11} and S_{22} are at 0 dB for probes in the air after calibration.

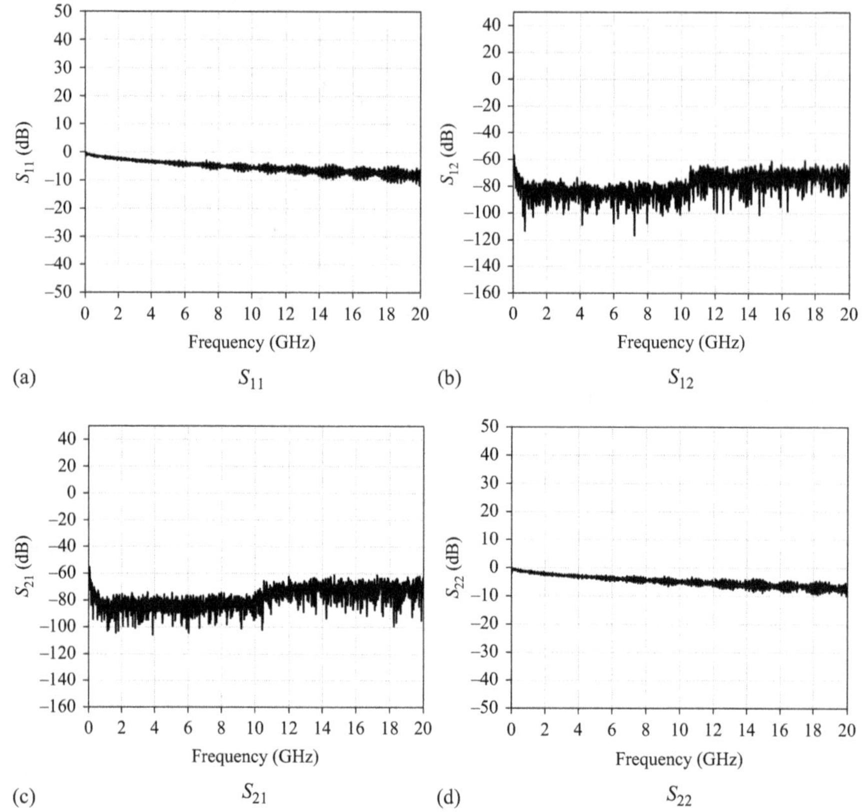

Figure 9.36 Scattering parameters for probes in air before calibration

After calibration is performed, one could also place the probes on the through a transmission line (which was used for calibration) to check if the S_{12} and S_{21} are at 0 dB for the calibrated range of frequencies, as shown in Figure 9.38.

For verification of on-wafer DUT de-embedding electromagnetic field solves could be employed to cross-verify the results after de-embedding. The DUT alone without any bond-pads added to it could be solved any EM field solver and the results obtained by on-wafer de-embedding are cross-verified against these values.

9.7 Calibration for off-chip measurements

If the MEMS device has to be tested as a standalone design in a PCB environment, then off-chip testing has to be performed. Off-chip measurements require connectors such as SMA, probe cables and a VNA. For calibration, standard kits are available which contain various standards in the form of discrete components as

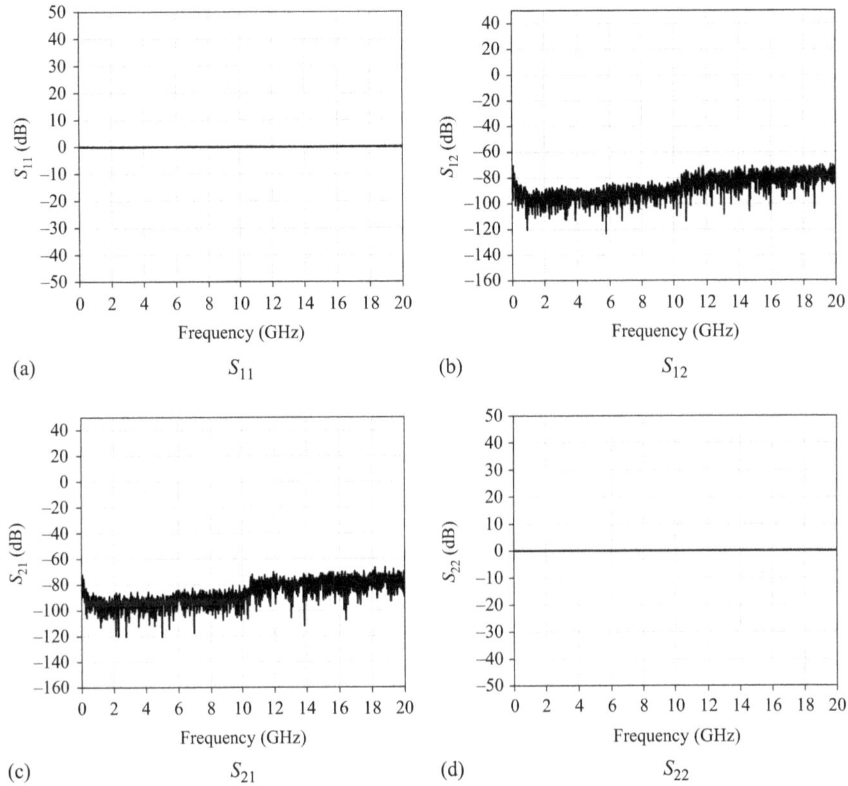

Figure 9.37 Scattering parameters for probes in air after calibration

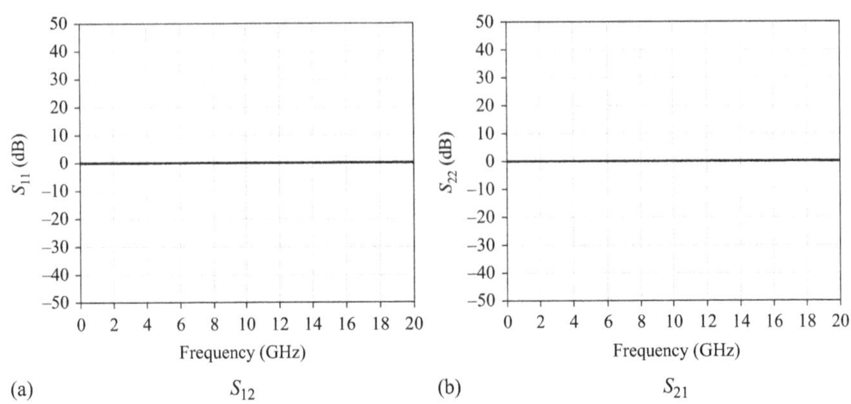

Figure 9.38 Transmission coefficients after calibration, when probes are placed on the through structure

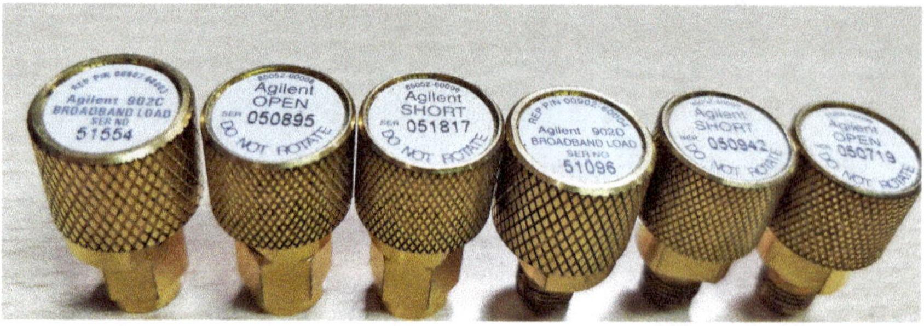

Figure 9.39 Discrete component calibration standards

shown in Figure 9.39. Both male and female standards are available depending on which connector it has to be connected to. These standards are used to remove the effect of connecting cables which are used to connect VNA and DUT printed circuit board.

The MEMS component is soldered on to PCB and the necessary input and output ports are bonded from the on-chip pads to PCB bond pads with bond wires. For the reduction of parasitic effects, the bond wire length has to be minimal. The parasitics introduced by bond pads and bond wires can be obtained by EM Simulation. In most of the designs, these parasitics are included for the design of MEMS components. For example, a bond wire loop provides series inductance across the bond pads which is included in the design.

9.8 Other testing methodologies

Apart from the electrical actuation scheme, there are piezoelectric, thermoelastic, optical, magnetomotive actuation schemes available to excite the resonator. In this section, we discuss techniques other than electrical detection, employed to test MEMS resonators.

9.8.1 Optical detection

Optical detection techniques based on interferometry and beam deflection are employed to test MEMS resonators. Detection based on the interferometric principles relies on measuring the phase difference between two superimposed waves of the same frequency. There are several techniques of which Michelson interferometry and Fabry–Perot interferometry are popular [16–19]. In Michelson interferometry (Figure 9.40), the beam reflected off a mirror and the beam reflected off the surface which is resonating are combined. The combined beam is detected by a photodetector. In Fabry–Perot interferometry (Figure 9.41), a cavity could be formed between the resonant structure and the substrate beneath. This substrate has to be reflective and resonant structure semi-transparent. The change in the intensity of light, detected by photodetector when in action gives us the details of resonance.

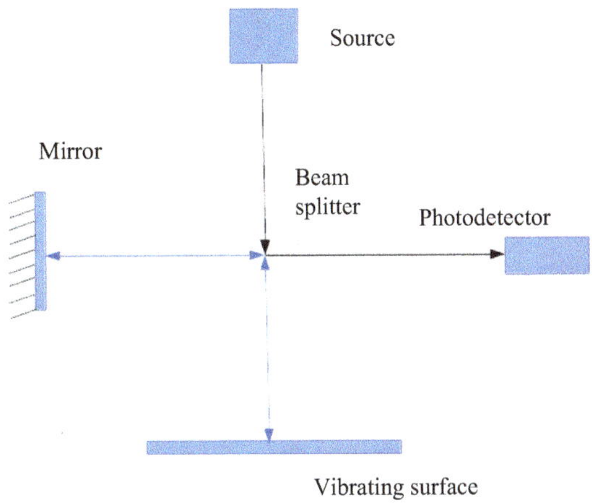

Figure 9.40 Michelson interferometer

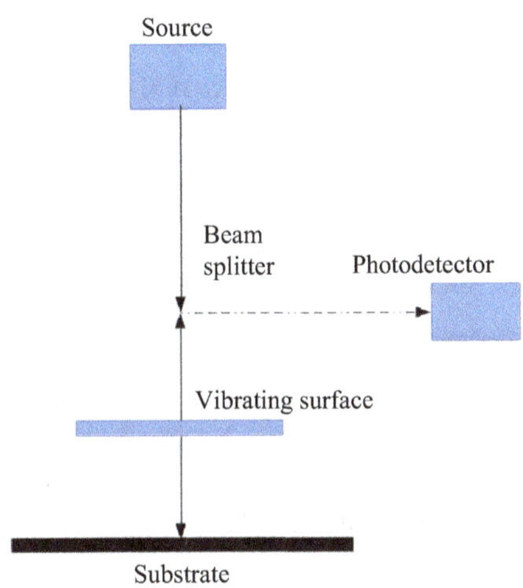

Figure 9.41 Fabry–Perot

For beam deflection detection (Figure 9.42) [20,21], the laser beam is focused on the surface of the resonator and the motion of the surface under resonance interferes with the otherwise stable reference beam. Change in reflectance due to this interference is detected by using a photodetector.

Optical detection techniques are efficient compared to the conventional measurements minimizing the interference between the input actuation and output detection. Coupling losses, energy losses due to surface scattering, the requirement of reflective and anti-reflective coatings, frequency limitations are few challenges of these techniques.

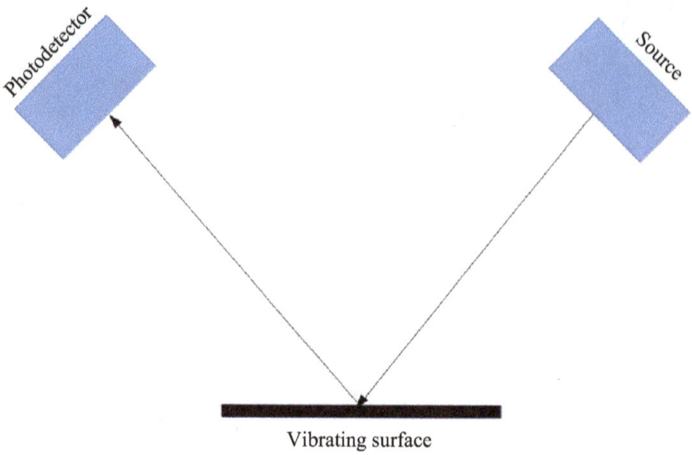

Figure 9.42 Beam deflection

9.9 Measurement of intermodulation distortion in MEMS resonators

Due to nonlinear effects, MEMS resonators might generate harmonics and intermodulation products. Of these, the third-order intermodulation product [22] due to odd-order nonlinearities can lead to unwanted spectral components in the passband of the filter. Consider two fundamental signals at frequencies ω_1 and ω_2. The nonlinearity in the resonators might lead to cubic mixing of these two frequencies and the third-order intermodulation will give products of frequencies at $2\omega_1 \pm \omega_2$ and $2\omega_2 \pm \omega_1$. If ω_1 and ω_2 differ from the center frequency (ω_0) of the passband of the filter by $\Delta\omega$ and $2\ \Delta\omega$ respectively, that is if $\omega_1 = \omega_0 - \Delta\omega$ and $\omega_2 = \omega_0 - 2\Delta\omega$, then $2\omega_1 - \omega_2$ is in the passband of the filter at ω_0. This third-order intermodulation (IM_3) signal will end up corrupting the desired signal.

The intermodulation test is carried out by a two-tone test where two sinusoid signals with equal amplitudes are applied to the input. The frequencies of two tones could be selected with respect to the center frequency of the passband of the filter as described above. The third-order intercept point (IP_3) is the amplitude at which the IM products become equal to that of the fundamental tones of the output. The input level at which this happens is called input-referred third-order intercept point (IIP_3) and the output corresponding to the input is called output referred third-order intercept point (OIP_3). It should be noted that the third-order products rise three times faster than the desired components of the output.

To test the MEMS resonators, signal generators can be used to supply the two tones and the central frequency of the passband [23–25]. A spectrum analyzer could be employed to observe the output spectrum. Input power levels of all the input signals must be the same and varied in steps. At each input power level, output power levels of the fundamental tone and intermodulation products could be plotted. The third-order intercept point is calculated by extrapolating the third-order intermodulation power and the fundamental tone power and finding the point at which they cross. Higher the third-order intercept point, the lower the distortion caused by the third-order intermodulation.

9.10 Summary

The testing of MEMS resonators helps not only in verifying if the design specifications have been met but also in further optimization of the design, if necessary. In this chapter we have seen various methods to test MEMS resonator filters. As the frequency of operation of the resonators further increases, we may have to come up with specific calibration mechanisms to remove the parasitic effects.

References

[1] Rytting D. ARFTG 50 year network analyzer history. In: 2008 71st ARFTG Microwave Measurement Conference. IEEE; 2008. p. 1–8.

[2] Martens L. High-frequency characterization of electronic packaging. vol. 1. Springer Science & Business Media, New York; 2013.

[3] Rytting D. Appendix to an analysis of vector measurement accuracy enhancement techniques. Hewlett Packard. 1982;p. 1–42.

[4] Rytting D. An analysis of vector measurement accuracy enhancement techniques. In: RF and Microwave Symp. Exhibition; 1980.

[5] Rehnmark S. On the calibration process of automatic network analyzer systems (short papers). IEEE Transactions on Microwave Theory and Techniques. 1974;22(4):457–458.

[6] Ballo D. Applying error correction to network analyzer measurements. Microwave Journal. 1998;41(3):64–75.

[7] Keysight Technologies Applying Error Correction to Vector Network Analyzer Measurements. Keysight Technologies; [updated 2018 Feb 28; cited 2019 Jul 07]. Available from: http://literature.cdn.keysight.com/litweb/pdf/5965-7709E.pdf.

[8] BIPM I, IFCC I, IUPAC I. OIML. International Vocabulary of Basic and General Terms in Metrology Second. International Organization for Standardization, Geneva.

[9] Rumiantsev A, and Ridler N. VNA calibration. IEEE Microwave Magazine. 2008;9(3):86–99.

[10] Engen GF, and Hoer CA. Thru-reflect-line: An improved technique for calibrating the dual six-port automatic network analyzer. IEEE Transactions on Microwave Theory and Techniques. 1979;27(12):987–93.

[11] ISS Map 101-190. FormFactor; [updated 2017 Aug 14; cited 2019 Jul 07]. Available from: https://www.formfactor.com/download/iss-map-101–190/.

[12] van Wijnen PJ. A new straightforward calibration and correction procedure for 'on wafer' high frequency S-parameter measurements (45 MHz–18 GHz). In: Proceedings of the Bipolar Circuits and Technology Meeting, 1987; 1987.
[13] Cho MH, Huang GW, Chen KM, *et al.* A novel cascade-based de-embedding method for on-wafer microwave characterization and automatic measurement. In: 2004 IEEE MTT-S International Microwave Symposium Digest (IEEE Cat. No. 04CH37535). vol. 2. IEEE; 2004. p. 1237–40.
[14] Eisenstadt WR, and Eo Y. *S*-parameter-based IC interconnect transmission line characterization. IEEE Transactions on Components, Hybrids, and Manufacturing Technology. 1992;15(4):483–90.
[15] Kolding TE. A four-step method for de-embedding gigahertz on-wafer CMOS measurements. IEEE Transactions on Electron Devices. 2000;47(4): 734–40.
[16] Carr DW, Sekaric L, and Craighead HG. Measurement of nanomechanical resonant structures in single-crystal silicon. Journal of Vacuum Science & Technology B: Microelectronics and Nanometer Structures Processing, Measurement, and Phenomena. 1998;16(6):3821–24.
[17] Carr DW, Evoy S, Sekaric L, *et al.* Parametric amplification in a torsional microresonator. Applied Physics Letters. 2000;77(10):1545–47.
[18] Meyer C, Lorenz H, and Karrai K. Optical detection of quasi-static actuation of nanoelectromechanical systems. Applied Physics Letters. 2003;83(12): 2420–22.
[19] Kouh T, Karabacak D, Kim D, *et al.* Diffraction effects in optical interferometric displacement detection in nanoelectromechanical systems. Applied Physics Letters. 2005;86(1):013106.
[20] Czaplewski D, Sullivan J, Friedmann T, *et al.* Mechanical dissipation in tetrahedral amorphous carbon. Journal of Applied Physics. 2005;97(2): 023517.
[21] Sandberg R, Boisen A, and Svendsen W. Characterization system for resonant micro-and nanocantilevers. Review of Scientific Instruments. 2005; 76(12):125101.
[22] Razavi B. RF microelectronics. Pearson Education, India; 2012.
[23] Alastalo AT, and Kaajakari V. Third-order intermodulation in microelectromechanical filters coupled with capacitive transducers. Journal of Microelectromechanical Systems. 2006;15(1):141–48.
[24] Zuo C, Sinha N, Pisani MB, *et al.* Channel-Select RF MEMS Filters Based On Self-Coupled A1N Contour-Mode Piezoelectric Resonators. Departmental Papers (ESE). 2007;p. 318.
[25] Nordquist CD, and Olsson RH. Power handling and intermodulation distortion of contour-mode AlN MEMS resonators and filters. In: 2011 IEEE MTT-S International Microwave Symposium. IEEE; 2011. p. 1–4.

Chapter 10

3D packaging for the integration of heterogeneous systems

Pavani Vamsi Krishna Nittala[1], Prosenjit Sen[1], K.N. Bhat[1], and M.M. Nayak[1]

'Another direction of improvement (of computing power) is to make physical machines three dimensional instead of all on a surface of a chip (2-D). That can be done in stages instead of all at once; you can have several layers and then many more layers as the time goes on.' – an excerpt from a talk delivered by Richard Feynman titled 'Computing Machines in the Future' at the Gakushuin University in Tokyo on August 9, 1985. Three decades later, advances in nanofabrication have enabled us to imagine, design, create and stack microchips for 3D Integration.

Currently, heterogeneous 3D integration is faster than Moore (Figure 10.1). This thesis details the development of processing technologies for the 3D packaging for the integration of heterogeneous systems.

10.1 Three-dimensional integration

Due to the limitations of silicon material, further scaling [2–5] of devices is much more challenging. Hence, to keep up with the Moore's law [6] (miniaturization law), there is a need for an immediate technology driver [7], and 3D integration technology is believed to be a potential candidate. In 3D integration technology, the processed or to be processed wafers/chips are stacked in the third dimension [8–12]. This technology not only provides reduction in board footprint, high-speed and low-power devices but also allows the possibility to integrate multi-functional platforms by stacking [3,8,13–15]. The 3D integration technology enables heterogeneous integration with higher density [7] making it faster than Moore [1,16,17].

The use of TSVs (through-silicon via [7,18–20]) in 3D integration makes 3D integration different from 3D packaging [8]. TSV technology involves six main steps:

1. Etching/drilling the silicon via using deep reactive-ion etching (DRIE) tool
2. Dielectric layer deposition by the plasma-enhanced chemical vapour deposition (PECVD) technique

[1]Centre for Nano Science and Engineering, Indian Institute of Science, Bengaluru, India

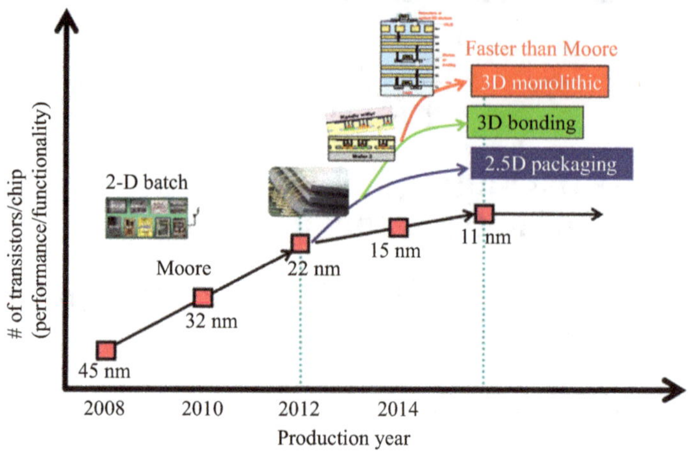

Figure 10.1 Semiconductor and package technology road map [1]

3. Barrier and the seed layer depositions by the physical vapour deposition (PVD) techniques
4. Filling of the vias using electroplating (generally copper is used to fill the vias)
5. Copper over-burden removal by the chemical mechanical polishing (CMP) technique and
6. TSV copper revealing.

TSV offers unique advantages and enables higher number of interconnections, reduced latency, lower inductance, lower capacitance, and permit higher-speed communications, and lower-power-level communication links between circuits [21].

10.1.1 3D integration: manufacturing methods

Out of the various 3D integration schemes followed, one of the more popular ideas is to categorize the process by die or wafer-level stacking: wafer-to-wafer (W2W), chip-to-chip (C2C) and chip-to-wafer (C2W). In the W2W stacking approach, processed wafers were thinned, aligned, bonded and further dies were singulated using dicing. In the C2C stacking approach, processes wafers were diced and known good dies were stacked individually by the bonding process. In the C2W approach, processed wafers were diced and known good dies were identified. Further, these dies were assembled on a new wafer for reconstruction, followed by further stacking and singulation. W2W integration technique offers wafer-level integration and hence has the highest throughput with the best alignment accuracy. However, only the identical size dies will be stacked and there is an inherent risk due to a defective die stacked to the good dies, thereby destroying the whole stack. C2C stacking approach is not a cost-effective solution due to single die processing. However, it is a very well-known technique in high-margin and high-performance device application areas like space. The C2W technique offers unique advantages

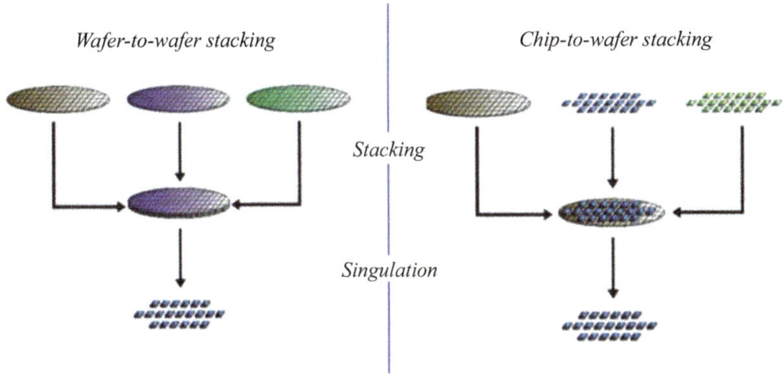

Figure 10.2 Difference between W2W and C2W stacking methodologies [22]

such as heterogeneous stacking, integration of non-identical-sized known good dies and integration of dies with dissimilar materials. Hence, this technique allows stacking of not only silicon-based semiconductor dies but also the other semiconductor devices manufactured at different foundries. Figure 10.2 shows the difference between W2W and C2W stacking approaches [22].

10.2 3D IC technology landscape

As compared to 2D technologies, 3D integration technology offers many advantages like devices with better electrical performance, wider bandwidth, lower power consumption, smaller form factor, higher density, lower cost, and lighter weight. Present-day 3D integration technology can be classified into four major platforms:

1. Package-level 3D integration
2. Chip-level 3D integration or 2.5D stacked IC technology (3D SIC)
3. Within-die 3D integration or 3D system-on-chip (3D SOC)
4. Monolithic 3D IC integration

10.2.1 Package-level 3D integration

Over the past two decades, many technologies for package-level 3D integration have been developed and are in high-volume manufacturing [4,7,12]. Among them, the most popular technology is to stack the processed dies and interconnect them to a printed circuit board (PCB) using wire-bonding technique, as shown in Figure 10.3(a) and (b). The applications include micro electromechanical systems (MEMS) die and application-specific integrated circuit (ASIC) die stacking for sensors [23], as shown in Figure 10.4(a) and the memory die stacking.

Another approach using package-on-package (PoP) stacking technology is shown in Figure 10.4(b). PoP is widely used in mobile devices, where packages

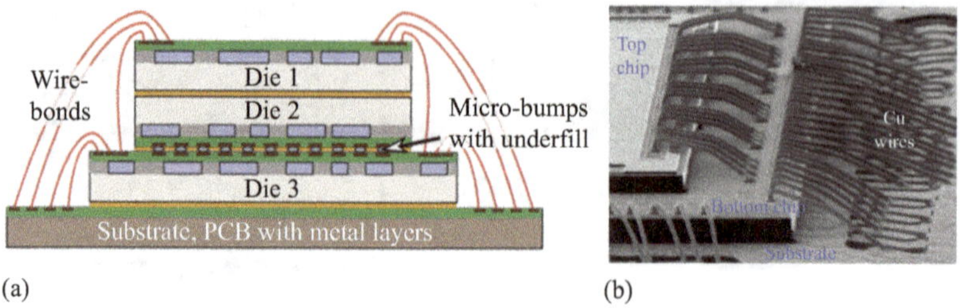

Figure 10.3 Package-level 3D integration: (a) traditional die stacking with wire bonding to the PCB [24] and (b) Amkor's 3D IC packaging with copper wires [8]

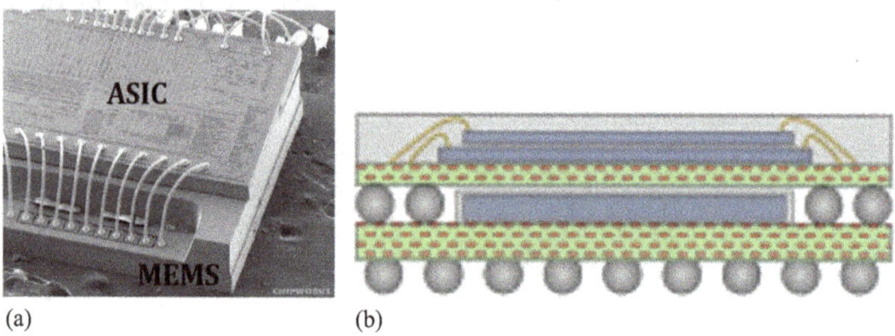

Figure 10.4 (a) MEMS and ASIC die stacking and interconnection using wire bonding [23] and (b) package-on-package stacking approach [7]

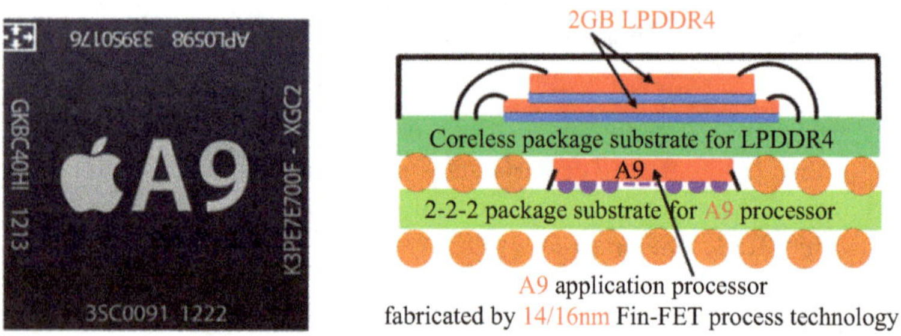

Figure 10.5 Package-on-package in Apple's iPhone 6s smartphone [25]

were stacked together to realize 3D integration. Currently, in mobile devices, memory, external LPDDR DRAM (low-power, double-data rate dynamic random-access memory) and non-volatile flash memories are stacked [7]. Using 3D PoP, memory die and its controller are packaged and stacked independently using the ball grid array (BGA) and interconnected above a PCB [4,25] (Figure 10.5). In many devices, application processor and DRAM were using 3D PoP [25].

Another well-known approach is embedded wafer-level ball grid array [26] (eWLB). This approach involves reconstituting a wafer (from the known good dies), by moulding individual die, placed in an array format into an artificial wafer [26,27].

10.2.2 Chip-level 3D integration

In this approach, the processed dies were stacked and interconnected, enabling high-density face-to-face connections using smaller connections (micro bumps [28]). The electrical connection to the active layer is realized through the silicon substrate of the active device. This methodology is popularly known as through silicon via (TSV) technology [4,7,12] which is the heart of this approach. TSVs are further used for interconnecting the dies or connecting the die to a package substrate using micro bumps, as shown in Figure 10.6(a).

In order to enable high-bandwidth [30] and low latency, multiple dies [31] are stacked side-by-side [32] onto a silicon carrier wafer in this approach. The silicon carrier wafer with TSVs [33] used in this approach is called as an interposer. The interposer wafer does not contain any active elements and hence this technology is popularly called as the 2.5-D technology or 3D-stacked IC technology (3D SIC) (Figure 10.6(a)). The main areas of applications include memory die stacking and interconnecting it with a logic SOC using high-density interposer bus [7,25], as shown in Figure 10.7. It is also used in FPGA die stacking allowing for high-speed connections. Intel's embedded multi-die interconnect bridge (EMIB) technology [29]

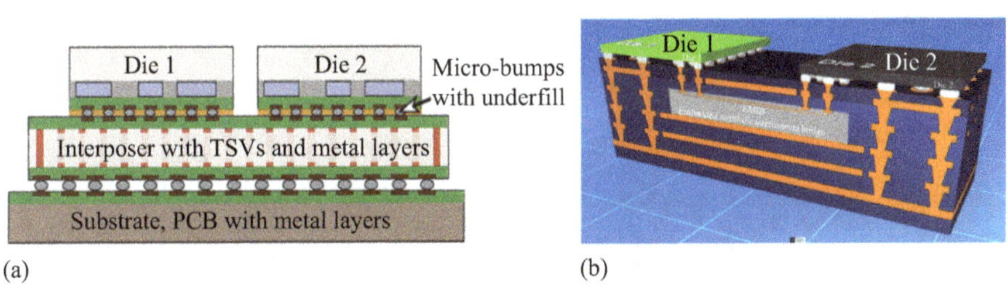

Figure 10.6 Interposer technology: (a) schematic of the interposer technology [24] and (b) Intel's EMIB technology [29]

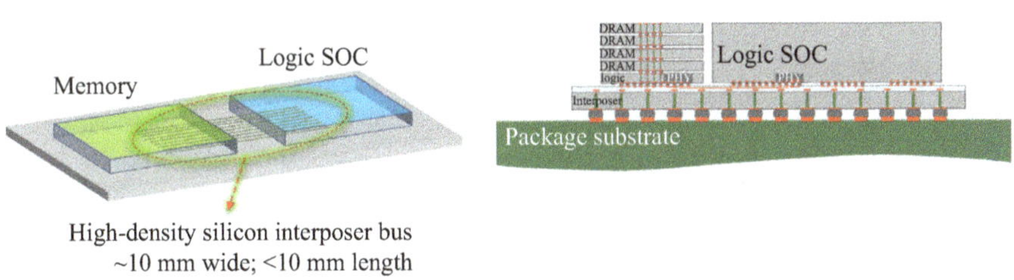

Figure 10.7 Typical interposer technology used in 3D memory stacks with logic SOC die [7]

is also an example of interposer technology wherein, silicon-embedded bridge in the organic package or interposer takes care of the lateral communication between the chips as shown in Figure 10.6(b).

10.2.3 Within-die 3D integration

Within-die 3D integration technology is also known as 3D system-on-chip (3D SOC) technology. In the VLSI system-on-chip technology, a collection of large functional circuit blocks (individual IP's) are connected at the global interconnect level. These functional blocks may have different functional requirements (digital, analogue, CMOS, memory – DRAM/SRAM, etc.) and need to be integrated to perform complex functions. Instead of having multiple functional blocks as individual ICs on a PCB, SoC is split into two or more parts and was realized with optimum 3D IC technologies at lower cost per function. Further, these functional circuit blocks were stacked and interconnected using 3D interconnect processes. An example of such technology is Amkor's Double POSSUM [25] multi-stacked die configuration enabling the integration of ASICs, MEMS, memory, microcontrollers, etc., as shown in Figure 10.8.

The fabrication methodology using this approach is illustrated in Figure 10.9. In this approach, W2W bonding methodology is used to align and bond the cleaned and plasma-treated processed wafers at room temperature (RT), as shown in Figure 10.9. The top wafer can be thinned, and the silicon back-side electrical contacts were created to interconnect top and bottom wafers. In this approach, W2W overlay bonding accuracies are very critical.

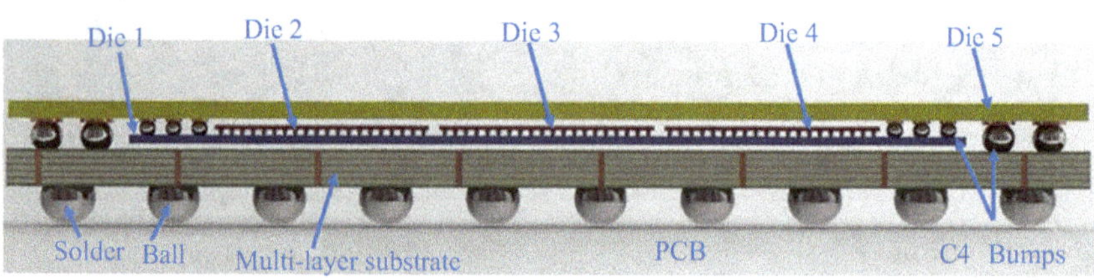

Figure 10.8 Amkor's multi-chip-to-chip interconnects [25]

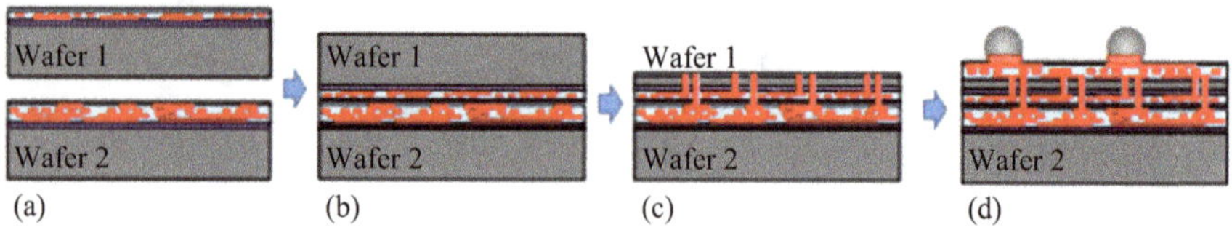

Figure 10.9 (a) Processed device wafers, (b) RT alignment and bonding of both the wafers, (c) thinning of the top wafer (wafer 1) and realization of TSVs and (d) back-end processing for bonding to a final PCB [7]

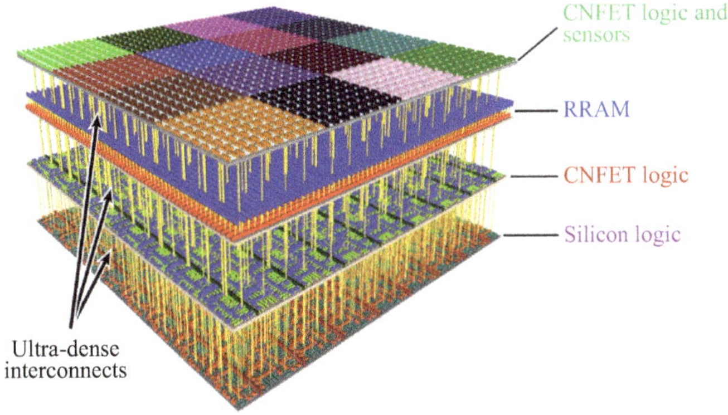

Figure 10.10 Monolithic integration of four functional layers interconnected through dense vertical inter-connects [34]

10.2.4 Monolithic 3D integration

The sequential 3D process of bonding an un-patterned semiconductor layer to processed wafer for sequential building of second layer of devices is referred to as monolithic 3D integration [7,34–42]. Even though the sequential processing of second device layer on top of processed layers technique offers higher yield, this process is severely impacted by thermal limits [7]. Hence, identical device layer stacking is not feasible using this sequential approach. This approach is appropriate for 3D heterogeneous integration technologies, where dissimilar material property layers need to be stacked and interconnected using variety of interconnection technologies (not limiting to only TSV technology). An example of monolithically integrated vertical layers is shown in Figure 10.10.

10.3 3D heterogeneous integration

In the past 50 years, there has been a consistent growth in the semiconductor industry which was guided by the Moore's law [6]. Due to the existing limitations in the silicon-processing technologies, further scaling of devices has become technically and financially challenging [43]. Consequently, the focus has shifted towards scaling of systems for which there is a need to investigate alternative technologies. With the advent of mobile computing and communications market [44], there has been a sudden increase in demand for miniaturized systems with extended functionality and better performance. To address these needs, researchers are focusing on comparatively less explored assembly and packaging technologies [45]. As discussed in previous reports [5,10,46,47,48], over the last 45 years silicon has scaled by a factor of 1,000, whereas the printed circuit board (PCB) trace pitch, ball grid array pitch and the bump pitch to packages have scaled only 4 times. One of the approaches to address this issue is to utilize the third dimension for scaling by heterogeneous integration [8–10]. Heterogeneous integration involves the integration and interconnection of various components manufactured separately in a single stack so that the integrated system can be more powerful, efficient and functionally diverse [45,49]. Currently, various technologies are being developed for heterogeneous integration. Such 3D integrated heterogeneous systems would provide complex functionality [50,51], for the upcoming plethora of market demands in Internet of Things (IoT) [17,52]-based smart devices with big data cloud applications [53] and healthcare-based applications [54].

Though there are several reports on 3D integration technologies, which have been for stacking and interconnection mechanisms, all these technologies are confined to silicon-based devices. With the existing processing technologies, it is difficult to integrate complex combinations of GaN-HEMTs, MEMS, microfluidics, optical devices and CMOS at the wafer-level. Moreover, due to the cross-contamination issues, most of these devices were never accepted in the standard CMOS foundries. These devices were very expensive compared to the existing CMOS-based devices. To address these challenges, there were efforts towards developing processing technologies for 3D heterogeneous integration by post fab vertical stacking techniques.

To address these challenges, in Sections 10.4 and 10.5 we demonstrate techniques to vertically stack ultra-thin layers of different devices from different technologies fabricated using different materials. In Section 10.4, we developed fabrication techniques for the transfer of ultra-thin silicon (UTSi) layers of thickness ~1.45 μm on to a foreign substrate using low-temperature bonding approach. In order to enable this transfer, we have analysed and resolved the associated stress and reported a scheme of mitigating the stress issues. This resulted in a crack-free transfer of ultra-thin silicon layer. Based on this technique, we have demonstrated three-layer stacking of the ultra-thin silicon layers with functional MOSFETs in each layer using low-temperature epoxy bonding processes. Electrical characterization results of NMOS/PMOS devices in each layer have been presented and compared before and after the transfer.

Further, 3D heterogeneous integration of miniaturised hybrid systems is discussed in Section 10.5. In this 3D packaging technology, we were able to integrate photodetector, optical filters, microfluidic chip with microheater and an LED. The functionality of the heterogeneous system was demonstrated by measuring the increase in photodetector current due to the fluorescence property of the Rhodamine B and Rhodamine 6G. Finally, we have demonstrated an approach for interconnecting the stacked layers using the non-conventional inkjet printing technique.

10.4 3D stacking of ultra-thin silicon layers with functional MOS devices

Fabrication techniques for the transfer of ultra-thin silicon (UTSi) layers of thickness ~1.45 μm on to a foreign substrate using low-temperature bonding approach are demonstrated. Using this technique, three-layer stacking of the ultra-thin silicon layers with functional MOSFETs in each layer using low-temperature transient liquid-phase bonding-based processes are demonstrated.

10.4.1 Transistor fabrication NMOS and PMOS

Self-aligned n-channel and p-channel metal-oxide-semiconductor field-effect transistors (MOSFETs) were fabricated on p-type and n-type SOI wafers, respectively, with 2-μm-thick device layer, 1-μm-thick buried oxide (BOX) layer and handle layer of thickness 450 μm. The fabrication process flows utilized for both the NMOS and PMOS layers are briefly shown in Figure 10.11. To determine the diffusion schedules for the source/drain regions as well as to achieve the required sheet resistance for the contact regions, TCAD Athena simulations were performed. The device layer thickness post transistor fabrication was ~1.4 μm.

10.4.1.1 Ultra-thin silicon transfer using epoxy and Au–In TLP bonding

The above transistors were passivated with ~90 nm low-temperature (140 °C) PECVD oxide, leaving only the contact pad areas open for interconnects, as shown in Figures 10.12(a) and (a). The device side of the SOI substrate was temporarily bonded to a carrier glass substrate using temporary bonding material Brewer Science HT: 10:10 as shown in Figures 10.12(b) and 10.13(b). During the temporary bonding process, the backside of the SOI substrate (i.e. handle layer) was protected using photoresist (PR). This was done to prevent contamination of the handle layer surface by HT: 10.10 residues, which was observed to interfere with the subsequent handle layer etch. The PR was removed and the handle layer of the SOI wafer was etched away using isotropic dry etch process in the deep reactive ion etching (DRIE) tool. The etch stopped at the BOX layer as seen in Figures 10.12(c) and 10.13(c).

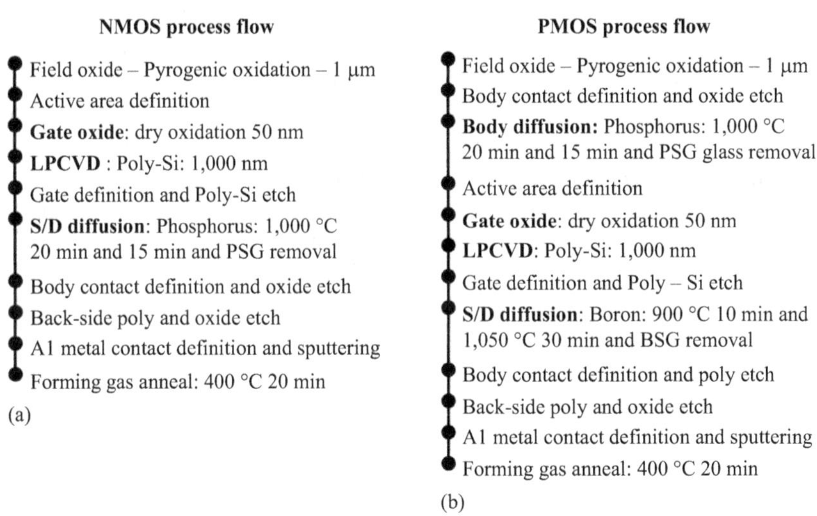

Figure 10.11 Process flow for the NMOS and PMOS MOSFET fabrication

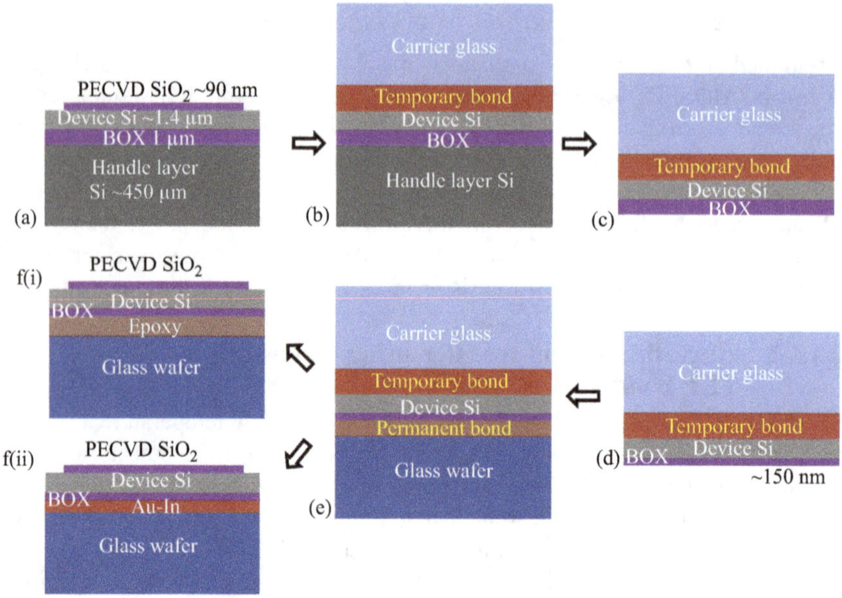

Figure 10.12 Schematic of the process flow for the ultra-thin silicon fabrication: (a) ~1.4-μm-thick Si device layer after transistor fabrication, (b) temporary bonding of glass carrier, (c) removal of the SOI handle layer, (d) controlled etch of BOX, (e) permanent bonding of glass wafer, f(i) final device with epoxy bonding and f(ii) final device with Au–In bonding

During further processing, cracks were observed on the thin device layer as shown in Figure 10.13(d). Further, it was also noted that these cracks were present only for the transfer of 2-μm device layers. Similar transfer of 10-μm device layers did not result in such cracks. Therefore, these cracks were attributed to the stress arising from the deposited layers on the top and the BOX at the bottom of the device silicon layer. By sequentially removing the different layers from a dummy SOI wafer and measuring the curvature using the kSA MOS stress measurement tool, the stress of the buried oxide layer was found to be ~ 290.5 MPa compressive. After handle layer removal, the compressive BOX relaxes by stretching. This leads to cracking of the silicon device layer. Hence, it was concluded that the solution for reducing the residual stress effect is to reduce the thickness of the BOX layer. Removing the BOX layer completely, however, led to the wrinkles after transfer, as shown in Figure 10.13(e). These wrinkles can be attributed to the remaining stress due to the bonding layer and the layers on the top side of the thin silicon device layer. Crack-free and wrinkle-free thin silicon, as shown in Figures 10.12(d) and 10.13(g), could be achieved when a controlled wet etching of the buried oxide layer was carried out to reduce it to ~150 nm. Stress cracks are observed when more than 200 nm of BOX was left as shown in Figure 10.13(f). We observed that leaving

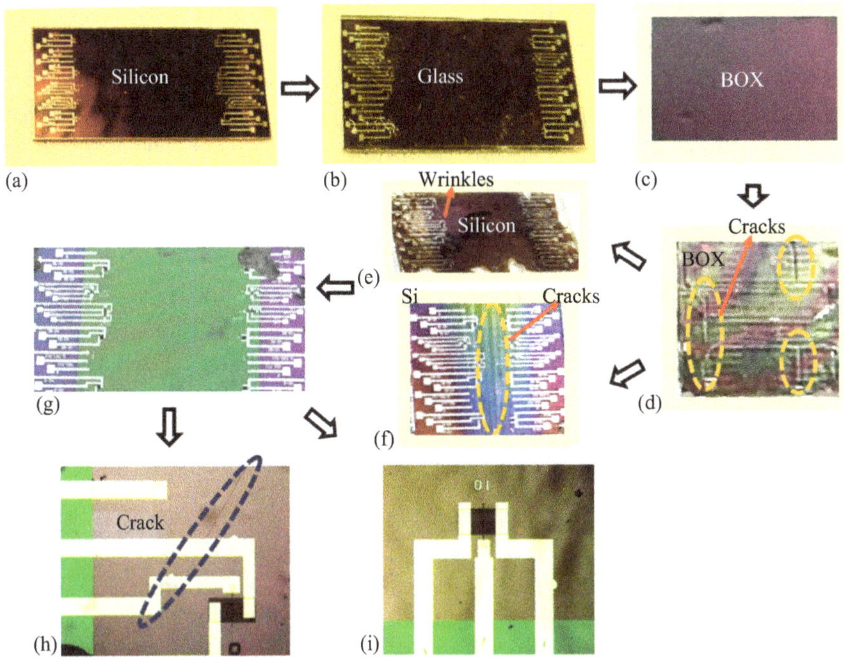

Figure 10.13 *(a) Top view of the as fabricated transistor SOI die, (b) glass carrier wafer temporary bonding over the SOI die, (c) image showing the BOX after handle layer removal using DRIE, (d) stress cracks were observed after the transfer process upon further processing, (e) wrinkles were observed on the transferred devices after the complete removal of BOX, (f) cracks were observed on the transferred devices with partial etch of the BOX, (g) photograph of a decently bonded sample without any wrinkles or cracks, (h) observed micro-cracks on a transistor after transferring to a foreign substrate using Au–In bonding and (i) micrograph of the epoxy-bonded sample without any cracks*

between 100 and 200 nm BOX on the silicon device layer balances the stress induced by the bonding layer and the different layers present on the front side of the device layer.

The ultra-thin silicon device layer was transferred to the permanent package substrate (e.g. glass), using two approaches as shown in Figure 10.12(e). For transfer, we explored: (I) RT or low temperature epoxy bonding and (II) low temperature Au–In bonding as described in the following sections.

Method (I)

In this epoxy bonding approach, EPO-TEK UJ1190 was spin-coated at 1,600 rpm on a cleaned glass substrate. The ultra-thin Si device layer on a temporary substrate was brought in contact with the glass substrate and exposed to UV light using the MJB4 lithography tool for 60 s. The exposure was completed in 15 cycles at intervals of 30 s at RT. This resulted in a permanent bond between the ultra-thin silicon and the glass substrate, as shown in Figure 10.12(f(i)). The carrier substrate was then separated using thermo-mechanical debonding on a hot plate at 150 °C as recommended by the manufacturer. The remaining temporary bonding material was thoroughly cleaned using the wafer bond remover from Brewer Science.

Method (II)

For Au–In bond, Cr/Au (20/100 nm thickness) was sputtered on the back of ultra-thin silicon and 1 μm indium was evaporated on the foreign substrate (in this case glass substrate). These two substrates were brought together and subjected to 600 N force at a temperature of 150 °C for 30 min in a bonding tool under vacuum. Prior to the bonding, oxide was removed from the indium surface using a dilute HCl dip just. After the bonding process, the carrier wafer was separated from the stack by the thermo-mechanical debonding method. This was followed by cleaning the remaining bonding material. The final device is shown in Figure 10.12(f(ii)).

From the SEM/FIB pictures in Figure 10.14(a), the post-bonding thickness of Au–In alloy was 1.3 μm whereas, the epoxy thickness was ~25 μm, as shown in Figure 10.14(b). For Au–In bond, the SEM/FIB image clearly shows voids. The incomplete bonding is attributed to the low bonding temperature. Upon careful observation under the microscope, micro-cracks on the ultra-thin silicon layer were observed at random location for the case of the indium-bonded sample, as shown in Figure 10.13(h). These cracks could result from large bonding forces used in the bonding process. On the other hand, in the case of epoxy bonding cracks were absent (see Figure 10.13(i)) and the yield was almost 100%. Absence of cracks for epoxy bonding can be expected due to lower stiffness of epoxy and lower bonding force. Further, the epoxy bonding process is relatively cheap compared to the TLP bonding which involves two metal deposition steps and specialized bond tool requirements.

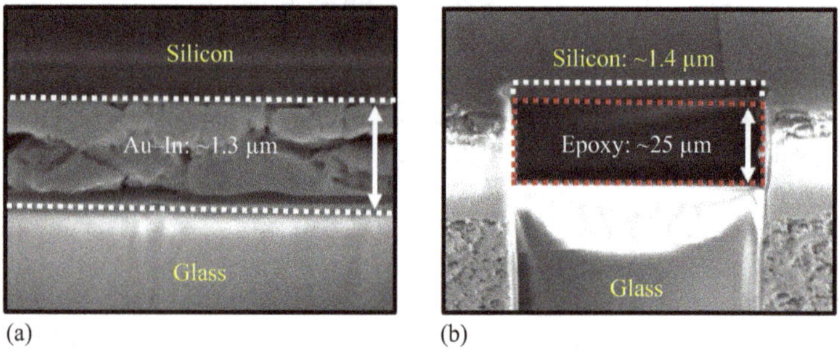

Figure 10.14 FIB images of the bond interface using (a) gold–indium transient liquid-phase bond with Au–In alloy thickness of ~1.3 μm and (b) epoxy-bonded sample with epoxy thickness of ~25 μm

10.4.2 Vertical stacking process flow

Based on the better results obtained with the epoxy bonding, we adopted this approach for vertical stacking of multiple layers of ultra-thin silicon substrates. The stacking scheme remains the same as discussed in the previous section and is shown in Figure 10.15. It involves temporary bonding of a die to the carrier wafer, backside silicon removal and controlled oxide etching followed by permanent bonding, as shown in Figure 10.15(b) and (c). For further stacking, a SU-8 2002 layer was used to planarize the stack and provide dielectric isolation. Using a photolithography step, the contact pad areas were opened, all the other areas were left with ~1–2-µm-thick SU-8. Before bonding the second layer, SU-8 surface was

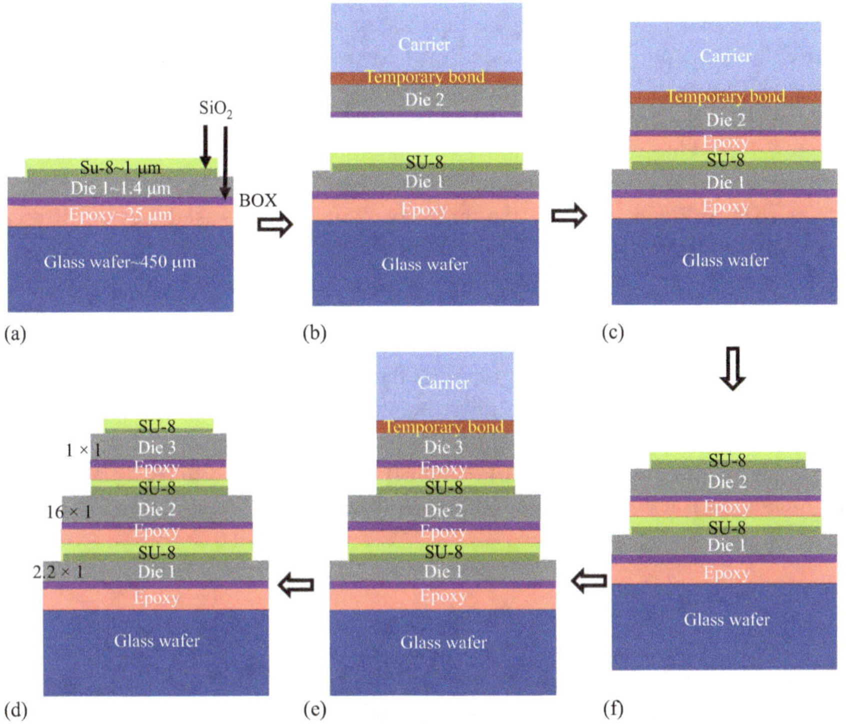

Figure 10.15 Schematic of the process flow for vertical homogenous stack: (a) transistor fabrication and transfer to the glass wafer using epoxy bonding methodology, (b) preparation of second die for the transfer over the first one – temporary bonding of carrier glass to SOI, etching of handle layer silicon and partial etch of oxide, (c) permanent bonding of the device stack using epoxy, (d) separation of the glass carrier and thorough cleaning, (e) another ultra-thin silicon permanent bonding to the existing stack and (f) final stack

treated with low-power oxygen plasma. This step is required for SU-8 surface activation which allows proper spreading of epoxy and better adhesion.

The second layer of ultra-thin Si device layer was prepared using the same technique as described above. Silicon being not transparent to the UV light, UJ1190 epoxy cannot be used for further stacking (i.e. second and third layer). Hence, a thermally curable epoxy (EPO-TEK 377) was explored and used for further bonding/stacking. The selected epoxy has very low viscosity and low outgassing, both properties being beneficial for the stacking process. The epoxy was spin-coated on to the backside (BOX) of the ultra-thin Si device layer at 1,800 rpm for a minute. This was followed by bonding on the hot plate at 135 °C for 30 min. After bonding, the carrier wafer was removed using thermo-mechanical debonding method.

In preparation for the transfer of third layer, 1-μm-thick SU-8 was spin-coated on the stack and patterned over the device region, as shown in Figure 10.15(d). Finally, the third die was also bonded to the existing stack with devices facing to the top as shown in Figure 10.15(e). Thermo-mechanical debonding of the carrier wafer leads to the completion of the stacking process, as seen in Figure 10.15(f). Successful stacking was also achieved by bonding the first layer using thermally cured epoxy (EPO-TEK 377) instead of UV-cured epoxy (UJ1190).

As seen in Figure 10.15(f), the three layers in the stack-used dies of different dimensions are as follows: die 1 was 2.2 cm × 1 cm, die 2 was 1.6 cm × 1 cm and die 3 was 1 cm × 1 cm. The lengths of the dies were designed to be different to ensure that both sides of the die can be accessed from above to probe the device contact pads, as seen in Figure 10.16(a). In the fabricated stack, the bottom-most layer consisted of PMOS devices, whereas the middle and the top layer consists of NMOS devices. Figure 10.16(a) shows the photograph of the stack, where the top third layer with devices and the interconnect pads on both sides of all the three layers can be seen. Figure 10.16(b) demonstrates the cross-sectional FIB image of the three-layer stack with epoxy and silicon layers.

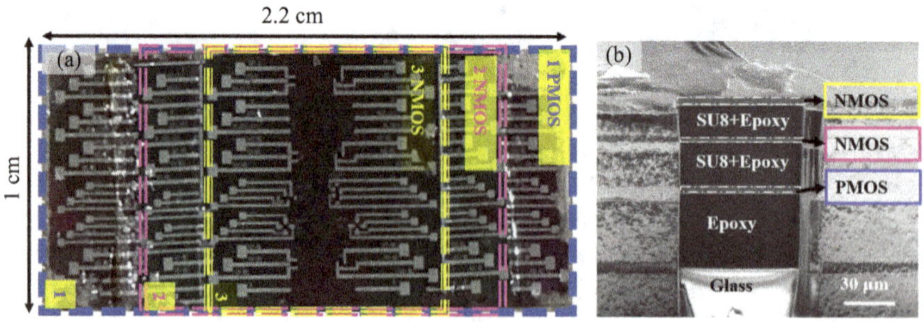

Figure 10.16 Image of the three-layer stack: (a) top view photograph showing all the three bonded layers and (b) cross-sectional FIB image showing all the three functional NMOS and PMOS layers

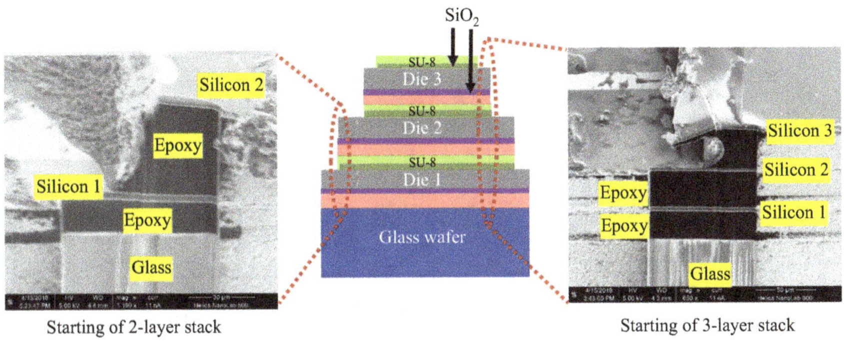

Figure 10.17 Epoxy spread at the bond interfaces to the subsequent layers. Keep-out zone in both the layers was ~ 40 μm

10.4.3 Keep-out zone

FIB cross-sections at various positions on the stack can be seen in Figure 10.17. Image in the left shows the edge of the two-layer stack and the image on the right shows after stacking the three layers. From the images it can be observed the effect of the epoxy on the edges of the bottom and the top die due to spin coating and bonding is ~40 μm. Hence, the keep-out zone for the devices on the edges of both the stacked silicon layers would be ~40 μm from the respective edge.

10.4.4 Characterization of the transferred devices

10.4.4.1 Diffusion measurements

The PMOS and NMOS transistor (gate width = 200 μm, length = 20 μm) structures were simulated using TCAD Athena tool to identify the right diffusion parameters for fabricating the experimental devices. Based on the simulation study, phosphorus pre-deposition and drive-in both were performed at 1,000 °C for 20 min and 15 min, respectively, to realize the S/D regions in the NMOS transistor. For the PMOS, the boron pre-deposition was performed at 900 °C for 10 min followed by drive-in at 1,050 °C for 30 min. The sheet resistance (SR) of fabricated NMOS and PMOS were ~4.6 Ω/□ and ~30 Ω/□, respectively.

10.4.4.2 DC electrical measurements

Electrical DC measurements of the NMOS transistors were performed before and after the transfer to foreign substrate using both the 'Au–In' and 'epoxy' bonding approaches. I_d–V_d (output) measurements were performed at $V_{gs} = 0$ to 7 V in steps of 1 V and $V_d = 0$ to 5 V. I_d–V_g (transfer) measurements were performed at $V_{ds} =$ 7 V and $V_{gs} = -2$ to 5 V. Figure 10.18(a) and (c) summarizes the output characteristics of both the bonding approaches. It can be observed that at lower V_g before and after transfer, the I_d values were identical in both the bonding cases. However, in both cases, reduction in current can be observed for $V_{gs} > 3$ V. This

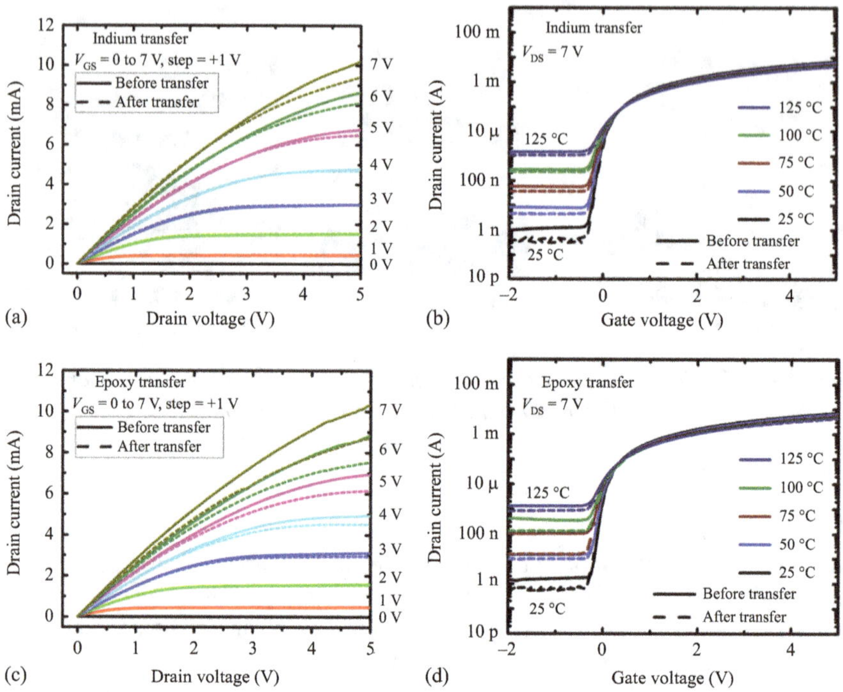

Figure 10.18 Electrical characteristics of the 20 μm channel length NMOS device before and after transfer to foreign substrate. Indium bond (a) output characteristics at RT and (b) transfer characteristics at various temperatures. Epoxy bond (c) output characteristics at RT and (d) transfer characteristics at various temperatures

could be attributed to the poor thermal conductivity of the glass substrate. Also, it can be observed that reduction in current is lower in the case of Au–In bond as compared to the epoxy bond at higher gate voltages; This can be attributed to better lateral dissipation of heat through the Au–In alloy. Whereas the epoxy being a poor conductor of heat, this will lead to higher device temperatures and hence the observed current reduction.

Figure 10.18(b) and (d) summarizes the transfer characteristics (I_d versus V_g) at various temperatures (25 °C to 125 °C, in steps of 25 °C), respectively, for the indium bonding and epoxy bonding cases. The off-currents after the transfer are compared with those obtained before transfer for both cases. As expected, the off-currents are observed to increase with temperature. For both the bonding approaches, the transferred devices showed lower off-currents than the device on the SOI substrate. In the case of indium bonding, the extracted threshold voltage (V_t) before stacking was −86.01 mV and after stacking it was −67.4 mV. In the case of the epoxy bonding, the extracted V_t before stacking was −53.9 mV and after stacking it is equal to −43.3 mV. These results indicate about 20% shift in the threshold voltage for both the approaches.

10.4.4.3 Negative differential resistance

To confirm that the reduction in on-current is due to the temperature rise in both the bonding techniques, we carried out the I_d–V_d measurements at elevated temperature and up to higher drain voltages (V_d = 15 V) for V_g = 10 V on both Au–In and epoxy-bonded NMOS device samples. These measurements were carried out between RT (25 °C) and 125 °C at an interval of 25 °C. Similar measurements were also carried out between 25 °C and 200 °C for NMOS devices fabricated on the SOI wafer and retained without transferring them on to foreign substrate (we designate them as the 'Original NMOS SOI device'). Figure 10.19(a) and (b) shows the I_d–V_d measurements of the NMOS transistor transferred to glass using Au–In and epoxy bonding approaches, respectively. Figure 10.19(c) shows the results obtained on the

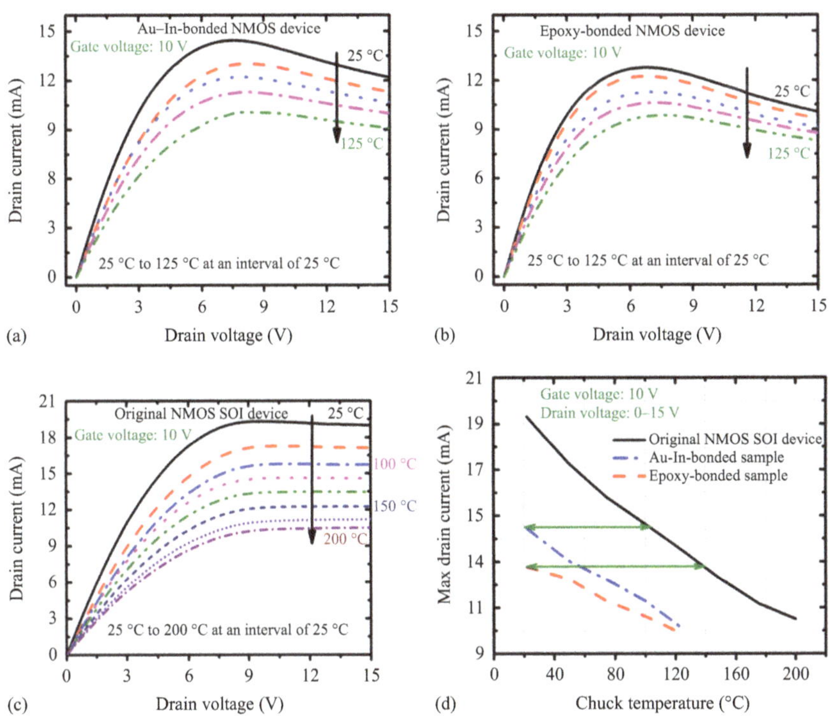

Figure 10.19 Device behaviour at elevated temperatures: (a) output characteristics of the Au–In-bonded sample, (b) output characteristics of the epoxy-bonded sample (c) output characteristics of the original NMOS SOI sample and (d) mapping the temperature of the Au–In bond and the epoxy with respect to the as fabricated sample using the maximum current obtained

original NMOS SOI device. As expected, in these devices, due to the rise in the applied chuck temperature the drain current (saturation current) is reduced at higher temperatures. The decrease in the saturation current occurs because of the decrease in the mobility of charge carriers in the channel caused by the increased electron–phonon scattering at higher temperatures. For the transferred substrates we also observe a region of negative differential resistance (NDR) region [55,56]. Very minimal effect of the same can be observed in the original NMOS SOI device, as shown in Figure 10.19(c).

Once the MOSFETs are transferred to a glass substrate through Au–In or epoxy bonding, the effect of poor thermal dissipation on the output characteristics becomes prominent, even at RTs. In both cases, with increase of drain voltage, the drain current increases to a maximum and then starts to decrease leading to an NDR region, as can be seen in Figure 10.19(a) and (b). This happens as the power through the device increases the device channel heats up due to the poor thermal dissipation through the bonding material and the glass substrate. On comparing the MOSFET characteristics in case of Au–In bonding with those of epoxy bonding, the epoxy being an even poorer thermal conductor, the reduction in device current is further enhanced.

To measure the rise in temperature, the maximum current obtained in all the three cases was recorded and plotted in Figure 10.19(d). It may be noted from the graph, that the maximum current for RT measurements in the Au–In-bonded devices approximately correspond to those obtained at 100 °C in the original NMOS SOI device. Whereas, for the epoxy bonding approach, the maximum current value at RT corresponds to ~150 °C in the original NMOS SOI device, as shown in Figure 10.19(d). Hence, the Au–In-bonded sample would be at 100 °C and the epoxy sample would be 150 °C at the maximum current-carrying conditions. As V_d increases further, the power dissipation in the channel increases leading to further rise in temperature and hence, reduction in current with an increase in V_d.

To understand the NDR effect on the post-transfer devices, TCAD Atlas simulations were conducted on the NMOS SOI transistor. Steady-state I_d–V_g simulation results at higher drain voltages on SOI wafer (with no vertical stacking) are shown in Figure 10.20(a). Simulated measurements were identified to be in close match with the original fabricated NMOS SOI device measurements as shown in Figure 10.20(a). The Au–In- and epoxy-bonded devices results (after fabrication) were also plotted and at higher drain voltages where a significant negative differential resistance can be observed. The same has been verified through thermodynamic MOSFET simulations using TCAD Atlas as shown in the same figure. We suspect the significant NDR effect in the post-transferred devices is due to the increase in the channel temperature during operation. This could be due to the lack of proper thermal dissipation through the substrate. To verify the same, we have simulated the increase in the channel temperature with increase of the drain voltages for all the devices at RT. as shown in Figure 10.20(b).

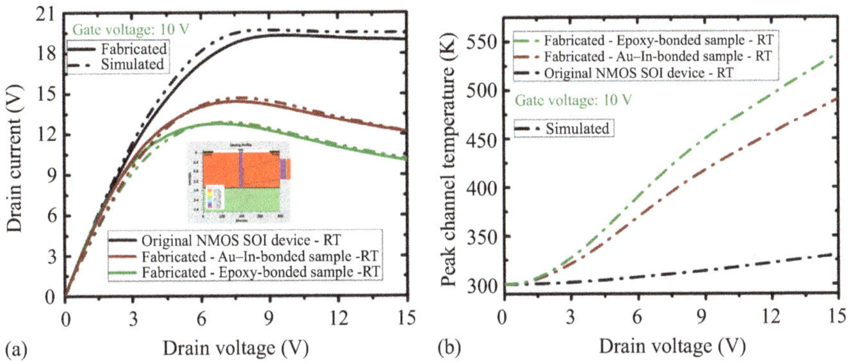

Figure 10.20 (a) Comparison of the simulated and fabricated output characteristics of pre- and post transfer of the NMOS SOI device to glass substrate using Au–In bond and the epoxy bond and (b) channel temperature simulation measurements with change in drain voltages, indicating significant channel temperature rise post transfer

10.4.4.4 3-Layer stack with functional devices and DC measurements

Having established the impact of both the bonding processes on the device characteristics, due to the low temperature and cost-effectiveness of the epoxy-bonding process, it was further utilized for three-layer stacking. Each of these pre-fabricated layers has NMOS or PMOS devices. The first layer with PMOS devices, output and the transfer characteristics before and after transfer can be seen in Figure 10.21(a) and (b). The second and the third NMOS layer results can be seen, respectively, in (i) Figure 10.21(c), (d) and (ii) Figure 10.21(e) and (f). Also, the stack was subjected to elevated temperatures and recorded the transfer characteristics to understand the behaviour of the devices upon stacking. No significant changes have been observed in the device behaviour upon stacking the functional MOSFET layers.

10.4.4.5 5-Layer ultra-thin silicon stack

In this section, we demonstrate the feasibility of extending this bonding process to a greater number of layers. Figure 10.22 shows the FIB images of functional 5-layer stack with three 10-μm-thick layers and two 2-μm-thick layers. However, due to the step design, the number of layers would be limited by the die size. To address this issue and to improve the silicon real estate, the die sizes can be replaced with equal-sized dies incorporating metal-filled TSV-based interconnects. Silicon thickness being very low (~1.4 μm thick), the 'via' opening can be made smaller and lesser metal and barrier fills would be needed. However, the epoxy thickness is around 25 μm, which is still to be reduced further to get the full benefit of this process.

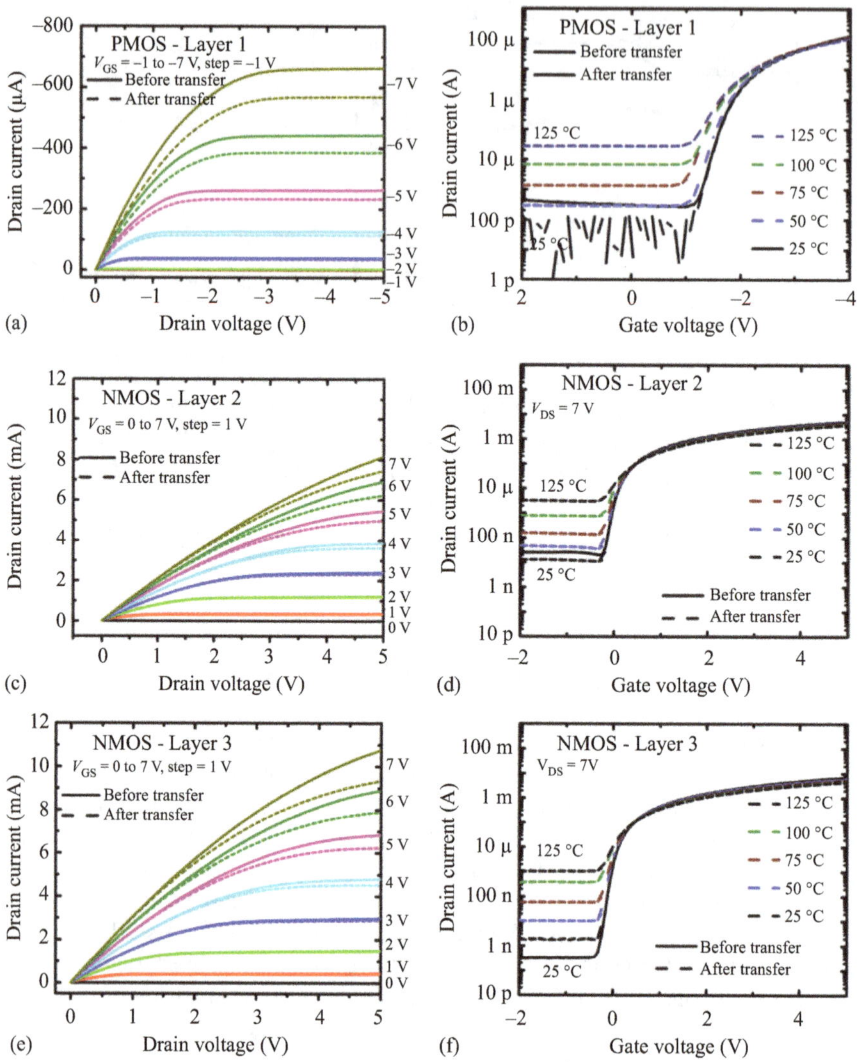

Figure 10.21 Electrical characteristics of the three-layer stack before and after stacking comparison: (a) output characteristics of the first/bottom (PMOS) layer in the stack, (b) transfer characteristics of the first/bottom (PMOS) layer in the stack at RT and elevated temperatures, (c) output characteristics of the second (NMOS) layer in the stack, (d) transfer characteristics of the second (NMOS) layer in the stack at RT and elevated temperatures, (e) output characteristics of the third/top (NMOS) layer in the stack and (f) transfer characteristics of the third/top (NMOS) layer in the stack at RT and elevated temperatures

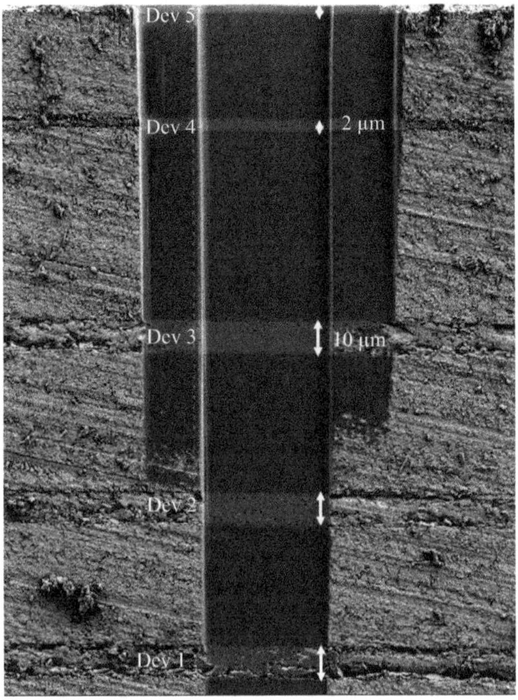

Figure 10.22 Five-layer stack with functional ultra-thin Si layers. First two layers of 2 μm thickness and rest three layers of 10 μm thickness

10.4.5 Reliability measurements on the ultra-thin silicon stack

10.4.5.1 Fabrication of two-layer stacks with functional devices

Two-layer stacked devices were fabricated as shown in Figure 10.23 using the process flow discussed in the earlier sections. The first stack has PMOS devices in both the layers, whereas the second stack has NMOS in one and the PMOS in the other layer.

10.4.5.2 Reliability tests

The fabricated stacks were subjected to the industry standard Joint Electron Device Engineering Council (JEDEC) standard electrical reliability tests JESD22-A120A and JESD22-A104E-G. The first test JESD22-A120A for the removal of moisture diffusivity involves the exposure to 85 °C for 24 h and for the water solubility and diffusivity the stacks were exposed to 85 °C with 60 % RH for 65 h. (However, the actual test is supposed to be done for 168 h.) The second test (A104E-G) is a temperature cycling test to determine the device capability in withstanding extreme thermal shocks. This test involves sudden alternating low

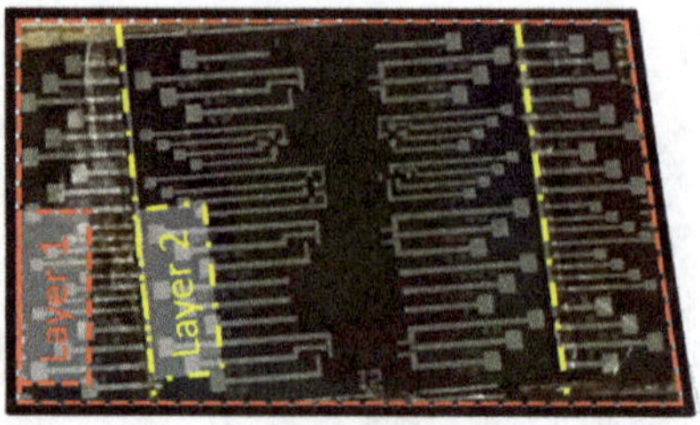

Figure 10.23 Photograph of the two-layer stacked die with layer 1 of 1.6 cm × 1 cm and layer 2 of 1 cm × 1 cm

(−40 °C) and high (+125 °C) temperatures for 10 cycles each for a period of 30 min at each temperature. After subjecting to each of these tests, the device behaviour has been tested by measuring the electrical DC characteristics of fabricated transistors.

10.4.5.3 DC electrical measurements

Fabricated NMOS and PMOS transistors before and after the transfer to glass substrate were electrically characterized using Agilent B1500 device analyser and compared, as shown in Figures 10.24 and 10.25. Identical devices were measured and compared after each stage of fabrication and the reliability tests. Figure 10.24 shows a typical set of DC measurements of the two-layer stack with both the layers having PMOS devices.

The channel lengths of the fabricated PMOS Layer 1 and layer 2 were 6 and 20 μm, respectively. The output characteristics with $V_d = 0$ to -5 V and V_g from -3 to -7 V in steps of -1 V and the transfer characteristics with $V_d = 0$ to -5 V were measured for both the layers. After the first part of the JESD22-A120A thermal stability (85 °C for 24 h) test and the second JESD22-A104E-G test, the devices were characterized and compared as shown in Figure 10.24. It may be noted that after these intensive tests there is no significant change in the device behaviour. However, at higher gate voltages slight reduction in the drain current has been observed. This could be attributed to using glass as the permanent substrate, which is a poor thermal conductor.

Images in Figure 10.25 show another two-layer stack DC measurement with Layer 1 NMOS and the layer 2 PMOS devices. The NMOS and PMOS in Layer 1 and 2 are with channel lengths of 10 μm. The NMOS output characteristics with $V_d = 0$ to 5 V and V_g from 0 to 4 V in steps of 1 V and the transfer characteristics with $V_d = -2$ to 5 V were measured and presented in Figure 10.25. The PMOS output characteristics with $V_d = 0$ to -5 V and V_g from -3 to -7 V in steps of -1 V and the transfer characteristics with $V_d = 2$ to -5 V were performed. After the second part of the JESD22-A120A (water solubility test 85 °C with 60 % RH for 65 h) test, the devices were measured and compared as shown in Figure 10.25. After each of these intensive tests, there it may be noted that there is no significant change in the device behaviour.

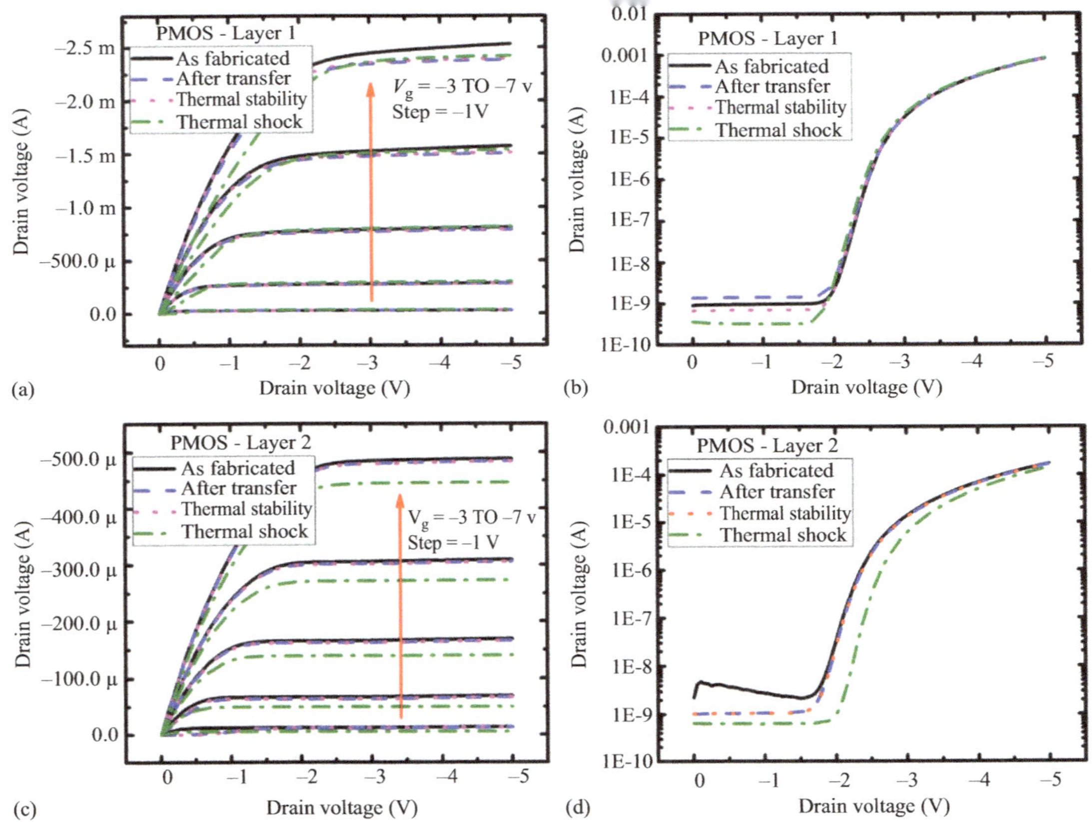

Figure 10.24 *DC measurements of before and after transfer and after the reliability tests (Thermal stability – JESD22-A120A and Thermal shock – JESD22-A104E-G) of both PMOS devices in Layer 1 and 2: (a) output characteristics with varying gate voltage for Layer 1, (b) transfer characteristics for Layer 1, (c) output characteristics with varying gate voltage for Layer 2 and (d) transfer characteristics for Layer 2*

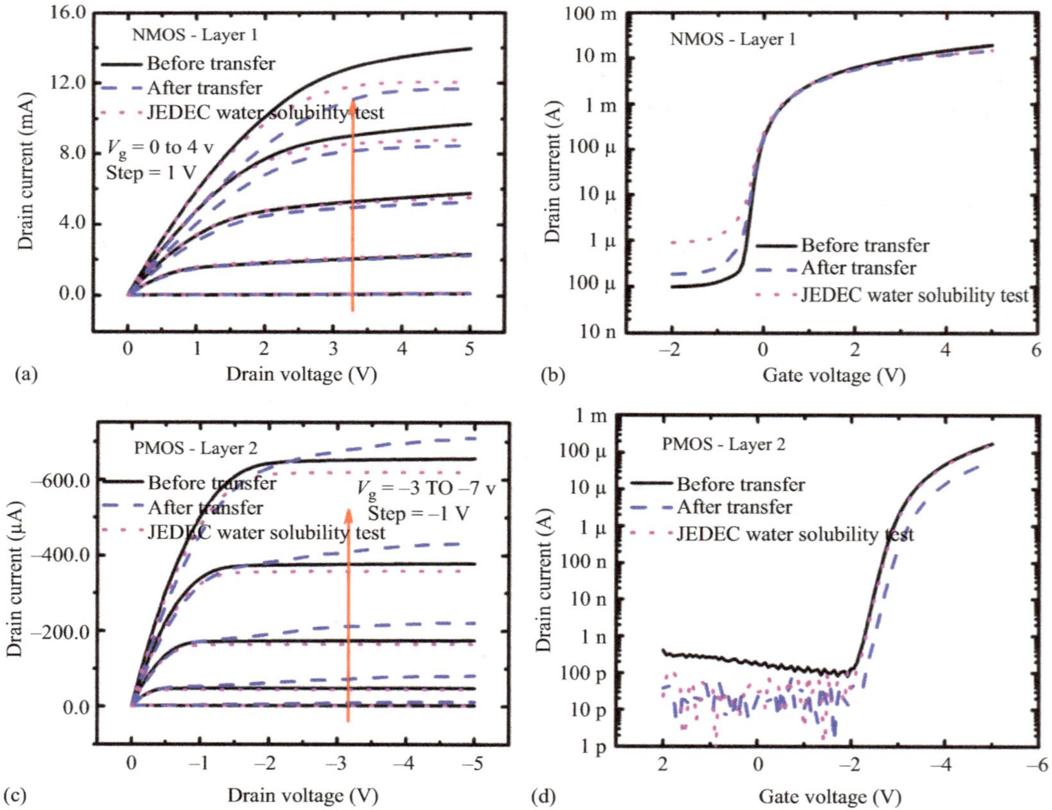

Figure 10.25 *DC measurements of before and after transfer and after the reliability tests (Solubility test – JESD22-A120A) of the NMOS device in Layer 1 and PMOS device in layer 2. (a) Output characteristics with varying Vg for Layer 1 (b) Transfer characteristics for Layer 1 (c) Output characteristics with varying Vg for Layer 2 (d) Transfer characteristics for Layer 2*

10.5 3D integration of heterogeneous dies for fluorescent detection

Scaling by 3D integration of various heterogeneous components enables miniaturized systems. However, heterogeneous system integration is challenging due to the dissimilarities in materials and process used in the fabrication of individual components. In this chapter, we demonstrate a simple 3D integration method for miniaturisation of systems. Various components of the system were stacked using SU-8-based planarization and epoxy-based bonding. Spacer dielectric (SU-8) was patterned using photolithography for the formation of interconnect vias. Electrical interconnects over the large topography between the layers was formed by screen-printing of silver nanoparticle epoxy. Using this integration technique, we demonstrate a fluorescence sensing platform consisting of silicon photodetector, plastic optical filters, commercial LED and a glass microheater chip. This chapter resolves several fabrication challenges of planarization, stacking and interconnection of these divergent chips. For example, process incompatibility of the plastic optical filters was resolved by additional passivation using Parylene-C. The functionality of the demonstrated system is verified by detecting the fluorescence property of Rhodamine B and Rhodamine 6G dyes. Rhodamine B's sensitivity to temperature was also demonstrated using the on-chip microheater. This process flow can be scaled to stack a larger number of layers for demonstrations of more complicated systems with enhanced functionality and applications.

There have been reports [34,42,57] and products on the silicon-based 3D integration of homogeneous memory and logic devices [29]. In contrast to homogeneous integration [58], building 3D systems by stacking the heterogeneous dies of MOS [59,60], MEMS [61], microfluidics [54], optical devices [59], etc. is significantly more difficult. Heterogeneous devices are fabricated on different substrates using different processing technologies. Hence, their integration into a single 3D stack requires addressing fundamental issues related to processing compatibility [62], different thermal budgets [63], different coefficients of thermal expansion and other differences in their mechanical properties. Also, technologies developed for 3D integration of CMOS devices are not easily translatable for 3D integration of heterogeneous devices. For example, through-substrate via (TSV) technology is still not mature for non-silicon substrates and hence using TSV for 3D integration of heterogeneous dies is currently not feasible. Further, high-temperature metal-based bonding techniques [64] cannot be used for the integration of temperature-sensitive components such as plastics.

In this section, we demonstrate a 3D stack consisting of a silicon device, two plastic filters, one LED and a glass microheater device. Various complementary metal oxide semiconductor (CMOS)-based systems were reviewed in [54,65–69], with electrical, electrochemical, optical, thermal and magnetic sensing capabilities. In addition, optical filtering mechanisms have been discussed in detail in [70]. Polymer-based filters were also explored [71]. Out of the optical, electrochemical and micromechanical detection mechanisms in microfluidic systems, fluorescence-based bio sensing [72–79] is more prevalent due to its ease of implementation. The main benefits of this would be high selectivity, very low detection boundaries and the availability of markers for tagging the bio markers in the bio-sensing platforms.

In this chapter, we present a cost-effective method of integrating heterogeneous components in 3D stack using SU-8 [80], by integrating diverse components with the aim of extreme miniaturization of a spectro-fluorometer. A spectro-fluorometer consists of (i) an excitation source, (ii) a detector and (iii) filters. Filters block wavelengths associated with excitation spectrum while allowing only the fluoresced light from the test sample to reach the detector. Further, to be able to perform chemical reactions like polymerase chain reaction (PCR) [81,82], a microheater is required. In this chapter, we report a detailed fabrication flow for heterogeneous stacking of a silicon-based photodetector, two optical plastic filters, a tiny commercial blue LED and a glass-based chip with microheater and interconnect pads. Using the developed heterogeneously integrated system in this work, we demonstrate temperature-dependent fluorescence detection from Rhodamine B and Rhodamine 6G.

10.5.1 Individual components fabrication

The process flow developed for 3D integration of heterogeneous dies is shown in Figure 10.26. In summary, the photodetector was fabricated on a p-type silicon substrate as shown in Figure 10.26(a). The details of design and fabrication of the photodetector are presented in Section 10.5.1.1. Connections to the external world were planned on one side for all the dies (see Figure 10.26(b)). Commercially available plastic optical absorption filters were used in this fluorescence detection system. The filters were aligned and bonded manually over the photodetector (see Figure 10.26(b)). After bonding the filters, the stack was planarized using spin-coated polymer SU-8 to enable further stacking as seen in Figure 10.26(c). Platinum-resistive heater and interconnections for the LED were designed and fabricated on a thin glass substrate (150 μm thick). This glass substrate was aligned and bonded using epoxy over the stack (Figure 10.26(d)). After bonding, the stack was again planarized using spin-coated polymer. The planarizing polymers were patterned using a lithography step to open vias for the interconnections, as shown in Figure 10.26(e). These vias were subsequently filled with silver epoxy using the screen-printing technique which enabled the electrical interconnection between the top glass wafer to the bottom silicon. Finally, a commercial blue LED was bonded over the glass wafer using silver epoxy (as shown in Figure 10.26(f)). Manually coloured SEM image of the device with all the components can be seen in Figure 10.27. The demonstrated system has been scaled to the size of a USB stick as shown in Figure 10.28. The process flow summarised above involves three main challenges: stacking, planarization and formation of interconnects. The fabrication details of the photodetector and glass fluidic chip with micro-heater are presented in Sections 10.5.1.1, 10.5.1.2 and 10.5.1.3.

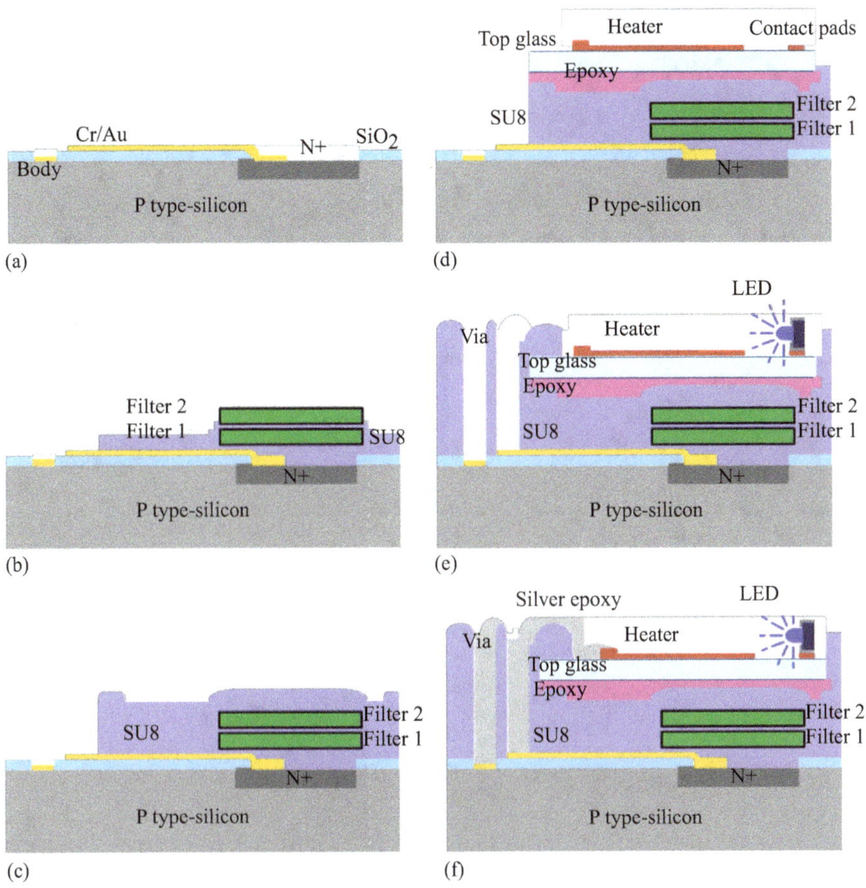

Figure 10.26 Schematic of the process flow for 3D heterogeneous stack: (a) silicon photodetector fabrication, (b) filter bonding using SU-8, (c) planarization of the filter stack, (d) bonding of the microfluidic chip (microheater on glass) using epoxy, (e) planarization and via opening for interconnects and (f) screen-printed interconnects

10.5.1.1 Photodetector fabrication

In this work, a simple silicon p-n junction photodetector was simulated and fabricated for fluorescence detection. The choice of the junction depth for the fabricated detector was based on the light penetration depth calculated from the optical absorption coefficient of the silicon as a function of wavelength [83]. From the intrinsic absorption coefficient for silicon as a function of wavelength [83], we estimate that the photons of ~500 nm wavelength will be absorbed beyond a depth of ~200 nm in the silicon. Hence, a shallow junction photodetector with a junction depth of 200 nm was designed using process simulator TCAD Athena. Even though this work is based on a simple photodetector, for better sensitivity this

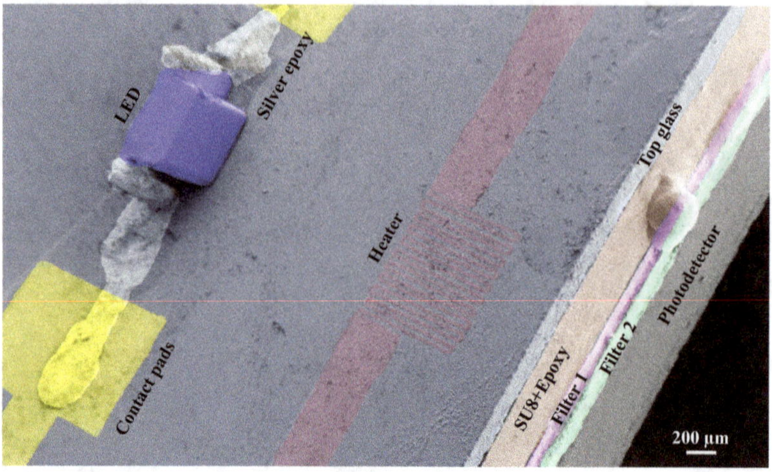

Figure 10.27 False-coloured SEM image of the device showing all the components in the system

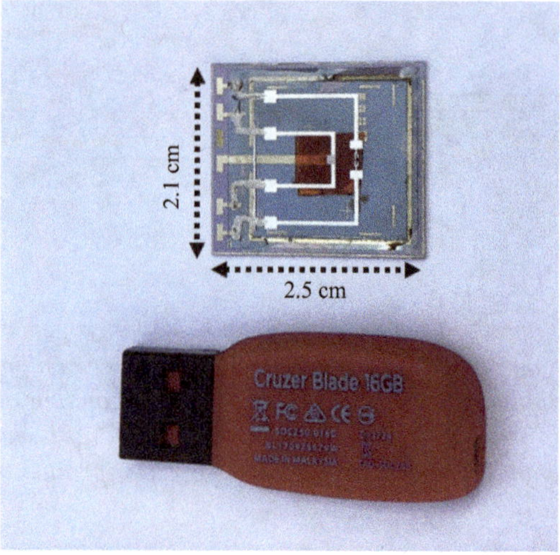

Figure 10.28 Size of the proposed heterogeneous system in comparison with a commercial USB stick

photodetector can be replaced with an avalanche photodetector without any significant modifications in the integration process.

Silicon photodetector die fabrication involves three lithography steps. In this fabrication scheme, 1-μm-thick pyrogenic silicon dioxide was grown on a thoroughly cleaned 1–10 Ω-cm P-type <100>silicon wafer. The oxide diffusion mask was defined using lithography and oxide wet etch. During wet etching of oxide, the backside of the wafer was protected using photoresist. Phosphorus (n+) diffusion was performed in a $POCl_3$ diffusion furnace at 900 °C for 15 min (pre-deposition), followed by 10 min of annealing in the nitrogen ambient (drive-in) without removing the phospho-silicate glass (PSG), which was formed during the pre-deposition step. After the diffusion process, PSG was removed in a dilute HF bath. In comparison with the simulated photodetector sheet resistance 16.14 Ω/□, the measured sheet resistance after fabrication was 17.33 Ω/□. The sheet resistance measurements were performed using the four-probe measurement technique. This was followed by contact metallization using 20 nm chrome and 100 nm gold. Finally, after the metal patterning, the wafers were annealed in the forming gas ambient for 15 min at 400 °C to passivate the dangling bonds and interface states. The fabricated photodetector die can be seen in Figure 10.29(a).

10.5.1.2 Selection of optical filter

As explained in the previous report [70], an ideal filter for fluorescence detection application blocks (i.e. 0% transmission) the excitation wavelengths but transmits 100% of the emitted wavelengths allowing only the fluorescent light to reach the detector. The selection of the filter was based on the excitation and emission wavelength of the fluorescent dye and the radiant intensity of the LED. In this work, blue LED with a peak wavelength of 430 nm was used as an excitation source. The excitation peak for the Rhodamine B (for fluorescence from Sigma-Aldrich) used was 553 nm [84]. The filter was chosen to block the light having a wavelength below ~ 550 nm as the emission peak for the dye used was 627 nm. Hence, this fluoresced light (627 nm) will be *detected* by the photodetector, and the filters will *block* anything below 550 nm. For this application commercially available 105 Orange filter from LEE Filters was chosen. Attaining lossless transmission and perfect blocking is however very difficult. Based on the attenuation of the blocked wavelengths, we decided to stack two filters to achieve sufficient rejection of the excitation LED.

10.5.1.3 Fabrication of the glass fluidic chip with microheater

A 2 kΩ platinum resistive heater was designed and fabricated on the ~150-μm-thick glass coverslip. Piranha cleaned coverslips were patterned lithographically followed by the sputtering of Ti/Pt and photoresist lift-off in acetone to form the heater. On the coverslip along with resistive heater, bond pads for the excitation LED were designed as shown in Figure 10.29(c). The dimensions of the LED used in this work (from Kingbright) were 1.6 mm × 0.8 mm × 1.1 mm (thick). It is a blue colour GaN LED with a peak wavelength of 430 nm. Rather than mounting the LED horizontally with the LED facing upwards, it was mounted facing sideways as shown in the SEM image Figure 10.29(f). This was done to illuminate the biological sample droplet which will reside in the same horizontal level. The LED was bonded manually, viewing under the microscope using silver epoxy as shown in Figure 10.29(f). The SEM image of the microheater can be seen in Figure 10.29(g).

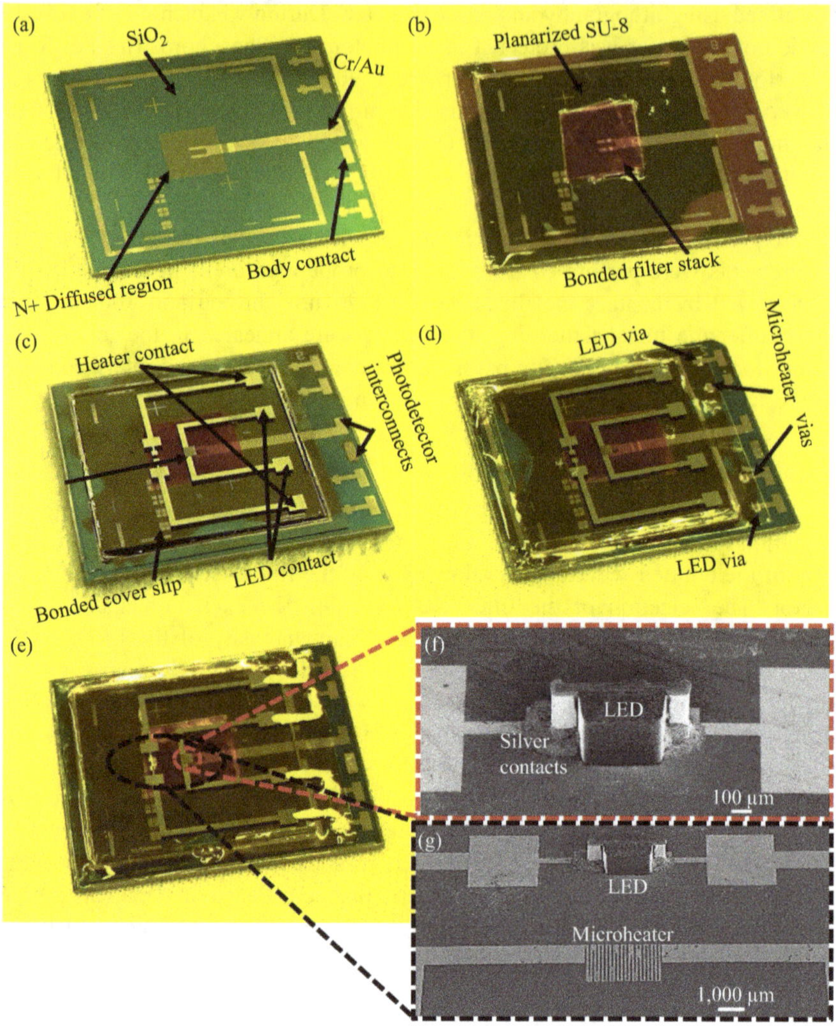

Figure 10.29 Micrographs at various stages of the fabrication process: (a) silicon photodetector, (b) filters bonding and SU-8 planarization, (c) glass microfluidic chip (with microheater) bonding using epoxy, (d) planarized stack and open vias for metallization, (e) final device with bonded LED and interconnections, (f) SEM image of LED and the silver epoxy used for bonding LED, and (g) SEM image of microheater and LED on the microfluidic chip

10.5.2 Hybrid integration

10.5.2.1 Device stacking by epoxy bonding

Stacking these heterogeneous components in 3D and interconnecting them electrically was the main objective of this work. As the individual dies of the 3D stack were on heterogeneous substrates, using through-substrate via (TSV) was not feasible. Instead our approach uses interconnects at the edge of the dies as presented in previous chapters. To enable the formation of electrical routing and interconnects at the edges, the stack needs to be planarized after each bonding step. For planarization spin-coated SU-8 was used in this work. SU-8 is available in various ranges of viscosities allowing formation of planarizing layers with varying thickness. This allows stacking and planarization of devices having different thickness without significant changes in the process. An additional benefit of choosing SU-8 was its ability to be lithographically patterned even for very thick layers which makes formation of high aspect ratio vias for interconnects very simple.

Various epoxies including SU-8, EPO-TEK UJ1190, 377, 353-ND, Fevicol Fevikwik and M-Bond were investigated for bonding the components to the stack. The key properties of these bonding materials which effect the stacking are coefficient of thermal expansion (CTE), viscosity, glass transition temperature, outgassing, adhesion, curing methodology, colour (transparent or opaque) and optical properties. Properties of the bonding epoxies are compared in Table 10.1. Considering the low viscosity, very low outgassing and better optical transmission properties, EPO-TEK 377 and UJ1190 (shown inside the dotted line box in the table) were selected for bonding of the microfluidic device. To keep the process simple, SU-8 was used to bond the filters on the photodetector. The filters were aligned manually using the alignment marks structured on the photodiode chip. The alignment marks (+) can be seen in Figure 10.29(a) and (b).

10.5.2.2 Bonding plastic filters to silicon photodetector

During the integration process, it was observed that the optical filters were reacting with SU-8 and other solvents which made them incompatible with the fabrication process. In order to make them compatible with the integration steps, we decided to conformally coat them with a process compatible polymer. Parylene was selected due to its good thermal endurance, zero outgassing and strong resistance to various solvents, acids and alkalis. A 1–1.5-μm-thick conformal coating of Parylene-C was obtained by a vacuum-coating approach, using PDS 2010, SCS Labcoater. Parylene-coated filters were found to be compatible with the fabrication processes. To enhance the bonding strength, the hydrophobic Parylene-coated filters were treated with oxygen plasma before bonding. The 120 s oxygen plasma surface activation was carried out in an Oxford Instruments Reactive Ion Etching (RIE) tool (PlasmaLab100). The plasma process used ICP power of 300 W, platen power of 50 W, platen temperature of 15 °C, chamber pressure of 10 mTorr and oxygen mass flow rate of 10 sccm.

After plasma treatment, these filters were immediately subjected to SU-8 bonding as follows. Bonding process begins with cleaning the photodetector dies in acetone and IPA. The photodetector dies were next spin-coated with SU-8 2005 at 2,000 rpm to obtain a 6–7-μm-thick film. Parylene-coated filter was bonded manually above the active (n+) region of the photodiode followed by baking the sample on a hot plate at 95 °C for 3 min. During assembly, a flat Teflon tweezer was used to squeeze out any trapped air. Such an approach is feasible for flexible substrates only. For brittle substrates, this approach of mechanically squeezing out

Table 10.1 Key properties of the bonding materials

#	CTE 10^{-6} in/in°C Below T_g	Viscosity Centipoise	Tg Glass transition temp °C	Outgassing @ 200 °C	Adhesion kg/cm^2	Curable	Appearance	Comments	Optical properties @ 23°
Fevicol-Fevikwik	–	40–50	–	High	~70	RT	Clear	Non-patternable	–
M-Bond	–	2.038	–	High	~2–2.8	RT	Amber	Non-patternable	–
EPO-TEK 353 ND	54	3,000–5,000 @ 50 rpm	≥90	0.22%	>140.6	Thermal	Amber/Dark red	Non-patternable	≥50 % @ 550 nm and ≥98% 800–1,000 nm
EPO-TEK UJ1190	68	501 @ 100 rpm	100	0.04%	–	UV	Pale yellow	Non-patternable	≥94 % @ 520 nm–1,000 nm and ≥80% 380–1,000 nm
EPO-TEK −377	57	150–300 @100 rpm	≥95	0.06%	102.3	Thermal	Amber/Dark red	Non-patternable	≥90% @ 600 nm–1,000 nm
SU-8	52	2005:6 2015:1500 2,100:> 15,000	50 (not Cross link linked) 210 (Cross linked)	7.5% @ 95 °C	best	Thermal	Clear	Patternable	>90% @ 400 nm–800 nm

trapped air can lead to device breakage. To bond the second filter over the first one, same process of bonding using SU-8 has been repeated as shown in Figure 10.26(b). The alignment accuracy of the first filter to the photodetector active region was ~50 μm and was performed under the microscope. And the filter to filter alignment was ~150 μm. The same can be observed in the SEM image, as shown in Figure 10.30(d).

10.5.2.3 Planarization of the bonded filter stack

To enable further stacking of subsequent layers over the filters, the stack needs to be planarized. SU-8 2100 was used for planarizing the filter stack. SU-8 2100 was spin-coated over the samples at 3000 rpm for 1 min followed by a softbake at 60 °C for 10 min. The temperature was then ramped to 80° C at a ramp rate of 5 °C min^{-1}. Slow ramping reduces stress issues by ensuring uniform solvent evaporation over the SU-8 thickness. The sample was held at 80 °C for 60 min and then slowly cooled to RT at a rate of 5 °C min^{-1}. Using a photolithographic mask, the substrates were then exposed to UV light in the EVG 620 mask aligner. The total exposure was broken into three steps with each step having a dose of

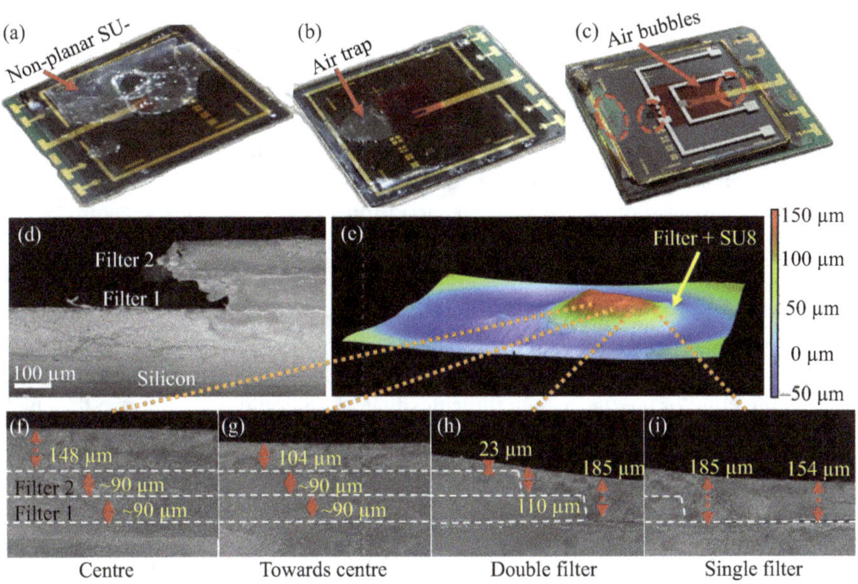

Figure 10.30 (a) Defects appearing due to non-planar SU-8 after filter stack planarization step, (b) air traps due to SU-8 outgassing, (c) air-traps due to SU-8 bonding, (d) SEM image after the filter stacking, (e) 3D mapping of the device after the first planarization step using the surface profiler, SEM image at the (f) centre of the device, (g) towards the centre of the device, (h) starting of the double filter and (i) starting of the single filter

100 mJ/cm^2. Between each step, a gap of 30 s was provided to relax and cool the substrate. Such an approach prevents the T-topping [85] of the resist, by reducing the surface heating of the SU-8 film. Post-exposure bake was performed at 60 °C for 10 min and 85 °C for 25 min. The temperature ramp up and ramp down were at a rate of 5 °C min^{-1}. A 10 min relaxation at RT was performed after the softbake, exposure and post-exposure bake steps. Post-bake relaxation helps the photoresist film to attain its optimum hydration levels by reabsorbing the lost moisture. The samples were then developed in SU-8 developer for approximately 22 min with mild agitation. The completion of the development process was judged through optical inspection. Following the IPA/DI water rinse, hard-bake was performed at 95 °C for 5 min. Figure 10.29(b) shows the photograph after successful bonding of the two-filters on the stack and subsequent planarization.

On some of the sample's planarization failed due to the formation of a wavy SU-8 film as shown in Figure 10.30(a). This failure was easily detected through optical inspection after the lithography process. Even though the reason for the formation of wavy SU-8 film is not very clear, the failure was observed particularly in the stacks where Parylene was damaged at the corners of the filters. If the coated filters were not handled properly, Parylene was observed to peel off from the edges of the filter. This issue was resolved through improving device handling and sorting out the filters with damaged Parylene coating through added inspection steps. Subsequent bonding to such wavy SU-8 layers leads to integration failure from incomplete bonding or appearance of trapped air.

SEM image after bonding the two filters using SU-8 2005 can be seen in Figure 10.30(d). After the planarization step, 3D topographical map was generated using the Dektak surface profiler. Cross-sectional SEM images show the SU-8 thickness at various points after the planarization. SEM image at the centre of the filters can be seen in Figure 10.30(f), which shows ~148-μm-thick SU-8 above the filter stack. As we move away from the centre, SU-8 thickness is reduced to ~104 μm as shown in Figure 10.30(g). Figure 10.30(h) and (i) shows the SEM image at the edge of the double and single filter stacks, respectively. From the 3D map and the SEM image, it was clear that the SU-8 was higher above the filter and reduces towards the edge of the die, which indicates that SU-8 could not planarize the stack completely. The current planarization step is ensuring a smooth transition of SU-8 from the top of the filters to the photodiode.

10.5.2.4 Bonding of the glass fluidic chip

To bond the glass wafer to the stack, initially SU-8 was explored. To enhance wettability, 1-μm-thick Parylene was deposited on the backside of the glass substrate. SU8-2035 was spin-coated above the existing stack consisting of the photodetector and the filters. During the pre-exposure softbake step while the sample was partially wet, coverslip was manually aligned and bonded under the microscope. However, after the lithography and during the post-exposure/hard-bake steps air traps under the glass substrate appeared as shown in Figure 10.30(b) and (c). This was attributed to the outgassing of the SU-8 and crosslinking related volume shrinking [86]. SU-8 outgasses at a rate of 7.5% at

95 °C (Micro Chem SU-8 2000 data sheet). Most of the stress issues in SU-8 happen due to this crosslinking related volume shrinking. With glass coverslip on the top and the photodetector stack at the bottom, the gas generated gets trapped and leads to bubbles. To address this issue, EPO-TEK epoxies UJ1190 and 377 were considered. EPO-TEK UJ1190 is a UV curable epoxy which has very low outgassing (0.04% at 200 °C). A 20 μl drop of epoxy was dispensed over the photodetector stack followed by aligning the glass wafer to the device under the microscope achieving an alignment accuracy in the order of 100 μm using the alignment marks on the photodetector die. The assembled stack was exposed to UV, with a dose of 150 mJ/cm^2 for 5 times at a regular interval of 20 seconds. The exposure was followed by relaxation of 50 min. Figure 10.29(d) shows the epoxy-bonded sample without any trapped gas. As epoxy wetted the glass surface well, the Parylene coating step was not required. Using EPO-TEK 377 also gave similar results. The remaining non-planarity after the SU-8 step is removed by the epoxy bonding step, as seen in Figure 10.30.

10.5.2.5 Planarization of the fluidic chip and polymer via opening

Like the first planarization step, SU8-2100 was used again for planarizing the microfluidic device layer. This planarization reduces the step height at the microfluidic device edge and allows formation of electrical routing and interconnects between the photodetector device and the microfluidic device. SU-8 photolithography was performed to open vias to the underlying photodetector layer. The UV exposure energy used in this step was slightly higher, with a dose of 110 mJ/cm^2 for 3 times and 30 s gap between each exposure step. A longer relaxation of 60 min was provided after exposure to ensure minimal thermal stresses in the stack. Finally, the sample was hard-baked for 10 min at 110° C. Figure 10.29(d) shows the stack after planarization and via opening. Open vias for metal interconnection between the top microfluidic device and bottom photodetector device can be observed in Figure 10.31.

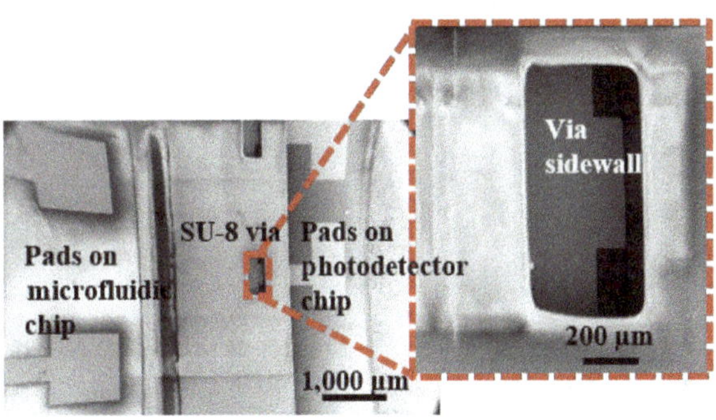

Figure 10.31 Open vias after SU-8 lithography

10.5.2.6 Interconnecting the components in the stack

To electrically connect the top microfluidic chip with the bottom photodetector chip, metallic interconnects are required. Due to the large contour arising from the multiple stacking, bonding and planarization steps, using conventional metal deposition and lithography techniques for forming interconnects is not feasible. Hence, non-conventional techniques like inkjet printing and screen printing were explored. Diamatix 2831 inkjet printer with 10 pL cartridge was used for printing. Silver ink with <50 nm particles was used in this process (silver dispersion from Sigma-Aldrich). Samples were treated with oxygen plasma (300 W – ICP and 50 W – RF, O_2–10 sccm, and 10 mT) for 60 s to enhance adhesion of the silver ink to the glass and SU-8. To increase the thickness of the silver interconnects, the printing was performed 5 times with a gap of 30 min between printing the layers, as shown in Figure 10.32(a) and (b). This interval between printing subsequent layers helped the silver ink to get dried as the printer platen was kept at 50 °C. Using the camera on the printer tool, the alignment accuracy achieved was in the order of 40 μm in filling the vias.

It was however observed that in some of the samples the connections were not proper. This was due to the high aspect ratio of the SU-8 vias. With a single layer thickness of ~440 nm for the inkjet-printed silver nano-ink, getting a continuous film on the straight sidewall of the SU-8 via was difficult. Incomplete coverage of silver ink on the SU-8 via sidewall was verified by measuring the local electrical connectivity. SEM image (Figure 10.32(d)) shows the multiple z-height differences in the device. In the cases where the electrical continuity was broken, vias were manually filled with silver epoxy under the microscope as shown in Figure 10.32(b).

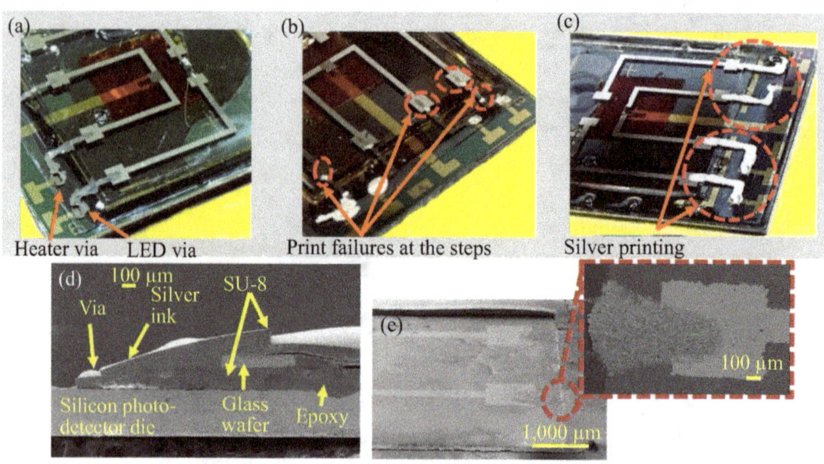

Figure 10.32 (a) Inkjet-printed device, (b) inkjet-printed device with defects at the various contours, (c) screen-printed device without any defects, (d) SEM image of the inkjet-printed device showing various contours in the device and (e) SEM image after the screen printing

Due to the limitations of the inkjet process as discussed above, screen-printing technique was explored which allows the formation of thicker metal lines in a single step. Screen printing allows the use of ink with significantly higher nano-particle loading. Such thicker (higher viscosity) inks are incompatible with the inkjet technique. Nylon mesh with the required pattern was purchased from local sources. Silver epoxy was squeezed through the mesh using a hard rubber blade. Micrograph and SEM image of the screen-printed device can be seen in Figure 10.32(c) and (e), respectively. The screen-printed epoxy thickness was ~13.9 μm. Screen printing was successful in covering all the bond interfaces as shown in Figure 10.29(e) and was found to be repeatable with almost 100% yield in filling the vias. The alignment accuracy was in the order of 150 μm which is slightly poor as compared to the inkjet printing technique.

10.5.3 Device component testing

10.5.3.1 Effect of stacking on photodetector sensitivity

Two Parylene-coated plastic filters were stacked above the photodiode active region using SU-8 2005. This was followed by planarization using SU-8 2100 and stacking of the glass substrate using epoxy bonding technique. To understand the effect of these various materials used in the stacking process spectral response of the device was verified at each stage of the process. For spectral responsivity measurements, a quantum efficiency setup (QE, Sciencetech) was used. The spectral response of the as fabricated photodetector was measured for wavelengths varying from 400 to 700 nm. The sensitivity was found to be ~0.1 A/W as shown in Figure 10.33(a). To confirm the functionality of the filters and effect of Parylene on them, spectral responsivity of the photodetector was measured with filters placed on them. Figure 10.33(a) shows the measured spectral responsivity of the photo-detector at different wavelengths for different configurations. Better blocking of the blue light below 550 nm was observed when a two-filter stack was used instead of one as shown in the measurements (Figure 10.33(a)). The slope in the transition region also becomes sharper improving the rejection of the light from the excitation LED. This guided our design to use a two-filter stack. Figure 10.33(b) shows the spectral response at the various stages of system integration. The measurements were performed on different combination of filter stacks and the final device. From the measurements (Figure 10.33(b)) we observe that the 1.5 μm Parylene coating has minimal effect on the optical path. It is clear from the graph that the spectral response of the final device is slightly less than the filters bonded, but the difference is not very high and can be neglected.

10.5.3.2 Effect of microheater proximity on photodetector sensitivity

The fabricated microheater is in close proximity to the photodetector. To understand the effect of increased temperature on the photodetector sensitivity, spectral responsivity measurements using quantum efficiency setup (QE, Sciencetech) were performed at various heater voltages, as seen in Figure 10.34. The maximum

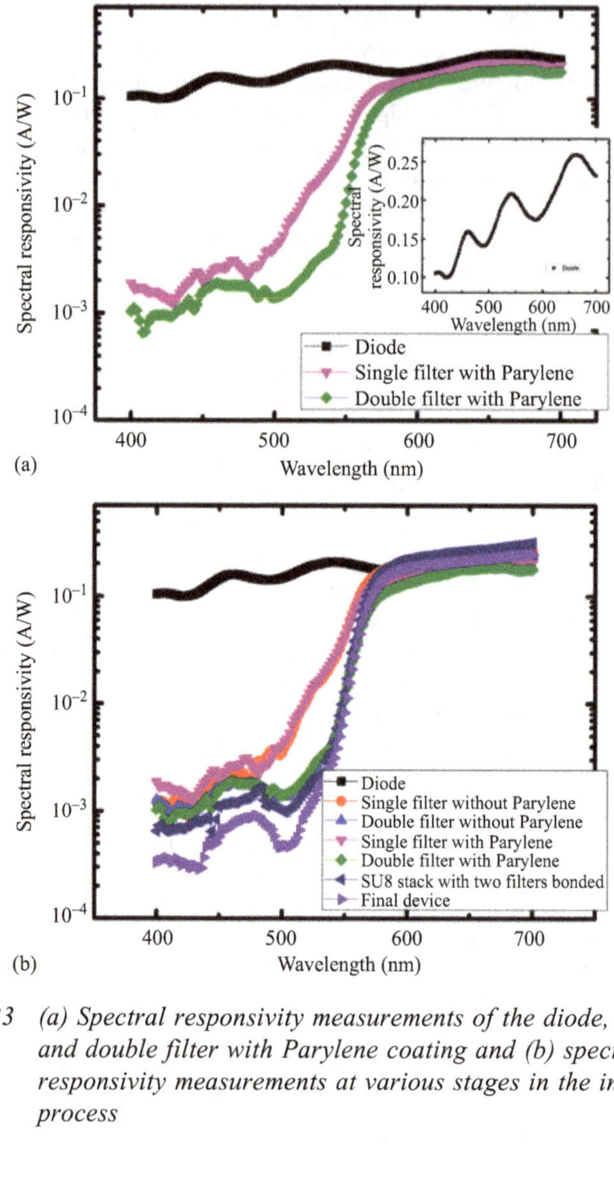

Figure 10.33 (a) Spectral responsivity measurements of the diode, single and double filter with Parylene coating and (b) spectral responsivity measurements at various stages in the integration process

voltage used translates to a heater temperature of 50 °C (see Figure 10.35). At lower wavelengths where most of the light is rejected by the filters, the increase in noise due to higher temperature is visible in the spectral response. However, at higher wavelengths spectral responsivity results show that the increased temperature has a negligible effect on the photodetector response. This is because at higher wavelengths the photodetector current is significantly higher than the noise currents.

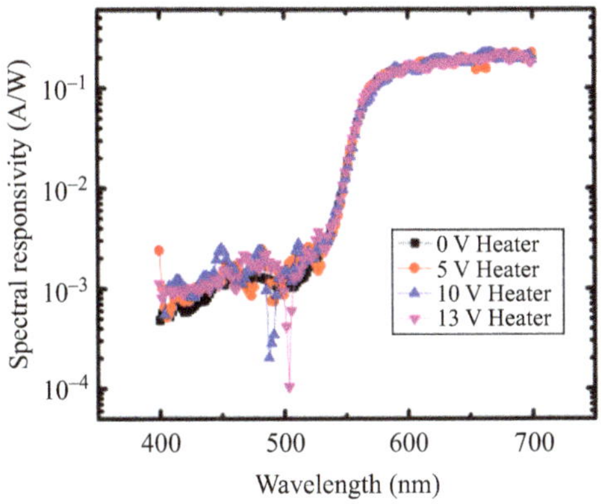

Figure 10.34 Spectral responsivity measurements of the final device at different heater conditions

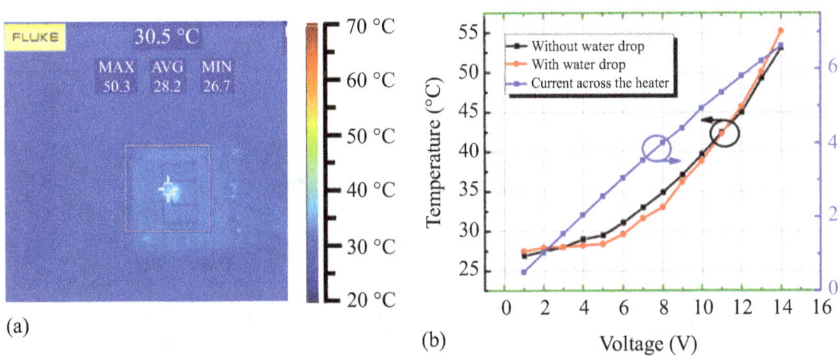

Figure 10.35 (a) IR image of the device under test and (b) temperature vs. voltage plot of the on-chip microheater

10.5.3.3 Platform testing

The platinum microheater was calibrated using an IR camera (Fluke). The fabricated die was kept under the camera and the temperature was measured by gradually increasing voltage between the microheater terminals. The microheater voltage vs. substrate temperature is plotted in Figure 10.35(b). Measurements of the microheater were performed with and without a 1 μL water droplet. Also, the change in current with respect to the applied voltage across the heater is recorded as shown in Figure 10.35(b). Based on this graph, it is clear that to obtain heater temperature of ~50 °C, a voltage of 14 V has to be applied across the heater. Figure 10.35(a) shows the IR camera image of the device.

Ray path of the complete setup can be seen in Figure 10.36. Intensity of the blue LED light source was controlled by the source measure unit (SMU) and the light beam from the blue led was used as an excitation source. It is important to

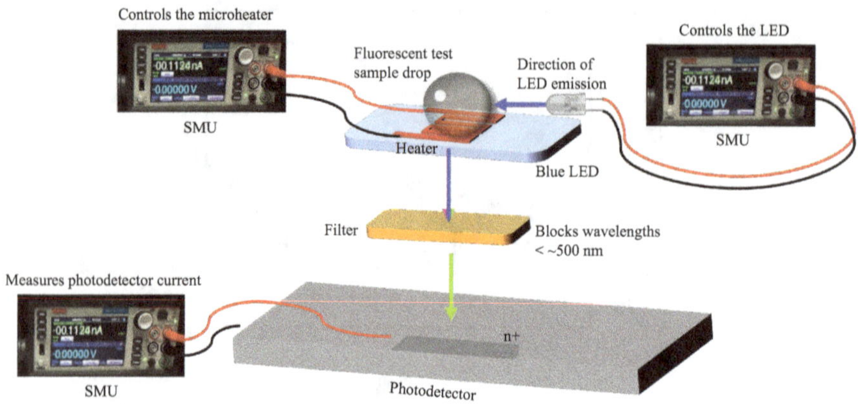

Figure 10.36 Ray path of the complete setup: the 430 nm peak wavelength blue LED was used as an excitation source and was directed towards the test sample. The light beam falling on the filter stack would allow only the fluoresced light and directs the same towards the photodetector detector. The change in the current of the photodetector was recorded using the SMU

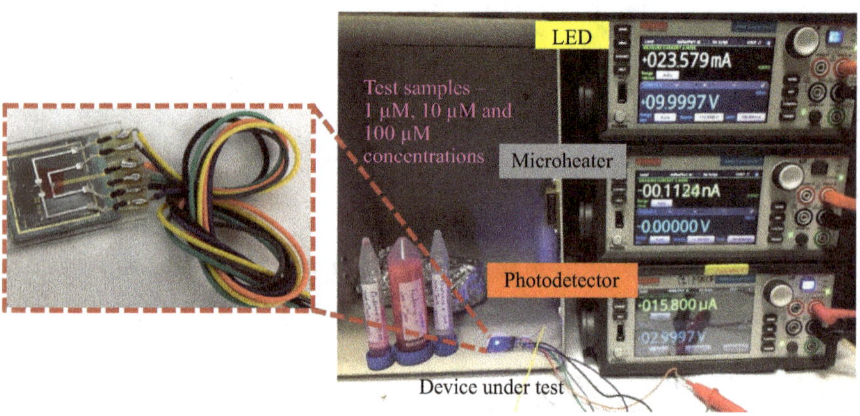

Figure 10.37 Test setup with three source meters for the LED, microheater and photodetector, device under test and the Rhodamine dye

note that the integrated LED is facing sideways on the fluidic chip. This makes the emission axis of the LED parallel to the chip surface and hence, allowing illumination of a droplet placed on the fluidic chip. The test sample (Rhodamine) was placed on the top surface of the microheater. The light is emitted in all directions from the dye droplet. The filters block the wavelengths associated with excitation spectrum while allowing only the fluoresced light from the test sample. The measurements were performed in a closed box as shown in Figure 10.37. The fluoresced light will give rise to a change in the current which was recorded using another SMU. The effect on Rhodamine with temperature was also studied by providing voltage to the microheater. The test sample was placed directly on the top surface of the microheater assembly, to ensure the supplied temperature reach the test sample. After each measurement, the surface of the sample was wiped with IPA.

An image of the experimental setup is shown in Figure 10.37, which consists of wire-bonded test sample, three sourcemeters (Kiethley 2460) to supply voltages to the photodetector, LED and microheater. Rhodamine B and 6G were purchased as powder from Sigma-Aldrich. Rhodamine B has an excitation peak wavelength of $\lambda_{ex} = 553$ nm and an emission peak wavelength $\lambda_{em} = 627$ nm. Rhodamine 6G [87] has an excitation peak wavelength of $\lambda_{ex} = 526$ nm and an emission peak wavelength of $\lambda_{em} = 557$ nm. Solutions of 1, 10 and 100 μM concentration were prepared in DI water. In our experiments the photodiode measured the intensity of the light emitted by the test samples with and without heating to demonstrate the functionality of the 3D heterogeneous microfluidic system.

10.5.4 Experimental results

A 1 μL drop of Rhodamine 6G dye was placed over the microheater, which is right above the photodetector's active region. Voltage was applied to the LED through a 10 KΩ resistor; this is to ensure there is no high current to the LED. A 3 to 10 V was applied to the LED using sourcemeter and the photodetector output was recorded at each step for 1, 10 and 100 μM concentrations of the solutions. As a control experiment the response was also measured without any dye. The same experiments were repeated for the Rhodamine B fluorescence solution as well. These measurements were performed without applying any voltage to the heater. Figure 10.38(a) and (b) shows the results of the Rhodamine 6G and Rhodamine B, respectively. These results show that there is a consistent increase in output current of the detector for increasing supply voltage to the LED. Also, with increasing concentration of Rhodamine in both the 6G and B cases, increase in the current has been observed.

The response of these dyes is sensitive to the solution temperature. Figure 10.38(c) and (d) shows the graph of the detector output current for the increasing temperatures of Rhodamine 6G and B, respectively. In this case, the LED voltage was fixed at 10 V and the experiments were performed for 1 μM and 10 μM concentration solutions. Results in Figure 10.38(c) show the increase in fluorescence with increasing voltage to the heater for the Rhodamine 6G dye. For Rhodamine B, contrary to other reports [88,89], we observed an increase in fluorescence with increasing temperature.

The final isometric view SEM image of the device can be seen Figure 10.39 and the cross-section of the stack with the silicon photodetector, both filters, SU8 and epoxy, glass wafer and the final SU-8 which was used to fill the silver epoxy in the vias. The integration scheme leads to a significant reduction in the volume of the overall system as compared to previous demonstrations [67,72] and hence achieving scaling of overall system through 3D integration.

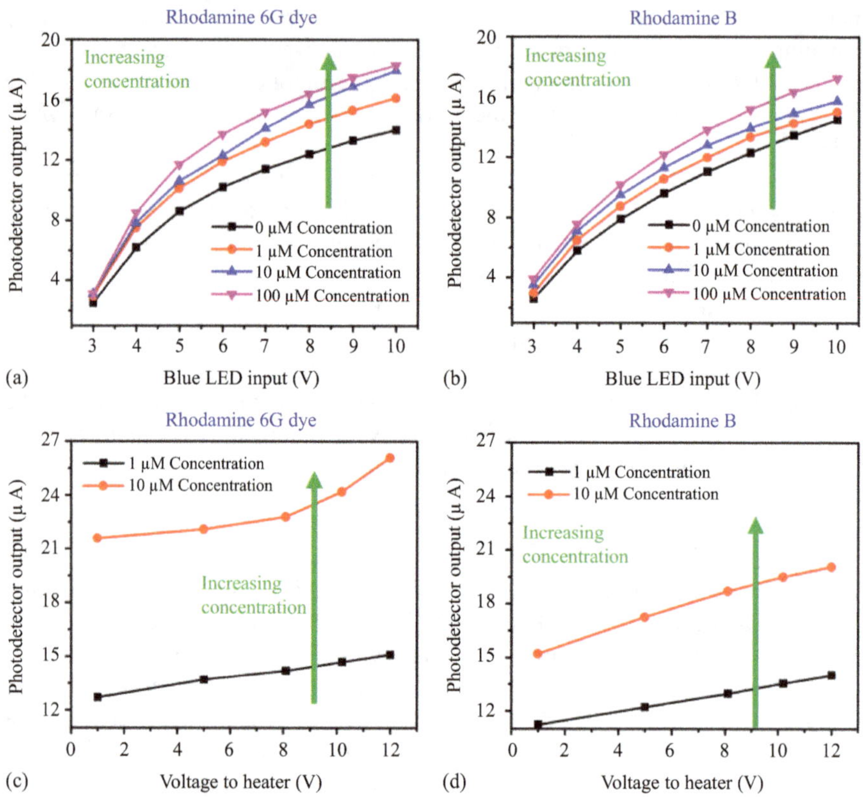

Figure 10.38 Photodetector output current vs. LED input: (a) Rhodamine 6G, (b) Rhodamine B: photodetector output current vs. voltage to the heater, (c) Rhodamine 6G and (d) Rhodamine B

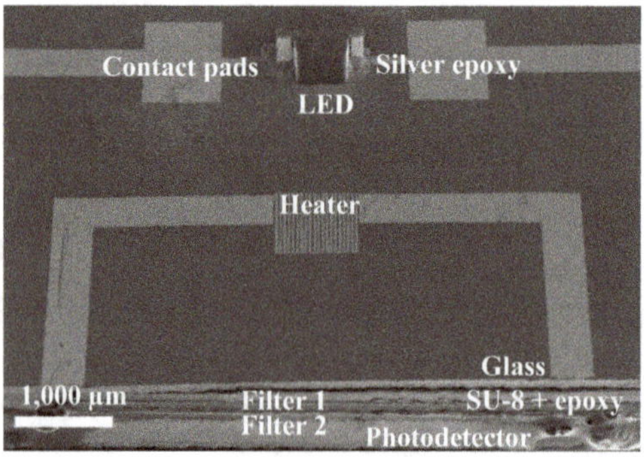

Figure 10.39 Isometric and the cross-section SEM image of the device with all the components in the system

10.6 Summary

We have discussed innovative processing technologies that would allow 3D packaging by the post-fab vertical stacking technique, suitable for the packaging industry. These novel simple processes may pave way towards 3D-stacked ultra-thin devices.

References

[1] Saraswat K. Silicon compatible optical interconnects. 2016 IEEE Int Interconnect Technol Conf/Adv Met Conf IITC/AMC 2016. 2016
[2] Dennard RH, Gaensslen FH, Hwa-Nien YU, Leo Rideout V, Bassous E, and Leblanc AR. Design of ion-implanted MOSFETs with very small physical dimensions. Proc IEEE. 1999;87(4):668–78
[3] Iyer SS, and Kirihata T. Three-dimensional integration: A tutorial for designers. 2015;7(4):63–74
[4] Knickerbocker, Andry PS, Dang B, Horton RR, Interrante MJ, Wright CSPRJPKSRSEJSSMS a MS a WTCKTBCWSL, *et al.* Three-dimensional silicon integration. Int Bus. 2008;52(6):553–69
[5] Iyer SSS. Heterogeneous integration for performance and scaling. IEEE Trans Components, Packag Manuf Technol. 2016;6(7):1–10
[6] Moore GE. Cramming more components onto integrated circuits (Reprinted from Electronics, pp. 114–117, April 19, 1965). Proc IEEE [Internet]. 1965 [cited 2017 Aug 7];86(1):82–5. Available from: papers3://publication/uuid/8E5EB7C8-681C-447D-9361-E68D1932997D
[7] Beyne E. The 3-D Interconnect technology landscape. IEEE Des Test. 2016;33(3):8–20
[8] Lau JH. 3D IC Packaging and integration. 2015; Available from: http://s3.amazonaws.com/sdieee/1817-SanDiegoCPMTDL_Lau_advancedpackaging.pdf
[9] Christiaens W, Bosman E, and Vanfleteren J. UTCP: A novel polyimide-based ultra-thin chip packaging technology. IEEE Trans Components Packag Technol. 2010;33(4):754–60
[10] Iyer SS. Three-dimensional integration: An industry perspective. MRS Bull [Internet]. 2015;40(03):225–32. Available from: http://www.journals.cambridge.org/abstract_S0883769415000329
[11] Fischer AC, Forsberg F, Lapisa M, *et al.* Integrating MEMS and ICs. Microsystems Nanoeng [Internet]. 2015 May 28 [cited 2017 Aug 7]; 1(April):15005. Available from: http://arxiv.org/abs/1604.04843
[12] Farooq MG, and Iyer SS. 3D integration review. Sci China Inf Sci. 2011; 54(5):1012–25
[13] Govaerts J, Bosman E, Christiaens W, and Vanfleteren J. Fine-pitch capabilities of the flat ultra-thin chip packaging (UTCP) technology. IEEE Trans Adv Packag. 2010;33(1):72–8
[14] Bleiker SJ. Heterogeneous 3D Integration and packaging technologies for nano-electromechanical systems. 2017

[15] Ma Q, Wang Z, and Pan L. Monolithic integration of multiple sensors on a single silicon chip. 2016;0–3
[16] Shukri DD, Co-advisor S. 3D ICs interconnect performance modeling and analysis.
[17] Association SI. 2015 ITRS Section 5: More Moore. 2015;1–52
[18] Shen W-W, and Chen K-N. Three-dimensional integrated circuit (3D IC) key technology: through-silicon via (TSV). [cited 2017 Aug 30]; Available from: https://nanoscalereslett.springeropen.com/track/pdf/10.1186/s11671-017-1831-4?site=nanoscalereslett.springeropen.com
[19] Motoyoshi BM. Through-silicon via (TSV).pdf. 2009;97(1):43–8
[20] Ma M. TSV in IC packaging: now and future. Semicon Taiwan
[21] Eroglu SEK, Cho WY, and Leblebici Y. A chip-level post-CMOS via-last Cu TSV process for multi-layer homogeneous 3D integration. 2016 12th Conf PhD Res Microelectron Electron PRIME 2016. 2016;3–6
[22] Pangracious V, Marrakchi Z, and Mehrez H. Three-dimensional design methodologies for tree-based FPGA architecture. 2015;350. Available from: http://link.springer.com/10.1007/978-3-319-19174-4
[23] Fischer AC, Forsberg F, Lapisa M, *et al.* Integrating MEMS and ICs. Microsystems Nanoeng [Internet]. 2015;1(April):15005. Available from: http://www.nature.com/articles/micronano20155
[24] Knechtel J, Sinanoglu O, and Elfadel I (Abe) M, Lienig J, and Sze CCN. Large-scale 3D chips: challenges and solutions for design automation, testing, and trustworthy integration. IPSJ Trans Syst LSI Des Methodol [Internet]. 2017;10(0):45–62. Available from: https://www.jstage.jst.go.jp/article/ipsjtsldm/10/0/10_45/_article
[25] Lau JH. Recent advances and new trends in flip chip technology. J Electron Packag [Internet]. 2016;138(3):030802. Available from: http://electronicpackaging.asmedigitalcollection.asme.org/article.aspx?doi=10.1115/1.4034037
[26] Meyer T, Ofner G, Bradl S, Brunnbauer M, and Hagen R. Embedded wafer level ball grid array (eWLB). 10th Electron Packag Technol Conf EPTC 2008. 2008;994–8
[27] Lau JH, Li M, Li QM, *et al.* Reliability of fan-out wafer-level packaging. 2018;1–12
[28] Tummala RR. Fundamentals of microsystems packaging [Internet]. 2001. 967 p. Available from: http://books.google.com/books?id=nf1vwR9VYGEC&pgis=1
[29] Mahajan R, Sankman R, Patel N, Kim DW, Aygun K, Qian Z, *et al.* Embedded multi-die interconnect bridge (EMIB)-A high density, high bandwidth packaging interconnect. Proc - Electron Components Technol Conf. 2016;2016–August:557–65
[30] Wang C, Cheng H, Chung M, Pan P, Ho C, and Chiu C. High bandwidth application on 2.5D IC Silicon Interposer. 2014;568–72
[31] Hellings G, Scholz M, Detalle M, *et al.* Active-lite interposer for 2.5 & 3D integration. IEEE Symp VLSI Circuits, Dig Tech Pap. 2015;2015–August (April 2013):T222–3

[32] Ho SW, Ding MZ, Lim PS, *et al.* 2.5D through silicon interposer package fabrication by chip-on-wafer (CoW) approach. Proc 16th Electron Packag Technol Conf EPTC 2014. 2014;(65):679–83
[33] Wang MJ, Hung CY, Kao CL, *et al.* TSV technology for 2.5D IC solution. Proc - Electron Components Technol Conf. 2012;284–8
[34] Shulaker MM, Hills G, Park RS, *et al.* Three-dimensional integration of nanotechnologies for computing and data storage on a single chip. Nature [Internet]. 2017;547(7661):74–8. Available from: http://dx.doi.org/10.1038/nature22994
[35] Lee KH, Wang Y, Wang B, *et al.* Monolithic integration of Si-CMOS and III-V-on-Si through direct wafer bonding process. 2017;6:571–578, 2168-6734(001)
[36] Park J. Physics and technology of low temperature germanium Mosfets for monolithic three dimensional integrated circuits. Jin-Hong Park [Internet]. 2009;(June). Available from: http://cis.stanford.edu/~saraswat/Thesis/Jin-Hong Park Thesis.pdf
[37] Deshpande, V, Djara, V, Connor, E. O. *et al.* Three-dimensional monolithic integration of III–V and Si (Ge) FETs for hybrid CMOS and beyond technology three-dimensional monolithic integration of III–V and Si (Ge) FETs for hybrid CMOS and beyond. Jpn J Appl Phys. 2017;56(4s):04CA05
[38] Ren J, Member S, Liu C, and Tang CW. A novel Si–GaN monolithic integration technology for a high-voltage cascoded diode. IEEE Electron Device Letters. 2017;38(4):501–4
[39] Kerr AJ, Lee H, Palacios T, *et al.* Monolithic integration of silicon CMOS and GaN transistors in a current mirror circuit. J Vac Sci Technol B. 2017; 30(2):02B101 (2012)
[40] Lee KHKEK, Bao S, Kohen D, *et al.* Monolithic integration of III-V HEMT and Si-CMOS through TSV-less 3D wafer stacking. Proc - Electron Components Technol Conf. 2015;2015–July (March 2016):560–5
[41] Fitzgerald EA. Novel integrated circuit platforms employing monolithic silicon CMOS + GaN devices. Trans ECS, Soc TE. 2016;75(12):31–7
[42] Shulaker MM, Wu TF, Sabry MM, Wei H, Wong HP, and Mitra S. Monolithic 3D integration: a path from concept to reality 2015. 2015;1197–202
[43] Koyanagi M. Challenges in 3D integration. Trans ECS, Soc TE. 2013;53(3): 237–44
[44] Koyanagi M. Heterogeneous 3D integration - Technology enabler toward future super-chip. Tech Dig - Int Electron Devices Meet IEDM. 2013;8–15
[45] ITRS. International Technology Roadmap for Semiconductors 2.0 2015 Edition outside system connectivity. 2015;1–37
[46] Garrou P, Lu JJQ, and Ramm P. Three-dimensional integration. Handbook of Wafer Bonding New York: Wiley. 2012;(December):301–28
[47] ITRS. International Technology Roadmap for Semiconductors, Edition 2015, Beyond C-MOS. Int Technol Roadmap Semicond. 2015
[48] Shulaker MM, Hills G, Park RS, *et al.* for Computing and data storage on a single chip. Nat Publ Gr [Internet]. 2017;547(7661):74–8. Available from: http://dx.doi.org/10.1038/nature22994

[49] Lee K, Kanno S, Kiyoyama K, Fukushima T, Tanaka T, and Koyanagi M. A cavity chip interconnection technology for thick MEMS chip integration in MEMS-LSI multichip module. 2010;19(6):1284–91

[50] ITRS. International Technology Roadmap for Semiconductors 2.0 2015 Edition Heterogeneous Integration. 2015;1–92

[51] Tummala R, Wolter KJ, Sundaram V, Smet V, and Raj PM. New era in automotive electronics, a co-development by Georgia tech and its automotive partners. 2016 Pan Pacific Microelectron Symp Pan Pacific 2016. 2016;3–6

[52] ITRS. International Technology Roadmap for Semiconductors 2.0 2015 Edition System Integration. 2015;1–21

[53] INTERNATIONAL TECHNOLOGY ROADMAP FOR SEMICONDUCTORS 2.0. Factory Integration. 2015;1–80

[54] Khan SM, Gumus A, Nassar JM, and Hussain MM. CMOS Enabled microfluidic systems for healthcare based applications. Adv Mater [Internet]. 2018;30 (16):1705759. Available from: http:https://doi.org/10.1002/adma.201705759

[55] Chen S, Griffin PB, and Plummer JD. Negative differential resistance circuit design and memory applications. IEEE Trans Electron Devices. 2009;56 (4):634–40

[56] Hull AW. The dynatron: a vacuum tube possessing negative electric resistance. Proc Inst Radio Eng. 1918;6(1):5–35

[57] Shulaker MM, Van Rethy J, Wu TF, *et al.* Carbon nanotube circuit integration up to sub-20 nm channel lengths. ACS Nano. 2014;8(4):3434–43

[58] Kim SK, Xue L, and Tiwari S. Low-temperature polymer-based three-dimensional silicon integration. IEEE Electron Device Letters. 2007;28 (8):706–9

[59] Zimmer F, Lapisa M, Bakke T, Bring M, Stemme G, and Niklaus F. One-megapixel monocrystalline-silicon micromirror array on CMOS driving electronics manufactured with very large scale heterogeneous integration. J Microelectromech Sys. 2011;20(3):564–72

[60] Zhang X, Jo PK, Zia M, May GS, and Bakir MS. Heterogeneous interconnect stitching technology with compressible microinterconnects for dense multi-die integration. IEEE Electron Device Lett. 2017;38(2):255–7

[61] Temiz Y, Zervas M, Guiducci C, and Leblebici Y. Die-level TSV fabrication platform for CMOS-MEMS integration. 2011 16th Int Solid-State Sensors, Actuators Microsystems Conf TRANSDUCERS'11. 2011;1799–802

[62] Abgrall P, and Gué a- MM. Lab-on-chip technologies: Making a microfluidic network and coupling it into a complete microsystem – A review. J Micromechanics Microengineering. 2007;17(5):R15–49

[63] Ghoneim MT, Fahad HM, Hussain AM, *et al.* Enhanced cooling in monocrystalline ultra-thin silicon by embedded micro-air channels. AIP Adv. 2015;5(12):127115

[64] Welch WC, Chae J, and Najafi K. Transfer of metal MEMS packages using a wafer-level solder transfer technique. IEEE Trans Adv Packag. 2005;28 (4):643–9

[65] Datta-Chaudhuri T, Smela E, and Abshire PA. Calhoun: The NPS institutional archive Dspace repository system-on-chip considerations for heterogeneous integration of CMOS and fluidic bio-interfaces system-on-chip considerations for heterogeneous integration of CMOS and fluidic bio-interfaces. IEEE Trans Biomed Circuits Syst [Internet]. 2016 [cited 2017 Aug 7];10(6). Available from: http://hdl.handle.net/10945/52401

[66] Temiz Y, Kilchenmann S, Leblebici Y, and Guiducci C. 3D integration technology for lab-on-a-chip applications. Electron Lett [Internet]. 2011;47 (26):S22. Available from: http://digital-library.theiet.org/content/journals/10.1049/el.2011.2683

[67] Datta-Chaudhuri T, Abshire P, and Smela E. Packaging commercial CMOS chips for lab on a chip integration. Lab Chip [Internet]. 2014;14(10):1753. Available from: http://xlink.rsc.org/?DOI=c4lc00135d

[68] Rae BR, Muir KR, Gong Z, *et al.* A CMOS time-resolved fluorescence lifetime analysis micro-system. Sensors. 2009;9(11):9255–74

[69] Chung J, Hwang HY, Chen Y, and Lee TY. Microfluidic packaging of high-density CMOS electrode array for lab-on-a-chip applications. Sensors Actuators B Chem [Internet]. 2018 [cited 2017 Aug 7];254:542–50. Available from:http://ac.els-cdn.com/S0925400517313291/1-s2.0-S0925400517313291-main.pdf?_tid=e3daf52a-7b44-11e7-9dde-00000aacb362&acdnat=1502092327_eb88c9e103d3a5e1277a0c8d9b27fb46

[70] Dandin M, Abshire P, and Smela E. Optical filtering technologies for integrated fluorescence sensors. Lab Chip [Internet]. 2007;7(8):955. Available from: http://www.ncbi.nlm.nih.gov/pubmed/17653336%5Cnhttp://xlink.rsc.org/?DOI=b704008c

[71] Smela E, Dandin M, and Abshire P. Polymer filters for ultraviolet-excited integrated fluorescence sensing. J Micromechanics Microengineering [Internet]. 2012;22(9):95018. Available from: http://stacks.iop.org/0960-1317/22/i=9/a=095018

[72] Nelson N, Sander D, Dandin M, Prakash SB, Sarje A, and Abshire P. Handheld fluorometers for lab-on-a-chip applications. IEEE Trans Biomed Circuits Syst. 2009;3(2):97–107

[73] James TD, Scullion MG, Ashok PC, Di Falco A, Dholakia K, and Krauss TF. Valve controlled fluorescence detection system for remote sensing applications. Microfluid Nanofluidics. 2011;11(5):529–36

[74] Lim J, Gruner P, Konrad M, and Baret J-C. Micro-optical lens array for fluorescence detection in droplet-based microfluidics3. 2013 [cited 2017 Aug 7];13. Available from: www.rsc.org/loc

[75] Myers FB, and Lee LP. Innovations in optical microfluidic technologies for point-of-care diagnostics. Lab Chip [Internet]. 2008;8(12):2015–31. Available from: http://pubs.rsc.org/en/content/articlehtml/2008/lc/b812343h%5Cn/Users/flisi/Google Drive/7.0 Virus sensors/2008,Rev,LoC-Innovations in optical microfluidic technologies for PoC diagnostics.pdf%5Cnhttp://dx.doi.org/10.1039/B812343H

[76] Yao M, Shah G, and Fang J. Highly sensitive and miniaturized fluorescence detection system with an autonomous capillary fluid manipulation chip. Micromachines. 2012;3(2):462–79

[77] Long F, Zhu A, Zhou X, *et al.* Highly sensitive and selective optofluidics-based immunosensor for rapid assessment of bisphenol A leaching risk. Biosens Bioelectron [Internet]. 2014;55:19–25. Available from: http://dx.doi.org/10.1016/j.bios.2013.11.069

[78] Zhang R, Zhao BB, Huang K, *et al.* The 2018 GaN power electronics roadmap. Appl Phys Lett [Internet]. 2017;8(5):999–1002. Available from: http://xlink.rsc.org/?DOI=C7TC02221B

[79] Lafleur JP, Jönsson A, Senkbeil S, and Kutter JP. Recent advances in lab-on-a-chip for biosensing applications. Biosens Bioelectron. 2016;76:213–33

[80] Mata A, Fleischmann AJ, and Roy S. Fabrication of multi-layer {SU-8} microstructures. J Micromech Microeng [Internet]. 2006;16:276\,-\,284. Available from: http://dx.doi.org/10.1088/0960-1317/16/2/012

[81] Armani M, Rodriguez-Canales J, Gillespie J, *et al.* 2D-PCR: a method of mapping DNA in tissue sections. Lab Chip. 2009;9(24):3526–34

[82] Norian H, Field RM, Kymissis I, and Shepard KL. An integrated CMOS quantitative-polymerase-chain-reaction lab-on-chip for point-of-care diagnostics. Lab Chip [Internet]. 2014;14(20):4076–84. Available from: http://xlink.rsc.org/?DOI=C4LC00443D

[83] Sze CCN. Physics of Semiconductor Devices [Internet]. Vol. 10, America. 2014. 739-751 p. Available from: http://link.springer.com/10.1007/978-3-319-03002-9

[84] Hammond PR. Spectra of the lowest excited singlet states of rhodamine 6G and rhodamine B. IEEE J Quantum Electron [Internet]. 1979;15(7):624–32. Available from: http://ieeexplore.ieee.org/xpls/abs_all.jsp?arnumber=1070050

[85] Lee SJ, Shi W, Maciel P, and Cha SW. Top-edge profile control for SU-8 structural photoresist. Proc 15th Bienn Univ Ind Microelectron Symp (Cat No03CH37488). 2003;389–90.

[86] Del Campo A, and Greiner C. SU-8: A photoresist for high-aspect-ratio and 3D submicron lithography. J Micromechanics Microengineering. 2007;17(6):R81

[87] Tapia Estavez MJ, Arbeloa FL, Arbeloa TL, and Arbeloa IL. Absorption and fluorescence properties of rhodamine 6G Absorbed on aqueous suspensions of Wyoming montmorillonite. Langmuir. 1993;9(14):3629–34

[88] Ali M a., Moghaddasi J, and Ahmed S a. Temperature effects in rhodamine B dyes and improvement in CW dye laser performance. Laser Chem [Internet]. 1991;11(1):31–8. Available from: http://www.hindawi.com/archive/1991/763535/abs/

[89] Ahmed RM, and Saif M. Optical properties of rhodamine B dye doped in transparent polymers for sensor application. Chinese J Phys. 2013;51(3):511–21

Chapter 11

Reliability issues of MEMS resonators

Poorvi K. Joshi[1] *and Meghana A. Hasamnis*[1]

This chapter introduces the terms and concepts needed to describe and evaluate MEMS resonator reliability. It is interesting to note that the reliability theory concepts are also used by actuaries in calculating life insurance premiums. Moreover, the life span of human provides a useful analogy for component reliability as well. This text describes some of the main points of various reliability issues of MEMS resonators arising due to its process technology, operation, and transduction method, the failures due to the packaging of these devices, aging, and frequency drift.

Lord Kelvin stated, "When you can measure what you are speaking out and can express it in numbers, you know something about it. But when you cannot measure it, when you cannot express it in numbers, your knowledge of the subject is of a meager and unsatisfactory kind".

So, we will first discuss the quantification of reliability for the system.

11.1 Introduction

Reliability is very much related to availability, which can be described as the ability of a component or system to perform at a specified moment or at a particular interval of time.

Reliability underlines the dependability of a product during the lifecycle management. In terms of availability, reliability can be defined as a probability assumed or worked out from reliability, testability, and maintainability. In terms of the frequency of failure, reliability can be defined as the probability of the required output of a component or system [1].

In nutshell, reliability means how long a component or system will perform its intended function without a breakdown or degradation of performance over time. It is the probability that something will work when you want it to.

Reliability requires a worm's view study of both qualitative and quantitative parameters of the component or systems. The assessment of reliability is difficult; it is usually based on available mathematics and statistics relying on validating

[1]Department of Electronics Engineering, Shri Ramdeobaba College of Engineering and Management, Nagpur, India

quantitative methods for prediction and measurement, to find the root cause for the failure of component and or system. In the absence of validated quantitative and qualitative, it is easy to portray the probability of success based on equations, however considering the massively multifaceted nature of probability the exact magnitude of success is difficult to predict [1].

Reliability, safety, and quality all go hand in hand for predicting the probability of failure. Reliability focuses on the durability aspect, safety focuses on preserving life and quality focuses on safeguarding the system requirement. High-reliability levels can also be achieved from good engineering and from paying attention to details, rather depending too much on using reliability calculations and statistics [2].

11.2 MEMS reliability

"Reliability" is an essential part of a product. No reliability means no business. Reliability is one of the critical drivers for acceptance and growth of MEMS. Emerging technologies depend on the acceptance of the market and as such needs to be designed into high volume and critical applications. As such, the approach for reliability physics must be handled from the root level at the time of evaluating and predicting field performance considering the lifetime of the product [1].

The portion of the lifetime prediction of the reliability program is seen in Figure 11.1. Reliability testing is one of the important factors which is required to accelerate the lifetime of the MEMS part with the help of acceleration factors which helps to identify proper lifetime prediction.

For understanding the behavior of the device under various operating conditions and environment, reliability and sustainability play a pivotal role in commercialization of MEMS. We must consider reliability for the whole life cycle of Reliable MEMS from fabrication to wear out stage for getting the same accepted widely in the market.

We generally found that reliability often dominates the device designs and as such its maximum performance may be traded off in return for reliability. Further,

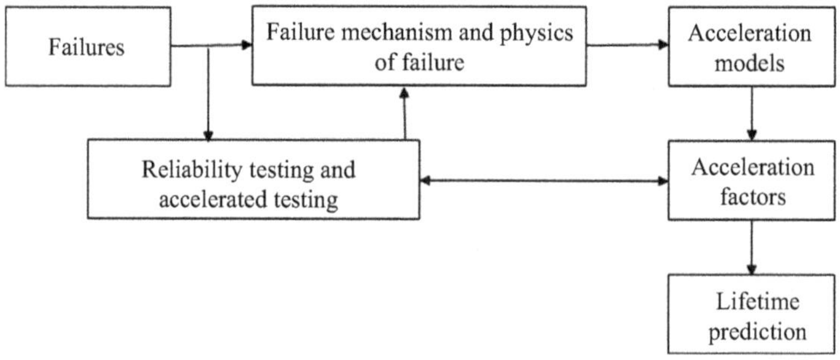

Figure 11.1 Lifetime prediction diagram [4]

reliability evaluation and control also contributes to the improvement of device performance. In the current ever-growing scenario, since the products are distributed all over the world and used in a wide range of environments reliability evaluation of the products took over the front seat and plays a more important role than ever [2].

MEMS devices, which are micromechanical devices fabricated using semiconductor fabrication technologies, are finding applications as core products for integrated systems which concentrate on miniaturization and advanced functionalities. Compactness and portability of MEMS devices paved the way for utilization even in mobile applications. Due to the size effect with the miniaturization of MEMS devices' dimensions and the precision of microfabrication, the reliability is considered to be high. Over a period of time, it is observed that high reliability of the MEMS devices has been established by the frequent development effort of many researchers and engineers. However, answers for the several reliability issues and their related phenomena still need to be identified. This opens a field for spending considerable effort to find a generalized theory on MEMS reliability which concentrates both in devices and materials. It seems that in the near future, the ideas and innovations concerning the MEMS reliability will be organized into a whole and every engineer will be able to design MEMS device of high performance and high reliability [1].

Micro materials for MEMS structures need to be concentrated upon for the reliability which is an entirely new aspect. It is observed that the reliability of Micro materials was not as important by far in classic microelectronic devices. Due to the advancement in evaluation methods, experimental trials and analysis of results observed have taken a giant step, which resulted in much significant new knowledge. The scope for increased reliability studies for advancing MEMS applications operating in harsh environments such as high temperature, high pressure, and high radiation also emphasizes the need for increased reliability studies and not only limited to the MEMS products that are already running in the market [1].

Given the fact that the reliability of MEMS devices can be extremely sensitive to the environmental conditions as such stringent demands for the design, the materials used, and the package is envisaged. Reliable MEMS device can only be accepted in any applications when reliability is taken into account from the very start of the development process. The same will be possible taking into consideration a very close interaction between all the concerned involved in the process, viz. designers, processing engineers, packaging specialists and simulation, test and reliability engineers. The purpose of "reliability tests" is, for this reason, better defined as: "To investigate and predict the effect of design, processing, packaging and use in different environments and conditions on the functioning and the lifetime of MEMS and define corrective actions" [3].

As we want to measure reliability, we need a standard or a unified series of standards against which we can compare our results. In case of reliability studies, there are standards which specify the operating conditions in which reliability is measured. In the field of reliability, there are two such yardsticks, which are typically referred to are the International Electrotechnical Commission (IEC) specifications and the United States Military Standards (MIL-STDs) [2,3].

The United States and most of the other industrial nations is a participant to the IEC documents. However, these documents are seldom referenced or utilized within the United States. This paves the way for the United States Military Standards as primary reference documents. Frequently referring to these military documents did not go down well with many industrial and commercial manufacturers, However, it looks like that there have been little other choices and must out of necessity reference some set of standards, if for no other reason than of comparison [3].

One of the major advantages that the military specifications offer is that documentation of a selection of practical tools is made that can be used to assess the reliability for military, industrial or commercial equipment. A change in philosophy in the military standards is allowing reliability tasks to be custom-made to explore the unique aspect and characteristics of a given program and not only depending on the letter of the specification. The usefulness of military standards and practical implementation in the field of reliability has taken a leap due to this change in allowance of philosophy thereby enhanced the possibility of their use in nonmilitary applications as well [3].

In the following portion, bathtub curve, one of the major philosophies, being used for reliability distribution, is discussed.

11.3 The bathtub curve

The important factor for the MEMS reliability is to know the distribution of failures over the lifetime of the product cycle. With the help of these concepts, distribution functions can be developed and used for predictive purposes. Bathtub curve reflects a hazard rate that changes over the lifetime of the product which is observed as high at the starting then subsequently reducing and increasing towards the end of the product life. The term "bathtub curve" emerges due to the behavior of product that will fail within the first few weeks to months of the product lifetime which is known as the Infant Mortality Portion. In general, an ideal failure behavior is to identify and eliminate the failures due to defects in the initial stage only, i.e. in the infant mortality portion of the curve through burn-in and/or defect reduction programs and avoid operating the product into the wear-out phase. The operational life is identified within the typically constant hazard rate section of the curve [3].

11.3.1 Failure rate over the life of a product

The failure rate is expected to vary over the life of a product – "Bathtub Curve"

Equipment reliability is frequently described by the familiar bathtub curve shown in figure 11.2. The bathtub curve is a composite curve formed by the addition of three separate curves: the first curve or phase is that of infant mortality, the second curve or phase is of useful life and the third is of wear-out. (These three curves are analogous to the three major factors in human lifespan statistics.) Each

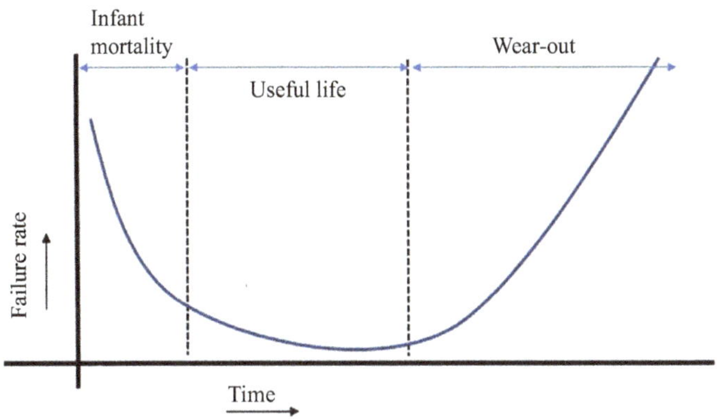

Figure 11.2 Bathtub curve [1]

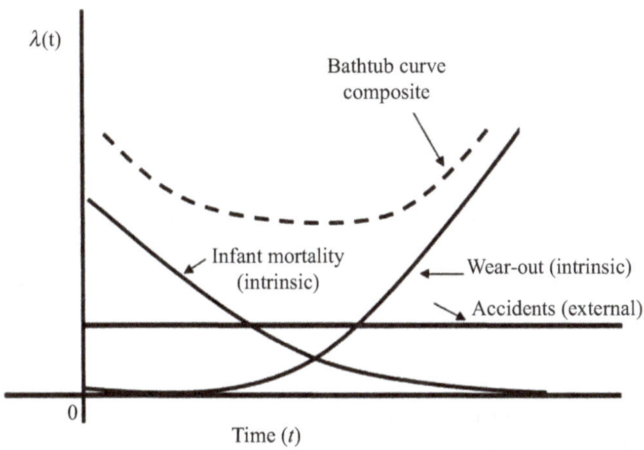

Figure 11.3 Illustrates how the bathtub curve can be the composite of three failure rates: infant mortality, wear-out, and externally induced failure [4]

of these curves also relates to the phase during which it is the principal contributor to an individual or an equipment's hazard rate. Distinctly different phenomena dominate during each of these three phases of the equipment's life [3].

Figure 11.3 illustrates how the bathtub curve can be the composite of three failure rates: infant mortality, wear-out, and externally induced failure.

Bathtub curve or life-cycle curve evaluates [3] and brought to notice the following:

- typical life history curve for infinite number of items
- comparison of failure rate with time
- three distinct phases – debugging, chance failure, and wear-out phase

Table 11.1 Bathtub curve summary

Phase	Failure rate	Possible causes	Possible improvement actions
Burn-in	Decreasing failure rate	Manufacturing defects, welding, soldering, assembly errors, part defects, poor quality control, poor workmanship, etc.	Better quality control, acceptance testing, burn-in testing, screening, highly accelerated stress screening, etc.
Useful life	Constant failure rate	Environment, random loads, human errors, chance events, "Acts of God", etc.	Excess strength, redundancy, robust design, etc.
Wear-out	Increasing failure rate	Fatigue, corrosion, aging, friction, etc.	Derating, preventive maintenance, parts replacement, better material, improved designs, technology, etc.

During the first phase, debugging (infant mortality) phase, the weaker parts have not yet been removed from the total part population. As these weaker parts are identified and removed from the population, either by a formal screening program or through normal equipment operation, the instantaneous failure rate, or the hazard rate, decreases rapidly to more or less residual level.

The residual value from the above stage persists through the second and most important phase, the useful life period. During the useful life period, failures still occur occasionally but they are random in nature and are randomly distributed with respect to time. The statistical frequency of these failures can be predicted with fairly good accuracy, but the exact time and location of actual failures cannot be pinpointed and predicted.

The third and final phase of equipment life is wear-out. During this phase, the accumulated damage due to the applied stresses begins to take its toll. The parts generally become weaker, more prone to failure and thus they fail with increasing frequency. For most types of electronic parts, the equipment itself becomes obsolete and is retired long before the parts enter the wear-out period. Thus, for the vast majority of electronic parts (but unfortunately not for all), the wear-out period can be safely ignored [3,4].

The abstracts of the bathtub curve portions are listed in Table 11.1.

To express lifetime behavior statistically, there are various distributions can be referred to. The exponential, the Weibull and the Lognormal are the known statistical reliability distributions.

11.4 Reliability evaluation methodologies

Mathematical statistics used in reliability shows significant results for lifetime behavior. The survivor or reliability function, cumulative distribution function, probability distribution function, hazard function, and the bathtub curve concept are to be used to measure failure distributions and predict reliability lifetimes.

Reliability is the probability of the product performing properly under typical operating conditions for the expected lifetime intended, and an expression to define reliability is [3,4]:

The probability of the device to perform as per its specifications and function under different operating conditions for an intended lifetime is term as reliability and is express as equation (11.1). It is defined as the probability of device operating without fail in time t.

$$R(t) = 1 - F(t) \tag{11.1}$$

where

- $R(t)$ is the reliability function or the survivor function.
- $F(t)$ is the cumulative failure distribution function (CDF). (Probability that device will fail if chose randomly by time t). A lifetime distribution model $f(t)$ is the probability density function (PDF) over the time range 0 to ∞ (infinity).

The relationship between the CDF and PDF is shown below.

$$f(t) = \int_0^t f(t')\,dt'$$

$$f(t) = \frac{d}{dt}F(t)$$

The hazard rate $h(t)$ is also known as the instantaneous failure rate. The probability of operating without failure up to that time interval.

$$h(t) = \frac{f(t)}{1 - F(t)} = \frac{f(t)}{R(t)}$$

Also written as

$$h(t) = -\frac{1}{R(t)}\frac{dR(t)}{dt}$$

which is equivalent to

$$h(t) = -\frac{d}{dt}(\ln R(t))$$

The integral of the hazard rate is the cumulative failure rate (cumulative hazard rate).

$$H(t) = \int_0^t h(t')\,dt = -\ln R(t)$$

The hazard rate $h(t)$ or instantaneous failure rate has the dimension time^{-1}. Since $R(0) = 1$ (no failures at time zero), the reliability rate over a time period t is the exponential of the cumulative hazard rate in that same period t.

$$R(t) = e^{-\int_0^t h(t')\, dt'}$$

An important quantitative reliability concept is how long the device will survive without failure. This is also termed meantime to failure (MTTF), more specifically, the mean-time to the first failure.

$$\text{MTTF} = \bar{t} \equiv \int_0^\infty tf(t)\, dt$$

11.5 Acceleration factors

Squeezing of the period for getting lifetime data of reliability testing at accelerated conditions is critical, which is made possible in a much shorter period. To get the answers to the performance in a reasonable time, the accelerated method is to be adopted. Using higher-than-normal stress, failure can be induced earlier than usual. This concept is used for releasing reliable product to market. For accelerated testing, stresses experienced in-use environment are "accelerated" or amplified to a level to accelerate the time to failure of an individual failure mechanism. The point to note for the reliability resting in the accelerated condition is to create the same failure mechanism as occurs in-use conditions. The know-how of the physics of failure helps in developing an acceleration model. An acceleration factor is calculated as compared to the use conditions. The known accelerated model which is the outcome of the method adopted can be used to predict the lifetime under normal condition. Highest reliability rating devices though limit the likelihood of faulty components causing product failure in the field, but one must ensure that parts have been adequately tested for reliability [3,4].

Ever-growing customer requirements and harsh market demands for improved reliability will never end. Steady improvement is a must to stay in the business. New product development and technology development should always put reliability as a priority. The continuous cycle of design, qualification, mass production, evaluation, and customer feedback enable to further improve the reliability of the product.

The field of MEMS does have a limited history of known failure models and acceleration factors are also not available in numbers when compared to the more seasoned semiconductor industry. With the approach of accelerated testing, the different acceleration factors are applied keeping in mind that the factors are not over applied on the product or system. The mapping of data plays a pivotal role during the accelerated testing which translates into the actual requirement of the product during the normal lifetime.

The operating environment, storage environment and the lifetime of the product are the prerequisites for the apt utilization of acceleration models while comparing to use the conditions. This will vary for each product development effort and depends on the common requirements of the customer and supplier [4].

A summary of some known MEMS failure mechanisms and accelerating stresses is provided in Table 11.2.

Table 11.2 Examples of failure mechanisms and accelerating factors [4]

Failure mechanism	Accelerating factors
Cyclic fatigue	No. of cycles, maximum applied strain, humidity
Creep (plastic deformation)	Temperature, applied strain
Stiction	Humidity, shock, vibration
Shorting and open circuits	Electric field, temperature, humidity
Arcing	Electric field, gas pressure, gas composition
Dielectric charging	Electric field, temperature, radiation, humidity
Corrosion	Humidity, voltage, temperature
Fracture due to shock and vibration	Acceleration, frequency (resonance), vacuum

Following are the prerequisites for lifetime predictions [4]

- Knowledge of operating and nonoperating environmental, the lifetime of the end product, and use conditions during manufacturing
- Packaging and application of end product
- Customer's acceptable failure rate over the lifetime of the product
- Necessary stress conditions to classify failure mechanisms
- Acceleration testing and models for lifetime prediction
- Failure distributions data, viz. statistical manipulation in reliability testing.

The point to assume that the shape of the curve is the same in the accelerated condition as in the use condition while using acceleration data to predict the lifetimes with acceleration models.

11.5.1 Lifetime units

FITS and ppm (parts-per-million) failure are the two popular units in which failure rates are reported. The unit of FITS is defined as the number of failures in 10^9 device-hours. The ppm unit is always given over a stated time interval. The FITS unit is a rate of failure, while the ppm is a cumulative amount of failures out of a known population over a specific period [4].

11.6 Failure modes and mechanisms

The process in which the expected or actual failure modes of a product are analyzed and suitably their effect is identified to reduce or mitigate is termed as reliability engineering. A product or a process could be potentially failed to achieve the desired outcome and the failure modes explain the reasoning behind the same. A failure mode has different scenarios viz. It can come between a cause and an effect or it is also possible that the cause or effect themselves might be the failure mode or for a single event to be a cause, effect and failure mode. Failure modes can be broadly categorized into two types, viz. design failure modes and manufacturing failure modes, which depend on their origin in the product development phase [5].

11.6.1 Design phase failure modes

The failure modes that originate in the design phase are distinguished by their origin in the design phase and can identify into two subcategories of failures—functional and material modes [4].

11.6.1.1 Functional failure modes

Inadequate design leading to a deviation from the specification due to degradation, loss or absence of desired performance is identified through functional failure modes. The overall reliability of the part gets affected due to the functional failure occurs due to the non-functionality of the device in the field. Insufficient designs are the factors which lead to the loss of function of the part which may occur at the beginning or later in life. It is essential to get to know the prediction well in advance thereby minimizing the risk of field failures and improving the reliability of the part. The functional failure modes can be divided between MEMS element design, system-level design, and package design.

Elemental design failures due to mask data faults, design rule violations, and engineering analysis faults lead to failures where the MEMS element does not perform as expected.

System-level design relates to the field failures due to system complexity. System complexity is quite visible in the MEMS product which comprises MEMS element, electronics, and package. This system complexity is a significant challenge to prepare a model which predicts overall system behavior. It is a daunting task to model without segregating the subcomponents to a sufficient level of abstraction without loss of accuracy. It is observed that the increase in system-level model abstraction decreases the failure predictability [4].

Package design of MEMS devices presents a unique challenge for MEMS designers and reliability engineers. If we compare with the conventional IC packaging wherein it is clear in the sense that the package must reject certain inputs like moisture, contamination, etc. and tolerate common forces such as temperature, shock, handling or tester forces but packaging it is quite evident that the MEMS element must interact directly with the outside world in order to perform their design function. MEMS products which sense pressure, temperature, chemical species, or control light or sound need direct interaction between the sensing element and the input that needs to be measured. Due to the varying requirement of MEMS application, it is observed that each MEMS application usually requires a new package design to optimize its performance or to meet the needs of the system which is also the primary reason why the cost of packaging a MEMS device remains high. Out of several categories of MEMS packages including metal packages, plastic packages, and thin-film multilayer packages that are similar to standard semiconductor packaging, there are several other factors that envisage MEMS packaging more complicated. As such, there is a need for comprehensive MEMS package design integration [4].

11.6.1.2 MEMS material failure modes

MEMS material failure modes specific are commonly divided into the following categories: thermo-mechanical, electrical, and environmental failures.

The failures resulting from thermo-mechanical forces which include the most common MEMS stress failures, i.e. residual stress, are known as thermo-mechanical failures.

Static or dynamic charge transfer within materials or across gaps or surfaces leading to several potential failure modes is categorized as electrical failure.

Environmental failures occur due to various reasons depending on the MEMS applications in which the sensing element is directly exposed to the operating environment which could, in some cases, be quite aggressive. The interactions between environmental forces and the materials within the device can result in several failure modes in MEMS [4].

11.6.2 Manufacturing failure modes

Failures related to manufacturing are due to the specific cycle of processing that has typical characteristics which are usually unavoidable. Manufacturing failure modes can be segregated into two types depending on where they originate in the manufacturing process viz. front-end processing and back-end processing. Figure 11.4 identifies some of the main manufacturing-related steps where defects may be encountered. Front-end processing can also be referred as local, which primarily leads to contamination and any form of voids or stringers; whereas back end, which can be referred as global defects, includes a broader spectrum of defects, from those due to wafer-level variations and handling to those related assembly.

11.6.2.1 Front-end process defects

Front-end process-related defects are divided into the following categories, viz. material transport failures—those due to deposition and etch steps, wafer-bonding failures—hermiticity, and tribological failures-stiction.

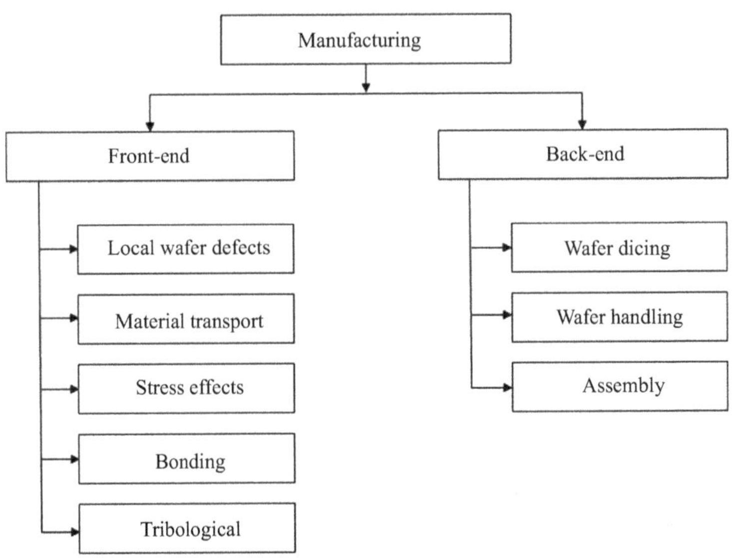

Figure 11.4 Manufacturing process-related defects [4]

11.6.2.2 Back-end process failures

Back-end process steps during the MEMS product development include all the steps after the final wafer fabrication step such as dicing, assembly process steps, and ATE20 testing. These steps can introduce a variety of failure modes [4].

11.6.3 In-use failures

Physics of failure puts the onus on the reason for failure of a well-designed, fabricated and packaged device. It is observed that there is an interconnection between the design, manufacturing and in-use failures. Know-how of the physics of failure, the properties of materials used and the link to the process flow enables in making improved design rules for reliable operation of the device in the expected operating environment. In-use failures can be segregated based on the origin, viz. mechanical origin (shock, fatigue, creep), electrical origin (dielectric charging, ESD) and related to the environmental effects (radiation, anodic oxidation).

Various predominant failure modes are found out in different devices. As the device gets older, this mode will change as successive changes in the process of development of the device thereby increasing the device reliability. To keep a pace and expectation of rapid time to market requires adaptation of concurrent engineering practices for efficiently designing a MEMS device to be reliable. Knowing the details of the physics of failure ensures high confidence to perform accelerated testing which is one of the key elements [4].

11.6.3.1 Mechanical failure modes

Fracture

One of the important failure modes for MEMS device is fracture. These generally include micromachined mechanical components. Given the property of silicon as a brittle material, it is presumed outside MEMS filed that silicon MEMS are the fragile device. However, if designing of the device is made in such a way that device operates well below fracture strength then silicon is suitable mechanical material, as long as one designs the device to operate well below the fracture strength. Elastic behavior of suitably engineered silicon suspensions designed with large safety factors can also be considered [4].

Mechanical shock resistance

Sudden acceleration defines a shock. Unit for shock is referred to as "g", where 1 g = 9.81 ms^{-2}, the acceleration in Earth's gravity. MEMS devices need to withstand large shocks to be reliable as the devices can accidentally face shock due to occasional dropping, falling of the pocket, falling on the floor during assembling, and rough handling during sports activity. Around 500 g for a 1.5 m fall on a hard surface is encountered [4].

The failure modes arisen due to shock [4] are as follows:

- Exceeding the yield strength of the material because of a large shock-induced deflection, resulting in fracture
- Parts coming in contact resulting in stiction

- Delamination
- Broken parts or particulates being generated or being displaced leading to short circuits and/or mechanical blockage.

A few points that make MEMS more shock-resistant are mentioned [4] below:

- The small mass of MEMS devices enables devices to easily accommodate shocks of a few g.
- Spring design also enables to deal with shocks
- Shocks of higher magnitude need more careful design as well as packaging.
- "Stoppers" is one of the techniques to limit the motion of beam as it minimizes displacement and kinetic energy
- Surfaces which might come into contact to be at the same electrical potential to avoid a current spike
- Utilization of package to dissipate the shock load.
- Stiction is also equally responsible for failure mode in addition to fractures.

Vibration

Failure due to vibration needs to concentrate on to determine a lower limit on vibration level which is expected for failure as a function of frequency. Estimation of the vibration acceleration which brings parts into contact or to reach fracture stress is to be known before vibration testing to avoid surprises. It can serve as a tool for the designer [4].

The lower limit on maximum safe vibration levels can be determined with the following three steps [4]:

- Possible failure modes due to vibration are to be identified
- Dynamics of the device is to be determined.
- To generate a plot of a safe lower limit of vibration thereby avoiding contact.

Creep

Plastic deformation under applied strain is called creep, which occurs only in ductile materials affecting metal thin films and not silicon (except at temperatures above 600 °C). Creep is a consequence of dislocation motion which depends on Temperature, Stress and Time. It is expressed as d€/dt, where € is the strain and t the time. With regard to atomic diffusion and dislocation mobility, temperature plays a key role. The homologous temperature (ratio of operating to melting temperature T_{melt}) provides insights related to dislocation mobility [4].

Three temperature regimes which are often used are:

- $0 < T < 0.3\ T_{\mathrm{melt}}$: no creep observed
- $0.3\ T_{\mathrm{melt}} < T < 0.9\ T_{\mathrm{melt}}$: dislocation motion leading to creep
- $0.9\ T_{\mathrm{melt}} < T < T_{\mathrm{melt}}$: diffusion creep.

By increasing creep resistance of a MEMS device in the following path helps in reducing creep in MEMS:

- Operating temperature to be reduced
- Applied stress levels to be reduced
- Material to be changed.

Few points regarding creeps allowing MEMs device more reliable [4] are:

- As long as there is no metal on the suspension, silicon MEMS are not affected by creep roughly below 600 °C. This is observed for SiN_x and SiO_x flexures too.
- Creep can be one of the main failure modes for metal MEMS
- For metal MEMS, creep can be mitigated by a few means, viz. reducing the applied stress, reducing operating temperature, or a change of material.

Fatigue

Initially, fatigue was considered to be an important failure mode since MEMS has moving parts, but it does not turn out to be a life-limiting factor in any commercial MEMS device. For silicon, fatigue occurs only for applied stresses greater than half the single-cycle fracture strength and thus any reasonable design will not have stress levels too high for fatigue to be relevant. For metal MEMS, fatigue can occur at lower relative stress levels but is generally much less problematic.

By controlling the maximum stress and the relative humidity, it is possible to avoid fatigue. Metal MEMS are more susceptible to fatigue because of their lower melting point [4].

Fatigue in brittle and ductile materials

The decrease in yield strength which is cycle-dependent and relates to slow crack growth leading to failure due to a periodically applied stress results in fatigue. The crack grows at each cycle of alternating stress thereby reducing the strength of the material and eventually leading to failure.

It is observed that fluctuating loads can be a factor for failure while the monotonic loads do not. Fatigue data is often plotted as a stress-life (S/N) curve between the maximum applied cyclic stress vs. the number of cycles to failure

Fatigue behavior in ductile materials (e.g., most metals) and brittle materials (e.g., silicon, ceramics) is very different. Fatigue generally occurs due to plastic deformation while brittle material does not deform at ambient temperature. Fatigue in ductile material occurs over a large range of stresses while in brittle material fatigue only occurs for stress levels near the yield strength [4].

Silicon MEMS

Fatigue has never been observed in the air at room temperature for bulk samples of Silicon. Moreover, dislocation activity is not observed at low homologous temperatures ($T_{ambient}/T_{melt} < 0.3$), and there is no sign of extrinsic toughening mechanisms, nor of stress, corrosion cracking. As such it is observed that fatigue was not expected in silicon in air room temperature.

Relative humidity plays a key role in fatigue lifetime for silicon.

High-cycle fatigue of silicon parts will not occur in the following cases [4]:

(a) The maximum cyclic stress is less than 20% of single-cycle fracture and if humidity is not controlled, or
(b) Maximum cyclic stress is less than 40% of single-cycle fracture and the device is hermetically packaged in an ultra-dry ambient.

Since silicon is a brittle material exhibiting a range of fracture strength, the conservative MEMS designer will limit the maximum designed stress to below 20% of yield strength, thereby avoiding fatigue.

11.6.3.2 Electrical failure modes

As we know that MEMS devices have moving parts unlike other integrated devices, it is presumed that their failures are often to be mechanical in origin. However, this might have been true in the initial days of MEMS but is no longer the case. As the design approach for MEMS has progressed, electrical failures have taken a major role in MEMS lifetime when compared to mechanical failures.

Electrical failure modes [4] happen due to the following reasons:

Dielectric charging

Principally, electrostatically driven or sensed MEMS devices are the ones which got affected with dielectric charging. MEMS having electromagnetic or thermal actuation or sensing principles are not sensitive to dielectric charging.

Bulk and surface leakage currents, and associated trapped charge, are a strong function of voltage, temperature, and relative humidity. In MEMS devices, the stress due to dielectrics must be carefully taken care of because it often serves a structural role.

Electrostatic forces which are undesired and difficult to predict time-varying rising from charging of the dielectrics in MEMS structures is the reason a serious performance issue for a wide range of electrostatically driven or sensed MEMS devices including microphones, displays, micromirrors, and RF switches.

The gradual shift in actuation, or a gradual change in rest or actuated position, or a gradual shift in release voltage is observed due to the failure modes related to dielectric charging. Drift in applied electrostatic force as a function of time is the main reason behind this.

The ways to solve or minimize the "charging" problem [4] are as mentioned below:

- Bipolar AC drive voltage
- Geometry changes: to minimize the area of exposed dielectric, or pattern the dielectric
- Shield-movable parts (sense mass, actuators) from electric fields due to trapped charge, selectively remove dielectric to avoid charging
- Charge dissipation layers to remove the surface charge and provide shielding
- Change dielectric or change composition to reduce the amount of trapped charge or
- Decrease discharging time constants (e.g., SiO_x instead of SiN_x)
- Reduce electric fields (e.g., thicker dielectric or with higher dielectric constant,
- Redesigned springs to operate at lower voltages)
- Optimized drive voltage (multi-level: one to actuate, one to hold), or charge monitoring
- Control of packaging ambient to minimize humidity and contaminants.

Electrical breakdown and electrostatic discharge event

Electrical breakdown is to be studied through a solid dielectric (e.g., an insulator such as silicon nitride or silicon oxide film) and through a gas (e.g., arcing between neighboring electrodes). The need to observe the electrical breakdown happens due to the sudden voltage and due to the current pulse from electrostatic discharge (ESD) event. The study of the lower but longer-lasting voltage from the normal drive signal for a MEMS device is also to be worked out. The approach to study this breakdown is passing through the following sequence, viz. breakdown through a gas, then though a solid, and finally the effect of ESD discharge on MEMS, for which the mechanical time constant is generally much longer than the pulse duration [6].

Electromigration
Migration of metal atoms under an applied electric field is called electromigration (EM). It is to be noted that one must distinguish electrolytic (metal ions in solution) from solid-state (atom motion in a metal wire or trace due to electron momentum transfer) electro migration. Electrolytic EM leads to failures by dendritic bridging which takes place on the printed circuit board level when enough moisture is present to allow surface conduction between neighboring conductors by ionic transport. Solid-state EM leads to failures such as voids and dendrites which occurs in microfabricated wires, in which the momentum from electrons can cause atomic displacement of the conductors when the current density and temperature are sufficiently high. Since very high current densities are encountered in IC circuits, Solid-state electromigration is an important failure mode for microelectronics which does not depend on ambient moisture [4].

11.6.4 Environmental failure modes

Failures due to factors external to the MEMS device also play an important role. Packaging plays a vital role when dealing with the interplay between environment and device.

Environment failure modes affecting all microelectronic devices [4] are categorized as

- Due to the effect of radiation on MEMS
- Due to the anodic oxidation and galvanic corrosion of poly-silicon, as examples of MEMS-specific corrosion issues
- Metal corrosion due to airborne humidity or ionic contaminants or atmospheric pollutants.

11.6.4.1 Radiation

Suitability of electronics components for use in high-radiation environments and design techniques to develop radiation-tolerant electronics and optics have adopted worm's eye views over a period of time. There exist well-established test procedures for space applications taking into consideration the physics of how different energetic particles interact with matter, the types of damage that are caused, and the influence on most electronic devices, optical components, and mechanical parts. There is no standard test procedure for the effect of radiation on MEMS due to the vast range of materials, technologies and applications MEMS have found out into. There are some proposed approaches though to test the effect of radiation on MEMS [4].

The point concerning the high end of space mission doses' effect on the mechanical properties of silicon and metals is mostly unchanged as yield strength is not significantly affected. Silicon as a structural material can be viewed as intrinsically radiation hard. This feature of silicon makes most MEMS devices mechanically radiation-tolerant by default.

Accumulation of charge in dielectric layers due to ionizing radiation is the main reason for MEMS devices failure. The trapped charge leads to device failure which happens due to the instance of large changes in calibration of capacitive accelerometers, or device failure due to stiction initiated by electrostatic forces from the trapped charge. As such, the devices are needed to be shielded or built with radiation-tolerant technologies.

The main types of radiation encountered near-earth consist of [4]:

- Trapped radiation: energetic electrons and protons magnetically trapped around
- the earth (Van Allen belts)
- Solar energetic particles: mostly highly energetic protons
- Galactic cosmic-rays: the continuous low flux of highly energetic
- Secondary radiation: radiation generated when the above radiation interacts with materials in the spacecraft, notably with shielding.

Accelerated radiation testing methods have been explored despite the complexity of the actual space radiation environment. Accelerated radiation testing methods have been developed with the utilization of mono-energetic particles whose relevance and suitability has been suitably demonstrated for microelectronic devices [7].

11.6.4.2 Electrostatic MEMS sensors and actuators

The main reason for failure in electrostatic MEMS devices at high radiation doses is the accumulation of charge in dielectric layers. It seems that failure occurred due to Mechanical issues however the root cause is the Electrical one. For a given device, total ionizing dose (TID) is the main radiation parameter that quantifies the amount of charging [4].

11.6.4.3 Radiation-hardening MEMS

The different impact that trapped charge in dielectric layers has on different actuation schemes and geometries relates directly to the difference in sensitivity of MEMS devices to radiation. MEMS operating on electrostatic principles are the most sensitive to charge accumulation in dielectric layers. However, MEMS operation on thermal and electromagnetic principles is much more radiation-tolerant.

Following are the techniques that eliminate or minimize the charging effects [4]:

- avoid undesired electrostatic forces (due to the charging of conductors)
- Change of dielectric material to one with lower trap density
- Geometry changes to eliminate the dielectric from between moving surfaces, and from under moving surfaces and to minimize the exposed area of dielectric, or replacing the dielectric films with arrays of dielectric posts
- Electrical shielding, by covering exposed dielectric with a conductor as at well-defined potential

- Since electrothermal and electromagnetic actuation principles are intrinsically more radiation tolerant than electrostatic operation, these actuation principles should be considered for applications where high radiation doses are expected.

11.6.4.4 Anodic oxidation and galvanic corrosion of silicon

Poly-silicon has become the preferred material in surface-micromachined MEMS due to its electrical and mechanical properties as well as its relative ease of processing. For bulk micromachined devices, single-crystal silicon (often SOI) is the preferred material. In dry ambient, the atmosphere found inside a package hermetically sealed in a dry and inert environment as well, poly-Si and single-crystal electrodes show truly impressive longevity. There are no signs of degradation or corrosion after several months at fields close to dielectric breakdown. It is seen that many commercial MEMS devices are designed to operate in an ultra-dry ambient in a hermetically sealed package so that the chip operates in a dry ambient even if the package is subjected to high relative humidity. There can be leakage currents on the dielectric between neighboring electrodes if the package is non-hermetic [4].

Following are the measures for mitigating anodic oxidation [4]:

- minimizing the operating voltages
- careful choice of voltage polarity
- hermetic packaging or environmental control.

11.6.4.5 Galvanic corrosion during release of HF

Final release step in processing MEMS devices is HF, which is very commonly used because of its excellent selectivity toward silicon, and its rapid etch rate of silicon oxides. It is known that 49% of HF does not significantly affect the morphology and materials properties of single-crystal silicon. However, there have been reports of HF affecting the material properties of polysilicon, particularly when metallic layers are present on top of the polysilicon. Metal traces or coatings on poly-silicon are required for electrical interconnections, as optical reflectors, RF antenna, or for stress control. Galvanic corrosion occurs due to the potential differences between the metal devices and the silicon. This corrosion occurs only in the HF bath and is therefore limited to the release step [4].

Few of the solutions to minimize galvanic corrosion are:

- Only metallize the polysilicon after the HF release
- Perform the release in vapor HF instead of liquid HF
- Ensuring reliable substrate contacts to minimize redox potentials.

11.6.4.6 Metal corrosion

Metal corrosion happens either chemically or electrochemically. Electrochemical corrosion is enabled by conduction through water monolayers in which contaminants have dissolved. It is driven by potentials which are either externally applied or galvanic. Corrosion is of particular concern for microelectronics and MEMS because of the μm or sub-μm scale of metal traces. It is to note that very small amounts of corrosion can lead to device failure. MEMS devices as sensors may need to operate in harsh or corrosive environments, for instance, pressure sensors for engine management or for turbine monitoring.

Corrosion has three accelerating factors, viz. [4]:

- Temperature, as temperature accelerates both diffusion and the kinetics of chemical reactions.
- relative humidity as it provides the water surface film and
- concentration of trace contaminants and airborne pollutants of reactive compounds such as sulfides and chlorides that dissolve in the water surface film giving rise to high levels of corrosive ions.

Most common packages for MEMS are plastics (e.g., over-molding on a lead frame) through which water can slowly permeate on the time scale of weeks. As such, most MEMS devices in the current scenario use several adsorbed monolayers of water on the surface.

We have seen many possible failure modes of silicon and metal MEMS devices, along with the physics of failure. The physics of failure is one of the important factors for determining the accelerated testing conditions, once the root cause has been identified. Specific test structures are often designed to investigate the failure mode which helps in a redesign, or a materials change, process change, etc., thereby ensuring enhanced reliability [4].

The device reliability can be determined from accelerated tests and can be compared to customer requests or system needs. The package is a key aspect of the MEMS reliability, as it can control the ambient the chip experiences, especially moisture. Acceleration of charging, corrosion, fatigue, and stiction are the factors causing failure. Packaging varies from device to device and must be attended in the early stage to ensure the expected failure modes can be controlled thereby enhancing the reliability.

11.7 Root cause and failure analysis

It is a daunting task to identify the root cause and find out suitable corrective action of reliability field failures. The points to be kept in mind for reliability program are

- to strategize for identifying potential failure modes
- failure mechanisms
- to identify risk areas in the design and process
- to be aware of containment strategies.

Containment strategies for achieving low failure rate by identifying the root cause and by taking suitable corrective action ensure the effectiveness of implementing for production.

Failure mode and effects analysis (FMEA) methodology is a stepping stone. It is experienced that a structured and well-planned approach is adopted to utilize the optimum proven methodologies for identifying the potential failure modes and mechanism. The FMEA methodology identifies the sections to be evaluated and filled by the. The purpose is to include all the failure modes that can occur using the cause-and-effect approach, [4], viz.

- How the failure occurs
- What parameters are changed
- What prevention measures are safeguarded
- What actions are taken to mitigate or eliminate the failure
- How the corrective actions in the form of reliability testing, inspections, data gathering and analysis are taken for eliminating the failure.

11.7.1 Failure mode and effects analysis

It is found that FMEAs are traditionally used for manufacturing processes and design. The process failure mode and effects analysis (PFMEA) and the design failure mode and effects analysis (DFMEA) are already in demand by many MEMS customers. A standard that defines the FMEA is the SAE J 1739 out of many others available. The reliability failure mode and effects analysis (RFMEA) approach is to be adopted for MEMS reliability.

The sequence to be followed in RFMEA is shown in a flow chart used (Figure 11.5). The goal of the project is to be well defined which enables the course of flow of RFMEA. Then comes the requirement of a team of engineers and scientists to generate the FMEA within the defined scope. It is to note that the input for RFMEA will be found within the PFMEA and DFMEA documents. There will be unique failure mechanisms only identified in the reliability FMEA [4].

11.7.2 RPN (risk priority number) levels

The format of the reliability FMEA (RFMEA) can match the format of process and design FMEA. One of the formats that includes a risk priority number (RPN) quantitative value for each identified potential failure mode on the RFMEA. The RPN value is obtained by multiplying three values [4].

- The first value is for severity. For each failure mode, a severity (S) value is assigned. This value is scaled between 1 and 10, where 1 is typically defined as a failure that is not noticeable and does not affect the product performance, while the severity of 10 is the most extreme failure and can result in death.

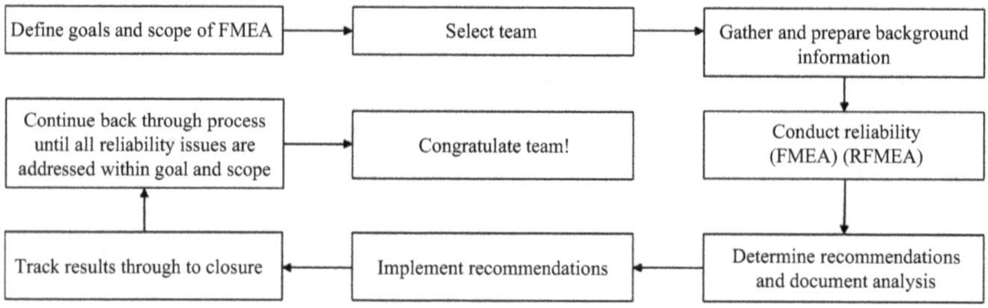

Figure 11.5 An example flowchart for the generation of reliability FMEA [4]

- The second value is an occurrence, which is assigned via the "effect" of the failure mode. The likelihood of an occurrence (O) value is how often the effect is predicted to be observed. Again, a 1–10 scale is given to occurrence where a 1 is very unlikely and a 10 is inevitable.
- The third value is detection (D). This is a function of the effectiveness of the prevention or mitigation measures adopted. If the outcome of detection by the inspection and test within the manufacturing facility, or reliability testing before shipment to the customer cannot bring out a failure mode and if it spills up to the customer then the detection level is 10. It means that the failure mode cannot be detected. Detection level will be 1 if there is a 100% surety that controls are in place which will result in catching the failure before shipment to the customer and the customer will not experience this particular failure mode.

11.8 Analytical methods for failure analysis

Analytical methods for failure analysis or characterization of MEMS parts are the most important aspects for reliability and those are too numerous to present, thus, a few key techniques are covered like dynamic analysis (laser Doppler vibrometry), structural techniques (interferometry, SEM, EBSD, FIB, TEM, AFM) and chemical/compositional techniques (Auger).

11.8.1 Laser Doppler vibrometry

The dynamic information of the MEMS devices is provided by laser Doppler vibrometer (LDV). It is a very powerful technique. In LDV, when the MEMS device is driven with a periodic AC signal, it results in surface movement and an interferometric technique detects the path-length shift of the reflected laser beam. The laser vibrometer also provides animation of the recorded data. The mesh which is adjustable one can cover a single device or entire array, and the laser beam of the laser vibrometer scans set-points within the mesh and records data from different devices within a few seconds. LDV tools exist that can measure out-of-plane single-point vibration, out-of-plane differential vibration, in-plane vibration, rotational vibration, 3D vibration, and both 2D and 3D surface vibration mapping [4].

11.8.2 Interferometry (ZYGO optical profiler)

Optical profiler is interference microscopes and is used to measure the height variations—such as surface roughness—on surfaces with great precision using the wavelength of light as the ruler. Optical interference profiling is a well-established method of obtaining an accurate surface.

Optical profiling uses the wave properties of light to compare the optical path difference between a test surface and a reference surface. Inside an optical interference profiler, a light beam is split, reflecting half the beam from a test material which is passed through the focal plane of a microscope objective, and the other half of the split beam is reflected from the reference mirror (Figure 11.6) [8].

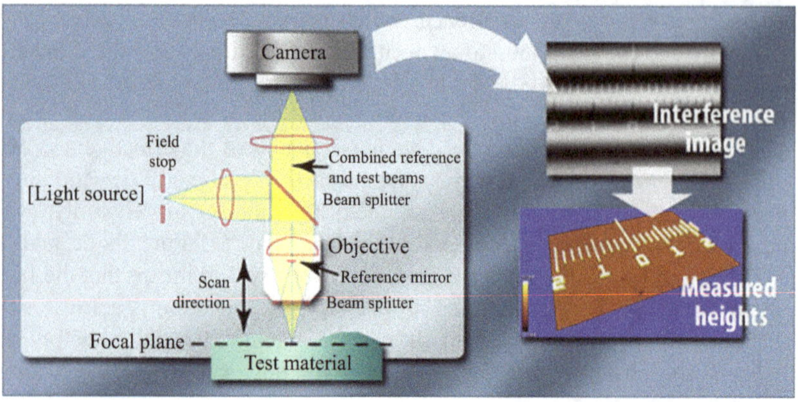

Figure 11.6 Basic operation of optical profiler [8]. Image courtesy of ZYGO

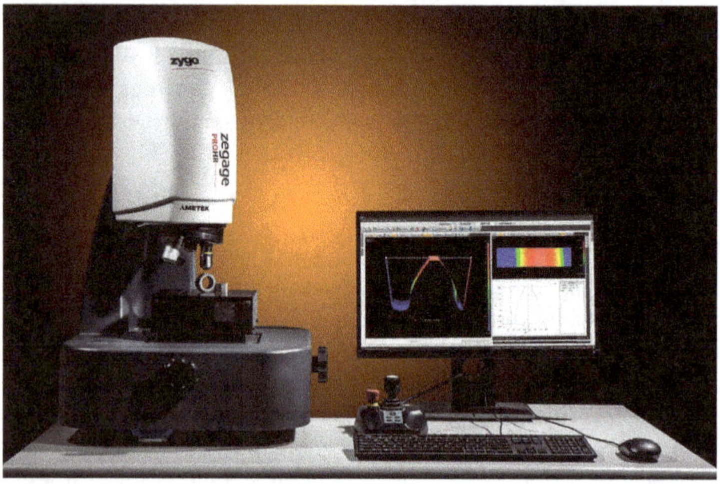

Figure 11.7 ZeGage™ Pro Instrument [8]. Image courtesy of ZYGO

11.8.2.1 ZeGage™ Pro optical profiler

The ZeGage™ Pro and ZeGage™ Pro HR 3D optical profilers (Figure 11.7) provide noncontact measurement and characterization of micro- and nano-scale features of virtually any surface and set the new standard for performance, ease of use, flexibility, and precision of benchtop-class industrial non-contact surface profilers.

The base ZeGage™ Pro model delivers surface mapping at the nanometer scale, meeting the needs for most industrial surfaces. For smoother samples, with a much finer surface finish, the ZeGage™ Pro HR offers more than 20× higher precision with 0.15 nm vertical precision [8].

11.8.2.2 NewView™ 9000 3D optical surface profiler

The NewView™ 9000 3D (Figure 11.8) optical surface profiler provides powerful versatility in non-contact optical surface profiling. With the system, it is easy and fast to

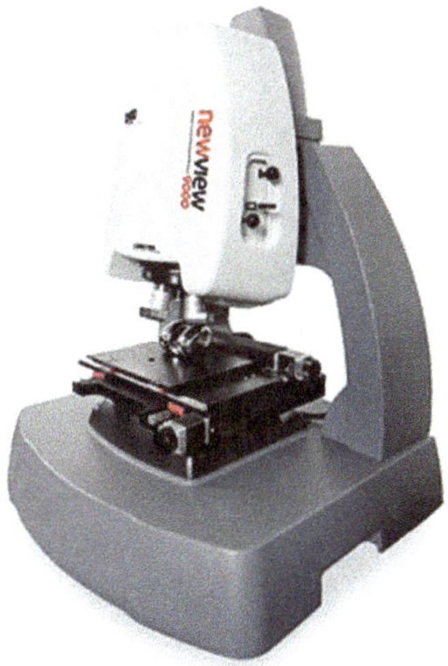

Figure 11.8 NewView^{TM} 90003D optical surface profiler [8]. Image courtesy of ZYGO

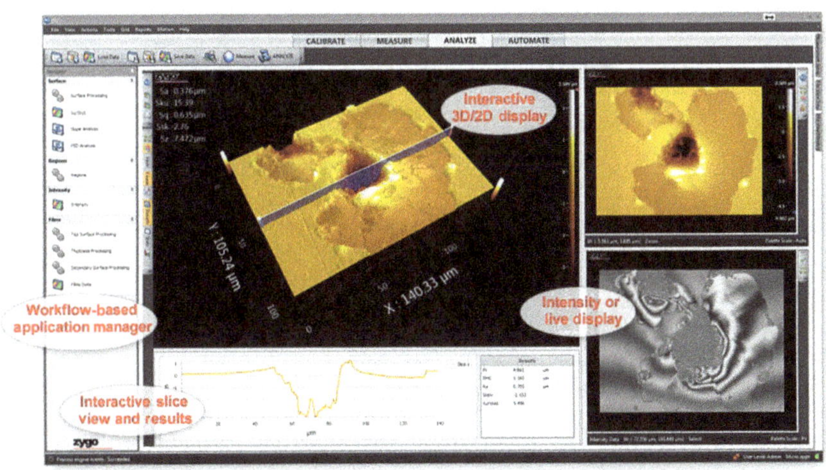

Figure 11.9 TM software screen [8]. Image courtesy of ZYGO

measure a wide range of surface types, including smooth, rough, flat, sloped and stepped. All measurements are non-destructive, fast, and require no sample preparation.

At the core of the system is Zygo's Coherence Scanning Interferometry (CSI) technology which delivers sub-nanometer precision at all magnifications and measures a wider range of surfaces faster and more precisely than other commercially available technologies, thus optimizing your return on investment (Figure 11.9).

All NewView 9000 profilers (Figure 11.10) are equipped with a triple-zoom turret which can be populated with discrete zoom optics tailor-made for the system.

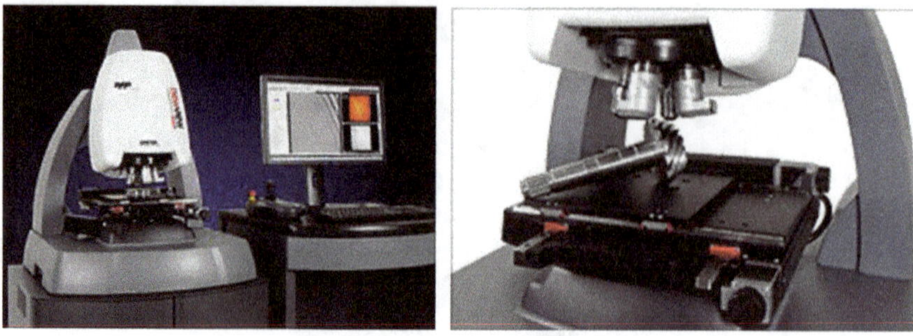

(a) (b)

Figure 11.10 (a) NewView 9000 optical profiling system [8]. Image courtesy of ZYGO. (b) Open work area features clear lines of sight [8]. Image courtesy of ZYGO

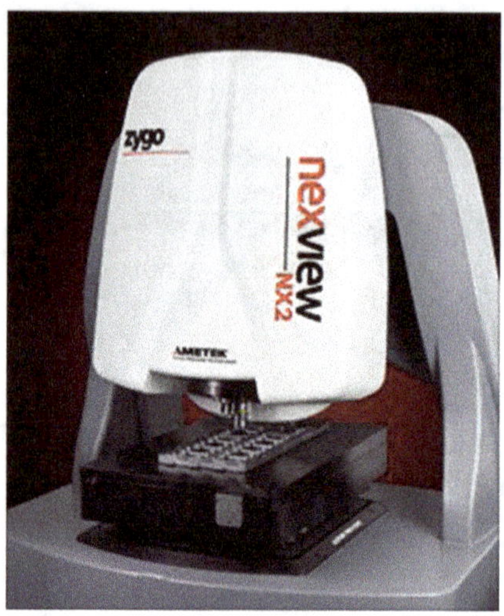

Figure 11.11 Nexview™ NX2 3D optical surface profiler [8]. Image courtesy of ZYGO

Sample staging configurations range from completely manual to fully automated with encoded travel [8].

11.8.2.3 Nexview™ NX2 3D optical surface profiler

Designed for the most demanding applications, the Nexview™ NX2 3D optical profiler (Figure 11.11) combines exceptional precision, advanced algorithms, application flexibility, and automation into a single package that represents Zygo's most advanced coherence scanning interferometric (CSI) profiler.

The completely non-contact technology optimizes the return on investment by delivering sub-nanometer precision at all magnifications and measuring a wider range of surfaces faster and more precisely than other comparable technologies commercially available. With applications as varied as flatness, roughness and waviness, thin films, step heights and more on virtually any surface and material, Nexview NX2 truly is the no-compromise profiler [8].

11.8.3 Scanning electron microscopy

For MEMS analysis, scanning electron microscopy (SEM) is one of the fundamental tools. Sizes of MEMS devices are in the order of a micron and the wavelength of light is too large for detailed delineation of the structure and surface during imaging. For this, a magnetic lens is used to direct the electron beam via a magnetic lens to a sample. As a result of this interaction of the beam with the sample, secondary electrons are emitted or backscattered which is detected by a secondary electron detector or backscattered detector, respectively. The collected electrons from the sample provide a surface image which is topographical in nature that is secondary electron image with atomic number and orientation information (backscattered electrons) [4].

11.8.4 Electron beam scatter detector (EBSD)

A detector is added to the SEM, to gain crystallographic quantitative information of polycrystalline and it is commonly referred to as EBSD (electron beam scatter detector). A tilted sample surface comes in contact with the electron beam; due to this, electrons are diffracted and suddenly hit the fluorescent screen of the detector. The orientation, grain boundary, and defect information are given by the diffraction pattern. A map of a polycrystalline sample is obtained as a result of scanning the electron beam, which results in quantitative information about the microstructure of the sample [4].

11.8.5 Transmission electron microscopy

Like SEM, transmission electron microscopy (TEM) uses an electron beam, but the electron beam is available at much higher accelerating voltages. The image is collected by thinning the sample when the beam moves through the sample. The images produced by TEM are at very high magnifications and can provide crystalline material crystallographic information as well as for individual grains in polycrystalline materials. For this powerful technique, careful sample preparation is required which is generally performed with the focused ion beam technique [4].

11.8.6 Focused ion beam (FIB)

For the investigation of MEMS failures, focused ion beam (FIB) sectioning is a well-adopted method. FIB is a dual-beam system with SEM in one tool. Figure 11.12 illustrates this system for the LEO 1500 cross beam system [9]. This allows FIB sputtering, deposition, and ion imaging, with secondary electron imaging in one tool.

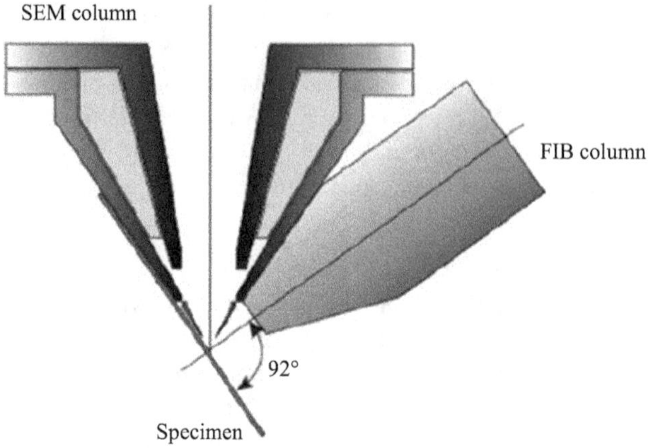

Figure 11.12 LEO 1500 cross beam FIB SEM system [10]. Reprinted with permission copyright 2003 SPIE

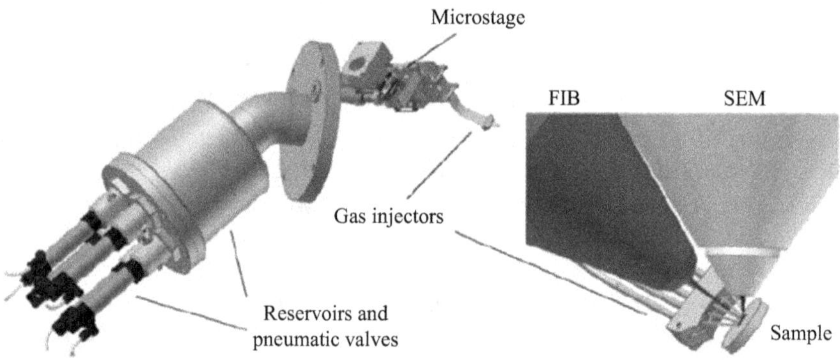

Figure 11.13 Gas injection system for FIB dual-beam system. [10] Reprinted with permission copyright 2003 SPIE

A gallium ion beam is used in FIB to sputter the sample for cross-section and also it can deposit material (typically tungsten) to samples. Figure 11.13 shows the gas injection system of the FIB. The beam starts in the coarse sputtering mode, and gradually reduces the sputter rate to a final polish phase [9].

11.8.7 Atomic force microscopy

One of another powerful and critical analytical tools for MEMS is the atomic force microscopy (AFM). A single atomic step height and sub-angstrom RMS surface roughness measurements are provided by AFM. It is found that more fine surface detail area was available with AFM than with the interferometer. At the end of a probe, a fine cantilevered tip scans the surface of the MEMS structure and a feedback system maintains a constant force between the probe and the surface at a constant deflection. This is known as "contact mode" which is also termed as a static model. This is an imaging mode. In dynamic mode AFM, the tip oscillates close to the resonance frequency and the oscillation changes at the surface area compared to a reference oscillation which provides surface information. In non-contact mode AFM, the tip aperture above the resonance frequency at a low amplitude oscillation. In non-contact mode, the soft samples are better imaged [4].

To achieve high measurement accuracy, optical readout technique and external mechanical excitation were implemented while keeping the samples simple. The latter is very important to identify and characterize the different degradation processes on the samples. On-chip excitation and detection require more complicated, multi-layer structures making the evaluation more difficult. To combine optical readout and external mechanical driving techniques, AFM is implemented. The AFM utilized in the field of reliability is described in detail in the previous study [6]. Park Scientific Instruments' AutoProbe M5 and Digital Instruments' MultiMode SPM are used. The sample chip with the cantilevers is mounted in the AFM head. The cantilever is excited mechanically at swept-frequency, while the deflection of the free end is measured continuously with a deflected laser beam. The typical excitation amplitude of the sweep is a few nanometers. The maximum displacement of the cantilever's free end in resonance in the air is 4–5 times greater than the excitation amplitude. This deflection is in the elastic range, as the thickness of the used cantilevers is 180–8,000 nm. The measured peak gives the resonant frequency of the cantilever beam. The samples were driven in resonance for a long time to characterize the long-term stability. Mechanical shock tests and large deflection test were also conducted in the AFM head. The built-in piezoelectric actuator in the AFM head applied mechanical shock to the sample. The large deflection was realized by resonant driving at higher excitation amplitudes or externally by an AFM cantilever. Repeated shock tests were conducted with 5 s resonant driving at a ten times higher excitation level than the one used for the resonance frequency determination. The resonance frequency is monitored constantly during the shock tests. The AFM operated in an environmental chamber filled with air, humid air, nitrogen-rich and argon rich atmosphere to study the influence of the environment on the resonant characteristics and on the shock response. All the silicon nitride, silicon carbide, and crystalline silicon samples were tested with the AFM method [10].

11.8.8 Auger analysis

Auger analysis is a technique in which an electron beam is used to excite the surface. The emitted Auger electron having much lower energy than the X-rays detected, and it is the characteristic of energy for each element. As such, the information gained is more surface sensitive as the Auger electron does not have a high enough energy to travel through and be emitted from the bulk of the sample. This technique can be used at a depth on the order of nanometers. In the vacuum chamber, the surface can be sputtered and upon completion of sputtering spectra can be collected. To obtain elemental data in a "depth profiling" manner, sequentially sputtering is performed.

The beam can be reduced to less than 1 μ in diameter which allows localized elemental surface information [10].

11.8.9 Electron beam-induced current

Electron beam-induced current (EBIC) is a SEM. For the assessment of defects in semiconductors, SEM-based mapping and non-destructive fault segregation technique are used extensively. EBIC technique relies on a semiconductor analysis technique that uses an electron beam to induce a current within a sample which is used as a signal for generating images that show characteristics of the sample. It is a well-known fact that a SEM is a convenient source of the electron beam for this purpose as such most EBIC techniques are performed using a SEM. Holes and electrons that are free to move around are created by the impact when the primary electron beam of the SEM penetrates the semiconductor itself. In a material that is free of any electric field, these holes and electrons will be in random motion which allows them to find each other and recombine. If electric fields within the semiconductor device do exist, then the holes and electrons will be separated and swept off to differently charged areas. These electric fields may either come from external sources or local to the sample. Ensuring the proper electrical contact with the sample, the movement of the holes and electrons generated by the SEM's electron beam can be collected, amplified, and analyzed in such a way that variations in the generation, drift, or recombination of these carriers can be displayed as variations of contrast in an EBIC image. EBIC imaging is very sensitive to electron-hole recombination, and this is the reason why EBIC analysis is very useful for finding defects that act as recombination centers in semiconductor materials [11]. The resulting current (EBIC) is imaged as the focused electron beam scans across the semiconductor device. The image contrast visualizes defects of the device as dark areas while areas free of defects appear bright. This image provides the qualitative assessment of subsurface electrical and physical properties within the semiconductor with the resolution of secondary electron imaging. An EBIC analysis is also very useful in isolating electrical shorts through several layers of metallization [11].

EBIC technique is shown in Figure 11.14. While referring plan view, the p-n junction is perpendicular to the electron beam, whereas in cross-section-view the p-n junction is parallel to the SEM beam. When an electron beam strikes a

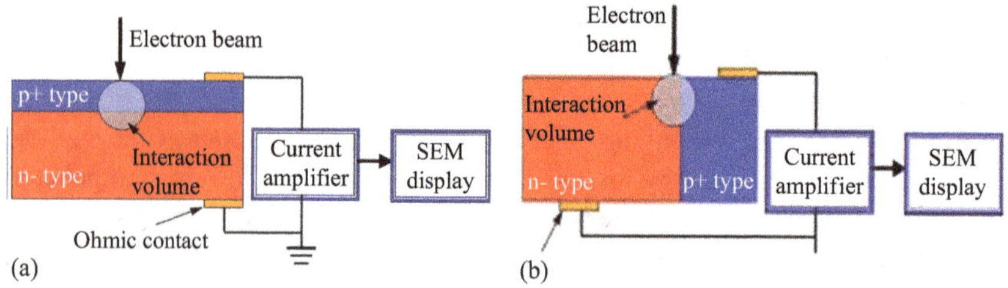

Figure 11.14 (a) Schemes of EBIC plan-view identification [13] (b) Schemes of EBIC cross-section-identification [13]

semiconductor to be tested, it will create electron-hole pairs in an interface volume. They will diffuse to the junction near to the p-n junction. Electrons will be drift to the n-type and holes will be drift to the p+ type. An EBIC will be produced when the p+ and n-types are connected to a Picoammeter or current amplifier [12].

This current can be used as the imaging signal for a SEM or scanning transmission electron microscope (STEM). EBIC image is obtained by connecting the EBIC signal scanned in the observation region. For the case of plan-view, defects such as dislocations tend to strongly decrease the minority carrier in their vicinity. As such it is observed that the EBIC signal is strong in areas without the defects and weak in areas around the defects. Therefore, it should figure out the defects in silicon in terms of the contrast change in its image [12].

A typical EBIC system consists of the following:

- a SEM
- an external low-noise current amplifier
- high-vacuum electrical feed-through that connects electrically the external current amplifier to the sample inside the SEM chamber and
- A sample holder with the required electrical contact points to facilitate the electrical connection of the sample to the feed-through.

It is found that EBIC currents are usually in the range of nano ampere to microampere while the primary beam current absorbed by the sample is in the pico ampere range. In areas of the p-n junction electron-hole recombination is enhanced where physical defects exist thereby reducing the collected current in those "defect" areas. The current through the junction is used to produce the EBIC image which results in the areas with physical defects appearing to be darker in the EBIC image than areas with no physical defects. As such, EBIC imaging is a convenient tool for finding sub-surface and other "difficult-to-see" damage sites. EBIC imaging is an effective analysis tool for bipolar circuits but it is not that effective in analyzing MOS circuits. This is due to the resulting false failures which occur when the gate oxides of MOS transistors tend to trap charges from primary beam charge injection [12].

Following are the applications of EBIC analysis [12]:

- Detection of crystal lattice defects, which show up as dark spots or lines
- location of p-n junction defects
- Detection of collector pipes that result in collector-emitter leakage currents in bipolar transistors
- Detection of extra junctions or doping layers
- Identification of ESD/EOS failures sites and
- Measurement of depletion layer widths and minority carrier diffusion lengths/lifetimes.

11.9 Reliability study of resonator

Since MEMS resonators are relatively new devices, reliability studies have just begun and some of the results obtained are presented in this section. The device used for reliability studies is ring resonators. Physical design specifications considered for ring resonator are ring diameter 6 mm, ring width 120 μm, ring thickness 100 μm, and resonant frequency 14,000 Hz. The frequency split between mode1 and 2 is desirable for close to zero since for getting high sensitivity the resonant motion is used on this sensor. The acceptable range of the frequency split depends on its quality factor (Q-value). It is observed that a high Q requires a small frequency split, whereas a low Q brings a wide allowance to the frequency split. However, a high Q-value is suitable for high sensitivity. To develop a more accurate gyro sensor for meeting the desired output for ESP application, a high-Q value with a vacuum package is adopted. As such, a ring resonator having an asymmetrical shape with high accuracy is required. To accomplish this, MEMS process technology is to be followed. One of the technologies, viz. photolithography with the capability of precise two-dimensional patterning and other technology DRIE having the capability of anisotropic vertical etching were used. The frequency difference between the two modes was finally controlled to less than 1 Hz with a nominal resonant frequency of 14,000 Hz [13].

11.9.1 Process and materials

The single-crystal silicon wafer is used for the ring and legs and photolithography and the DRIE technique is used. Anodic bonding is done on the silicon wafer which is bonded with a pedestal glass wafer. It is observed that Magnetic parts are automatically assembled on a spacer glass wafer in parallel.

The stress of silicon should be considered initially. The typical amplitude of the ring is designed as 20 μm. The maximum stress of the moving portion is 65 MPa. As the fracture strength of silicon is more than 500 MPa, which helps to ensure that the maximum stress during device operation is small enough and it has also been reported that less than 50% of the fracture strength does not show any fatigue [13].

After stress, the creep of the aluminum alloy film patterned on the ring should be considered. Conventional processes in IC fabrication is followed as the aluminum electrode is formed on the silicon dioxide film using physical vapor deposition (PVD), photolithography, and chemical etching. It is experienced that Aluminum has the potential for creeping at not very high temperatures [14]. It is widely recognized as the stress migration of aluminum alloys in IC chips. One of the prerequisites in IC technology is the low resistance where there are long wirings with a narrow width is encountered. The soft metal is also preferred for easy wire bonding. However, it is to ensure for MEMS that the mechanical characteristics of the metal layer on the resonator are stable when applying periodic stress by oscillation.

It is observed that the mechanical properties of aluminum alloy film that had been metalized by PVD changed on oscillation, resulting in a slight change in the resonant characteristics. The process conditions, such as pressure, power, temperature, and composition of the metal are the factors controlling the metal properties obtained when using PVD. Proper attention should be paid to creep and fatigue when metal material is used in MEMS [14].

The vacuum should be considered next, to obtain a resonator with a high Q-value the resonator is to be kept in the vacuum. This means that the ease of the oscillation is closely related to the pressure in the package. As such, it is necessary to guarantee that the inside pressure is maintained for life because the resonator should always be oscillated with a constant amplitude by the driving force within a limited range. It is easy to determine the pressure around the resonator device by measuring the Q-value, but it is difficult to measure fine leakage from the package. The helium bomb method using a helium leak detector is a standard technique to check for leakage from a hermetically sealed package of an electrical device [15]. However, it is observed that the detection sensitivity is two orders of magnitude less than the level required in this application. Therefore, this sealing process has been verified by accumulating a lot of process technology and know-how for having a suitable design to obtain a stable sealing process.

Then contamination should be considered. Since this device is capacitive as well as inductive, there is no narrow gap in the design. The minimum gap size is 100 μm, which is between the upper/lower pole and ring surface. This is large enough compared with the particle size in the wafer process, but it is still a critical size for the particles existing in the can-type package, therefore, several special cleaning processes for the assembly parts and a further clean assembly process were developed. A highly reliable sensor can be a reality by such contamination control in the manufacturing stage [15].

Finally, fracture at a high-g shock should be considered. The Bosch DRIE process is a relatively stable technique once the process conditions have been optimized. Several modes of unusual wall shape occur if the DRIE process is performed under critical conditions.

Due to the unusual shape which generates some portion, stress is concentrated which results in fracture of the silicon structure and thereby ending with a potential failure. The defect has various shapes that are unpredicted and as such control of the etching process to avoid unusual shapes is required.

This breakdown of the layer occurs when the ratio between deposition and etching is not optimized. Especially in the case of high aspect ratio and/or penetrating etching it has often been observed because of the ratio between deposition and etching changes as etching proceeds. A method to maintain the deposition to etching ratio during the etching by the installation of parameter lumping is an effective measure. Notching is one of the unusual sides etch conditions at the bottom area [16]. Charging of the insulation film lying on the bottom of the etched silicon layer causes a notch and several measures to ensure suitable etching conditions have been proposed. These measures are very effective in avoiding notching, hence this is not an issue assuming that the etching time can be controlled within a safe range. The modes were observed in the case of a higher etching rate with higher r.f. power. Photoresist process which is used for the optimization of the process conditions can be adopted to avoid these issues. It is necessary for the establishment of a stable process that design should have a sufficiently wide process window compared with the range of the total variation of the process [16].

11.10 Long-term stability

11.10.1 Stiffening effect

Information about the long-term stability of the resonator can be received by monitoring the resonance frequency. The silicon nitride and silicon carbide cantilever beams exhibited a stiffening effect during the long-term AFM resonant tests in air. As seen the resonance frequency increased gradually in air. This behavior is independent of the mechanical driving; the samples that stood still between the measurement points showed the same characteristics. The shapes of the curves are consistent. However, curves measured on samples with different shapes and sizes have different slopes. The thickness of the cantilever and the age of the cantilever are the two dominating factors which determine the curve. By designing more robust, thicker resonators, the stiffening problem can be overcome. Figure 11.15 shows the stiffening behavior on a several months old sample and after a buffered hydrogen fluoride (BHF) dip. The BHF removed the native oxide from the surface creating a fresh sample again. It is observed that old samples have weaker stiffening effect than fresh ones. The stiffening effect is a long-term degradation process. The resonant mode device needs frequent recalibration. Accelerated burn-in techniques can be a solution as the resonator is considerably stable after a few months of burn-in time [17].

11.10.2 Shock response

Another interesting phenomenon is the shock and large-deflection response of the thin SiN and SiC cantilevers. The shock and the large deflection generate an abrupt drop in the resonance frequency, up to a few percent. According to the stiffening effect, the resonance frequency increases again following the mechanical shock.

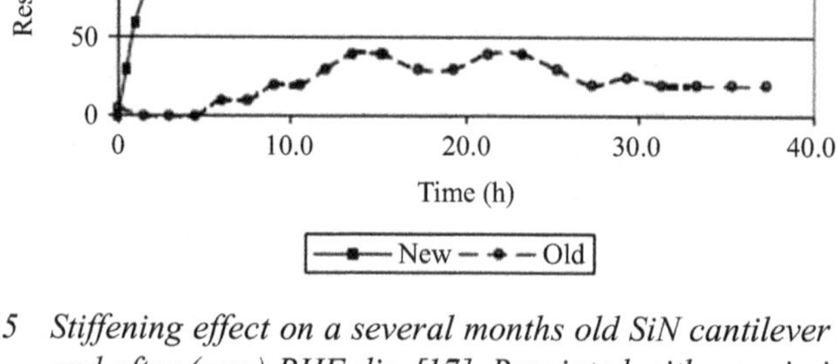

Figure 11.15 Stiffening effect on a several months old SiN cantilever before (old) and after (new) BHF dip [17]. Reprinted with permission copyright 2002 SPIE

The shock response is stronger in the case of thin cantilevers. The shock sensitivity is a major stability issue of the resonant mode devices. By implying robust design shock sensitivity decreases but this left with losing of measurement sensitivity as well [17].

11.10.3 Environmental influence

It is observed that the height of the negative frequency step is proportional to the magnitude of the shock or large deflection. Small shocks generate resonance frequency drops, and which fully recover in a few minutes. The shock tests were conducted with the AFM in various environments repeatedly. It is found that the thin silicon carbide and silicon nitride cantilevers behaved similarly. The curves in Figure 11.16 show the results of repeated shock tests conducted on a SiN cantilever beam in various environments. It is to note that each curve is an average of 8–10 shock tests and the deviation of the measurement points is 10%–20%. We can conclude from the tests that the resonance frequency drop and recovery effects are stronger in a humid environment while comparing in air. In addition, nitrogen and argon rich environments weaken the effects. The results suggest atmospheric packaging solutions overcome the shock response problem. It is to note that atmospheric packaging is much cheaper than vacuum packaging. It is found that packaging in pure argon or nitrogen atmosphere is satisfactory if vacuum packaging is not required to obtain high-quality factor. Moreover, the shock response is stronger on thinner cantilevers [17].

11.10.4 "Flycatcher" effect

Crystalline silicon paddle-beams and cantilever beams are subjected to cycle test in the air. Though no stiffening or shock response effects were found on these robust structures, the resonators showed another failure process. Because of the high-frequency vibration in the air (30–80 kHz), electrostatic charging occurs. The beams

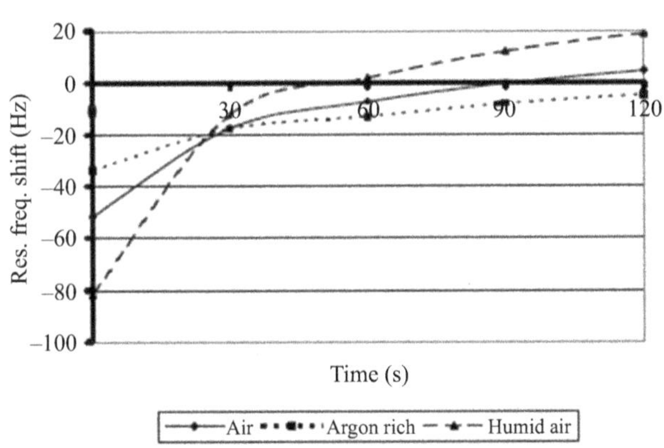

Figure 11.16 Repeated shock tests on a SiN cantilever in various environments [17]. Reprinted with permission copyright 2002 SPIE

gather airborne particles from the ambient air, like a flycatcher. It is understood that proper cleaning of the chips is crucial before packaging because most of the contamination on the resonating beam comes from the chip itself. The sticking gold/photo-resist clusters remaining from the lift-off process could not be removed with acetone. It comes off the surface during the vibration and sticks on the vibrating beam as shown in Figure 11.17 left [17].

A gold-coated paddle-beam was driven in resonance in air. Fine dust was introduced 20 h later in the measurement chamber. The resonance frequency dropped abruptly as shown in Figure 11.18. The dust particles settle in the chamber in approximately 10 h and then it is found that the resonance frequency reaches its minimum at this time. Then the resonance frequency increases again indicating that the beam releases slowly some of the dust particles. Subsequently, it happens again only the resonating beam gathers the airborne dust particles and the rest of the chip

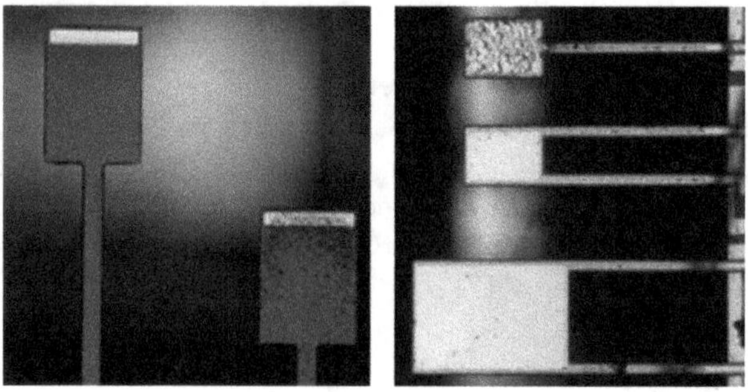

Figure 11.17 Si paddle-beam after a few hundred hours vibrating in the air (left), gold-coated paddle-beam vibrating industry environment (right) [17]. Reprinted with permission copyright 2002 SPIE

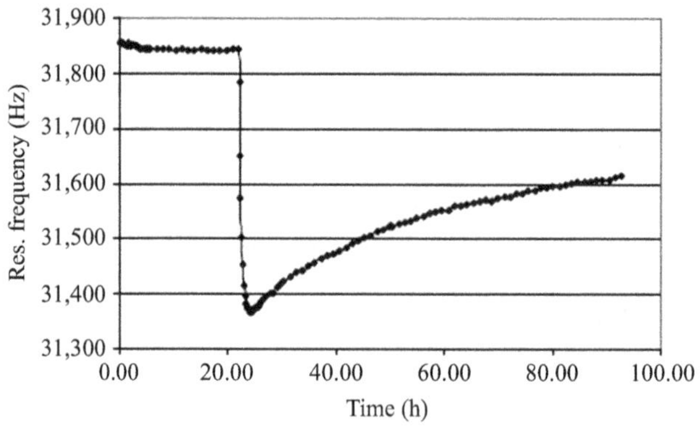

Figure 11.18 Cycling test on gold-coated paddle-beam. The introduction of dust particles decreases [17]. Reprinted with permission copyright 2002 SPIE

remains clean, as shown in Figure 11.17 right. These results point out that for ambient applications to maintain stable resonance frequency the protective package is required against dust and not necessarily the vacuum package [17].

11.11 Reliability of wafer-level vacuum package

11.11.1 Autoclave test

Autoclave test, or popularly called as pressure cooker test (PCT), or pressure pot test (PPOT), is a reliability test performed to assess the ability of a product or component to withstand severe temperature and humidity conditions. It is used to accelerate corrosion in the metal parts of the product, including the metallization areas on the surface of the die. It subjects the samples to the high vapor pressure generated inside the autoclave chamber. In this case, a micromechanical resonator is shown in Figure 11.19. It subjected to this test. It is seen that the bonding material seals the resonator cleanly. To test the hermetic reliability of the package, a batch of 53 resonators is tested before the autoclave test. JESD22-A102 standard is referred for the testing condition. The devices were placed under 121±2 °C, the relative humidity of 100%, and 2 atm (205 kPa) for 24 h, 96 h, and 168 h. The testing data shows the Q's of the resonators did not degrade and the frequency shift is within measurement error. It is observed that based on the permeation data and pumping capacity of getter material embedded inside the cavity, this package has no problem maintaining the vacuum within 100mTorr for 10 years [18].

11.11.2 High-temperature storage life test

For electronic components, there are recommended environmental conditions for their storage which allows them to perform at proper functionally. However, it is well

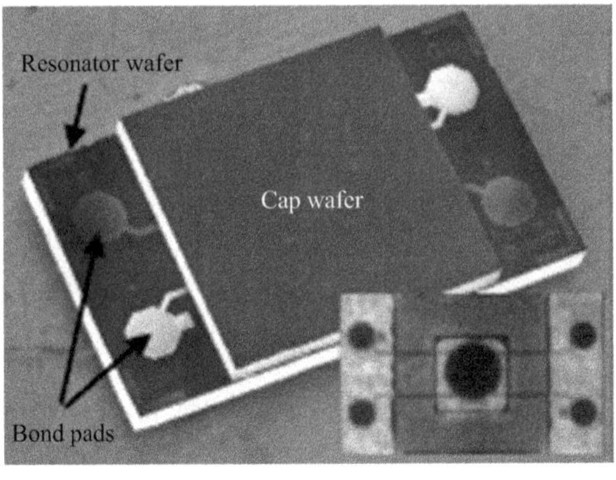

Figure 11.19 SEM picture of a singulated, vacuum-sealed micromechanical resonator. The inset shows the IR image of the bonding ring [18]. Reprinted with permission copyright 2006 IEEE

known that storage may not always be close to the recommended standard. This is where the high-temperature storage life (HTSL) test comes into play, testing the parameters of the device so that it can be determined what damage will result from being stored in less-than-ideal conditions. The HTSL test in response to how many businesses and consumers may put away an electronic device for a considerable length of time in conditions that may include exposure to high temperature [18].

This type of test is used to screen, monitor, qualify, or evaluate all electronic components or systems. The test itself is used to determine the overall effects of temperature and the passage of time for devices that are stored. This means that the device is tested to see how it reacts to being stored in high or low-temperature environments along with the time it may spend in storage before being used. This includes data retention failure mechanisms or non-volatile memory devices. This test also indicates that you cannot buy the electronic components or systems and store them for an infinite amount of time [18].

In the case of resonators, a batch of 64 resonators was stored under 150 ± 10 °C up to 1,000 h based on JESD22-A103 high-temperature standard. It is to note that Neither Q degradation nor frequency shift was observed. The yield of this particular wafer is as high as 97.3% even after the HTSL test. As such this indicates that there is no stress degradation within the resonator package.

Initially, five temperature-compensated MEMS oscillators were tested. After the testing, oscillators were placed into an oven at 150 °C for high-temperature storage. Parts are taken out of the oven for a thermal cycle test. As shown in Figure 11.20, the frequency deviation across the temperature did not change with time. As we know that the packaged resonator passed the HTSL, this data indicates that the resonator/oscillator package did not degrade with high-temperature storage [18].

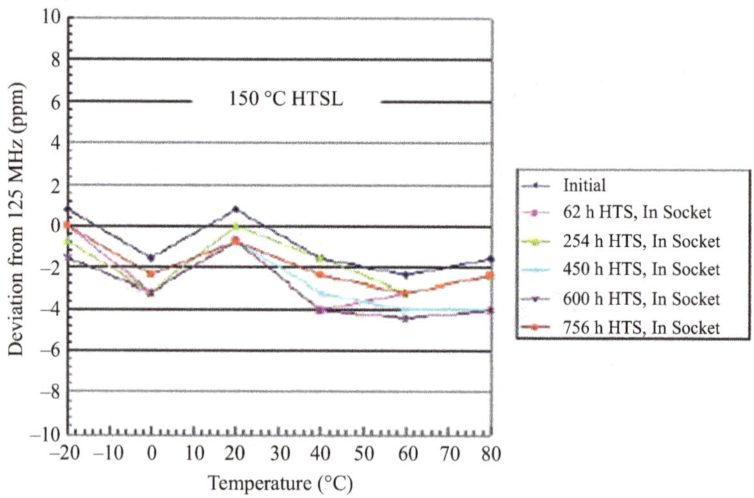

Figure 11.20 High-temperature storage life + thermal cycling [18]. Reprinted with permission copyright 2006 IEEE

11.11.3 Mechanical strength of bonding

This test is also popularly known as a bond test. This test is used to determine the ability of an adhesive to remain in contact with a surface or material while under stress or the adhesive's ability to hold together two materials as they are stressed. The bond itself is tested after it has been applied and allowed to cure by either applying a force directly to the adhesive in an attempt to remove it from the material or by attempting to separate to materials to have been connected by the adhesive. The force is then either steadily increased or remains constant and is applied to the bond until it fails [18].

Bond tests determine an adhesives' bond strength. The bond strength of a material may be considered to be its overall "stickiness" and will depend on the type of stress the bond experiences and the temperature at which the test is run. This strength will also depend upon the direction of the force applied to the bond. For example, a bond may require more force to be broken in shear than in direct tension.

There are many different types of bond testing, but the most commonly performed are tensile, shear and peel testing with torsional, impact, and pull-off tests performed less often. Each bond test is used to determine the adhesives bond strength in a given direction or under a given type of stress. Nearly all bond testing will take one of two forms. Either the adhesive will be applied to the surface of a material as a type of coating and the adhesives' bond strength is determined to be its ability to remain in contact with the material while under stress or the adhesive is applied between the two rigid substrates and its bond strength is determined as its ability to hold the two substrates together while under stress [18].

In case of resonators, it is to ensure that the cap will not be delaminated from the MEMS substrate during assembly as well as over-molding processes and to achieve this the bonding needs to be mechanically strong enough. This bonding usually tested by standard methods. Typically, the backend process requires 5 kgf of bonding strength to ensure a reliable assembly. During the experimentation, it was found that the average of glass frit bonding is as high as 9.3 kgf. No trace of delamination is observed for this type of device, which has gone through the severe QFN assembly process [18].

All these tests indicate that MEMS-based silicon oscillators have entered into an era of productization. The following reliability tests of these oscillators support this claim.

11.11.3.1 Aging

An aging test, usually also called accelerated aging, is a testing method that uses a higher level of heat, humidity, oxygen, sunlight, vibration, etc. to speed up the normal aging process of items. In case of MEMS, aging under mechanical stress is usually very important as compared to electrical failures viz. dielectric breakdown. Mechanical parts are run at very high speed, far in excess of what they would receive in normal usage. Aging is also one of the major concerns of overall MEMS devices, as a mechanical device that vibrates tens of millions of cycles per second. The requirement of aging based on timing and clock applications is generally ±5 ppm the first year, which is equivalent to the resonator traveling 6,000 km a year with ±30 m of accuracy.

Instead of resonators alone, the aging testing was conducted with real temperature compensated oscillator products. In the test, the aging characteristic at 85 °C, which is the maximum operating temperature based on the specification. The oscillators have +1/−4.5 ppm for the first year and +1.2/−5.2 ppm for 10 years. At room temperature, on the other hand, the samples show +1/−2 ppm over 18 months [18].

11.11.3.2 Shock resistance

One of the features that micromechanical resonator oscillators additionally have over quartz crystal is the shock resistance. MEMS resonator typically has mass as small as 10^{-14} kg and its stiffness is usually as high as tens of thousands of N for high-frequency structures. The resonator structure only bends 23 Å with 100,000 g shock. The resonator itself will survive the g-shock very well.

However, the impact of the g-shock on the packaging could be severe since the packaging contains most of the mass. In this particular evaluation, we flip the resonator on top of ASIC with solder balls as shown in Figure 11.21(a), which probably represents a worse case compared to ceramic package or plastic overmold package. The failure mechanism is that the smaller solder balls on the resonator break out from the UBM adhesion layer due to the stress generated along the *z*-axis as well as the shear stress. FEM results show that the smallest stress and smallest shear stress that can pull the solder off the ASIC are 100 MPa and 40 MPa, respectively. If the oscillator experience 30,000 g shock along the *x*-, *y*-, and *z*-axis, the *z*-axis stress and two shear stresses are listed in Table 11.3. Therefore, this type of oscillator will survive 30,000 g of shock. A recent test has shown that MEMS oscillators survived the g-shock of an air gun. Three out of three 125 MHz MEMS oscillator did not show any performance degradation [18].

11.11.3.3 Vibration operation

Under random vibration ranging, silicon MEMS oscillators were tested from 2.84G to 50G. Table 11.4 lists the power spectral density for different vibration amplitudes. The output power spectral density is much lower than what is required in mil-std-883 [18].

11.11.3.4 Thermal cycling

Thermal cycling also known as temperature cycle testing (TCT), or simply temperature cycling or temp cycling, is used to determine the ability of components to resist extremely low and extremely high temperatures. It is also used to evaluate their ability to withstand cyclical exposures to these temperature extremes. In this testing usually, mechanical failure occurs. The mechanical failure resulting from cyclical thermomechanical loading is known as a fatigue failure, so temperature cycling primarily accelerates fatigue failures. Thermal shock testing closely resembles TCT, which also accelerates fatigue failures. Figure 11.22 shows the frequency variation of an oscillator while the temperature switches between 40 °C and 100 °C. The temperature cycles every 5 min and frequency measurement is taken every minute. As shown, the frequency variation is within ±5 ppm [18].

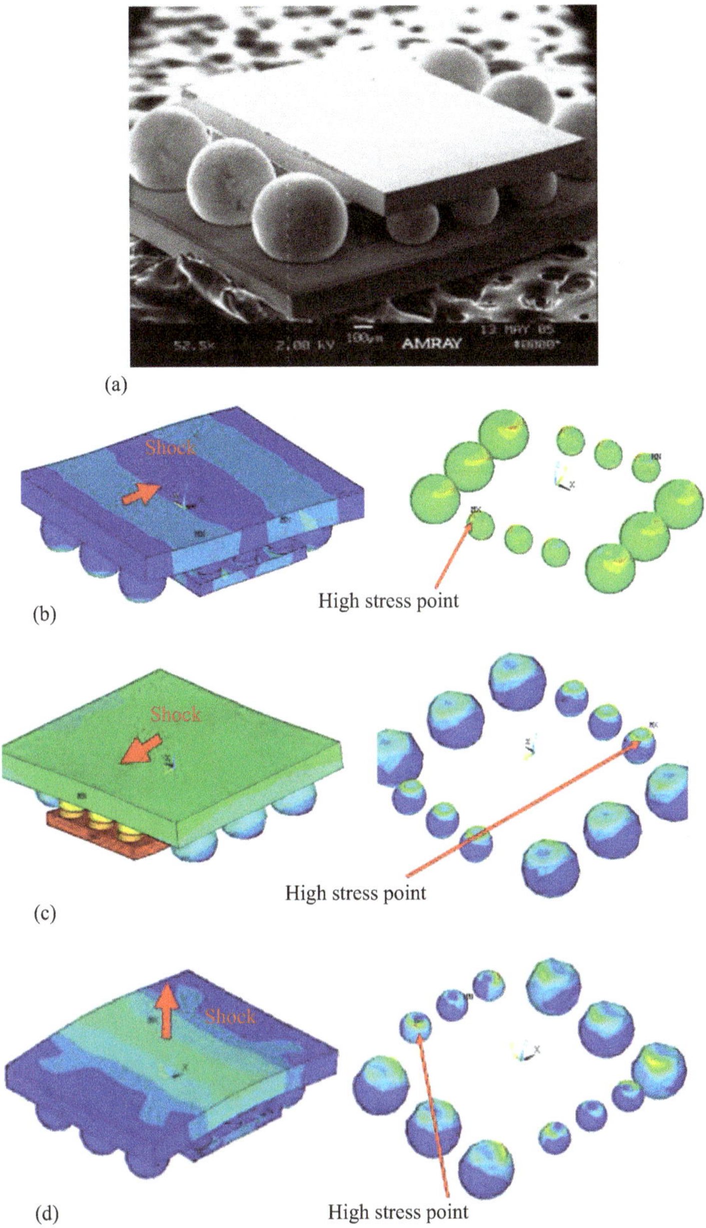

Figure 11.21 (a) Packaged MEMS for g-shock evaluation [18]. Reprinted with permission copyright 2006 IEEE. (b) Plot of stress due to a shock force at 30,000 g at the x-axis [18]. Reprinted with permission copyright 2006 IEEE. (c) Plot of stress due to a shock force at 30,000 g at the y-axis [18]. Reprinted with permission copyright 2006 IEEE. (d) Plot of stress due to a shock force at 30,000 g at the z-axis [18]. Reprinted with permission copyright 2006 IEEE

Table 11.3 Stress generated due to 30,000 g shock on a BGA-assembled MEMS oscillator [18]

	Description	*X*-shock	*Y*-shock	*Z*-shock
SEQV	von Mises stress [MPa]	43.2	46.8	27
SZ	Stress in the *Z*-direction [MPa]	60.9	72.6	47.1
SXZ	Sheer stress (*XZ*) [MPa]	15.3	N/A	7.8
SYZ	Sheer stress (*YZ*) [MPa]	N/A	15	8.7
UZ	Displacement [μm]	N/A	N/A	0.45

Table 11.4 Vibration power spectral density [18]

MIL_std_883, Method 2026		MEMS oscillator test results	
Overall RMS	**Output power spectral density**	**Overall RMS**	**Output power spectral density**
5.2	0.02	2.84	0.02
7.3	0.04	14.53	0.06
9.0	0.06	19.75	0.13
11.6	0.1	25.06	0.2
16.4	0.2	30.13	0.2
20.0	0.3	35.00	0.4
23.1	0.4	40.13	0.5
28.4	0.6	42.25	0.7
36.6	1.0	50.13	
44.8	1.5	x	x

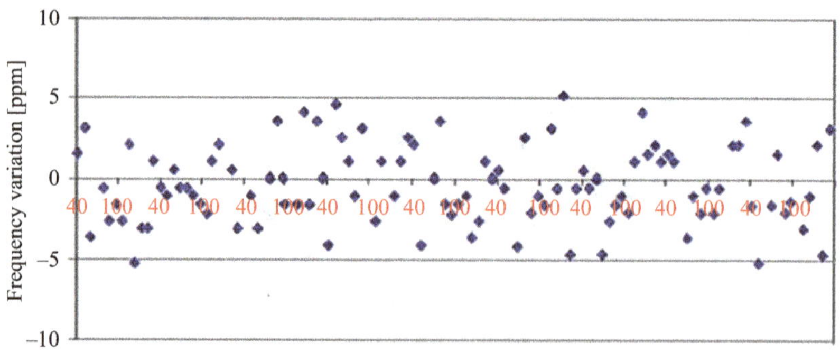

Figure 11.22 Frequency stability during thermal cycling [18]. Reprinted with permission copyright 2006 IEEE

11.12 Summary

In this chapter, we discussed the determination of the reliability of systems and components. The chapter presented a brief about the importance of reliability study concerning various failures associated throughout the life cycle of a component. Failure and its causes during the life cycle of the mechanical component are discussed with the help of bathtub curve analysis. Reliability determination in a reasonable time can be done with accelerated testing but the failure mode and its mechanism must be known in advance. Various failure modes and mechanisms occurred during designing, functioning, the packaging is discussed. Apart from the mathematical modeling of accelerated testing, it is important that the exact reason and nature of the failure is ascertained. The instruments required for the determination of failure are also discussed. The electromechanical components have to sustain the electrical and mechanical forces leading to additional modes of failures. The chapter highlights the root cause analysis and various analytical failures analysis. The reliability issues related to mechanical components and in particular to resonators are also discussed.

References

[1] W. van Spengen. 'MEMS reliability from a failure mechanisms perspective'. Microelectronics Reliability. 2003; 43(7): pp. 1049–60.

[2] Fuqua Norman B., Reliability Engineering for Electronics Design, New York: Marcel Dekker, 1987.

[3] Paul A. Tobias, and David Trindade, Applied Reliability, 3rd Edition, Chapman and Hall/CRC, Boca Raton, FL, 2011

[4] Hartzell A. L., Silva M. G. and Shea H. R., MEMS Reliability, New York: Springer, 2011

[5] Document SAE J 1739: Potential Failure Mode and Effects Analysis in Design (Design FMEA) and Potential Failure Mode and Effects Analysis in Manufacturing and Assembly Processes (Process FMEA) Reference Manual, SAE, 400 Commonwealth Drive, Warrendale, PA 15096-0001.

[6] J. Walraven. 'Failure mechanisms in MEMS'. Proceedings International Test Conference (ITC '03). 2003. pp. 828–32

[7] L. P. Schanwald, J. R. Schwank, J. J. Sniegowsi, D. S. Walsh, N. F. Smith, K. A. Peterson, *et al.* 'Radiation effects on surface micromachined comb drives and microengines'. IEEE Transactions on Nuclear Science 1998, 45(6), pp. 2789–98.

[8] ZYGO optical Metrology Instruments. Available at, 'https://www.zygo.com/?/met/'. Access on 10-10-2019.

[9] Gnauck P., and Hoffrogge P. 'A new SEM/FIB crossbeam inspection tool for high resolution materials and device characterization'. Proc of SPIE. 4980, Reliability, Testing, and Characterization of MEMS/MOEMS II. Micromachining and Microfabrication, Jan 2003, San Jose, CA, United States. 2003. pp. 106–13

[10] Mowat I., Moskito J., Ward I, Kawayoshi H., Winter D., Strossmam G., *et al.* 'Analytical methods for nanotechnology'. Technical Proceedings of the 2007 NSTI Nanotechnology Conference and Trade Show, Volume 4. May 2007.

[11] Available on, http://eesemi.com/ebic.htm. Access on 23 March 2019.

[12] Vu Le Huy, and Shoji Kamiya. 'A Direct Evidence of Fatigue Damage Growth Inside Silicon Mems Structures Obtained with Ebic Technique'. Vietnam Journal of Mechanics, VAST. 2014; 36(2):109–18

[13] T. Tsuchiya, A. Inoue, J. Sakata, M. Hashimoto, A. Yokoyama, and M. Sugimoto, 'Fatigue test of single crystal silicon resonator'. Technical Digest of the 16th Sensor Symposium, 1998; Japan. 277–80

[14] D. S. Gardner, H. P. Longworth, and P. A. Flinn, 'Detection of reactions and changes in thin-film morphology using stress measurements'. Journal of Vacuum Science and Technology A. 1992; 10(4): pp. 1428–41

[15] United States Department of Defense. MIL - STD - 833F, Method 1014.11, a United States Defense Standard, 2004.

[16] I. W. Rangelow, 'Critical tasks in high aspect ratio silicon dry etching for microelectromechanical systems'. Journal of Vacuum Science and Technology A. 2003; 21(4): pp.1550–62

[17] Kazinczi R., Mollinger J. R., and Bossche A. 'Environment-induced failure modes of thin film resonators'. Journal of Micro/Nanolithography, MEMS, and MOEMS. 2002; 1(1): 63–69

[18] Hsu W., 'Reliability of Silicon Resonator Oscillators'. International Frequency Control Symposium and Exposition. Miami, FL, USA, June 2006. IEEE; 2007. pp. 389–92

Index

www.ingramcontent.com/pod-product-compliance
Lightning Source LLC
LaVergne TN
LVHW011828200926
848484LV00002B/340